Introductory Horticulture

Delmar Publishers' Online Services

To access Delmar on the World Wide Web, point your browser to:

http://www.delmar.com/delmar.html

To access through Gopher: gopher://gopher.delmar.com
(Delmar Online is part of "thomson.com", an Internet site
with information on more than 30 publishers
of the International Thomson Publishing organization.)

For information on our products and services:
email: info @ delmar.com
or call:
800-347-7707

Introductory Horticulture

FIFTH EDITION

H. Edward Reiley
&
Carroll L. Shry, Jr.

Delmar Publishers™

I(T)P An International Thomson Publishing Company

Albany • Bonn • Boston • Cincinnati • Detroit • London • Madrid
Melbourne • Mexico City • New York • Pacific Grove • Paris • San Francisco
Singapore • Tokyo • Toronto • Washington

NOTICE TO THE READER

Cover art courtesy of Don Almquist
Cover Design: Christina Almquist

Delmar Staff
Publisher: Tim O'Leary
Acquisitions Editor: Cathy L. Esperti
Developmental Editor: Cathy L. Esperti
Senior Project Editor: Andrea Edwards Myers
Production Manager: Wendy A. Troeger
Marketing Manager: Maura Theriault

COPYRIGHT © 1997
By Delmar Publishers
a division of International Thomson Publishing Inc.
The ITP logo is a trademark under license

Printed in the United States of America

For more information, contact:

Delmar Publishers
3 Columbia Circle, Box 15015
Albany, New York 12212-5015

International Thomson Publishing Europe
Berkshire House 168 - 173
High Holburn
London WC1V 7AA
England

Thomas Nelson Australia
102 Dodds Street
South Melbourne, 3205
Victoria, Australia

Nelson Canada
1120 Birchmount Road
Scarborough, Ontario
Canada M1K 5G4

International Thomson Editores
Campos Eliseos 385, Piso 7
Col Polanco
11560 Mexico D F Mexico

International Thomson Publishing Gmbh
Königswinterer Strasse 418
53227 Bonn
Germany

International Thomson Publishing Asia
221 Henderson Road #05 - 10
Henderson Building
Singapore 0315

International Thompson Publishing - Japan
Hirakawacho Kyowa Building, 3F
2-2-1 Hirakawacho
Chiyoda-ku, 102 Tokyo
Japan

Library of Congress Cataloging-in-Publication Data
Reiley, H. Edward.
    Introductory horticulture / H. Edward Reiley & Carroll L. Shry, Jr.—5th ed.
        p. cm.
    Includes index.
    ISBN 0-8273-6766-X (textbook)
    1. Horticulture.   I. Shry, Carroll L.   II. Title.
SB318.R43   1995   95–15151
635—dc20   CIP

❧ CONTENTS ❧

One of the most rapid growth areas in the field of agriculture is that of horticulture. This new-found interest is due in large part to an increasing awareness by the general public, as people look inwardly at a need to be closer to nature and outwardly at a world in which efficient methods of food production and conservation of land are increasingly important.

This awareness and the resulting growth of the field have generated a need for a broad-based first level text which explores the basic principles of horticulture and methods of practical application of these principles. INTRODUCTORY HORTICULTURE, Fifth Edition, an established text in Delmar's Agriculture Series, was written to fill this need.

Foremost in the text is an examination of fundamental horticulture principles, from careers in the field to the cultivation of specific crops. Special care has been taken to treat highly technical subjects, such as plant propagation and taxonomy, in a comprehensive yet understandable manner. A section on pesticides includes up-to-date information on the subject, with special emphasis on personal safety and the protection of human beings and the environment. Use of biological control of pests whenever possible is advocated and up-to-date information presented. Certain topics, such as the creation of holiday arrangements, basic floral design, the construction of terrariums and bonsai plantings, and the cultivation of houseplants, have been added especially for student appeal.

Many changes and additions have been made to the fifth edition of INTRODUCTORY HORTICULTURE by Reiley and Shry. The authors took this opportunity to incorporate most of the ideas and suggestions that users of this text have requested during its first ten years in print. Among the major changes, you will find:

- ◆ More plant science material and emphasis.
- ◆ Math problems added in some units.
- ◆ Greenhouse management section added.
- ◆ New color pictures to update older ones.
- ◆ Update on material as needed.
- ◆ Revised Instructor's Guide.

In addition to these specific changes, each unit has been appropriately updated to reflect changes in the horticulture industry. Many new photographs and illustrations have also been included in the fifth edition of INTRODUCTORY HORTICULTURE.

The format of the text complements the authors' direct approach to the subject matter. A general objective and list of competencies at the beginning of each unit tell what students should accomplish before proceeding to the next unit. Review material and activities, also included in each unit, allow the students to put into practice skills they learn through study of the text material. Numerous line drawings, photographs, and charts clarify and summarize text content. A glossary at the end of the book lists terms with which horticulture students should be familiar.

An extensive Instructor's Guide accompanies the text. A pretest and post-test are included so that individual comprehension of subject matter may be gauged before and after study of the text. Also contained in the guide are answers to review questions, additional student activities, suggestions for instructor demonstrations, supplemental reading lists, and, when appropriate, special activities for the advanced student as well as masters for overhead transparencies.

The authors of INTRODUCTORY HORTI-CULTURE, H. Edward Reiley and Carroll L. Shry, Jr., are both involved in agriculture education in Maryland. Mr. Reiley is the retired Director of Vocational Education for the Frederick County Board of Education. He received a Bachelor of Science degree in Horticulture and a Master of Science degree in Agriculture Economics from the University of Maryland. Before working in a supervisory capacity, Mr. Reiley taught agriculture in secondary school and continuing education classes. He is a member of various professional organizations, including the American Rhododendron Society, and is a life member of the FFA Alumni Association. He was named Vocational Educator of the year in Maryland for 1978, and received the Honorary American Farmer degree in 1979. In his spare time, Mr. Reiley breeds new varieties of rhododendrons and azalea hybrids and manages a nursery.

Carroll L. Shry, Jr., is presently a landscaping/nursery production instructor at the Frederick County Vocational-Technical Center, Maryland. He holds a Bachelor of Science degree in Agricultural Education/Horticulture from the University of Maryland and a Master of Science degree in Agricultural Education/Horticulture from West Virginia University. He has taught vocational agriculture classes at the comprehensive high school, vocational technical center, community college, and continuing education levels. Mr. Shry was named Vocational Educator of Maryland in 1980, and Outstanding Vocational Educator of the Year in Region 1—American Vocational Association, 1981. He is also active in many professional organizations, including the National Vocational Agriculture Association, the American Horticultural Society, the Associated Landscape Contractors of America, the American Nurserymen Association, and the National FFA Alumni Association. Mr. Shry operates a landscape business.

The Fifth Edition of INTRODUCTORY HORTICULTURE has been carefully designed to enhance the study of horticulture in agriscience programs. For best results, you may want to become familiar with the features incorporated into this text and accompanying learning tools.

ADDED FEATURES

◆ Pedagogical use of FULL COLOR to accelerate learning through highly illustrated material.

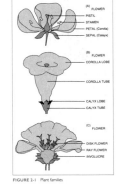

◆ Competency-based level of achievement needed for success is spelled out in front of each unit.

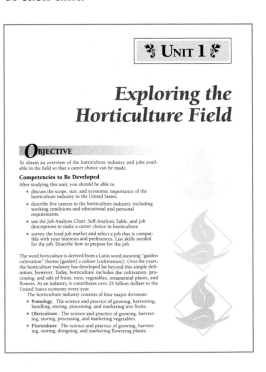

◆ FULL-COLOR interior plant identification chart.

ENHANCED CONTENT

The updated and enhanced content addresses the evolving agriscience curriculum:

◆ Broad applications to science and math, providing the appropriate balance of applied theoretical knowledge for the horticultural curriculum.

◆ Expanded coverage of micropropagation, fertilizers, and pesticides.

◆ Expanded coverage of integrated pest management. This coverage provides modern concepts in pest control to minimize use of chemicals by emphasis on biological and cultural control.

◆ Extensive coverage of skills needed by first-level horticulture students to move on to a specialized horticultural field of study.

EXTENSIVE TEACHING/ LEARNING PACKAGE

The complete supplement package was developed to achieve two goals:

1. To assist students in learning the essential information needed to continue their exploration into the exciting field of horticultural science.

2. To assist instructors in planning and implementing their instructional program for the most efficient use of time and other resources.

INSTRUCTOR'S GUIDE TO TEXT

The Instructor's Guide provides answers to the end-of-chapter questions and additional material to assist the instructor in the preparation of lesson plans.

LAB MANUAL

Order # 0-8273-6915-8

This comprehensive lab manual reinforces the text content. It is recommended that students complete each lab to confirm understanding of essential horticultural science content. Great care has been taken to provide instructors with low cost, strong science-based labs.

LAB MANUAL INSTRUCTOR'S GUIDE

Order # 0-8273-6916-6

The Instructor's Guide provides answers to lab manual exercises and additional guidance for the instructor.

TEACHER'S RESOURCE GUIDE

Order # 0-8273-6768-6

This new supplement provides the instructor with valuable resources to simplify the planning and implementation of the instructional program. It includes transparency masters, motivational questions and activities, answers to questions in the text, and lesson plans to provide the instructor with a cohesive plan for presenting each topic.

COMPUTERIZED TESTBANK

Order # 0-8273-7489-5

An all-new computerized testbank (IBM and MAC compatible) with more than 500 new questions that will give the instructor an expanded capability to create tests.

AGRISCIENCE TESTMASTER CD-ROM

Order # 0-8273-7496-0

This testbank contains over 3,000 questions correlated to the following Delmar titles:

Cooper/AGRISCIENCE: Fundamentals and Applications, 2E

Cooper/AGRICULTURAL MECHANICS: Fundamentals and Applications, 3E

Reiley/INTRODUCTORY HORTICULTURE, 5E

Herren/THE SCIENCE OF AGRICULTURE: A Biological Approach

Ricketts/LEADERSHIP PERSONAL DEVELOPMENT AND CAREER SUCCESS

Camp/MANAGING OUR NATURAL RESOURCES, 3E

Gillespie/MODERN LIVESTOCK AND POULTRY PRODUCTION, 5E.

The CD-ROM also contains ESAgrade Electronic Gradebook and ESAtest III. These enable the instructor to create and administer on-line tests and keep track of grades.

DELMAR'S AGRISCIENCE LASERDISC PACKAGE

Order # 0-8273-7300-7

Delmar's Agriscience Laserdisc Package offers instant accessibility to interactive video segments on four double-sided Level I laserdiscs directly corresponding to all the major content areas in agriculture. The discs can be obtained as a package or individually:

Disk 1: Animal Science
Order # 0-8273-7301-5

Disk 2: Plant Science
Order # 0-8273-7302-3

Disk 3: Business and Mechanical Technology
Order # 0-8273-7303-1

Disk 4: Forestry and Natural Resources Management
Order # 0-8273-7304-X

In addition, a correlation guide complete with bar codes correlates all bar codes contained on the laserdiscs with all of Delmar's top agriscience titles.

COLOR TRANSPARENCY PACKAGE

Order # 0-8273-6911-5

Forty color transparencies based on illustrations from the text provide the instructor with yet another means of promoting student understanding of horticulture concepts.

✤ ACKNOWLEDGMENTS ✤

Appreciation is expressed to the following for their special technical assistance:

Frances Mary Dewey, Instructor, Junior College of Albany, Division of Russell Sage College, Albany, New York

Jack E. Ingels, Chairman, Department of Plant Science, State University of New York Agriculture and Technical College, Cobleskill, New York

TECHNICAL REVIEWERS

Special appreciation is granted to the following reviewers who provided their input on the revision:

MeeCee Baker, Greenwood High School, Mifflin, Pennsylvania

Richard Ledford, Bradley High School, Cleveland, Tennessee

Mack Sanders, South Doyle High School, Knoxville, Tennessee

A.S. Azcona, Lyman High School, Longwood, Florida

M. Theresa Nowicki, Pembroke, Massachusetts

The following reviewers assisted with the first edition:

Ronnie A. Thomas, Park View High School, South Hill, Virginia

Mack Sanders, South Doyle High School, Knoxville, Tennessee

John Dawley, Groesbeck High School, Groesbeck, Texas

Keith Zamzaw, Texas A & M University, College Station, Texas

Edie Coleman, Fauquier County High School, Warrenton, Virginia

Dana Griffin, Corsicana High School, Corsicana, Texas

Dr. Alan McDaniel, Virginia Polytechnic Institute and State University, Blacksburg, Virginia

Richard Stotler, West Orange High School, Winter Garden, Florida

Bill Dodd, M. B. Smiley High School, Houston, Texas

Dr. Phillip L. Edgecomb, County College of Morris, Randolph, New Jersey

Thanks are extended to the following for their special assistance:

Dr. Conrad Link, Department Head, Horticulture, University of Maryland

Dr. Francis Gouin, Professor, Horticulture Department, University of Maryland

Jeffers Nursery Inc., Slingerlands, New York

Shemin Nurseries, Inc., Burtonsville, Maryland

Hewitt's Lawn and Garden Center, Latham, New York

Westminster Nursery, Westminster, Maryland

Dutch Plant Farms, Frederick, Maryland

Stadler Nursery, Catonsville, Maryland

Treeland Nursery, Inc., Frederick, Maryland

Carroll Gardens, Westminster, Maryland

Bluemont Nurseries, Inc., Monkton, Maryland

J. Frank Schmidt & Son Co., Boring, Oregon

Frederick County Vocational-Technical Center, Frederick, Maryland

Netherlands Flowerbulb Information Center, Brooklyn, New York

American Nurseryman, Washington, DC

Robert Jackson, Westminster Nursery, Westminster, Maryland

Landscape Contractors Association, Rockville, Maryland

National Landscape Association, Washington, DC

Chapel Valley Landscape, Inc., Lisbon, Maryland

Mary Reiley

Judy Shry

A large portion of the material in this text was classroom tested at Linganore High School, Frederick, Maryland. Unit 3 was classroom tested at Wilco Area Career Center, Lockport, Illinois.

Horticulture: An Introduction

Exploring the Horticulture Field

OBJECTIVE

To obtain an overview of the horticulture industry and jobs available in the field so that a career choice can be made.

Competencies to Be Developed

After studying this unit, you should be able to

- ◆ discuss the scope, size, and economic importance of the horticulture industry in the United States.

- ◆ describe five careers in the horticulture industry, including working conditions and educational and personal requirements.

- ◆ use the Job Analysis Chart, Self-Analysis Table, and job descriptions to make a career choice in horticulture.

- ◆ survey the local job market and select a job that is compatible with your interests and preferences. List skills needed for the job. Describe how to prepare for the job.

The word *horticulture* is derived from a Latin word meaning "garden cultivation" (*hortus* [garden] + *culture* [cultivation]). Over the years, the horticulture industry has developed far beyond this simple definition, however. Today, horticulture includes the cultivation, processing, and sale of fruits, nuts, vegetables, ornamental plants, and flowers. As an industry, it contributes over 25 billion dollars to the United States economy every year.

The horticulture industry consists of four major divisions:

- ◆ **Pomology.** The science and practice of growing, harvesting, handling, storing, processing, and marketing tree fruits.

- ◆ **Olericulture.** The science and practice of growing, harvesting, storing, processing, and marketing vegetables.

- ◆ **Floriculture.** The science and practice of growing, harvesting, storing, designing, and marketing flowering plants.

◆ **Landscape and Nursery Industry**. The
science and practice of propagating,
growing, installing, maintaining, and using
grasses, annual plants, shrubs, and trees in
the landscape.

Additional specialized areas include:

◆ **Seed Production**. The science and practice of producing, processing, and selling
high-quality seed crops for use in growing
a wide range of horticultural crops that
grow true from seed.

◆ **Related Occupations**. Those areas that
provide products and services necessary in
the production of horticultural crops. Such
resources include goods such as fertilizers,
pesticides, containers, machinery, and
services such as education and research.

JOBS IN HORTICULTURE

There are various sources of employment for individuals trained in horticultural practices. These
businesses include greenhouses, nurseries, garden
centers, golf courses, parks, orchards, floral design shops, grounds maintenance operations, and
vegetable and fruit growers. Figure 1-1 is a job
analysis chart which explains positions in these
areas of employment. The following job descriptions, adapted from the *Handbook of Agricultural
Occupations* (published by the United States Department of Agriculture), provide a more detailed
account of the most common positions in the
industry.

JOB TITLE: GREENHOUSE EMPLOYEE

JOB DESCRIPTION A greenhouse employee
grows plants in a heated glass or plastic greenhouse, figure 1-2. The plants that are grown may
be vegetables or flowers. The work involved may
include the propagation of trees or ornamental
shrubs from seed or cuttings. Greenhouse activities include the production of out-of-season vegetables such as tomatoes, cucumbers, or leaf
lettuce; the growing of cut flowers, potted plants,
and bedding plants in preparation for sale; and
the starting of plants for outdoor nursery beds.
When a greenhouse is part of a nursery operation,
a worker may have the duties of a nursery employee as well as those of a greenhouse worker.
Employees in greenhouses screen, mix, and ster-

FIGURE 1-2 A greenhouse worker grooming plants in
the greenhouse. (Ed Reiley, Photographer)

ilize soil and place it in containers for growing
plants. They sow seed, start cuttings, and transplant seedlings and plants. They water, weed, thin,
prune, fertilize, and spray growing plants. They
are also responsible for maintaining the greenhouse structure and equipment.

WORKING CONDITIONS Most of the work
of a greenhouse employee is done indoors. Outdoor jobs are usually done during favorable
weather. This work requires a great deal of walking, stooping, and bending over plants or seedbeds. Hands and clothing may be stained from soil
and plant sap or juices.

A job in a greenhouse involves a great deal of
manual labor, but the labor is not usually difficult. Working hours are usually regular, with work
generally steady throughout the year. Seasonal
demands sometimes require working overtime.

**PERSONAL AND EDUCATIONAL QUALI-
FICATIONS** For individuals to really enjoy
greenhouse work, they must be interested in
watching plants grow and develop. Good health
is important, but certain types of physical disabili-
ties will not prevent an interested individual from
entering the field. A high school education with a
course in vocational agriculture or horticulture is
preferred for one interested in this vocation.

Closely related to greenhouse work is flower
arranging, figure 1-3.

JOB TITLE: NURSERY EMPLOYEE

JOB DESCRIPTION The nursery employee
grows seedlings and plants for landscaping, fruit
production, and replanting in forests, figure 1-4.
The employee may work in one of several different

JOB TITLE	Does it entail year-round work?	Are there regular hours?	Is most of the work outdoors or indoors?	Does it offer variety?	Is the work in one place?	Are there fringe benefits?	Does the job involve working with others?	What are the educational requirements?	Is there an opportunity for promotion?
greenhouse worker	yes	Generally, but some overtime is usually required.	indoors	yes	yes	yes	Yes, to some extent.	high school diploma with a course in agriculture or horticulture	yes
nursery worker	in many cases	Yes, but there are peak seasons.	mostly outdoors	yes	yes	some	yes	high school diploma with a course in agriculture or horticulture	yes
garden center employee	yes	Yes, but there are peak seasons.	both indoors and outdoors	yes	yes	some	yes	high school diploma with a course in agriculture or horticulture	yes
golf course employee	no	Yes, during the golfing season.	outdoors	yes	yes	not to a great extent	yes	high school diploma with a course in agriculture or horticulture	yes
assistant grounds keeper	yes	Yes, some overtime is required on occasion.	mostly outdoors	yes	yes	There may be some.	not necessarily	high school diploma with a course in agriculture or horticulture	yes
park employee	yes	yes	outdoors	yes	yes	not to a great extent	yes	high school diploma with a course in agriculture or horticulture	yes
vegetable grower	depends upon grower	no, seasonal	outdoors	yes	yes	some	yes	high school diploma with a course in agriculture or horticulture	yes
orchard employee	depends upon grower	no, seasonal	outdoors	yes	yes	some	yes	high school diploma with a course in agriculture or horticulture	yes
employee of small fruit grower	depends upon grower	no, seasonal	outdoors	yes	yes	some	yes	high school diploma with a course in agriculture or horticulture	yes
employee of floral design shop	yes	yes, with some overtime	indoors	no	yes	yes	yes	high school diploma with a course in agriculture or horticulture	yes

FIGURE 1-1   Job analysis chart. (Adapted from the Job Profile Chart, Ohio State University)

FIGURE 1-3 A flower designer finishes a sympathy arrangement. (Ed Reiley, Photographer)

FIGURE 1-4 Nursery workers load azaleas to a flat-bed wagon. (Courtesy of Chesapeake Nurseries)

types of nurseries. Some nurseries specialize in producing fruit trees and small fruit transplants, some in ornamental trees and shrubs, and some in forest replanting materials. Some nursery employees operate greenhouses and produce their own seedlings and plants from cuttings. Some produce planting stock of two or more major lines of plants, trees, or shrubs.

Individuals employed in nurseries prepare seedbeds; plant seedlings; prepare cuttings for rooting; and weed, cultivate, water, and prune plants. They also perform other cultural practices such as spraying and grafting. They dig, grade, and pack plants for shipment. They may cut, lift, and lay sod. They also transplant shrubs and trees, and, in a tree nursery, gather and process forest tree seeds. Aiding in the maintenance and repair of buildings and equipment is also usually a part of the job.

WORKING CONDITIONS Most of the nursery worker's time is spent outdoors. If a greenhouse is a major part of the employer's business, the employee will naturally spend a considerable amount of time in the greenhouse. Planting and cultivating must be accomplished when weather conditions are good, but much outdoor work at a nursery can be done in rather bad weather.

Most of the work is considered manual labor; some of it involves heavy lifting. Working hours are regular. Employment may be for the entire year, depending upon the skills of the worker.

PERSONAL AND EDUCATIONAL QUALIFICATIONS The nursery employee should enjoy working with plants and being outdoors. A truck driver's license may be required. The ability to identify plants and the knowledge of how plants are used in the landscape are very desirable. A high

school education with a course in agriculture or horticulture is preferred.

JOB TITLE: GARDEN CENTER EMPLOYEE

JOB DESCRIPTION The employee of a garden center has many jobs, including caring for ornamental plants, moving plants and supplies into selling areas, arranging plants and supplies for display purposes, and selling the various products handled by the center. A garden center may be a part of a large retail store, a part of a nursery or greenhouse operation, or a retail store operated independently of other business.

The work of garden center employees includes cleaning, stocking, and arranging garden supplies on shelves and counters and in windows. They water, spray, and trim ornamental plants and control environmental conditions. They unload and unpack supplies as they arrive from wholesalers, make deliveries, and load orders onto trucks and into customers' cars. They also give information and advice to customers concerning plants and lawns and their care.

WORKING CONDITIONS Garden centers are built and arranged to attract customers. As a result, a garden center employee usually works in a clean, pleasant, and comfortable environment. Parts of the work area are usually heated during cold weather. Other areas are unheated but protected from rain, snow, and wind. Some of the work may be outdoors during the season in which ornamental shrubs and trees are sold. The work is fairly regular, but has seasonal peaks. Some garden centers close completely or operate with only a small crew during the winter months.

PERSONAL AND EDUCATIONAL QUALIFICATIONS Employees of garden centers should enjoy working with people and caring for ornamental plants. They must be good salespeople and must therefore be able to talk easily and in a friendly way. A high school education with a course in vocational agriculture or horticulture is desirable.

JOB TITLE: GROUNDS MAINTENANCE EMPLOYEE

JOB DESCRIPTION A grounds maintenance employee cares for the area surrounding an industry, business, church, school, airport, apartment building, private estate, cemetery, or shopping center. These employees plant and care for lawns and ornamental plants. The work entails mowing grass, reseeding, controlling weeds, and planting and spraying ornamental plants. They also rake grounds and dispose of leaves and other refuse. A year-round job is provided through maintenance and repair of walks, driveways, and equipment. The work may involve making minor repairs to buildings and providing for snow removal.

WORKING CONDITIONS The grounds maintenance employee works outdoors and deals mainly with ornamental plant materials. The work does provide variety, but is not extremely strenuous. The environment in which the employee works is usually very pleasant, although certain jobs must be done under a variety of weather conditions. The work is steady throughout the year. Employees work regular hours, but there are some peaks in the work load.

PERSONAL AND EDUCATIONAL QUALIFICATIONS The grounds maintenance employee should not mind working alone and should enjoy working with plants, tools, and small garden equipment. The demand for high school graduates specializing in this area is increasing as job opportunities expand in the landscape industry.

JOB TITLE: GOLF COURSE EMPLOYEE

JOB DESCRIPTION Golf course employees are responsible for the overall maintenance of golf courses. They care for the turf on both the greens and fairways. They install and use irrigation and drainage equipment, clean and maintain sand traps, change the locations of cups, and aerate the soil. They may also prune shrubs and trees, replace soil as needed, and repair equipment and buildings.

WORKING CONDITIONS The work of the golf course employee is usually done outdoors. Most of it is done during fairly good weather. The workday usually consists of eight hours. In the southern part of the United States, employment is steady throughout the year. In sections of the country with cold winters, employment is from March or April through October or November.

PERSONAL AND EDUCATIONAL QUALIFICATIONS Because a great deal of walking is done in golf course maintenance, the employee should have good health, but certain physical handicaps will not interfere with one's success. The employee should enjoy working outdoors and be able to get along well with others. A high school education with a course in agriculture or horticulture is desirable.

JOB TITLE: CITY, STATE, OR NATIONAL PARK EMPLOYEE

JOB DESCRIPTION The park employee does whatever work is needed for the proper maintenance of parks. This includes maintaining the trees, shrubs, flowers, and lawns that make up the planting area. The city park worker deals mainly with formal flower beds and lawn areas, while the state or national park worker is usually concerned with the care and maintenance of natural woodlands or forests.

The work of the park employee includes mowing grass; trimming the edges of walks and driveways; planting, pruning, and caring for trees, shrubs, hedges, lawns, and flower beds; controlling insects, diseases, and weeds; and caring for the soil. It also includes jobs such as the removal of trash and snow, maintenance of swimming pools, care of boating facilities, general maintenance of buildings and equipment, and repair of roadways and drives.

WORKING CONDITIONS The park employee is outdoors a great deal of the time. Most of the work is manual labor, and is accomplished in a healthy, pleasant environment. At times, park improvement jobs must be done in bad weather conditions. This occupation usually gives year-round employment, and working hours are regular. In certain types of state park work, there may be peak periods.

PERSONAL AND EDUCATIONAL QUALIFICATIONS A high school education with a course in vocational agriculture or horticulture is desirable.

Average hourly rate for city, state, or national park employees varies widely from region to region. An approximation as a result of some research (1992) would indicate the following:

Unskilled workers	$5.50
*Skilled workers	6.80
*Supervisors	12.50

*Note: These two positions may require one or two years of post-secondary education or years of on-the-job training.

OPPORTUNITIES FOR COLLEGE GRADUATES

The jobs listed and described above are entry-level jobs requiring only a high school diploma, not the highly skilled jobs requiring college degrees. College graduates are in great demand, and for the student interested in pursuing a degree in horticulture jobs are plentiful. Consider the following information from a recent survey:

1. Approximately 32 percent of all teachers and scientists in the U.S. Department of Agriculture and various colleges of agriculture will retire in the next few years.

2. Each year there are about 14 percent more jobs in agriculture than there are graduates to fill them.

3. A four-year college graduate will earn approximately one million dollars more during a lifetime career in horticulture than a high school graduate. Starting salaries range from $18,000 to $30,000.

What are these opportunities? *Scientific jobs* are available, such as food, soil, or water chemists, plant breeders, and forest pathologists. There are also *sales and service jobs*, including marketing specialists, sales representatives, buyers, and landscape contractors.

Many other opportunities include agriculture production, farm management consulting, finance, manufacturing, processing, toxicology, engineering, entomology, communications, education, extension work, public relations, and photography. The opportunity for employment is excellent for all of these jobs requiring college degrees.

If you have an interest in any of these employment opportunities, check with your teacher or school counselor to be sure you get the necessary high school math and science courses for college and a career in your chosen area.

Many of these professional careers will require more paper work and much of the work will be done indoors. There are good fringe benefits and retirement plans with most of them, as well as high salaries during the working years. The college graduate in horticulture has a bright future.

SELECTING A JOB

To simplify the task of choosing a specific career in the horticulture industry, first study the questions posed in figure 1-5, Self-Analysis Table. When the questions have been answered, match the information with the job analysis chart found in figure 1-1. This procedure may be followed for jobs other than those listed in the job analysis chart. Simply collect the information concerning the job and match it with your self-analysis.

Career opportunities may also be broken down according to the type of work they involve: production, management, and sales, figure 1-6. If outside work is preferred, an area in production might be considered. If inside work without the company of other people is appealing, a professional career, such as research, should be considered.

There are other factors to consider before selecting a career:

◆ Will the job be challenging?

◆ What types of duties does the work involve?

◆ What skills does the job require?

◆ Are jobs in the field available?

Using figures 1-1 and 1-5, check several of the listed jobs to determine which comes closest

In a career, do you desire a job that involves:
working outside?
working inside?
year-round work?
regular working hours?
a variety of tasks?
working in one place?
the security of fringe benefits?
working alone?
working with others?
specific educational requirements?
the guarantee of promotion?

FIGURE 1-5 Self-analysis table

JOBS THAT REQUIRE WORK WITH THE HANDS, TOOLS, AND EQUIPMENT. MUCH OF THE WORK IS DONE OUTSIDE, EXCEPT FOR THE GREENHOUSE WORKER.	
greenhouse worker	nursery grower
propagator	bedder
tree surgeon	pruner
turf worker	grounds keeper
greens keeper	landscape gardener
orchardist	small fruit grower
vegetable crop grower	farm chemicals sales worker

JOBS THAT MAY REQUIRE ADVANCED TRAINING, PLANNING, AND PAPER WORK. MUCH OF THE WORK IS DONE INSIDE.	
landscape architect	consultant
teacher	researcher
plant breeder	office supervisor
sales person	plant disease specialist
greenhouse manager	nursery manager
garden center manager	pest control specialist
pesticide specialist	landscape contractor

Note: This represents only a partial list. It is suggested that a survey of jobs available in the local job market be made. These jobs may then be analyzed using the process outlined in figure 1-1.

FIGURE 1-6 Job list

to satisfying your personal job expectations. Draw up a blank form as in figure 1-1 and fill in the information for the job you select. (See student activities at the end of this unit.) Research the job selected. If the job is available in your community or work area, arrange to discuss your interest with someone involved in that career, such as the manager or owner of a local business. A visit to the class by the owner or a visit to the business may be arranged.

THE FUTURE OF THE HORTICULTURE INDUSTRY

The horticulture industry is currently in a state of rapid growth. It is an industry full of career opportunities and is making use of the newest technologies in mechanized equipment and computers. As people become increasingly concerned about their surroundings, world resources, nature, conservation, and controlling pollution, horticulture becomes more important. More plants will be used to beautify surroundings, to conserve soil and control pollutants, and to protect and feed wildlife. As food costs increase, more

families will plant gardens. Employment, which involves working with plants or providing goods and services related to the horticulture business, will be widely available.

PRODUCTION OF HORTICULTURAL CROPS

As the population increases and people have more leisure time, more gardening of all types will be done. The demand for fruits, vegetables, flowers, ornamental plants, and landscaping services will grow. Increase in the demand for vegetable plants by home gardeners means a substantial increase in the production business.

SERVICE INDUSTRIES

As more horticultural goods are produced, more and more seeds, fertilizers, insecticides, fungicides, growth-regulating chemicals, hand tools, and equipment are also needed. These service industries generally seek to employ persons with a background in horticulture.

MARKETING

More people are engaged in marketing and distributing horticultural products than in their production. As the volume of products increases, so does the need for employees in marketing with a background in horticulture.

INSPECTION

Government agencies employ inspectors of both fresh and processed horticultural crops and products. Private processing firms interested in a quality product also employ field inspectors to spot disease and insect outbreaks as well as to determine when the crop is at peak quality for harvesting.

TEACHING AND EXTENSION WORK

More and more students need to be trained in horticulture at secondary and post-secondary school levels. This requires more teachers to fill the need. A quickly changing technology also requires a field force of specialists, such as extension agents and county agricultural agents, to work with producers. The homeowners and those who garden as a hobby also require the advice of professionals. Educators often use the news media as a mass communication method to reach the public more easily.

RESEARCH

As demands for greater production are placed on the horticulture industry, researchers will play an important role, figure 1-7, and will be needed in increasing numbers in the future. Better varieties of plants must be developed; more effective insect and disease control established, both through resistant varieties and through biological and chemical control; and better ways of growing, harvesting, preserving, packaging, storing, and marketing must be developed. The new technique of tissue culture propagation still needs research for application to a wider range of plants.

The horticulture industry has met the challenge in the past through the development of new products and procedures such as:

- hybrid varieties
- better greenhouse climate control
- mist propagation
- propagation by tissue culture
- automatic watering and fertilizing
- control of the blooming date of plants by regulation of the length of darkness or light
- plant growth regulators
- rooting hormones
- weed control
- resistant varieties of plants and biological pest control
- extended storage life of foods
- mechanical harvesting

FIGURE 1-7 Sophisticated laboratories and highly trained personnel continue to produce new and exciting plants and products for horticulture. (Courtesy of pbi/ Gordon Corporation)

- dwarfing rootstock for tree fruits
- better methods of packaging and processing products such as concentrated juices and frozen foods
- improved cultural practices
- gene splicing

As the horticulture industry becomes more complex and more technical in nature, more training will be required to hold the available jobs. The future holds great promise for the person willing to learn and to work in horticulture. There is every reason to believe the challenge will be met in the future. You can be part of this existing industry.

 # Student Activities

1) Using a telephone book, make a list of horticulture businesses in the local area.

2) Using the local newspaper, list the horticultural jobs advertised in the classified section.

3) Select one business in the community and interview the owner as well as several employees. Report to the class on available jobs and the type of duties and working conditions involved in each job.

4) Write to your local agricultural college and ask about jobs offered to its graduates in horticulture and about its placement service.

Self-Evaluation

Select the best answer from the choices offered to complete the statement or answer the question.

1) The science and practice of growing and harvesting tree fruits, small fruits, and tree nuts is called

a) ornamental and landscape horticulture.   c) olericulture.
b) floriculture.   d) pomology.

2) The science and practice of growing vegetable crops is called

a) ornamental and landscape horticulture.   c) olericulture.
b) floriculture.   d) pomology.

3) The science and practice of growing and harvesting flowering plants is called

a) ornamental and landscape horticulture.   c) olericulture.
b) floriculture.   d) pomology.

4) The science and practice of propagating, growing, maintaining, and using grasses, annual plants, shrubs, and trees in the landscape is called

a) ornamental and landscape horticulture.   c) olericulture.
b) floriculture.   d) pomology.

5) A greenhouse employee works

a) mostly outdoors.   c) with nursery stock only.
b) mostly indoors.   d) alone.

6) The greenhouse worker has a job that consists of

a) very irregular and seasonal work.
b) very strenuous manual labor.
c) regular hours that are steady throughout the year.
d) constant outside work.

7) Nursery workers

a) spend a great deal of time outdoors.
b) spend most working hours in the greenhouse.
c) can always depend on steady work.
d) do not need to be able to identify plants.

8) To find out what jobs are available locally, the student should

a) make a survey of local horticultural businesses.
b) check with the Department of Labor.
c) check with a local employment agency.
d) all of the above

9) A garden center employee should

a) enjoy meeting people.
b) enjoy a job that is mostly paper work.
c) enjoy working alone.
d) not be concerned about finishing high school.

10) A grounds maintenance employee should

    a) enjoy working outdoors and working alone.
    b) enjoy mowing grass as part of the job.
    c) enjoy repairing equipment and walks.
    d) all of the above

11) As the population grows and more people build homes, the horticulture industry will

    a) remain at the present level of employment.
    b) grow also as the demand for plants increases.
    c) get smaller, since homeowners will grow their own plants.
    d) none of the above

12) Which one of the positions listed below is never related to horticulture?

    a) researcher
    b) teacher
    c) marketing specialist
    d) stock broker

13) Most of the jobs in horticulture require

    a) a high school diploma.
    b) excellent physical condition.
    c) an ability to meet people.
    d) a person who enjoys being alone.

14) Three jobs in horticulture that require outdoor work and that are done with hand-operated tools and equipment are

    a) tree surgeon, researcher, landscape gardener.
    b) teacher, orchardist, turf worker.
    c) tree surgeon, turf worker, small fruit grower.
    d) nursery worker, pruner, plant breeder.

15) Horticultural salespeople should have a background in horticulture because

    a) it helps them to know the product they are selling.
    b) it helps them to talk to growers.
    c) it helps them to know which items to offer for sale.
    d) all of the above

16) The word *horticulture*, a Latin word, means

    a) "grower of crops."
    b) "plant cultivator."
    c) "lover of plants."
    d) "garden cultivation."

Plant Taxonomy: How Plants Are Named

OBJECTIVE

To differentiate between scientific and common plant names and explain the binomial system of naming plants.

Competencies to Be Developed

After studying this unit, you should be able to

- explain why scientific plant names are used.
- explain the difference between genus, species, and variety.
- list five plants by their common and scientific binomial names.
- give four examples of family names (both Latin and common).

MATERIALS LIST

- ✔ nursery catalogs
- ✔ issues of American Nurseryman Magazine
- ✔ *Hortus Third*, staff of the L.H. Bailey Hortorium, Cornell University

Most plants have more than one common name; some have several. For example, the trout lily is also known as the tiger lily, adder's-tongue, dog's-tooth violet, and yellow snowdrop. The Judas tree and redbud are the same tree, but are known by these different names in different parts of the United States. Common names can be confusing, since two totally different plants may have the same common name. The cowslip is one of these: in New York State, it is a marsh-loving, buttercup-like plant; in England, it is a primrose-like plant found on dry, grassy slopes. Both have yellow flowers, but, apart from their color, they have little else in common. Since common names can be so misleading, it is important to use the same name when plants from different areas are spoken of—to use common names when buying or selling could be disastrous. Common names (e.g., pea, maple, tomato, or petunia) are an important part of the horticulture industry, however, because these names are a part of the everyday language used by industry persons and laypeople.

THE BINOMIAL SYSTEM OF NAMING PLANTS

The early scholars always wrote in Latin or Greek, so, naturally, when they described plants or animals, they gave them scientific Latin or Latinized Greek names. However, this way of naming plants also caused problems; the names were often long and difficult. For example, *Nepeta floribus interrupte spicatis pedunculatis* was the name for catnip and *Dianthus floribus solitariis, squamis calycinis subovatis brevissimis, corollis, crenatis* the name for the common carnation.

The famous Swedish botanist, Linnaeus, simplified the matter by developing the binomial (two-name) system for naming plants. This system is still used today. He gave all plants just two Latin names as their scientific name. For example, he renamed the catnip *Nepeta cataria*. The first name is known as the *generic name*; this is the plant's group name. All plants having the same generic name are said to belong to the same *genus*. All plants belonging to the same genus have similar characteristics and are more closely related to each other than they are to the members of any other genus. The second name is the *specific name* or special name. All plants with the same specific name belong to the same *species*. (The Latin word *species* means "kind.") It is difficult to define exactly what a species is, but we can say that plants of the same species have the same characteristics and will consistently produce plants of the same type. Today, species are often subdivided into *varieties*. One variety of a species resembles that of another variety, but there are always one or two differences that are consistent and inherited. For example, the peach tree is known as *Prunus persica*; the nectarine is *Prunus persica* var. nucipersica. Another subdivision is termed a *cultivar* (cv.). The term culti-

var means "cultivated variety." For example, many people consider red to be the ultimate fall color, giving the brightest, most unique display of all. The most commonly known cultivars of *Acer rubrum* (red maple) are 'Red Sunset' and 'Autumn Flame,' which are the most reliable for brilliant reds and a long-lasting display of foliage.

The generic name is usually a noun and the species name an adjective. Sometimes, generic names are the names of early botanists, for example, *Buddleja* was named in memory of Adam Buddle. Some common generic names include *Acer* (maple), *Chrysanthemum* (mum), *Dianthus* (pink), *Hibiscus* (mallow), *Mimulus* (monkey flower), *Sedum* (stonecrop), *Papaver* (poppy), *Pinus* (pine), and *Pelargonium* (geranium). And in some cases the generic name has become the common name of the plant, such as *Euonymus*, *Rhododendron*, and *Forsythia*.

The species name, because it is an adjective, often gives important information about the plant. Sometimes, it tells us the color of the plant, for example, *Betula lutea* is the yellow birch (*lutea* means "yellow"); *Betula alba*—the white birch (*alba* means "white"); *Quercus rubra*—the red oak; *Juglans nigra*—the black walnut. Sometimes, the species name tells us if the plant is creeping or erect, for example, *Epigaea repens* is the scientific name for trailing arbutus.

Sometimes, the specific name gives geographical information about where a plant occurs naturally, for example, *Anemone virginiana* is the Virginia anemone; *Taxus canadensis* is the Canadian yew. *Macro* and *micro* are Greek words meaning "large" and "small." Therefore, a plant with the species name *macrophylla* could be expected to have large leaves, while a plant with the species name *microphylla* would have small leaves (*phyllus* means "leaf"), for example, *Philadelphus microphyllus* is the scientific name for the little-leaf mock orange.

When the meanings of the scientific names of plants are understood, the names are interesting and not difficult to learn. The easiest way to become familiar with them is to say them out loud; every time you transplant a seedling, dig a shrub, or sow some seeds, say the Latin name. When scientific names are used, the horticulturist is able to order plants from any part of the world; it is a universal language. The French, Dutch, the Texan, and the New Englander all use the same Latin names. Where common names are often misleading, the scientific name never is. For example, people often confuse the red maple with the Japanese red maple. They are two completely different trees. The red maple (*Acer rubrum*) grows to be over 75 feet tall; it has green leaves in the summer and red foliage in the fall. The Japanese red maple (*Acer palmatum*) has red leaves throughout the year and does not grow above 25 feet tall.

PLANT FAMILIES

Related genera (plural of *genus*) with similar flower structures are grouped together into major units known as *families*. For example, the rose family, known as Rosaceae, consists of several genera—*Prunus* (plum), *Fragaria* (strawberry), *Rubus* (bramble), and *Malus* (apple). All members of each

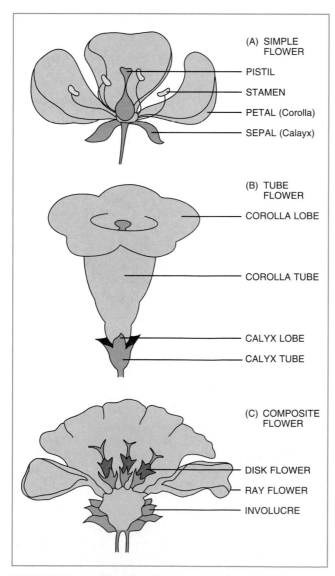

(A) SIMPLE FLOWER
— PISTIL
— STAMEN
— PETAL (Corolla)
— SEPAL (Calayx)

(B) TUBE FLOWER
— COROLLA LOBE

— COROLLA TUBE

— CALYX LOBE
— CALYX TUBE

(C) COMPOSITE FLOWER

— DISK FLOWER
— RAY FLOWER
— INVOLUCRE

FIGURE 2-1   Plant families

genus of the rose family have relatively simple flowers with separated petals, figure 2-1 (A). On the other hand, in the Solanaceae, the potato family, the petals are fused or joined to form a corolla tube, figure 2-1 (B). The Solanaceae includes the genera *Solanum* (potato), *Petunia*, and *Nicotiana* (tobacco). The Compositae (daisy family) is the largest of all the plant families. Members of the Compositae have two kinds of flowers packed together to form a single head or "flower," figure 2-1 (C). The outer flowers (known as ray flowers) may have large or small petals. The inner flowers (disk flowers) always have small petals. The daisy family includes these genera: *Aster* (aster), *Artemisia* (silver mound), *Achillea* (yarrow), *Helianthus* (sunflower), *Chrysanthemum* (mum), *Senecio* (cineraria), *Calendula* (pot marigold), and many others. Other important families are the Cruciferae (cabbage family), Umbelliferae (carrot family or umbellifer family), Papaveraceae (poppy family), Liliaceae (lily family), and Graminae (grass family).

EXPRESSING SCIENTIFIC NAMES

Notice that when the Latin names of plants are printed, they are expressed in italics. This is because when names and phrases are written in a language other than our own, it is conventional to print them in italics or underline them if they are typewritten or handwritten. Also, by convention, the generic name is always written first, then the species name, and last the cultivar (cv.) name. The generic name always begins with a capital letter; the species name with a small letter. The cultivar name is preceded by the letters *cv* or enclosed with quotes (i.e., 'Red Sunset') with the first letter of the cultivar name capitalized. Sometimes, when a number of species all belonging to the same genus is the subject, the generic name is abbreviated to only the first letter, for example, to express several different types of maples, the generic name for maple, *Acer,* is abbreviated to A. The red maple may be expressed as *A. rubrum* and the Japanese maple as *A. palmatum.* The term cultivar refers to a specific plant, as in *Pinus strobus* 'pendula,' weeping white pine, or *Fagus sylvatica* 'Atropurpurea,' copper beech, whose leaves are deep maroon color.

Taxonomy Chart: hierarchy of specification

Kingdom—Plant
  Division or Phylum—Spermatophyta (seed plants)
    Class—Angiospermae (seeds in fruit)
      Order—Acerales
        Family—Aceraceae
          Genus—*Acer*
            Species—*rubrum*
              Variety or cultivar—var.
              'October Glory'

THE TAXONOMIST

Scientists who identify and classify plants are known as *taxonomists*. An international set of rules has been drawn up to ensure that every different species has a different binomial name and that the scientific name assigned to that plant is the oldest binomial name ever used for that plant. This international set of rules is known as the *International Code of Botanical Nomenclature*.

Self-Evaluation

A) Select the best answer from the choices offered to complete the statement.

1) Scientific names are used to

  a) show the chemical makeup of plants.
  b) avoid confusion concerning the names of plants.
  c) increase the number of names of plants.
  d) all of the above

2) Scientific names of plants are expressed in Latin because

  a) it is a dead language.
  b) it is easy for all nationalities to pronounce.
  c) it is an international language and was used by the early scholars to express plant names.
  d) it is an easy language to learn.

3) The name of the person who developed the binomial system for naming plants is

  a) Linnaeus.
  b) Plato.
  c) Socrates.
  d) Hortus.

4) The generic name of the plant is

  a) placed last and begins with a small letter.
  b) placed first and begins with a small letter.
  c) placed first and begins with a large letter.
  d) placed in the middle of the name in parentheses.

5) A *genus* can be defined as

  a) a group of plants that have more in common with each other than they have with the members of any other genus.
  b) a group of plants that are all alike.
  c) a group of plants that have the same flower structure.
  d) none of the above

6) A *species* can be defined as

    a) a group of plants that are alike in almost every feature and consistently produce like plants.

    b) a group of plants that have more in common with each other than with the plants of any other group.

    c) plants that are all the same size.

    d) plants that all have the same flower color.

7) Latin names of plants are italicized

    a) to make them appear important.

    b) because it is conventional to italicize words and phrases that are expressed in a different language.

    c) because it helps to remember them.

    d) because all plant names are always italicized.

8) A person who identifies and classifies plants is known as a (an)

    a) taxidermist.   c) agronomist.

    b) taxonomist.   d) horticulturist.

9) For a plant with the scientific name of *Acer rubrum* cv. Red Sunset, 'Red Sunset' is the _____ name.

    a) genus   c) family

    b) species   d) cultivar

B) Match the correct genus with its common name.

    a) geranium 4   (1) *Nicotiana*

    b) tobacco 1   (2) *Papaver*

    c) oak 6   (3) *Dianthus*

    d) yew 5   (4) *Pelargonium*

    e) poppy 2   (5) *Taxus*

    f) pink 3   (6) *Quercus*

    g) pine 8   (7) *Juglans*

    h) walnut 6   (8) *Pinus*

    i) birch 9   (9) *Betula*

    j) sunflower 10   (10) *Helianthus*

    k) maple 11   (11) *Acer*

    l) beech 12   (12) *Fagus*

C) Diagram the seven groups of the taxonomical hierarchy of specification.

SECTION 2

Plant Science

Parts of the Plant and Their Functions

OBJECTIVE

To recognize the main parts of a plant and describe the function of each.

Competencies to Be Developed

After studying this unit, you should be able to

- ◆ note two contributions of plants to the life cycle on earth.
- ◆ list and describe the purpose of the four main parts of plants.
- ◆ explain the process of photosynthesis.
- ◆ explain the major structural difference between dicot and monocot stems, and how the stems grow.
- ◆ describe the process of pollination.

MATERIALS LIST

- ✔ a complete plant with roots, stem, leaves, and flower or seeds
- ✔ a plant with a fibrous root system
- ✔ a plant with a tap root system
- ✔ a large flower for dissection
- ✔ transparencies of plant parts
- ✔ overhead projector
- ✔ knives for dissecting flower parts
- ✔ a small monocot and dicot plant

THE IMPORTANCE OF PLANTS

The importance of plant life on earth cannot be overemphasized. Without plants, life on earth could not exist. Directly or indirectly, plants are the primary source of food for humans and animals. Whether people eat plants or eat animals that feed on plants, plant life is vital as a food source.

Plants play another essential role by producing oxygen. Without oxygen, life on earth could not exist. Plants are the major producers of oxygen on this planet. All plant life, from the smallest plankton in the ocean to the giant redwood tree, works to produce oxygen.

In addition to supplying food and oxygen, plants help to keep us cool, renew the air, slow down the wind, hold soil in place, provide a home for wildlife, beautify our surroundings, perfume the air, and furnish building materials and fuel.

PARTS OF A PLANT

Most plants are made up of four basic parts: leaves, stems, roots, and flowers, which later become fruit or seeds, figure 3-1.

LEAVES

Leaves are the food factory of the plant, producing all food that is used by the plant and stored for later use by the plant or by animals.

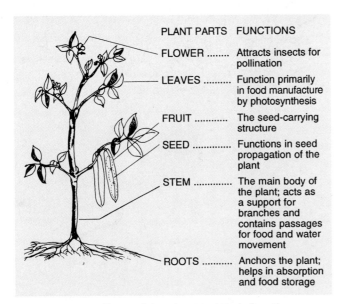

FIGURE 3-1   Parts of the plant and their functions

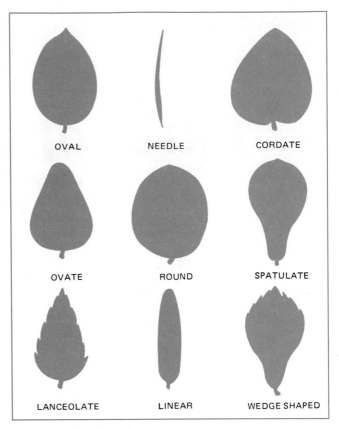

FIGURE 3-2   Leaf forms

Leaves vary a great deal in shape and size. Most leaves are flat. Some, such as the leaves of pine trees, are needlelike, while others, such as onion leaves, are cylindrical. The shape and size of leaves helps to identify plants, figure 3-2.

The arrangement of leaves on plants also differs. Some plants have leaves which alternate on the stem; some are positioned opposite one another. Others are *whorled* (arranged in a circle around the stem), figure 3-3.

EXTERNAL LEAF STRUCTURE   Leaves consist of the *petiole*, or leaf stalk, and the *blade*, a larger, usually flat part of the leaf, figure 3-4. Notice that the leaf blade has veins and a midrib. The *midrib* is the large center vein from which all other leaf veins extend.

The *veins* of the leaf form its structural framework. Most leaves have one of the forms shown in figure 3-2. Different leaf *margins* (edges of plant leaves) are shown in figure 3-5. Awareness of the different leaf margins assists in plant identification.

INTERNAL LEAF STRUCTURE   Internally, leaves have specialized cells which perform very important tasks, figure 3-6. The skin of the leaf,

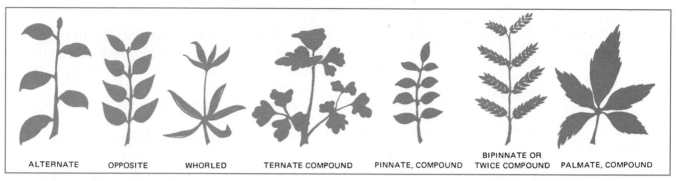

FIGURE 3-3   Leaf arrangements

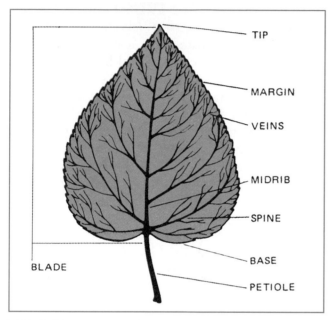

FIGURE 3-4   Parts of a leaf

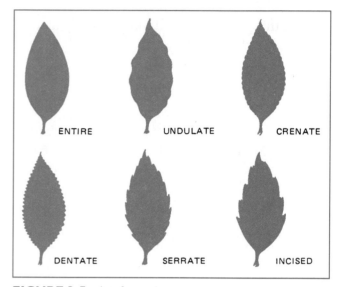

FIGURE 3-5   Leaf margins

called the *epidermis*, is a single layer of cells. Its chief function is to protect the leaf from loss of too much moisture. There are special cells in the leaf skin known as *guard cells*. These cells open and close a small space or pore on the underside of the leaf called a *stoma* to allow the leaf to breathe and *transpire*, or give off moisture and exchange gases such as oxygen and carbon dioxide. The guard cells are crescent shaped and the inner walls are thick. As the walls become turgid, due to water pressure in the cells, they open the stoma.

In the center of the leaf are food-making cells which contain *chloroplasts*. The green color of the chloroplasts, which gives green leaves their color, comes from the chlorophyll they contain. These cells, through a process called photosynthesis, manufacture food. *Photosynthesis* is the process by which carbon dioxide and water in the presence of light are converted to sugar and oxygen. This is

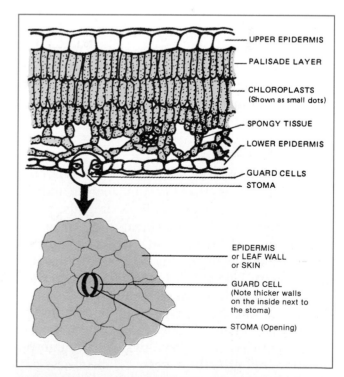

FIGURE 3-6   Cross section of leaf (top). Detail of guard cells and stoma (bottom).

the beginning of the food chain for all living things on earth. The chemical process is explained here in chemist's shorthand:

$$6\ CO_2 + 6\ H_2O + 672\ kcal = C_6H_{12}O_6 + 6\ O_2$$
carbon + water + light = glucose + oxygen
dioxide energy (sugar)

Thus, the equation states that 6 molecules of carbon dioxide plus 6 molecules of water are combined using 672 kilocalorie of light energy to produce 1 molecule of glucose, also known as dextrose or corn sugar, and 6 molecules of oxygen. *Note:* One kilocalorie is the energy required to heat 1,000 grams of water 1°C or 1 quart of water approximately 2°F.

LIGHT ENERGY

$$6\ CO_2 + 6\ H_2O = C_6H_{12}O_6 + 6\ O_2$$
carbon dioxide + water = sugar + oxygen

CHLOROPHYLL

The oxygen resulting from the photosynthesis process is used directly by animals. It is the vital ingredient in all forms of *oxidation* (combining with oxygen), such as burning, rusting, and rotting.

Food manufactured in the leaves moves downward through the stem to the roots. It is then used by the plant or stored in the stem or root in the form of sugar, starch, or protein. The leaves themselves are also used as food for various animals, including human beings. They are often the most nutritious part of the plant.

RESPIRATION

Plants also respire 24 hours a day just as animals do. In this process they consume oxygen and give off carbon dioxide the same as animals. Roots, stems, and leaves all use (breathe in) oxygen as they grow and breathe out carbon dioxide. An adequate supply of oxygen to all parts of the plant is necessary. Due to poor soil drainage and water pushing the air from the soil pore space, roots suffer most from short oxygen supply and possible plant death. Plants produce more oxygen through photosynthesis than they consume through respiration and growth processes. This excess oxygen is used by animals and by all other processes of oxidation.

STEMS

Stems have two main functions: (1) the movement of materials, such as the movement of water and minerals from roots upward to the leaves, and the movement of manufactured food from the leaves down to the roots, and (2) the support of the leaves and reproductive structures (flowers and fruit or seeds). Stems are also used for food storage, as in the Irish potato, and for reproduction methods which involve stem cuttings or grafting. Green stems manufacture food just as leaves do.

EXTERNAL STEM STRUCTURE The outside of the stem consists of *lenticels*, or breathing pores, bud scale scars, and leaf scars. *Bud scale scars* indicate where a terminal bud has been located; the distance between two scars represents one year of growth. *Leaf scars* show where leaves were attached, figure 3-7.

Some plants, such as the Irish potato and the gladiolus, have stems which differ greatly from the stems of other plants. The stems in these cases are used for food storage and plant reproduction. A

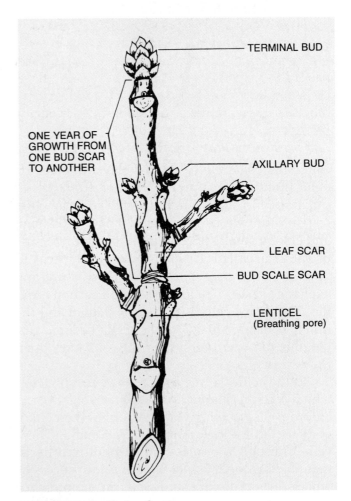

FIGURE 3-7 Parts of a stem

detailed discussion of this type of stem and how it differs from others is given in Section 3, Plant Propagation.

INTERNAL STEM STRUCTURE   In all stems, water and minerals travel up the xylem and manufactured food travels down the phloem. In *dicots*, plants which have two seed leaves (cotyledons), xylem and phloem form two layers separated by the *cambium,* which produces all new cells, figure 3-8. Dicot stems may continue to increase in diameter because the cambium builds new phloem cells on the outside and new xylem cells on the inside. Trees are good examples of dicots that have stems which continue to expand. In woody dicots, sapwood consists of new active xylem; heartwood is old, inactive xylem; and tree bark is old, inactive phloem (plus other cells).

This theoretically unlimited continuation of growth through the production of new cells is one of the major differences between plants and animals. For example, all the members of an animal's body are differentiated before birth. A human usually stops growth at maturity. Yet, a dicotyledonous tree continues to grow new branches and roots, and to grow taller as long as it lives.

Some plants, such as grasses, have a different stem structure, figure 3-9. These plants are referred to as *monocots* because they have only one cotyledon (seed leaf). Corn is an example of a monocot. Monocots have vascular bundles which contain both phloem and xylem tissue in each small bundle. Notice that the monocot stem has no outside circling cambium. All cells are formed in the initial stage of stem growth and merely enlarge to create the size of the mature stem. There is no further enlargement of stem size by formation of new cells, however, some stems form additional vas-

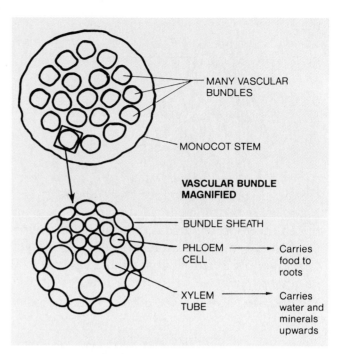

FIGURE 3-9   Cross section of a monocot stem

cular bundles as they grow. These plants are limited in size much as animals are.

ECONOMIC IMPORTANCE   The stems of some plants, such as the Irish potato and asparagus, are used as food. Others are used as building materials, such as the lumber from tree trunks.

ROOTS

Roots are usually underground and, therefore, are not easily visible. Roots function to

- ◆ anchor the plant and hold it upright.

- ◆ absorb water and minerals from the soil and conduct them to the stem.

- ◆ store large quantities of plant food.

- ◆ propagate or reproduce some plants.

The first three functions are essential to all plants.

STRUCTURE   The internal structure of a root is much like that of a stem. Older roots of shrubs and trees have phloem (old outer layers form corklike bark) on the outside, a cambium layer, and xylem (wood) inside, just as stems do. The phloem carries manufactured food down to the root for food and storage and the xylem carries water and minerals up to the stem.

The external structure of the root is very different from that of the stem. Whereas the stem has a terminal bud which initiates new growth, roots have a root cap. Just behind the root cap are

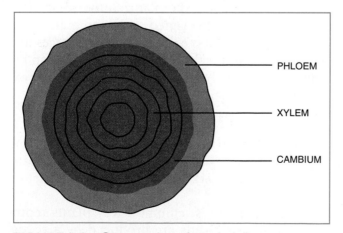

FIGURE 3-8   Cross section of a typical dicot stem

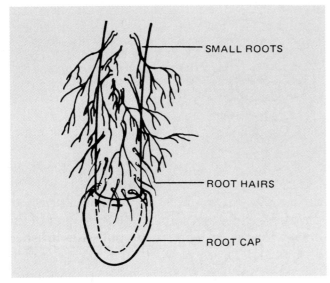

SMALL ROOTS

ROOT HAIRS

ROOT CAP

FIGURE 3-10   External view of a young root (greatly enlarged)

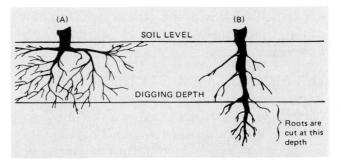

(A)   SOIL LEVEL   (B)

DIGGING DEPTH

Roots are cut at this depth

FIGURE 3-11   The fibrous root system (A) and the tap root system (B)

many root hairs. Side roots of increasing size form as the root grows older, figure 3-10. The root cap produces a continuous supply of new cells which rub off to lubricate a path and protect the cap and new root as the root pushes its way through the soil. The *root hairs* absorb moisture and minerals which are conducted to the larger roots and to the stem of the plant.

In addition to their function within the plant itself, many roots are important as cash crops for food. Vegetables such as carrots, beets, radishes, and sweet potatoes are all roots. Roots also serve in the process of propagation. Plants with tuberous roots, such as the dahlia, peony, and sweet potato, are propagated by separating the root clump or by rooting sprouts from the root, as is the case with the sweet potato.

TYPE OF ROOT SYSTEM   The ease with which nursery stock is transplanted or moved depends to a great extent on the type of root system that the plant possesses.

FIBROUS ROOT SYSTEM V. TAP ROOT SYSTEM   Plants with fibrous root systems are much easier to transplant than plants which have tap root systems. As illustrated in figure 3-11, when a plant with a fibrous root system and a plant with a tap root system are dug from the same depth in the ground, a greater percentage of the fibrous root system is saved. The roots are shorter, smaller, and more compact.

The tap root system has longer and fewer roots. Because of this, much of the root system is cut off

when a plant is dug. The ends of the roots which are lost in the cutting contain many root hairs necessary in the absorption of water and minerals from the soil. The larger roots serve only to conduct and store water, nutrients, and food. If too many of the small roots are lost, the plant may not be able to replenish the moisture lost by the leaves, and the plant will dry out and die.

FLOWERS, FRUITS, AND SEEDS

To most people, flowers are something of beauty meant to be seen and enjoyed. Those same people usually think of fruits and seeds as healthful foods. The parts that are admired and enjoyed by human beings, however, have an entirely different purpose for the plant.

The beauty of the flower, for example, is necessary to attract insects. In their visits for nectar or pollen, these insects fertilize the flower by means of a process called *pollination*. This is the beginning of fruit and seed formation. The fruits and seeds are made attractive to animals and birds so that they are collected, eaten, and spread. This, in turn, reproduces the plant. Some seeds have sticky coats and are carried on the animal's fur to be dropped in a new location. These devices and actions are not for the convenience of human beings or animals, but to ensure the continued existence of the plant itself.

PARTS OF THE FLOWER

Flowers differ in such features as size, shape, and color, but generally have the same basic parts. These basic parts are necessary for the production of seed.

Seed is the most common way plants reproduce in nature. Seeds are produced by a sexual

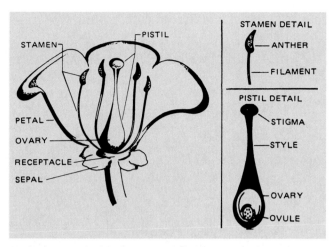

FIGURE 3-12   Parts of a complete flower

process with a male and a female parent involved. A *complete flower* has both male and female parts and only one parent is needed if the plant is *self-fruitful*, or can pollinate itself. The following is a description of a *complete flower*, figure 3-12. The complete flower contains four main parts: sepals, petals, stamens, and pistil.

The sepals are the green leaflike parts of the flower that cover and protect the flower bud before it opens.

The petals are actually leaves, but are generally known as the most striking part of the flower. The bright colors of the petals, which are usually present on flowers act to attract insects for pollination.

The stamens make up the male reproductive part of the flower. Each stamen consists of a short stalk called a *filament* and a saclike structure on top of the filament called an *anther*. The anther contains pollen, which is the male sex cell.

The pistil, located in the center of the flower, is the female part of the flower. It produces the female sex cells, the eggs (*ovules*). These eggs, if fertilized, become seeds. The pistil has three main parts: a sticky *stigma* on top to catch pollen and a *style*, a tube that leads to the third part, the *ovary*. The egg cells develop in the ovary. After fertilization, the ovary grows to become a fruit or a seed coat.

The flower is constructed so that insects attracted to it for nectar must first climb over the anther and brush pollen on the hairy surface of their bodies. As they climb onto the center of the flower for nectar, part of the pollen is brushed onto the stigma of the pistil. This allows the fertilization process to begin. The pollen grain sprouts like a seed and sends a long stalk down the style to the ovary and the egg cells. The pollen sperm cell then fertilizes the egg cells and seeds begin to develop. The ovary enlarges into a seed coat or fruit.

An *incomplete flower* has only the male parts or the female parts of the flower, but not both. Thus, a male flower has sepals, petals, and stamens, but no pistil. A female flower has sepals, petals, and pistil, but no stamens.

Flowers play a very important role in the florist and nursery businesses. Many plants are grown solely for the beautiful flowers they produce. Plants may grow flowers only to attract insects for pollination, but people grow them for their beauty and economic value. The floriculture industry (that part of the horticulture industry that deals with flowers) is a multimillion dollar business.

 # Student Activities

1) Dissect a flower. Sketch and label all parts.

2) In a notebook, draw and label a cross section of the four main plant parts.

3) Cut a cross section of a dicot stem and a monocot stem and draw a sketch, noting the differences.

Caution: Exercise care when handling knives or saws.

4) Obtain two plants common to the area and label a sketch of a leaf from each plant. Note the shape or form, type of margin, and arrangement of leaves. See text for information.

5) Start a monocot seed (corn) and a dicot seed (bean). As they develop, note that the monocot has one seed leaf and the dicot has two seed leaves.

 # Self-Evaluation

A) 1) Write the formula for photosynthesis in chemist's shorthand and in words.

2) Label the lettered parts of the plant and flower shown below.

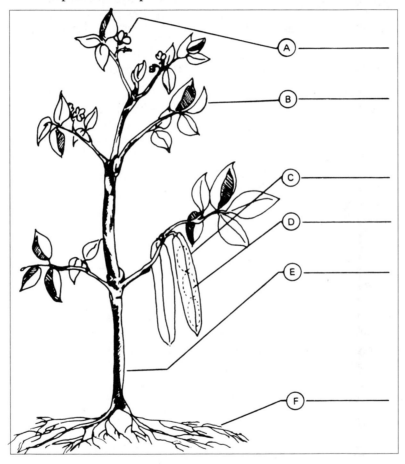

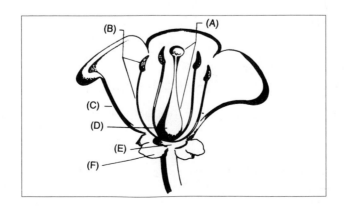

B) The column on the left contains a statement associated with plants and their functions. Select the correct term from the right-hand list and match it with the proper statement on the left.

1) A specialized stem that stores food and is used in plant propagation.
2) Water and minerals travel up this part of the plant stem (the wood of the stem).
3) Manufactured food travels down this part of the plant stem (the bark of the stem).
4) The part of the stem found in dicots but not monocots, which allows dicot plants to continue to produce new cells after initial cell division.
5) Plants with these root systems are easier to transplant than plants with tap root systems.
6) The flower is important to the plant only if it is fertilized and produces this plant part.

a) seed
b) phloem
c) Irish potato
d) food
e) xylem
f) fibrous
g) cambium
h) anchor
i) fertilize

C) Select the best answer from the choices offered to complete the statement or answer the question.

1) The two most important things that human beings receive from plants are

a) shade and food.
b) oxygen and food.
c) beauty and food.
d) food and clothing.

2) Plants also accomplish other purposes. They

a) keep us cool and clean the air.
b) slow the wind, prevent erosion, and provide beauty.
c) perfume the air and furnish fuel and building materials.
d) all of the above

3) The four basic parts of a plant are the

a) leaves, stems, roots, and flowers which produce seeds.
b) flowers, leaves, pollen, and fruit.
c) sepal, pistil, stamen, and ovary.
d) none of the above

4) Two main functions of stems are

a) to store food and convert it to starches.
b) to move materials and support plant parts.
c) to manufacture food and store it for future use.
d) to furnish food for human beings and other animals.

5) The stems of dicots, which have a cambium layer of cells,

a) produce all new cells quickly, then cease growing.
b) continue to produce new cells and grow larger.
c) are always green.
d) are cold resistant.

6) Monocot stems have the xylem and phloem

a) on each side of a cambium layer.
b) in small bundles scattered through the stem.
c) side by side.
d) near the outside of the stem.

7) The xylem, or wood, of a stem

    a) conducts manufactured food down to the roots.
    b) stores food.
    c) is green in color.
    d) conducts water and minerals up to the leaves.

8) Which of the following is not a function of roots?

    a) storage of food
    b) absorption of water
    c) anchoring of plants
    d) manufacture of food

9) Which three of the following vegetables are roots?

    a) carrots, beets, sweet potatoes
    b) celery, Irish potatoes, radishes
    c) Irish potatoes, carrots, beans
    d) kale, Irish potatoes, beets

10) The green color of leaves is caused by tiny particles in the food-producing cells called

    a) guard cells.
    b) epidermis.
    c) chloroplasts.
    d) starch.

11) Green plants are able to manufacture food only in the presence of

    a) light.
    b) carbon dioxide.
    c) water.
    d) all of these

12) The four main parts of a flower are the

    a) pollen, ovary, pistil, and stamen.
    b) sepal, petal, stamen, and pistil.
    c) sepal, pistil, ovary, and stigma.
    d) none of these

13) Which of the following is provided for a plant by its leaves?

    a) shade
    b) a food factory
    c) anchorage
    d) support

14) Plants are easier to transplant if they have a

    a) a tap root system.
    b) large root system.
    c) fibrous root system.
    d) single root.

15) An example of a plant with a specialized root that stores food and is used to propagate the plant is the

    a) Irish potato.
    b) carrot.
    c) radish.
    d) sweet potato.

16) The major function of root hairs on roots is to

    a) grow into larger roots.
    b) absorb water and minerals from the soil.
    c) protect the root as it pushes through the soil.
    d) keep the root warm.

17) Pollination is a sexual process in which pollen is deposited on the stigma of the plant. It starts the process of fertilization and

    a) growth of the pollen tube.
    b) seed formation.
    c) production of a fruit or seed coat.
    d) all of the above

18) The stamen is

    a) the male part of the flower.
    b) the part of the flower that produces pollen.
    c) the part of the flower that holds the anther.
    d) all of the above

19) The pistil is

    a) the female part of the flower.
    b) the male part of the flower.
    c) the showy part of the flower.
    d) the pollen-producing part of the flower.

20) When fertilized, the eggs in the ovary grow into

    a) fruit.   c) seeds.
    b) seed cases.   d) flowers.

21) One of the major differences between plants and animals is

    a) dicotyledonous plants continue to grow taller as long as they live.
    b) plants manufacture their own food.
    c) an animal's final growth size is determined at birth.
    d) all of the above

Environmental Requirements for Good Plant Growth

OBJECTIVE

To explain the basic needs of plants and the various factors that make up their environment.

Competencies to Be Developed

After studying this unit, you should be able to

- ◆ list four factors that affect the roots of plants.
- ◆ describe the differences between clayey, sandy, and loamy soils and identify a sample of each.
- ◆ explain three ways to improve soil drainage and two ways to increase moisture retention of soil.
- ◆ explain what is meant by the pH value of soil.
- ◆ compose a balanced fertilizer program for one plant that is grown commercially in the area.
- ◆ list four aboveground requirements for good plant growth.
- ◆ list the three major plant food elements and two functions of each.

MATERIALS LIST

- ✔ fertilizer bag of garden fertilizer and lawn fertilizer
- ✔ fertilizer bag or container of water soluble fertilizer with minor elements
- ✔ samples of clayey, sandy, and loamy soils
- ✔ complete plant
- ✔ three half-gallon jars
- ✔ four potted plants for light and temperature experiments

To properly grow into profit-making crops, plants require a certain environment. This environment can be divided into two parts: the underground environment in which the roots live and grow, and the aboveground environment in which the visible part of the plant ex-

CROWN
Trees increase each year in height and spread of branches by adding a new growth of twigs. This new growth comes from young cells in the buds at the ends of the twigs.

TRUNK
The tree trunk supports the crown and produces the bulk of the useful wood.

ROOTS
Roots anchor the tree; absorb water, dissolved minerals and nitrogen necessary for the living cells which make the food; and help hold the soil against erosion. A layer of growth cells at the root tips makes new root tissue throughout the growing season.

FIRE RUINS TIMBER
Disease and insects enter through fire scars.

Diseased or decayed wood   Insect damage

PHOTOSYNTHESIS
Leaves are the most important chemical factories in the world. Without their basic product, sugar, there would be no food for man or animal, no wood for shelter, no humus for the soil, no coal for fuel.

Inside each leaf, millions of green-colored, microscopic "synthetic chemists" (chloroplasts) manufacture sugar. They trap radiant energy from sunlight for power. Their raw materials are carbon dioxide from the air and water from the soil. Oxygen, a byproduct, is released. This fundamental energy-storing, sugar-making process is called photosynthesis.

What happens to this leaf-made sugar in a tree? With the aid of "chemical specialists" (enzymes), every living cell—from root tips to crown top—goes to work on the sugar. New products result. Each enzyme does a certain job, working with split-second timing and in harmony the others. In general, they break down sugar and recombine it with nitrogen and minerals to form other substances.

THINNING INCREASES GROWTH

35 years growth before thinning   16 years growth after thinning

INNER BARK (Phloem) carries food made in the leaves down to the branches, trunk, and roots.

SAPWOOD (Xylem) carries sap from roots to leaves.

HEARTWOOD (was sapwood, now inactive) gives strength.

OUTER BARK protects tree from injuries

CAMBIUM (a layer of cells between bark and wood) is where growth in diameter occurs. It forms annual rings of new wood inside and new bark outside.

ENZYMES
- Change some sugar to other foods such as starches, fats, oils, and proteins, which help form fruits, nuts, and seeds.
- Convert some sugar to cell-wall substances such as cellulose, wood, and bark.
- Make some of the sugar into other substances which find special uses in industry. Some of these are rosin and turpentine from southern pines; syrup from maples; chewing gum from chicle trees and spruces; tannin from hemlocks, oaks, and chestnuts.
- Use some of the sugar directly for energy in the growing parts of the tree—its buds, cambium layer, and root tips.

TRANSPIRATION
Transpiration is the release of water-vapor from living plants. Most of it occurs through the pores (stomates) on the underside of the leaves. Air also passes in and out.

FIGURE 4-1 The growth of a tree. (Courtesy United States Department of Agriculture)

ists. Figure 4-1 illustrates the two environmental systems of trees and how they relate to tree growth.

In the commercial production of plants, individuals must control the plant environment to obtain the *optimum* (best) return for the investment made, figure 4-2.

THE UNDERGROUND ENVIRONMENT

In a news article Richard W. Zobel, plant geneticist of the USDA, stated that outer space may be less a mystery than the 24 inches of soil just below the earth's surface. Unlocking the secrets of this area of the soil called the rhizosphere could result in the development of "crops that are more resistant or salt tolerant, have dramatically higher yields, are more nutritious and require less fertilizers and chemical pesticides," Mr. Zobel said.

We need to learn how micro-organisms interact with the plant, and with each other. Plant roots aren't all the same; they're functionally different and have different growth patterns. Some die back every two weeks, continually feeding the micro-organisms in the soil.

By understanding the rhizosphere a whole multitude of things will come down. For example, we could reduce chemical use on crops by knowing what biological elements in the soil will do the same thing. This affects ground water quality since you reduce the chances of chemicals moving into the water.

It is only recently that we have had the tools to study the rhizosphere and the interaction of all the elements in this area. Even now it is a very difficult task to try to study the interaction since

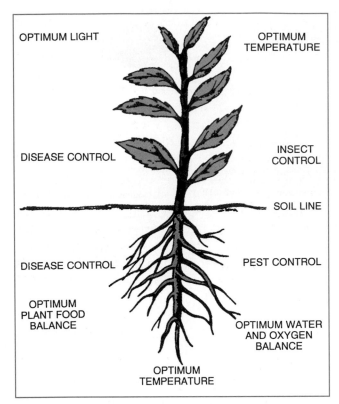

FIGURE 4-2 Requirements for optimum plant growth

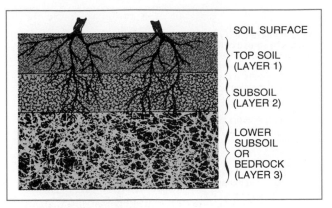

FIGURE 4-3 Soil profile

the sciences of soil physics and chemistry, plant biology and physiology, entomology and microbiology are all designed to study a single microorganism or element at a time. It is difficult when all are put together.

Although some plants require a more specialized underground environment, there are certain factors that affect the growth and development of all plants. The *medium* (soil or soil substitute) in which plants are grown is a very important factor. (*Note:* The plural of medium is *media* or *mediums.*) Through their roots, which anchor them in the soil, plants take in oxygen, moisture, and minerals—all vital to plant life. Many times, plant food is added to the soil to encourage better growth.

SOIL

Soil is made up of sand, silt, clay, organic matter, living organisms, and pore spaces which hold water and air. Soils are classified according to the percentage of sand, silt, and clay they contain.

Soil particles vary greatly in size. A sand particle is much larger than a silt particle. Clay particles are by far the smallest. These clay particles hold moisture and plant food elements much more effectively than larger particles. A certain amount of clay in all soil is important for this reason.

Soils vary greatly in general composition, depending on their origin. Some soils were formed as a result of rock breaking down over thousands of years; others developed as certain materials were deposited by water. A normal soil profile consists of three layers: (1) topsoil, (2) subsoil, and (3) soil bedrock or, if rock is not present, lower subsoil, figure 4-3. Topsoil represents the depth normally plowed or tilled, and contains the most organic matter or decaying plant parts. Deep-rooting plants send roots down in the subsoil, which is a well-defined layer immediately below the topsoil. If the soil is well drained, roots penetrate deeper into the subsoil since oxygen is available at greater depths. Roots may penetrate until rock, hard clay, or water prevents further growth.

The natural structure of soils is more important to the outdoor gardener, fruit grower, and nursery worker who plants outside than to the greenhouse operator or nursery worker who grows plants in containers. The worker who grows plants in containers can add ingredients to the soil to change its structure, moisture-holding ability, drainage ability, or fertility. In fact, most container plants are grown in completely soilless mixes.

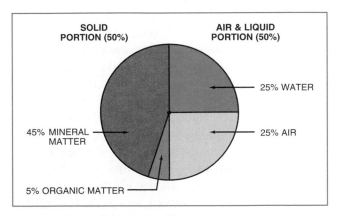

FIGURE 4-4 Soil composition

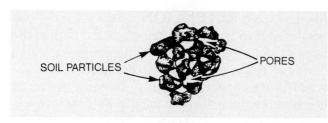

SOIL PARTICLES → ← PORES

FIGURE 4-5 Pore space in soil

An ideal soil is about 50 percent solid material, figure 4-4. This solid material consists mainly of minerals and a small percentage of organic matter. The other 50 percent of the ideal soil is pore space, figure 4-5. These pores are small holes between soil particles and are filled with water and air in varying amounts. After rain or irrigation, the pores may be nearly filled with water and the air is pushed out. As the soil dries, the amount of water decreases and the pores gradually fill with air again. The ideal water-to-air ratio in the pores is about half and half, 50 percent air to 50 percent water.

The amount of moisture and air a soil will hold depends on the soil structure and type of soil. Sandy soils with large particles also have large pore spaces. Water is lost more quickly from these large pores as the force of gravity drains the water out. These are called well-drained soils. As the clay content of the soil increases, more water is held. If soils contain too much clay, they may not drain well enough to allow enough oxygen in the pore space for good plant growth.

TYPES OF WATER IN SOIL *Gravitational water* is water that the soil is unable to hold against the force of gravity. It becomes a part of ground water or drains away in streams. It is of little value to the plant since it drains away quickly, carrying soluble plant food elements with it. Soils with larger pore spaces lose water faster to the force of gravity than do soils with smaller pore spaces.

Capillary water is held against the force of gravity. It is held in the small pore spaces of the soil and as a thin film of water around soil particles.

There are three types of capillary water. *Free moving capillary water* moves in all directions in soil just as it does in a glass capillary tube in a science lab. The soil must be saturated at low levels for water to continue to move upward in field soil. There is actually very little upward movement of water.

Available capillary water or field capacity is the water left after capillary movement stops. The soil surface is now dry and any water held is as a thin film around soil particles and in small capillary

tubes or pore spaces. Water does not continue to move through the soil at this point. Plant roots must continue to move through the soil in the search for water as the soil immediately surrounding them is dried out from root absorption of water. Roots will not continue to grow into air-dry soil where no moisture is available. When the soil is moist with available water, they will grow again. The field capacity for water is high (water is held) in heavy soils, with clay particles providing a lot of surface area and tiny pore spaces for the water to cling. This is the most important water for plants. They can extract and use about half of this water. As the film of water on soil particles gets thin more force is required to absorb it. When this force becomes about 15 atmospheres or about 200 pounds per square inch, plants can no longer absorb it. At this moisture level, plants will wilt and die. Desert plants can extract water at a force greater than 200 pounds per square inch.

Unavailable capillary water is not available to plants. It is held tightly as a molecular film around the soil particles and can only be moved as vapor. Soil this dry is called air dry because under normal drying conditions no more moisture will move.

SANDY SOIL *Sandy* (or *light*) soils include soils in which silt and clay make up less than 20 percent of the material by weight. These soils drain well, but have little capacity to hold moisture and plant food.

CLAYEY SOIL To be classified as a clayey soil, a soil must contain at least 30 percent clay. Such a soil is known as a heavy soil. Heavy soils have relatively poor drainage and aeration capabilities. (To *aerate* is to supply with air.) Because of this, heavy soils tend to hold more moisture than is good for plants. However, this type of soil also holds fertilizer and plant food well, which can be beneficial to plant growth.

LOAMY SOIL This is the most desirable soil for general use. *Loam* is a mixture of approximately equal parts of sand, silt, and clay. If it has more sand than silt or clay, it is known as a *sandy loam*; more clay, a *clayey loam*; more silt, a *silty loam*. The texture triangle shown in figure 4-6 is helpful in determining the names of soils.

SOIL IMPROVEMENT

Soils used for outdoor plant growth may be improved through increased drainage, irrigation

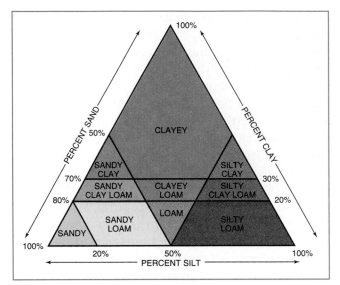

FIGURE 4-6 Texture triangle. This triangle may be used to determine the textural name of a soil by mechanical analysis (actually measuring the percent of sand, silt, and clay present in the soil). After the percentages of silt and clay are determined, these amounts are plotted on the soil triangle. This is done by projecting lines inward from the point on each side of the triangle which represents the percentage of that particular type of soil. The line drawn from the silt side of the triangle is placed parallel to the sand side of the triangle. The line projected from the clay line runs parallel to the silt line. The location of the point at which these two lines intersect indicates the name of the soil. The name of the section of the triangle in which the point is located is the name of the soil.

methods, and the addition of organic matter and plant food in the form of fertilizers.

Since it would be very expensive to change the percentage of sand, silt, or clay in a soil to improve the soil's drainage, aeration, or moisture-holding capacity, other practices must be used.

DRAINAGE AND AERATION   Drainage and aeration can be improved by changing the soil structure. One way this is done is by adding organic matter to encourage earthworms. Their tunnels and castings result in better soil structure through *aggregation*, the clinging together of soil particles in large crumblike shapes. Lime and gypsum (calcium sulfate) also aid in soil aggregation and improve structure in some cases. In effect, aggregation increases the size of soil particles much as if sand were added, that is, by the formation of larger spaces between particles. Another method is the use of tile drains to remove water from the soil. Raising planting beds and placing ditches between the beds are also methods used to control moisture in soil.

MOISTURE RETENTION   Often, sandy soils are not able to hold sufficient water for plants; the large particles and pore spaces encourage too much drainage. Adding organic matter improves the water-holding capacity of soils. Organic material holds many times its own weight in water, for example, peat moss holds fifteen times its weight in water. Organic matter also holds plant food effectively and allows the slow release of the food for plant use as the organic matter decomposes in the soil. Animal manure, green manure, peat moss, and sawdust are good sources of organic matter. (*Green manure* is a green herbaceous crop plowed under to improve soil.) *Mulches*, such as compost, wood chips, or bark, are placed on the surface of the soil to help retain soil moisture by reducing runoff, thereby allowing more rainwater to be absorbed into the soil. Mulches also keep soil cool, thus reducing evaporation loss. Irrigation methods are used to add water when rains do not supply enough.

Fertilizers should be used when necessary and according to recommendations of a soil test. Obtain directions on how to take a soil sample from a local extension agent or horticulture instructor. Send samples to the local soil test laboratory, asking for recommendations for the crop being grown.

Beyond these practices, the grower using field or nursery areas to cultivate crops has little control over the soil structure or its moisture content.

NUTRITIONAL DEFICIENCIES

Nutritional or plant food deficiencies often show up on the leaves of plants. Yellow or pale green leaves indicate a nitrogen deficiency. A phosphorus deficiency shows up as a purple color on the underside of the leaf. By the time these symptoms appear, damage has already been done to the plant. A soil test should have been used earlier to determine plant needs.

SOIL PESTS

Soils harbor certain diseases such as root rot and wilt. Nematodes, which are tiny animals, and insects may also damage roots. Resistant varieties of crops or natural controls such as crop rotation must be used to satisfactorily control these problems. Chemicals are sometimes used as a last resort to protect plants against diseases and insects.

Soil pasturization is needed any time soil or sand is used in a planting medium for container growing. The soil or sand should be heated thoroughly to 180°F for 30 minutes before being mixed with other materials.

PLANTING MEDIA MIXES

More and more growers are using planting mixes that contain little, if any, soil. In greenhouse operations and nurseries where plants are grown in flats, pots, or other containers, this is both convenient and economical.

ADVANTAGES   There are several advantages to using soilless media.

- The mix is uniform, that is, it does not vary in fertility, acidity (pH), or texture.

- The mixes are *sterile* (containing no disease organisms, insects, or weed seeds).

- Soilless mixes are lighter in weight, and thus easier to handle and ship.

- Good moisture retention and drainage are possible through the proper combination of ingredients.

DISADVANTAGES   Soilless mixes have some disadvantages.

- Since they are very lightweight, containers kept outdoors may be blown over.

- Since the mineral content in most mixes is low, minor plant food elements necessary to plant life may be missing, such as iron, sulfur, manganese, zinc, and calcium.

- Plants transplanted from mixes to soils often hesitate to extend roots into such a different growing medium. A problem develops when the growing medium remaining on the roots fails to blend with the soil into which the plant is transplanted. Moisture and other necessary nutrients are then unable to pass from the new soil to the roots. In addition, if the soil is heavy in clay, the plant's roots may never venture beyond the original root ball into the clayey soil. This situation may result in the death of the plant from lack of moisture. (The rhododendron and azalea, which are often grown in peat moss or sphagnum moss, are good examples of plants with this transplanting problem.) The problem may be solved by extending the root ends out into the soil when the plant is transplanted. Mixing sphagnum moss or pine bark into the soil at the permanent planting site also makes the transition easier.

CONTENT OF MIXES   Soilless mixes contain various combinations of the following materials.

Perlite, a gray-white material of volcanic origin, is usually used to improve aeration of the media. Horticultural grade perlite should be used because it has larger particles and provides better drainage and aeration than the finer builder's perlite. Sponge "ROK," an aluminum silicate, is used in the same manner as perlite.

Sphagnum moss is the dehydrated remains of acid bog plants, used in shredded form for seeds. Most horticulturists use shredded sphagnum moss to cover seeds because it is relatively sterile and lightweight, controls disease well, and has an excellent moisture-holding capacity. It is very acid.

Peat moss is partially decomposed vegetation that has been preserved underwater. The peat is collected from marshes, bogs, or swamps. It has a very high moisture-holding capacity, contains about 1 percent nitrogen, and is low in phosphorus and potassium.

Vermiculite is a very light, expanded material with a neutral pH (neither acidic nor alkaline). (When vermiculite is heated, the moisture in the mineral becomes steam and causes it to expand.) It has a very high moisture-holding capacity.

Limestone refers to ground natural limestone, also known as calcium carbonate ($CaCO_3$). It tends to raise pH.

Tree bark is usually the bark of pine or oak trees broken into small pieces and used in planting mixes. Fine (¼" or less in size) pine bark is now used extensively in container growing.

Slow-release fertilizers are fertilizers containing plant food which is gradually made available to plants over a period of time.

Mixes may contain two or more of the above ingredients. For plants requiring excellent drainage, coarse materials, such as bark or perlite, should make up a high percentage of the mix. Most mixes contain an organic material such as sphagnum or peat moss for moisture retention and a material such as perlite or bark for drainage and aeration. One of the most popular commercial mixes is 50 percent shredded sphagnum moss and 50 percent vermiculite, with a slow-release plant food added. Mixes are sold bagged and ready for use.

Soilless mixes may also be used as soil conditioners by digging them into the soil in varying amounts as needed. The mixes add organic matter, which increases moisture-holding capacity and improves soil drainage.

If soil is used as part of the planting mix, be sure it is sterilized or pasteurized. A mixture of one-third soil, one-third peat, shaving, or leaf mold, and one-third sand is suitable for most plants.

PLANT FOOD AND FERTILIZERS

Plants may be grown in soil, in soilless mixes, or in a combination of the two. In any case, plant food must be readily available to the plant. Fertilizer must be added as the plant requires food and the plant food elements must be in a water soluble form to be available to the plant.

WATER   Water is the most important plant food element by far. It makes up approximately 90 percent of the weight of plants and is the one most limiting factor in plant growth. All plant food elements are dissolved in water and move into and throughout the plant in a water solution. As noted earlier photosynthesis uses water in the manufacture of food and could not occur without it.

Only about 1 percent of the water absorbed is actually used by the plant. The other 99 percent is lost through the leaves and stems as water vapor in a process called *transpiration*. From 90 to 95 percent of the water lost from the plant is lost by transpiration through stomata in the leaves. Leaf temperature affects transpiration most. As leaf temperature rises, more water is lost. For each 10°C increase in temperature, the loss is doubled. A single corn plant can transpire 2 quarts of water per day. If it were not for this loss of water through transpiration, the water needs of plants would be greatly reduced. Loss of water from transpiration faster than the roots can pick it up results in wilting of the plant and plant death. Plants are unable to stop transpiration to conserve water.

Transpiration of water is high when soils are wet and the turgid guard cells of the stomata open wide to allow more water vapor and gas to escape. When water in the soil and plant is in short supply the stomata close and less water vapor is lost. It is estimated that there are 250,000 stomata in 1 square inch of the under-surface of an apple leaf. Stomata close at night and transpiration is greatly reduced. They open in response to light.

Transpiration is of little value to the plant. It cools the leaf slightly, but air movement and heat radiation are most important in leaf cooling. Transpiration is not needed for the absorption or movement of minerals in the plant. Perhaps its main value is in cooling and moistening the air around the plant through the process of evaporation of water. This natural air conditioning is beneficial to plant and animal.

Plant food may be divided into two groups: (1) *major elements*—nitrogen, phosphorus, and potassium (stated as *total nitrogen, available phosphoric acid,* and *soluble potash* when listed as fertilizer content), and (2) *minor elements*—calcium, magnesium, sulfur, iron, manganese, boron, copper, zinc, and chlorine. The elements are listed as major and minor according to the relative amounts of each element needed for good plant growth. Plants require relatively large amounts of the major elements and relatively smaller amounts of the minor elements. Other minor elements may also be needed in artificial mixes.

Commercial fertilizers and plant foods show the percentage or pounds per hundred weight of the three major elements in large numbers on the bag or container. This is the fertilizer analysis. If the container has the numbers 5-10-5 on the label, the mix is 5 percent elemental nitrogen, 10 percent available phosphoric acid, and 5 percent water soluble potash. The other 80 percent is filler material which makes it easier to spread the plant food evenly. The three main plant food elements, nitrogen, phosphorus, and potassium, are always listed in that order. It is easy to see how much actual plant food is being purchased on a percentage basis by reading the label, figure 4-7.

Minor elements may also be listed on the container. In parts of the country that have acidic soils, the reaction of acid content to the soil may be expressed in a statement such as "Acid equivalent to _____ pounds of limestone."

Just as human beings require a balanced diet, plants need a balance of food for best growth. Only a soil test can determine the amount of various plant food elements needed. If shortages exist, the plant will often show symptoms of deficiencies.

NITROGEN   Nitrogen is generally purchased in one of four forms. It is absorbed by plants in two forms—ammonium and nitrate.

- *nitrate of soda*, $NaNO_3$, which is highly soluble and quickly available. It also tends to lower soil acidity. It contains 16 percent nitrogen.

- *ammonium nitrate*, NH_4NO_3, which is not as soluble and available over a longer period of time. It contains 33 percent nitrogen.

FIGURE 4-7 The fertilizer bag shown on the left is sold as garden fertilizer. The analysis, in the bottom right-hand corner (10-10-10), indicates that the mix is 10 percent nitrogen, 10 percent available phosphoric acid, and 10 percent soluble potash, with a total of 30 percent actual plant food. The other 70 percent is filler, which aids in spreading fertilizer. The bag on the right, a lawn fertilizer, has a high ratio of nitrogen compared with the other elements. Notice that the fertilizer on the left contains the same percentage of all three elements. Grasses require more nitrogen than phosphorus or potash; this accounts for the higher ratio of nitrogen in the mix on the right.

- *ammonium sulfate*, $(NH_4)_2SO_4$, which becomes available more slowly and leaves the soil more acidic. It is good for plants that grow well in very acidic soil, and is 21 percent nitrogen.

- *urea formaldehyde*, which is an organic form of nitrogen and is more slowly available than the inorganic forms. It contains 38 percent nitrogen.

Of the three major plant food elements, nitrogen has the most noticeable effect on plants, with the effects showing soonest. Nitrogen encourages aboveground vegetative growth and gives a dark green color to leaves. It tends to produce soft, tender growth, a good quality for crops such as lettuce to possess. The tender growth makes the plant better tasting. Nitrogen also seems to regulate the use of the other major elements.

Because the addition of nitrogen quickly produces a visible effect, there is often a tendency to overuse it. Too much importance is often placed on this element without regard for a balanced plant food program.

Too much nitrogen may (1) lower the plant's resistance to disease, (2) weaken the stem because of long soft growth, (3) lower the quality of fruits, causing them to be too soft to ship, and (4) delay maturity or hardness of tissue and thus increase winter damage to plants.

Not enough nitrogen results in a plant being (1) yellow or light green in color, and (2) stunted in root and top growth.

Nitrogen is lost from the soil very easily by *leaching* (washing out). It is very soluble in water and is not held by the soil particles because of the charges of the particles involved. Soil particles have a negative charge; nitrogen also has a negative charge. Since like charges repel, nitrogen particles are not held in the soil. However, organic matter does tend to hold insoluble nitrogen which is released slowly into the soil. Nitrogen should not be used in excess for two reasons: (1) it is quickly lost from the soil through leaching, especially in sandy soils which lose water faster, and (2) it can damage plants if applied in too great an amount. Some plants such as legumes (beans, peas, etc.) manufacture their own nitrogen.

PHOSPHORUS Phosphorus is generally purchased in the following forms:

- Superphosphate—20 percent phosphate

- Treble superphosphate—46 percent phosphate

- Rock phosphate—25-35 percent phosphate

- Ammonium phosphate—48 percent phosphate

Phosphorus is present to some extent in all soils. Unlike nitrogen, it is held tightly by soil particles and therefore is not easily leached from soil. However, because it may not be in the water soluble form, it is usually not available to plants in the amount needed. This means that additional phosphate fertilizers should be applied. Whether or not additional phosphorus is needed can be determined by the use of a soil test.

Phosphorus affects plants in several ways.

- It encourages plant cell division.

- Flowers and seeds do not form without it.

- It hastens maturity, thereby offsetting the quick growth caused by nitrogen.

- It encourages root growth and the development of strong root systems.

- It makes potash (potassium) more easily available.

- It increases the plant's resistance to disease.

◆ It improves the quality of grain, root, and fruit crops.

Since phosphorus is held very tightly by soil particles, it does not usually cause damage to the field grown plants if excessive amounts are applied. However, container grown plants can be damaged by excesses of any soluble fertilizer element since it increases the soluble salt content present in the media. Fertilizers which are high in soluble salts *dehydrate* (dry out) plant roots by pulling water from the roots.

Insufficient phosphorus results in (1) purple coloring on the undersurface of leaves, (2) reduced flower, fruit, and seed production, (3) susceptibility to cold injury, (4) susceptibility to plant diseases, and (5) poor quality fruits and seeds.

POTASSIUM The most common sources of potassium are:

◆ muriate of potash—60 percent potash

◆ sulfate of potash—49 percent potash

◆ nitrate of potash—44 percent potash (also 13 percent nitrogen)

Potassium is rarely present in the soil in sufficient amounts to harm plants. It tends to modify both the fast, soft growth of nitrogen and the early maturity of phosphorus.

The presence of potassium is essential for several reasons.

◆ It increases the plant's resistance to disease.

◆ It encourages a strong, healthy root system.

◆ It is essential for starch formation.

◆ It is needed for the development of chlorophyll.

◆ It is essential for tuber development.

◆ It encourages the efficient use of carbon dioxide.

Since it is a major element, potash is generally added to soil. The amount is determined by soil tests.

Potassium deficiency appears as a marginal yellowing or scorch on the edges of leaves on the lower portion of the plant. This symptom is easily mistaken for moisture shortage during dry soil conditions.

LIME (CaCO₃) Lime acts as a plant food and as a material that affects soil acidity. Soil acidity, in turn, affects the availability of other plant food elements.

Lime furnishes calcium, one of the most important of the minor food elements. Calcium is important in the formation of plant cell walls, among other functions.

SOIL ACIDITY (pH)

Most plants grow best in soil with a pH of from 5.6 to 7.0. A pH of 7.0 is neutral; that is, the soil at pH 7 is neither acid nor alkaline (basic). *Alkaline* soil is the opposite of acid soil in pH rating. Hence, on a scale of 1 to 14, values lower than 7.0 indicate acid soils and values higher than 7.0 indicate alkaline soils, figure 4-8. Figure 4-9 gives the pH preferences of common flowers, ornamentals, vegetables, and small fruits.

In the United States, soils tend to be acid in areas where the parent material of the soil was acid and where the amount of rainfall exceeds evapo-

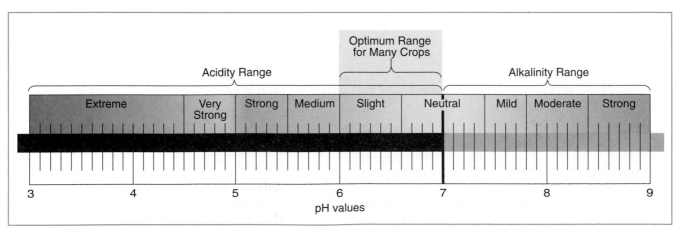

FIGURE 4-8 The lower the pH number, the more acidic the soil

FLOWERS AND ORNAMENTALS

Plant	pH	Plant	pH	Plant	pH	Plant	pH
African violet	5.5–7.5	Chrysanthemum	6.0–7.5	Gladiolus	6.0–7.0	Mock orange	6.0–8.0
Ageratum, blue	6.0–7.5	Clematis	5.5–7.0	Gloxinia	5.5–6.5	Morning glory	6.0–7.5
Alyssum	6.0–7.5	Clivia	5.5–6.5	Heather	4.5–6.0	Narcissus	6.0–7.0
Amaryllis	5.5–6.5	Coleus	6.0–7.0	Hens and chickens	6.0–8.0	Nasturtium	5.5–7.5
Anthurium	5.0–6.0	Columbine	5.5–7.0	Hibiscus, Chinese	6.0–8.0	Pansy	5.5–6.5
Arbor vitae	6.0–7.5	Coralbells	6.0–7.0	Holly, English	4.0–5.5	Peony	6.0–7.5
Arbutus, trailing	4.0–5.0	Cosmos	5.0–8.0	American	5.0–6.0	Petunia	6.0–7.5
Aster	6.5–7.0	Coxcomb	6.0–7.5	Hollyhock	6.0–8.0	Philodendron	5.0–6.0
Azalea	4.5–6.0	Crab, flowering	5.0–6.5	Honeysuckle bush	6.5–8.0	Phlox	5.0–6.0
Baby's breath		Crocus	6.0–8.0	Hoya	5.0–6.0	Pink	6.0–8.0
(gypsophila)	6.0–7.5	Cyclamen	6.0–7.0	Hydrangea, blue	4.0–5.0	Poinsettia	6.0–7.5
Baby's tears	5.0–5.5	Daffodil	6.0–6.5	pink	6.0–7.0	Poppy	6.0–7.5
Bachelor's button	6.0–7.5	Dahlia	6.0–7.5	white	6.0–8.0	Portulaca	5.5–7.5
Barberry	6.0–7.5	Daisy, shasta	5.0–6.0	Impatiens	5.5–6.5	Primrose	5.5–6.5
Bayberry	5.0–6.0	Daphne	6.5–7.5	Iris	5.5–7.5	Rhododendron	4.5–6.0
Begonia	5.5–7.0	Delphinium	6.0–7.5	Iris, Japanese	5.5–6.5	Rose	6.0–7.0
Bird of paradise	6.0–6.5	Deutzia	6.0–7.5	Ivy, Boston	6.0–8.0	Rose of Sharon	
Bleeding heart	6.0–7.5	Dieffenbachia	5.0–6.0	English	6.0–8.0	(althea)	6.0–7.5
Bloodleaf	5.5–6.5	Dogwood	5.0–6.0	Grape	5.0–6.5	Rubber plant	5.0–6.0
Bluebell	6.0–7.5	Eucalyptus	6.0–8.0	Jasmine	5.5–7.0	Salvia	6.0–7.5
Boxwood	6.0–7.5	Fern, asparagus	6.0–8.0	Lady's slipper	5.0–6.0	Scilla	6.0–8.0
Bridal wreath	6.0–8.0	bird's nest	5.0–5.5	Larkspur	6.0–7.5	Shrimp plant	5.5–6.5
Burning bush	5.5–7.5	Boston	5.5–6.5	Laurel	4.5–6.0	Snapdragon	6.0–7.5
Butterfly bush	6.0–7.5	maidenhair	6.0–8.0	Lilac	6.0–7.5	Sweet pea	6.0–7.5
Cacti	4.5–6.0	Forget-me-not	6.0–8.0	Lily, calla	6.0–7.0	Verbena	6.0–8.0
Cactus, Christmas	5.0–6.5	Forsythia	6.0–8.0	day	6.0–8.0	Wisteria	6.0–8.0
Camellia	4.5–5.5	Foxglove	6.0–7.5	Easter	6.0–7.0	Yew, Japanese	6.0–7.0
Campanula	5.5–6.5	Freesia	6.0–7.5	madonna	6.5–7.5	Yucca	6.0–8.0
Candytuft	5.5–7.0	Fuchsia	5.5–6.5	tiger	6.0–7.0	Zinnia	5.5–7.5
Canna	6.0–8.0	Gardenia	5.0–6.0	Lily of the valley	4.5–6.0		
Carnation	6.0–7.5	Geranium	6.0–8.0	Marigold	5.5–7.0		

VEGETABLES

Plant	pH	Plant	pH	Plant	pH	Plant	pH
Artichoke		Cauliflower	5.5–7.5	Lettuce	6.0–7.0	Radish	6.0–7.0
(Jerusalem)	6.5–7.5	Celery	5.8–7.0	Mushroom	6.5–7.5	Rhubarb	5.5–7.0
Asparagus	6.0–8.0	Chicory	5.0–6.5	Mustard	6.0–7.5	Rutabaga	5.5–7.0
Beans, lima	6.0–7.0	Chives	6.0–7.0	Okra	6.0–7.5	Sage	5.5–6.5
pole	6.0–7.5	Corn	5.5–7.5	Onion	6.0–7.0	Salsify	6.0–7.5
Beets, sugar	6.5–8.0	Cucumber	5.5–7.0	Parsley	5.0–7.0	Shallot	5.5–7.0
table	6.0–7.5	Eggplant	5.5–6.5	Parsnip	5.5–7.0	Spinach	6.0–7.5
Broccoli	6.0–7.0	Endive	5.8–7.0	Peas	6.0–7.5	Squash, crookneck	6.0–7.5
Brussels sprouts	6.0–7.5	Garlic	5.5–8.0	Peanut	5.3–6.6	Swiss chard	6.0–7.5
Cabbage	6.0–7.5	Horseradish	6.0–7.0	Pepper	5.5–7.0	Tomato	5.5–7.5
Cantaloupe		Kale	6.0–7.5	Potato	4.8–6.5	Turnip	5.5–6.8
(muskmelon)	6.0–7.5	Kohlrabi	6.0–7.5	Sweet potato	5.2–6.0	Watercress	6.0–8.0
Carrot	5.5–7.0	Leek	6.0–8.0	Pumpkin	5.5–7.5	Watermelon	5.5–6.5

SMALL FRUITS

Plant	pH	Plant	pH
Blackberries	5.0–6.0	Raspberries, black	5.0–6.5
Blueberries	4.0–5.5	red	5.5–7.0
Grapes	5.5–7.0	Strawberries	5.0–6.5

FIGURE 4-9 pH preferences of plants

ration of moisture from the soil. This is due to water draining through the soil and washing out the salts of sodium and calcium. This condition occurs in the eastern United States and along the West Coast down to about central California.

In areas where water evaporation from the soils is equal to or greater than the amount of rainfall, the salts of calcium and sodium tend to build up in the soils and increase the pH level. These salts may build up enough and thereby raise the pH so much that plants cannot grow in these soils. This occurs in many areas west of the Mississippi

River, some to the point of being too alkaline to grow crops. Irrigation with low-salt-content water can wash the alkali-producing elements out of some of these soils and greatly improve growing conditions.

LOWERING pH IN ALKALI SOILS Whenever it is necessary to lower soil pH for the best growth of plants, materials such as sulfur, iron sulfate, or aluminum sulfate may be used. This practice may be needed in alkali soil areas of the western United States. As mentioned, flushing these soils with low-salt irrigation water will also lower pH.

LIME, pH, AND OTHER PLANT FOOD ELEMENTS Lime serves a very important function in changing soil acidity or pH. When a soil test is made and the soil proves too acidic, lime is added to raise the pH.

Lime also affects the availability of other plant food elements to plants. For example, if a soil is acid with a low pH (5.5-6.5), phosphorus is tied up and not readily available. Adding lime releases phosphorus and makes it available to the plants. As another example, consider the fact that acid soils release iron and aluminum into soil. This poses a problem, since aluminum may be toxic or poisonous to some plants. The application of lime decreases the availability of aluminum and iron. On the other hand, liming such crops as blueberries, azaleas, and rhododendron may cause iron deficiency if the pH is raised above 6.0.

Liming to produce the proper pH also activates soil organisms and encourages the release of plant food. Soil structure is usually improved with the addition of calcium in the form of lime or gypsum. Gypsum does not change soil pH.

THE ABOVEGROUND ENVIRONMENT

Just as the entire plant is influenced by the underground environment of the plant roots, the entire plant is also affected by the environment surrounding the top of the plant. The aboveground environment is more changeable and may be more violent in its effect on plants.

The aboveground environment may be explained in terms of the factors affecting plants. These include (1) temperature, (2) light, (3) humidity, (4) plant diseases, (5) insects, and (6) gases or particles in the air.

TEMPERATURE

The temperature of the air has one of the strongest effects on plant growth. Some plants, such as lettuce, cabbage, and kale, grow best in cool temperatures. Others, such as corn, beans, and tomatoes, prefer hot weather. There are temperatures above which all plant growth stops. At the other extreme, temperatures near and below freezing also stop plant growth and, in fact, kill tender crops. For the best temperatures for raising specific crops, see figure 4-10. Generally the plant growth rate increases as temperature increases up to temperatures of about 90°F. This varies, but is a good general rule, providing that moisture is available to the plant and wilting does not occur.

LIGHT

Light must be present before plants can manufacture food. No green plant can exist for very long without light, whether that source is sunlight or light from an artificial source. Plants vary in the amount of light they require for best growth. Some plants prefer full sunlight; others prefer varying degrees of shade.

Light affects plants in other ways. Some plants, such as chrysanthemum, bloom only when the days begin to shorten. (Long nights are needed for flower buds to form.) This response to different periods of day and night in terms of growth and maturity of the plant is called *photoperiodism*.

Flowering is one way in which plants react to varying periods of light and dark. Plants may be classified in three groups according to this flowering reaction. *Short day* plants, such as chrysanthemum and Christmas cactus, flower when days are short and nights are long. *Long day* plants, such as lettuce and radishes, flower when days are long and nights are short. *Indifferent* plants are plants that do not depend upon certain periods of light or darkness to flower. The African violet and tomato are indifferent plants.

There are other ways in which plants react to length of days. For example, the black raspberry roots from cane tips in 5 to 10 days if the tips are covered with soil in September or October. In contrast, a rooting in midsummer may require up to six weeks. Dahlias develop a fibrous root system during the relatively long days of summer but as days shorten, the roots become thick storage organs.

Plants grow toward their source of light because the plant stem produces more growth hormones on the shady side, causing the stem on that side to grow to a greater length.

TEMPERATURE		SUNLIGHT	
Cool Temperature Plants 60°–80°F	Warm Temperature Plants 75°–90°F	Full Sunlight	Partial Shade

Vegetable Crops	*Vegetable Crops*	*Vegetable Crops*	*Vegetable Crops*
Beet Onion Broccoli Parsley Cabbage Pea Carrot Potato Chard Radish Endive Spinach Kale Turnip Lettuce	Bean Pepper Corn Squash Cucumber Sweet potato Eggplant Tomato Melon	Bean Melon Beet Onion Broccoli Pea Cabbage Pepper Carrot Potato Corn Spinach Cucumber Squash Eggplant Tomato Kale Turnip Lettuce	Beet Onion Broccoli Parsley Cabbage Radish Kale Spinach Lettuce Turnip
Flowering Plants	*Flowering Plants*	*Flowering Plants*	*Flowering Plants*

Arbutus
Azalea
Deciduous azalea
Primose
Rhododendron
	Aster
	Marigold
	Petunia
	(nearly all summer annual
	  bedding plants)
		Rose
		Marigold
		(most annual bedding
		  plants)
			Azalea
			Dogwood
			Lady's slipper
			Mountain laurel
			Rhododendron

FIGURE 4-10 Favorable temperature and light conditions for common vegetable crops and flowering plants

HUMIDITY

Most plants are not affected drastically by a minor change in *humidity,* the moisture level of air. Most plants grow best in the 40 to 80 percent relative humidity range. *Relative humidity* is the amount of moisture in the air as compared with the percentage of moisture that the air could hold at the same temperature if it were completely saturated.

Some plants are more sensitive to humidity than others. Provided that the roots are able to replenish moisture lost through plant leaves as fast as it is lost and that the plants do not wilt, low humidity is not a great problem for most horticultural crops. However, when hot, dry conditions cause plants to wilt, plant growth is slowed or stopped completely. If wilting is allowed to reach the extreme state of *permanent wilting*, death occurs. When the humidity is very high (80 to 100 percent relative humidity), other problems may arise. For example, high humidity may cause the spread of fungus disease.

PLANT DISEASES AND INSECTS

Any time a plant is suffering from disease or insect damage, production is reduced. The amount of reduction depends on how severe the damage is and what percentage or part of the plant is infected. For example, since leaf damage reduces the area available for producing food, the more leaves that are lost, the more severely total production is reduced. Stem injury may *girdle* (circle) or clog up a stem and kill the entire plant because the supply of water and minerals to the plant top and food to the roots is completely cut off.

Some diseases and insect damage may be prevented by the use of varieties of plants that are resistant to disease and/or insects, or by crop rotation or chemical sprays.

GASES AND AIR PARTICLES

Carbon dioxide (CO_2) is vital to plants for the production of food. There is rarely a severe enough shortage of carbon dioxide to cause damage to plants. However, greenhouse operators find that by adding carbon dioxide to the air, the growth rate of certain crops may be increased enough to more than pay for the added cost of the carbon dioxide. In field grown crops, there is no economical way to add carbon dioxide. Research is now being done to produce crops that can use carbon dioxide more efficiently. Yield increases of up to 50 percent may result, with no additional fertilizer use. Other growth-restricting factors such as lack of water are usually more important to outside crops and are therefore given more consideration.

Some air pollutants damage plants. Sulfur dioxide from coal furnaces and carbon monoxide from cars are known to reduce plant growth and, in severe cases, to kill plants. It is important to consider other toxic fumes in areas where concentrations are high enough to cause damage to crops.

Student Activities

1) Try this experiment to see how sand, silt, and clay particles are easily separated from soil. Take a half-gallon jar, fill it one-half full of soil, add one (1) teaspoon of table salt (NaCL), and fill with water to a point within 2 inches of the top. Replace the lid and shake until all the soil is *in suspension* (mixed very well). Set the jar aside and examine it the next day. The heavy sand particles should be on the bottom, silt above them, and clay on top. (Organic matter may be floating on top of the water.) Definite layers should be visible. Classify the soil based on the percent of sand, silt, and clay. Is it a sandy loam, silty loam, or clayey loam? (Use the texture triangle, figure 4-6).

2) Place a small amount of sand in one jar and clay in a second jar. Pour a cup of water into each. Record how long it takes the water to drain down through the medium in each jar. For best results, place 2 inches of coarse gravel in the bottom of each jar.

3) Examine the label on a complete plant food. List all the plant food elements and record the percentage of each available. Identify major and minor elements. Notice the small amount of the minor elements available.

4) Experiment with the effect of temperature on the growth rate of plants by placing two plants of the same species in atmospheres with a wide difference in temperature. Note the results.

5) Place a plant in complete darkness and record its growth rate and appearance as compared with a similar plant kept in normal lighting conditions.

6) Take a soil sample. Determine what type of fertilizer should be applied and the rate at which it should be applied for a specific crop.

7) Using a soil test kit, test a soil sample for pH level, nitrogen, phosphorus, and potassium. Make crop recommendations for specific crops.

Self-Evaluation

Select the best answer from the choices offered to complete the statement or answer the question.

1) By weight, how much salt and clay do sandy soils contain?

   a) less than 20 percent   c) 40 percent
   b) more than 30 percent   d) 50 percent

2) A soil having equal parts of sand, silt, and clay is called

   a) an aggregate.   c) a loam.
   b) a mixture.   d) none of these

3) The three layers of a normal soil profile are

   a) sand, silt, and clay.
   b) topsoil, subsoil, and root zone.
   c) root zone, topsoil, and subsoil.
   d) topsoil, subsoil, and bedrock or lower subsoil.

4) The pore space in a soil filled with _____ percent of water and _____ percent of air is the ideal ratio.

a) 25%, 25%
b) 20%, 30%
c) 40%, 10%
d) none of the above

5) The moisture-holding capacity of soil is sometimes improved by

a) adding organic matter.
b) using tile drains.
c) irrigation.
d) mulching.

6) Sandy soils, often called light soils, have

a) large soil particles and large pore spaces.
b) good drainage but little moisture-holding capacity.
c) good aeration.
d) all of the above

7) The four things that plants receive from soil are

a) anchorage of roots, food storage, air, and moisture.
b) anchorage of roots, oxygen, moisture, and nutrients.
c) nitrogen, phosphorus, photosynthesis, and minor elements.
d) all of the above

8) Certain plant food elements are called major elements because

a) plants use them in large amounts.
b) they are the first elements listed on the fertilizer bag.
c) they are plentiful in the soil.
d) none of the above

9) The advantages of a good artificial medium are

a) it is sterile and uniform in content.
b) it is lighter in weight and therefore easier to handle and ship than natural soil.
c) it has good drainage and moisture-holding ability.
d) all of the above

10) Nitrogen causes plants to

a) produce more flowers and seeds.
b) resist disease and develop strong roots.
c) harden off more rapidly.
d) grow rapidly and develop a dark green color.

11) Ammonium nitrate contains _____ percent nitrogen.

a) 45%
b) 10%
c) 33%
d) 20%

12) Phosphorus causes plants to

a) produce more flowers and seeds.
b) resist disease and develop strong roots.
c) mature more rapidly.
d) all of the above

13) Superphosphate fertilizer contains _____ percent phosphate.

a) 30%
b) 20%
c) 15%
d) 25%

14) Potassium causes plants to

    a) produce more flowers and seeds.
    b) resist disease and develop strong roots.
    c) grow much larger than they would otherwise.
    d) grow rapidly and develop a dark green color.

15) List two fertilizer sources of potassium for plants: _____ and _____ of potash.

    a) muriate and nitrate
    b) sulfate and ammonium
    c) nitrate and urea
    d) ammonium and muriate

16) Lime furnishes the plant food element

    a) nitrogen.
    b) phosphorus.
    c) potash.
    d) calcium.

17) Two plants that grow best in an acidic soil (pH 5.0 to 6.0) are

    a) azaleas and blueberries.
    b) corn and onions.
    c) corn and squash.
    d) onions and tomatoes.

18) To raise soil pH and lower soil acidity,

    a) more nitrogen is added to the soil.
    b) lime is added to the soil.
    c) a complete fertilizer is added to the soil.
    d) none of the above

19) To lower soil pH and raise soil acidity,

    a) sulfur is added to the soil.
    b) nitrogen is added to the soil.
    c) a complete fertilizer is added to the soil.
    d) lime is added to the soil.

20) The area of soil just below the earth's surface where most plant roots grow is called the

    a) subsoil.
    b) rhizosphere.
    c) top soil.
    d) active soil.

21) The classification of water in the soil that provides the most water to plants and is thus the most important is called

    a) gravitational water.
    b) unavailable capillary water.
    c) available capillary water.
    d) capillary water.

22) Water makes up approximately _____ percent of a plant's weight.

    a) 80
    b) 90
    c) 99
    d) 95

23) Green plants cannot live without light because

    a) it is necessary for the manufacture of food.
    b) they need light to breathe.
    c) light helps to warm them to the optimum temperature for growth.
    d) none of the above

24) Transpiration is a process in which plants

    a) lose water through stomata in the leaf.
    b) lose water through the leaf epidermis.
    c) breathe through the leaves.
    d) none of the above

25) As the outside temperature increases, plant growth normally

    a) increases if moisture is available.
    b) decreases because plants become too hot.
    c) decreases because the plant cannot receive moisture fast enough.
    d) increases because humidity always increases with the temperature.

Growth Stimulants, Retardants, and Rooting Hormones

OBJECTIVE

To list four types of growth regulators used in the horticulture industry.

Competencies to Be Developed

After studying this unit, you should be able to

- ◆ list one example of a substance used to stimulate plant growth.

- ◆ explain why chemical retardants are applied to floral crops and name two commonly used retardants.

- ◆ explain the use of rooting hormones on cuttings and list several rooting hormones.

- ◆ describe orally the source of one plant growth stimulant.

- ◆ demonstrate the proper application of a rooting hormone to cuttings.

- ◆ explain how rootstock is used in dwarfing fruit trees.

- ◆ define biostimulants.

MATERIALS LIST

- ✔ chrysanthemum and poinsettia plants
- ✔ rooting hormones (Hormodin #3 and #1)
- ✔ gibberellic acid
- ✔ B-Nine
- ✔ American and English holly cuttings. (Semi-hard cuttings from different shrubs may be used.)
- ✔ mist system or plastic enclosed pots or flats
- ✔ knife for making cuttings

Growth regulating substances in plants are called *hormones*. Hormones are organic chemicals which act and interact to affect growth rate. An example is seen is figure 5-1. The gardenia plant on the right was treated with the growth retardant B-Nine. A more compact, lower growing plant is produced. Prominent hormones are:

♦ *auxins* (Greek word meaning to increase), accelerate growth by stimulating cell enlargement.

♦ *gibberellins*, stimulate growth in stem and leaf by cell elongation. Also, stimulate premature flowering, growth of young fruits, and breaking of dormancy.

♦ *cytokinins*, stimulate cell division. Work along with auxins (will not work without auxins present).

♦ *inhibitors* (abscisic acid), inhibit seed germination, stem elongation, and hasten ripening of fruit (ethylene gas).

The auxins and gibberellins promote cell enlargement and the cytokinins stimulate cell division.

Growth hormones are organic chemicals principally produced by actively growing plant tissue such as short tips and young leaves. They move throughout the plant and may be found in many plant parts.

These chemicals react with one another in a very complex system in the plant. In some cases one concentration or amount of a hormone stimulates growth and a different concentration or amount restricts growth.

FIGURE 5-1 The growth retardant B-Nine used on the gardenia plant on the right results in a compact, low-growing plant. (Courtesy of Uniroyal)

APICAL DOMINANCE

A good example of regulating growth in plants is seen in the dominance of the terminal bud or shoot. Where *apical dominance* exists, the terminal bud secretes chemicals that inhibit or prevent the growth of axillary buds on the same shoot. Axillary buds are found in the axil or angle between a leaf and the stem. This causes the plant to grow tall and not send out side branches. Once the plant reaches flowering age and the terminal bud becomes a flower bud, the chemicals are no longer secreted. At this time, the plant starts sending out side branches. It appears that this is a genetic program directing the plant to grow above competing plants in its early years. Once height and access to sunlight is secured, the plant spreads over its competitors. Pinching out the terminal leaf bud of the shoot nullifies the effect on axillary buds, and branching occurs sooner than the plant would have branched if left alone. This is a method commonly used by nurseries to encourage early branching of shrubs and the production of more compact plants.

For years, human beings have been developing methods other than pruning to control the growth rate, size, and shape of plants. Some major achievements include the discovery of:

♦ chemical and natural stimulants that cause plants to grow taller or faster.

♦ chemical retardants that cause plants to grow slower or cease growing at a certain point.

♦ hormones that cause cuttings to root faster.

♦ dwarfing rootstock for tree fruits.

STIMULANTS

Applications of certain chemicals enable plants to grow taller. The most common chemical of this type is gibberellic acid (GA). An example of the effect this chemical has on plants is shown in figure 5-2. The chrysanthemum on the left, which had no chemical treatment, is growing at a normal rate. The plant on the right was treated with gibberellic acid. Notice how much taller the treated plant is.

Gibberellic acid causes the stems of plants to stretch out. The *nodes*, the joints at which buds,

FIGURE 5-2 A plant treated with gibberellic acid (right) compared with untreated plant. Notice how much taller the treated plant has grown. The treated plant also bloomed slightly sooner than the untreated plant; this may or may not have been caused by the gibberellic acid.

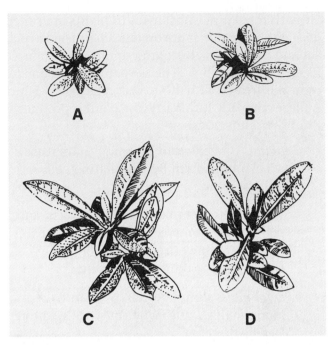

FIGURE 5-3 Growth increases on rhododendron using alfalfa as a stimulant. Examples A and B are untreated controls; C and D have been treated with alfalfa. (Courtesy of **American Rhododendron Society Quarterly Journal**, Spring 1979)

leaves, and branches grow out from the stem, are farther apart. Notice how much longer the stems which hold the blossom buds are on the treated plant. Gibberellins also occur naturally in plants.

A natural growth stimulant has been found in alfalfa, and while this stimulant has been known by nurserymen for over 30 years, interest in it has recently been renewed. The chemical that causes the stimulation of growth is the alcohol triacontanol.

Plant stimulation is brought about by mulching with alfalfa hay or watering plants with a tea made by soaking alfalfa feed pellets in water and using the tea to water the plants. The amount of Triacontanol needed is very small (4 milligrams per acre of crop). Too much seems to reduce the stimulating effect. This material is also sold commercially. Results with rhododendron seedlings were:

1. 55.1 percent increase in height
2. 27.6 percent increase in diameter
3. 27.6 percent increase in number of leaves
4. 53.5 percent increase in leaf length
5. 19.3 percent increase in leaf width[*]

[*]Research done by Dr. Hinderman of Ann Arbor, Michigan

This increased plant growth resulted with no added fertilizer. See figure 5-3.

If you wish to experiment with this material see the Student Activities section at the end of this unit for detail.

CHEMICAL RETARDANTS

At times, plants may grow too tall and open to be pleasing to customers. In these cases, chemicals are used to retard the growth of the plants, causing them to be shorter and more compact. Not only are these plants more attractive, but also the shorter stems are better able to support the flowers, thereby reducing the need for staking or tying. Today, chemical *retardants* (chemicals which *retard*, or slow down, growth) are used commercially as a normal part of the growing process of many plants. Figure 5-4 illustrates the effects of B-Nine used on azaleas.

One of the newest uses of plant growth retardants (PGR) is the application of a chemical to retard lawn grass growth. If applied at first "green-up," after the first mowing of grass in the spring, a chemical called "Limit" is absorbed by the roots and restricts grass growth for six to eight weeks. Another material called "Embark" is leaf-absorbed and also restricts grass growth.

FIGURE 5-4 The plant growth regulator (PGR) B-Nine was applied to the two azaleas on the right. Notice the compact growth and heavier flowering compared with the untreated plant on the left. (Courtesy of Uniroyal)

FIGURE 5-5 Two strips of turf grass. The strip on the left was treated with Embark. There is a tremendous reduction in growth rate compared with the untreated grass on the right. (Courtesy pbi/Gordon Corporation)

Figure 5-5 shows two strips of turf grass. Grass on the left side was treated with Embark, a PGR, which is absorbed through the leaves and moves to the growing point where it interrupts cell division, stem elongation, and seed head formation. One big advantage of this material is that roots continue to grow. Embark should be applied after the first or second mowing in spring.

Another use of retardants in the vegetable industry has proven valuable. Scientists at Pennsylvania State University experimented with soaking tomato seeds in a gel containing a growth retardant. The purpose was to keep the tomato plants from becoming too tall and spindly. The results were a 25 percent reduction in height and more vigorous plants, which remained compact until planted in the field. When the roots grew into soil

FIGURE 5-6 The shrubbery shown above was treated with the PGR Atrimmec. The treated area on the left has made very little growth compared with the untreated area on the right. (Courtesy pbi/Gordon Corporation)

not saturated with the chemical retardant, growth returned to normal. No reduction in yield resulted in this treatment and a compact, vigorous, easy-to-handle plant resulted.

A new plant growth regulator called Sumagic reduces the height of plants by inhibiting the natural production of the plant hormone gibberellic acid. Gibberellic acid causes stems to elongate or lengthen. This chemical is taken up by the leaves, stem, and roots and moves through the plant to the growing point where it inhibits the production of gibberellic acid.

Plant regulators are also being used in the landscape to control growth on shrubbery and hedges. One application of the PGR Atrimmic after pruning or some regrowth is claimed to last the entire season on plants that have only one flush of growth thus reducing or eliminating the need to prune. Plants that have a continuous growth habit would need a treatment again in 8 weeks.

Atrimmic is a systemic (a material that penetrates the plant, enters the plant sap, and moves through the plant) that works by blocking the plant hormones that stimulate growth. Applied as a foliar spray it is absorbed by the leaves and moves to the shoot tips where it temporarily stops shoot growth. Figure 5-6 shows an example of the use of Atrimmic on a hedge.

ROOTING HORMONES

When propagating plants from cuttings, it is important that a large percentage of the plants root, and that they root as quickly as possible. Some plants root very easily from cuttings without any chemical treatment. Geranium, azalea, and many soft, succulent houseplants require only a moist, well-aerated medium and high humidity to root. Many plants must have some assistance if any or very many of the cuttings are to root successfully.

Why is there a difference in ease of rooting? Early researchers discovered that some plants have more natural root-promoting chemicals than others. Indoleacetic acid (IAA) is a natural plant hormone that causes roots to form on plant stems. It is found in various plants in differing amounts.

Plants also root more easily or with more difficulty at different stages of maturity, which affects the hardness of the wood. Some root best when the wood is soft, others when it is almost hard.

The development of chemical rooting hormones made it possible to root certain plant cuttings that were considered impossible to root before. These chemicals also shortened the length of time required to root cuttings.

The chemical most commonly used for rooting cuttings is indolebutyric acid (IBA). Naphthaleneacetic acid (NAA) and alpha-naphthaleneacetic acid are also widely used. Indoleacetic acid, the first naturally occurring plant hormone to be used, is rarely, if ever, used today. It is generally not as effective as indolebutyric acid or naphthaleneacetic acid.

Rooting hormones are either mixed with talc and used as powders or dissolved in liquid and used as a wet dip. Some of the new liquid dips give better rooting percentages than do the powders. The strength of the active chemical varies from 0.1 percent to 0.8 percent, although concentrations as great as 2 percent of IBA have been used

on certain evergreens such as camellias and yews. Whatever their concentration, all rooting hormones should contain a fungicide (a chemical that kills fungi) such as captan to prevent rotting of the cutting. Captan also seems to help promote faster rooting.

The holly cutting on the right in figure 5-7 was treated with indolebutyric acid; the one on the left was untreated. After twenty-one days, the treated cutting was rooted. The untreated cutting had developed no roots.

DWARFING ROOTSTOCKS

As the cost of labor increased, orchardists began looking for smaller fruit trees that could be picked without the use of ladders. Research was started to discover a way to prevent apple trees from growing very tall. Pruning had been used to reduce growth, but the researchers realized that it was a time-consuming, expensive, and only partially effective method. It was then discovered that trees growing from certain types of roots did not grow as large as other trees. These trees also bore fruit at an earlier age. Using this principle, a complete series of rootstock known as the *malling rootstock* was developed in England. (*Rootstock* is a root or piece of root that is used for grafting.) It became possible to control the size and rate of growth of apple trees by selecting the proper rooting stock. Some varieties rooted from the malling rootstock grow close to normal size, while others are only one-quarter to one-half the normal size.

Since the original development of the malling stock, many other types have originated, including stock for apple, peach, and pear trees. The dwarf trees themselves can be purchased commercially. Check with your local agricultural experiment station to determine which rootstock is best for grafting or budding fruit trees in your area. Consult nursery catalogs for listings of dwarf fruit trees.

FIGURE 5-7 The holly cutting on right, treated with indolebutyric acid, developed roots in 21 days; the untreated cutting did not.

THE CHEMICAL BLOSSOM SET

One type of growth hormone is used on tomato blossoms early in the season to cause earlier development of fruit. Normally, the first blossoms

TRADE NAME	EFFECT	USED ON
A-Rest	shortens stem length	chrysanthemum lily poinsettia
B-Nine	shortens stem length	chrysanthemum hydrangea petunia azalea poinsettia
Cycocel	shortens stem on poinsettia induces heavier flower bud set on azalea	poinsettia azalea geranium
Phosfon	shortens stem length	lily chrysanthemum
Florel	induces basal breaks on roses retards growth on poinsettia increases number of branches on carnation speeds ripening of tomatoes assists in faster rooting of cuttings	rose poinsettia carnation tomato cuttings
Bonzi	shortens stem length	poinsettia geranium petunia chrysanthemum impatiens coleus pansy snapdragon celosia
Sumagic	shortens stem length	poinsettia chrysanthemum celosia coleus marigold periwinkle petunia ageratum dahlia

Growth retardants work best when the humidity is high, but plant foliage is dry. Always follow the directions on the label exactly.

FIGURE 5-8 Commercially used plant growth regulators. (**Note:** Wear protective clothing when called for on the label.)

on tomato plants often do not *set* (develop) fruit because of cool weather, lack of pollination, or for other reasons. The chemical *blossom set* causes seedless fruit to set on the blossoms and results in tomatoes ready to eat as much as ten days earlier than normal. Figure 5-8 illustrates a cluster of blossoms that was sprayed with the chemical and an untreated cluster. Notice that there are small tomatoes on the treated plant and only blossoms on the untreated plant. Experiment with some of the commercially used plant growth regulators in figure 5-8.

PLANT BIOSTIMULANTS

Biostimulants are natural products. They are organic, meaning from living organisms, and work to both stimulate soil microbial activity and improve soil cation exchange capacity, stimulate plant growth, and promote disease resistance. Humic acid, a product of humus or rotting organic matter, is an example of a soil microbial stimulant.

Root growth biostimulants improve water and nutrient uptake by the roots and also increase the number of fibrous roots. High levels of these growth stimulants are found in kelp seaweed. Fresh kelp is most effective and has been found to greatly increase growth and yields of plants. Processing such as drying, freezing, boiling, and chemical treatment decreases the stimulating effect. There is a product on the market that is processed to retain the stimulating effect of kelp.

Use of biostimulants can greatly reduce the need for fertilizers, especially nitrogen. This not only saves the grower money, but also reduces water pollution from excess nitrogen run-off or percolation into ground or stream water.

Methanol (a form of alcohol) used in small amounts appears to speed up plant growth. In tests by Dr. Andrew A. Benson (University of California) methanol (methyl alcohol) increased plant yield by 36–100 percent. A 10–30 percent solution of methanol was used.

Caution: Stay on the low side until plants' response or possible damage is observed.

The alcohol seems to work by blocking photorespiration. Plants use the water for growth rather than transpiring it into the air. This material works best on plants grown in full summer sun. Water use was reduced by as much as 50 percent from some plants.

This product is already on the market. Other plant stimulants such as triacontanol and DCPTA are being studied as possible growth regulators.

ALLELOPATHY

Allelopathy is the production of a chemical compound by one plant that slows down or stops the growth of another plant. This is a natural growth regulator some plants use to stop competition from other plants. Some green manure or cover crops grown for weed control, such as millet, have this effect on most weeds and prevent weed growth. The effect of these growth regulators may last for months or in some cases, such as the black walnut, years.

Some plants do not transplant well into beds where older plants of the same type are or recently were growing. Members of the rose family are good examples (hawthorn, apple, rose, peach, pear, etc.).

Student Activities

1) Apply gibberellic acid to an actively growing plant. Keep an identical plant which has not been treated as a check on the results.

2) Apply B-Nine to five poinsettias and five mum plants. Keep five untreated plants of each type as check plants. Record the differences noted in the plants' growing habits. (Be sure to follow directions on the label exactly.)

3) Start twenty holly cuttings. (If necessary, refer to Unit 7, Semi-hardwood Cuttings, for procedural suggestions.) Use a commercial rooting compound on ten of the cuttings. Keep ten untreated plants as checks. Be sure to follow the directions when using the rooting hormone. Record the results.

4) Contact a local orchardist or a state experiment station for details on dwarf fruit trees.

5) *Alfalfa growth stimulus experiment.* Take some potted plants just starting their growth cycle (about a week after transplanting) and treat as follows with alfalfa.

   a) Finely chop some good quality alfalfa hay.
   b) Mix ½ cup of the chopped hay with some hardwood mulch and place on top of the soil in a 6- to 10-inch pot; water it well.
   c) Compare to control plants with no alfalfa.

   An alternate method is to soak alfalfa pellets (cattle feed) in water (1½ cups of pellets per 5 gallons of water) for at least a day. Water the plants with ½ cup of this tea every 7 to 14 days.

Self-Evaluation

Select the best answer from the choices offered to complete the statement or answer the question.

1) Plants may be made to grow taller by applying the chemical

   a) naphthaleneacetic acid.
   b) gibberellic acid.
   c) phosphon.
   d) indoleacetic acid.

2) A naturally produced plant growth stimulant has been found in

   a) corn.
   b) the soil.
   c) alfalfa.
   d) hardwood mulch.

3) Plants are often treated with chemicals that dwarf or shorten them. This is done to

    a) make them more attractive and to eliminate the need for staking.
    b) make it easier to spray them.
    c) force them to bloom earlier.
    d) all of the above

4) B-Nine is a commercially used chemical that

    a) makes plants taller.
    b) helps cuttings root faster.
    c) causes plants to bloom sooner.
    d) shortens or dwarfs plants.

5) Gibberellic acid causes

    a) plants to grow taller.
    b) plant stems to stretch out.
    c) the development of a greater distance between nodes.
    d) all of the above

6) When applied to holly plants, the growth retardant phosphon

    a) injures the plant.
    b) shortens the plant and causes early berry formation.
    c) causes the plant to grow taller.
    d) none of the above

7) One rooting hormone occurring naturally in plants is

    a) indoleacetic acid.
    b) indolebutyric acid.
    c) naphthaleneacetic acid.
    d) talc.

8) Holly cuttings treated with indolebutyric acid should be well rooted in about

    a) ten days.
    b) twenty-one days.
    c) fourteen days.
    d) thirty-one days.

9) The chemical most often used commercially for rooting cuttings is

    a) indoleacetic acid.
    b) indolebutyric acid.
    c) naphthaleneacetic acid.
    d) talc.

10) The strength of the active chemical in the majority of rooting hormones on the market ranges from

    a) 0.1 percent to 0.8 percent.
    b) 1.0 percent to 8.0 percent.
    c) 5.0 percent to 15.0 percent.
    d) 0.01 percent to 0.08 percent.

11) A fungicide such as captan added to a rooting hormone

    a) helps prevent cuttings from rotting.
    b) makes the hormone last longer.
    c) causes the hormones to break down chemically.
    d) none of the above

12) Fruit trees may be dwarfed by

    a) not fertilizing.
    b) chemical sprays.
    c) budding or grafting using a special rootstock.
    d) none of the above

13) Blossom set is used to

    a) increase flower production.
    b) stimulate early fruit set on tomatoes.
    c) extend the life of flowers.
    d) none of the above

14) Biostimulants affect plants by

    a) improving water uptake.
    b) stimulating growth.
    c) increasing the number of fibrous roots.
    d) all of the above

Plant Propagation

Seeds

OBJECTIVE

To propagate at least one plant using seeds.

Competencies to Be Developed

After studying this unit, you should be able to

- ◆ identify the parts of a seed and the functions of each.
- ◆ differentiate between indirect and direct seeding methods.
- ◆ prepare a medium for seeds, sow seeds, and provide the proper conditions for germination.
- ◆ water, fertilize, and harden off seedlings before transplanting.
- ◆ transplant seedlings into flats or pots.

MATERIALS LIST

- ✔ planting container such as a pot or flat
- ✔ planting medium (sterilized)
- ✔ seeds of plants commonly grown in the area
- ✔ labels and marking pen or pencil
- ✔ leveling board or skew
- ✔ watering facilities for bottom watering
- ✔ transplanting containers
- ✔ watering can
- ✔ greenhouse or other growing space

WHAT IS PROPAGATION?

There are many ways of *propagating*, or reproducing, plants. The most common method of reproducing flowering as well as vegetable and cereal crops is through the use of seeds. This is a *sexual* process and requires the union of *pollen* (the male sex cell) with the *egg* (the female sex cell) in the ovary. Male and female cells may be from the same parent (*self-pollination*) or from separate parents (*cross-pollination*).

Seeds are a means of rapidly increasing the number of a certain plant. However, not all plants *come true to seed* (reproduce exact duplicates of the parent plant from seeds). Wheat and barley are examples of plants which do come true from seed. Others, such as rhododendrons, azaleas, and apples, do not come true from seed, causing the offspring to differ from either parent.

Plants that are not produced directly from seeds or do not produce seeds that will grow, such as some hybrids, must be propagated by another method to obtain exact duplicates. *Hybrids* are the offspring of two different varieties of one plant, each of which possesses certain characteristics that are desired in the new plants. A female plant of one variety and a male plant of another variety are *crossed* (bred) to produce offspring with the best characteristics of each parent. However, when hybrids themselves are reproduced from seed, their offspring do not have the same characteristics as the parents; rather, they have a variety of combinations of traits possessed by the plants originally used to produce the hybrids. To produce exact duplicates of these plants, asexual reproduction is used. This is not a sexual process, and no seeds are used in this method. Instead, the plant is propagated from one of its parts, such as the leaf, stem, or root.

Propagation may be accomplished by division of roots; by cuttings of leaves, stems, or roots; or by budding and grafting. The strawberry reproduces itself by runners; the lily by tiny bulblets. The reproduction method of the strawberry and lily occur naturally. Cuttings, budding, and graft-

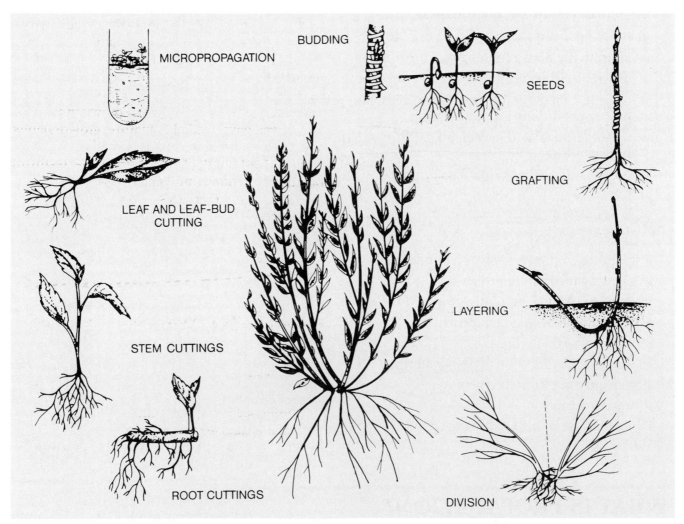

FIGURE 6-1　Commonly used methods of propagating plants. Except for seeds, these are all vegetative methods of propagation. (Courtesy Brooklyn Botanical Gardens)

ing, however, have been developed or improved upon by man.

Asexual reproduction is possible because each single cell of a plant contains all the characteristics of the entire plant and can regrow any missing part. Thus, a stem cutting removed from the roots of the plant develops new roots from cells along and at the base of the stem. Root cuttings develop new stem tissue in the same manner.

Plant propagation predates recorded history. Very early human beings planted seeds or divided plants to increase plant numbers and to carry them over the *dormant*, or resting, season. The quality of plants was improved by using seed from the best plants to produce other plants, or by separating naturally propagated plants of superior quality from the parent and planting them.

One of the newest approaches to plant propagation is a method called *tissue culture*, or *micropropagation*. This is an asexual method in which a growing shoot, tip, or bud of the plant is separated and grown on an agar gel or other nutrient medium. Through this method many thousands of new plants exactly like the parent plant can be produced very quickly. This is a good way to rapidly increase a new, superior plant and get it on the market years sooner. This method has been used for years with orchids and some other plants and is now being adopted as a commercial method in the propagation of many plants. Tissue culture will be discussed in greater detail in Unit 7.

Figure 6-1 shows the various kinds of plant propagation, which will be discussed in this section.

PROPAGATION OF PLANTS FROM SEEDS

In the horticulture industry, many plants are started from seed because it is a quick and economical method. For successful germination, the proper environmental and cultural conditions must be provided. These conditions include temperature, moisture, light, and medium.

Seeds of annual plants (plants that complete their life cycle in one year) from named varieties which are not hybrids generally come reasonably true from seed. *Named varieties* are specific individual strains of one type of plant which have been named to indicate their particular traits. The tomato plant has numerous named varieties such as Better Boy and Marglobe, each of which is different in size, days needed to mature, growth habits, and disease resistance. Seeds should not be saved from hybrids and planted, since the resulting plants would not come true from seeds. Many perennials are grown from seed, figure 6-2.

PLANT	GERMINATION TIME INDOORS (DAYS)
Achillea	5 to 7
Alyssum saxatile (basket-of-gold)	5
Anthemis (camomile)	5
Hardy aster	7 to 14
Campanula	10 to 24
Cerastium tomentosum (snow-in-summer)	10
Chrysanthemum	5 to 8
Dianthus (hardy pinks)	4 to 6
Doronicum (leopard's bane)	8
Eupatorium (perennial ageratum)	7
Gaillardia	5
Gloriosa daisy	7
Gypsophila (baby's breath)	7
Heliopsis	4
Hollyhock	7
Iberis (perennial candytuft)	6
Lupine	3 to 4
Myosotis (forget-me-not)	7 to 12
Papaver (Oriental poppy)	6 to 10
Phlox paniculata	21 to 28
Primula (primrose)	21 to 30
Salvia	5 to 21
Shasta daisy	5 to 7
Verbena	10

FIGURE 6-2 Perennials easily grown from seed

SPECIAL TREATMENT OF SEEDS FOR GERMINATION

Some seeds have a hard seed coat which must be soaked or scratched before the seeds are able to germinate. This process is called scarification. Examples of this type of seed are the red bud, Judas tree, and sweet clover. Sandpaper, scratching, or an acid bath may be used to weaken the seed coat.

Other seeds require a moist, cold rest period (dormant stage) at temperatures below 37°F (3°C) for eight weeks or longer. Examples of seeds which require a rest period are the apple, peach, pear, tree peony, maple, and yew. These may be buried or stratified in moist sand and kept cool.

Still other seeds must go through alternate wetting and drying. Some must have light to germinate; others must have darkness. These specific

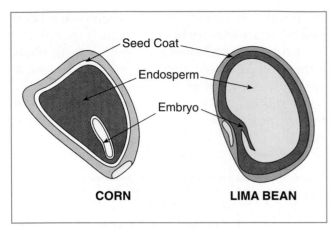

FIGURE 6-3 Composition of a seed

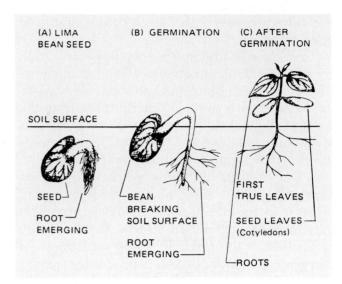

FIGURE 6-4 Lima been germination

requirements must be met when these seeds are planted. Always follow instructions on seed packets or enclosed planting directions.

COMPOSITION OF SEEDS

The basic parts of a seed are the seed coat, the endosperm (stored plant food), and the embryo, figure 6-3.

SEED COAT The *seed coat* is the outside covering of the seed which protects the embryonic plant. The seed coat makes it possible for seeds to be transported and stored for long periods of time.

ENDOSPERM (STORED PLANT FOOD) The *endosperm* is the food storage tissue which nourishes the embryonic plant during germination (the first start of growth in a seed).

EMBRYO (OR EMBRYONIC PLANT) The *embryo* is a new plant that is developed as a result of fertilization. During germination, it extends its roots and seed leaves to form a new plant, figure 6-4.

IDENTIFICATION AND SELECTION OF SEED

There are several important steps in the selection of seed.

◆ Identify which seeds are grown locally.

◆ Select seeds that have been tested for the current year's growing season for germination ability and purity. These tests must comply with state regulations.

◆ Purchase the seeds from a reliable dealer to assure that the variety is true (pure) and that germination ability is acceptable.

◆ Choose hybrid varieties for greater vigor, uniformity, and flowering ability.

◆ Select uniform heavyweight or primed seeds.

Some new things have been done with seeds to improve germination and uniformity of growth. First, seed *quality* has been improved and seeds have been sorted by weight. This results in all seedlings coming up at the same time. Heavier seeds tend to grow faster and produce larger plants. These seeds are usually worth more and are a good investment.

Second, seeds are *primed* or *enhanced*. This is done by soaking the seeds in salt solutions of KCL or ethyl alcohol. This treatment causes the growth hormones or enzymes to become active, and the seed radical or young root starts to grow. By regulating the amount of moisture and chemical, growth of the radical is stopped at a point where the seed may still be safely stored.

Seeds treated this way come up much sooner after planting. Where seeds are planted outside in the field, this is a great advantage. The sooner the seed comes up after planting, the less chance for soil insects and diseases to damage or kill the young seedling. Young plants become much more disease-resistant in just a few days and many more survive if growth is fast.

Plants from primed or enhanced seeds are more uniform in growth and more of them are ready for harvest at the same time. This is a great advantage where a once-over machine harvesting is done.

The selected and treated seeds as mentioned above cost more but outyield ordinary seeds and more than return the extra cost.

Small difficult-to-plant seeds, the ones you always plant too thickly, are now coated or pelletized to make them larger and easier to plant. Some companies even glue them to a tape with the proper spacing. All you do is stretch out the tape and cover it with soil.

GERMINATION MEDIUM

The best medium for germination has a favorable pH level and an adequate supply of plant nutrients. It is firm, porous, uniform in texture, sterile, and free of weeds, insects, and disease organisms.

A good germinating medium contains one or more of the following ingredients.

SOIL   The soil should be a loam composed of 45 percent mineral matter, 5 percent organic matter, 25 percent air, and 25 percent water.

CONSTRUCTION GRADE SAND   This is the best type of sand to use because it is more porous than some other sands, thereby allowing for better aeration and drainage. Sand particles have a negative charge; therefore, they do not hold plant nutrients in the medium as strongly as particles of soil or peat moss.

PEAT MOSS   Peat is partially decomposed vegetation which has been preserved under water. The peat is collected from marshes, bogs, or swamps. Peat has a very high capacity for holding water. It contains about 1 percent nitrogen and is low in phosphorous and potassium.

SPHAGNUM MOSS   Sphagnum moss is the dehydrated remains of acid bog plants. It is used in shredded form in seed germination. Sphagnum moss is relatively sterile and lightweight, controls disease well, and has excellent water-holding capacity.

HORTICULTURAL GRADE PERLITE   Perlite is a gray-white material of volcanic origin. It is most commonly used to improve aeration of media. Horticultural grade perlite consists of large particles, thereby providing good drainage and aeration.

VERMICULITE   Vermiculite is a very lightweight mineral which expands when it is heated. It is neutral (has a pH of 7) and has a very high water-holding capacity.

JIFFY MIX   Jiffy Mix is composed of equal parts of shredded sphagnum moss, peat, fine grade Terrlite vermiculite, and enough nutrients to sustain initial plant growth.

Although germination media are usually made up of one or more of the materials in the list above, good grade, sterilized topsoil provided with the proper drainage is sufficient. Sand or perlite should be added if the soil needs greater drainage and aeration. Peat, sphagnum moss, or vermiculite may be added to improve the moisture-holding capacity of the soil. For general purposes, a good mix consists of one-third soil, one-third sand or perlite, and one-third peat moss, sphagnum moss, or vermiculite. If the soil is heavy (with a high clay content), more sand or perlite may be needed; if it is sandy, less sand and perlite are required.

Soil used in any seed germination medium should be sterilized or pasteurized by heating it at 180°F for one-half hour. Sterilized or pasteurized soil may also be purchased at garden stores. It may then be mixed with other materials to form the desired medium.

INDIRECT SEEDING

Indirect seeding is a process in which seed is sown in a place separate from where the plants will eventually grow to maturity. The seedlings are transplanted one or more times before reaching the permanent growing area.

FLATS

When growing seeds, horticulturists often select a *flat*. Flats are made of plastic and come in many shapes and sizes. Size selection should be determined by the number of seeds to be sown. The medium selected is placed in the flat and leveled off to about ½ to ¾ inch below the top of the flat.

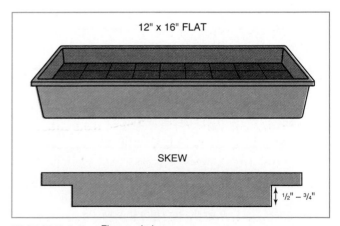

FIGURE 6-5   Flat and skew

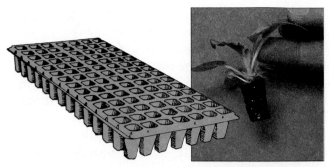

FIGURE 6-6   Flats with 98 separate cells for seedlings

This is done with a tool called a *skew*, sized to fit the flats. See figure 6-5.

After the flat is filled, rows are made in which to sow the seeds. A row marker may be used, or rows may be made one at a time using a straight board as a guide. It is best to sow seeds in rows because if disease strikes one row of seed, it can be removed without disturbing the others. Also, when several different varieties are in the flat, it is easier to label them by rows and much easier to transplant *seedlings* (young plants that have been germinated several days).

INDIVIDUAL CELLS IN CAVITY SEEDLING TRAYS

New seeding machines that can place one seed at a time in a series of small cell-like pots are now available. This allows the use of flats made up of many small pots. One or two seeds are planted in each cell, and are later thinned to one plant per cell. Cells are deep and narrow, making it possible to get many plants in a flat and yet have room for root growth. Plants from cells transplant with less root damage than from solid mat flats. Figure 6-6 shows a flat planted to begonia seedlings.

WHEN TO SEED

As business persons who must operate at a profit to survive, commercial plant growers must be sure

PLANT	TIME FROM SEEDING TO MARKET	GERMINATION TIME/DAYS	OTHER OBSERVATIONS
Ageratum	8 weeks in packs 11 weeks in pots	8	Sow seed on surface Germinate at 70°–80°F. (21°–27°C)
Annual Alyssum	7 weeks in packs 9 weeks in pots	8	Sow seed on surface Germinate at 70°F. (21°C)
Begonia	14 weeks in packs 16 weeks in pots	2–3	Sow seed on surface, need light to germinate Germinate at 65°–70°F. (19°–21°C)
Coleus	8 weeks in packs 9 weeks in pots	10	Cover seed 1/4 inch Germinate at 70°–80°F. (21°–27°C)
Impatiens	8 weeks in packs	8	Sow seed on surface, need light to germinate Germinate at 60°–70°F. (15°–21°C)
Marigold	9 weeks in packs 10 weeks in pots	7	Cover seed 1/4 inch Germinate at 70°–75°F. (21°–24°C)
Pansy	12 weeks in packs 13 weeks in pots	10	Sow seed on surface, cover with paper to keep dark Germinate at 70°F. (21°C)
Petunia	10 weeks in packs 11 weeks in pots	7–10	Sow seed on surface, need light to germinate Germinate at 70°–80°F. (21°–27°C)
Salvia	10 weeks in packs 11 weeks in pots	14	Cover seed 1/8–1/4 inch Germinate at 70°F. (21°C)
Verbena	10–11 weeks in packs	14–21	Cover seed 1/8 inch Germinate at 70°F. (21°C)
Vinca	13 weeks in packs 14 weeks in pots	14–21	Cover seed 1/8 inch Germinate at 75°–80°F. (24°–27°C)
Zinnia	5 weeks in packs 9 weeks for color in pots	5	Cover seed 1/8 inch Germinate at 80°F. (27°C)

FIGURE 6-7A   Seeding information for growing plants to market size in the greenhouse

that their plants are ready for sale at the correct time for outdoor planting and for holidays. Seeds must be planted on certain dates so that the resulting seedlings are ready for transplanting at the proper time. The chart in figure 6-7 gives planting information for seeds of various plants.

SOWING SEEDS

To sow seeds properly, use the following procedure. Take the packet of seeds and shake the seeds to the bottom of the packet. Hold the seed packet with the open end slightly lower than the rest of the packet and gently tap the packet. The seeds will move out slowly and gradually, making it easy to sow the seeds properly spaced in rows. Follow directions on the package for determining the distance apart the seeds should be sown. If flats with individual cells are used, place one or two seeds in each cell.

After the seeds are sown, cover them with shredded sphagnum moss, fine perlite, or fine sand. Cover with a layer of medium measuring about twice the thickness of the seed. (The seed package may have directions for depth of planting; if so, follow them for best results.) Some seeds, such as lobelia, petunia, and snapdragon, need light for germination and should not be covered when sown.

As soon as the seeds are sown, they are labeled with the name, variety, and date, figure 6-8. All labels should be printed clearly with pencil or waterproof marking pen. Ballpoint pen should not

PLANT	TIME TO SOW SEED	DAYS FROM TRANSPLANT IN THE GARDEN TO HARVEST	GERMINATION TIME IN DAYS	OTHER OBSERVATIONS
Beet	Seed outdoors	55–68	Depends on soil temp.	Cover seed 1/2 inch Germinate at 50°–60°F. (10°–15°C)
Broccoli	Seed outdoors	28–42	4–10	Cover seed 1/2 inch Germinate at 50°–85°F. (10°–30°C)
Bush or Snap bean	After soil is warm	65–90	7–10	Cover seed 1 inch
Cabbage	6–8 weeks before transplanting	80–100	4–10	Cover seed 1/2 inch Germinate at 50°–85°F. (10°–30°C)
Cucumber	3–4 weeks before transplanting date	50–65	6–10	Cover seed 1 inch Germinate at 70°–80°F (21°–27°C)
Eggplant	8 weeks before transplanting	80–100	14	Cover seed 1/4 inch Germinate at 70°–80°F. (21°–27°C)
Lima Bean	Seed outdoors after soil is warm	65–90	Depends on soil temp.	Cover seed 1 inch
Melons	3–4 weeks before transplant date	85–120	15–21	Cover seed 1 inch Germinate at 70°–80°F. (21°–27°C)
Onion	6–7 weeks before transplant date	100–115	7–10	—
Parsley	9 weeks before transplant date	60–75	15–21	Cover seed 1/4 inch Germinate at 50°–85°F. (10°–30°C)
Peppers	6–8 weeks before transplant date	57–75	8–14	Cover seed 1/4 inch Germinate at 70°–80°F. (21°–27°C)
Sweet Corn	After soil is warm	65–90 days from sowing seed	6–10	Cover seed 1 1/2 inches Germinate at 70°–75°F. (21°–24°C)
Tomato	4–6 weeks before transplant date	65–80	7–14	Cover seed 1/4 inch Germinate at 70°–80°F. (21°–27°C)

FIGURE 6-7B Seeding information for growing some vegetable crops. (See seed catalogs for others.)

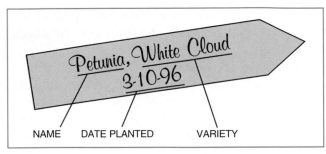

FIGURE 6-8 Plant label

FIGURE 6-9 A flat of begonia plugs ready for transplanting to pots. Notice the small holes in the flat where some plants have been removed. (Ed Reiley, Photographer)

be used since it washes out when the seedlings are watered.

WATERING THE SEED FLAT

Water aids in germination by making the seed coat soft so that the embryonic plant can germinate. Water itself is an important nutrient and also acts to dissolve other nutrients present in the medium, making them available to the growing plants.

To water freshly seeded flats, set the flats in a tub of water. This allows the seeds to be watered by *capillary action.* (The water is drawn up into the spaces between the soil particles to thoroughly moisten the medium.) After the medium is moistened, the flat is removed from the tub. If the flat is watered from the top, care must be taken to avoid washing the seeds out of the flat. Very small seeds should never be watered from the top because they may be washed too deeply into the soil.

CONDITIONS FOR GERMINATION

Seed flats containing seeds for germination should be located in a semishaded area of the greenhouse and receive a bottom heat of 65° to 70°F (18° to 21°C). (Some cool-season crops germinate well without bottom heat.) This may be accomplished in several ways. Location of the containers above the heat coil or hot water pipes or the use of a special propagating mat which is controlled with a thermostat are two possibilities. After the seeds are sown, the containers are covered with a pane of glass or clear plastic film to retain humidity. The covering should not touch the medium while the seedlings are germinating. Some seedlings, such as verbena, dusty miller, pansy, and portulaca, require a three-day period of darkness for germination. For these plants, newspaper is used instead of glass or plastic. The paper is removed as soon as the seeds germinate so that the new seedlings receive light immediately. After germination, the medium is kept moist but never wet. The flats are watered with a gentle

mist or from the bottom so that the small seedlings are not washed out.

The proper seedling medium is low in fertilizer elements. Seedlings should be fed weekly with a water-soluble fertilizer. As the seedlings approach transplanting size, a cooler temperature (55° to 60°F) (13° to 16°C) is provided to prepare the seedlings for the shock of transplanting. This is known as the *hardening-off* process. The process may also include a modest withholding of water to slow active growth.

Many seedlings are now sold as plugs, see figure 6-9. There is less transplant shock since the roots are not torn apart as they are when seedlings are transplanted from flats.

Seeds may be sown in pots, pans, or trays in a manner similar to that used for flats. The horticulturist should remember to provide drainage and uniform moisture. The containers should be watered from the bottom if possible. Plastic, newspaper, or glass is placed over the container to retain uniform moisture until the seeds germinate.

TRANSPLANTING SEEDLINGS

After seeds germinate, they develop seed leaves or *cotyledons,* the first leaves to appear on the plant, figure 6-10. The plant should be allowed to grow until the first true leaves are present before it is transplanted.

When handling seedlings, hold them by their true leaves using the thumb and forefinger. Do not hold by the stem; if the stem is badly bruised, the seedling could die. A bruised leaf is not nearly as serious an injury. While carefully holding the seed-

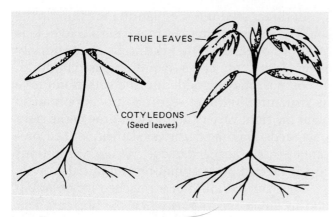

FIGURE 6-10 Tomato seedling with cotyledons and true leaves

TRUE LEAVES

COTYLEDONS
(Seed leaves)

ling, use a pot label to reach under the roots and lift, pushing the seedling out of the germinating medium. This exposes the roots. Do not shake the medium off the roots—exposure to air causes them to dry out. Plant the seedlings one by one about 2 inches apart in a flat. Using a *dibble* (a tool used to make the hole for transplanting seedlings), insert the seedlings to a depth slightly deeper than that at which they grew in the seedling flat. Gently press the medium around the roots. Water the seedlings at the soil surface with a gentle stream of water to settle the soil around the roots. The new seedlings are now ready to be grown to saleable size. (An alternate method involves using a trowel or fork when digging seedlings for transplanting.)

It is sometimes convenient to transplant seedlings into peat pots or market packs rather than flats. When a peat pot is used, one plant is placed in a pot and later transplanted directly to the garden with the plant remaining in the pot. Market packs generally hold six to twelve plants and are sold to the customer as such. The plants are then separated from the pack and planted individually in a permanent location.

Another indirect seeding technique is to plant a single seed in a Jiffy 7 peat moss pellet. These pellets, when soaked in water, expand to seven times their original size. They contain all the necessary nutrients to feed the small seedling until it is planted in a permanent site. This eliminates the seedling transplanting stage prior to permanent planting and allows the young seedlings to be transplanted to the garden with less transplant shock. The Jiffy 7 pellets are placed in plastic trays or flats for ease of handling. Manufacturer's directions should be followed when using these pellets.

Single seeds of squash, cucumber, and melon may also be planted in 2-inch peat pots in preparation for direct planting to the garden at a later time. These crops transplant with difficulty and the peat pot permits transplanting without disturbing the root system.

PROCEDURE

TRANSPLANTING TO POTS (Seedlings or Rooted Cuttings)

1. Work from a large table. Place the pots in a row on your right side.
2. Place all other material to your left side.
3. Cover drain hole with pottery or coarse sphagnum. (If using peat pots, this is not necessary.)
4. Hold the plant to be transplanted in the pot with the left hand.
5. Fill the pot to the rim with soil or soilless mix, gently holding the plant at the proper planting depth with the left hand.
6. Gently firm the planting medium around the plant by pushing down along the edge of the pot with the thumbs. Do not push hard enough to tear or bruise plant roots.
7. Label the pots or growing area according to variety, name, and date.
8. Move plants to growing area and water until water runs through the pot and out the drain hole.

PROCEDURE

TRANSPLANTING PLANTS FROM POTS

1. Work from a large table. Spread fingers of the left hand around the plant stem and over the top of the soil surface.
2. Invert the pot and gently tap the pot edge on the edge of the table.
3. Remove the pot from the root ball as it loosens.
4. If roots are pot bound, gently unwind them.

Seedlings of nursery stock plants such as trees and shrubs are generally planted in a flat and transplanted directly to the nursery row outdoors. Seeds of these plants may also be seeded directly in the nursery row.

DIRECT SEEDING

Many seeds are planted directly in the permanent growing area. This is referred to as *direct seeding.* It is the most economical method of seeding. Plants such as corn, melons, beans, beets, peas, lettuce, carrots, and most other vegetable crops are grown by this process. Some vegetables, such as tomatoes, peppers, cabbage, and eggplant, are generally transplanted as plants to the garden. Some trees and shrubs are also grown by direct seeding.

In direct seeding, the planting medium is the soil. The soil is prepared by removing all large clods or lumps of earth so that the seeds are uniformly covered. In some cases, manure, grass clippings, or compost may be added to the soil to improve its structure. In direct seeding, it is important to plant the right variety at the right time in the right soil. Seeds may need chemical treatment to prevent disease or insect damage. Weather conditions largely determine germination and initial growth.

The same conditions are needed for germination in direct seeding as when seeding in a flat: there must be sufficient moisture and aeration; the seedbed must allow firm contact with the seed; and the temperature must be high enough to support the germination process. Requirements for germination of specific seeds are given in the directions on the label of the seed packet.

 Student Activities

1) Plant seeds in a pot or flat. Care for the seeds through the transplanting stage. Keep a record of the date the seeds were planted, when they germinate, and when they are transplanted.

2) Collect various seed packets and compile a list of information given on the packets. Discuss the importance of such information.

3) Soak a bean seed in water for a few days. Compare the soaked seed with a dry seed. What difference do you notice in the seed coat?

4) Separate a soaked bean seed and see if you can identify the parts of the seed. Sketch and label the parts.

 Self-Evaluation

A) Select the best answer from the choices offered to complete the statement or answer the question.

1) Producing plants from seed is a type of _____ propagation.

   a) asexual   c) sexual
   b) bisexual   d) unsexual

2) Seeds are composed of the

   a) seed coat, endosperm, and embryonic plant.
   b) seed coat, root, and stem.
   c) eye, starch, and seed coat.
   d) root system, starch coat, and seed coat.

3) When two separate parent plants are involved in the pollination process, it is known as

a) self-pollination.   c) bisexual pollination.
b) cross-pollination.   d) asexual pollination.

4) Many horticultural crops are started from seed because

a) there are very few people who know how to propagate them any other way.
b) it is not possible to propagate them in any other way.
c) it is quick and economical.
d) both b. and c.

5) A good growing medium in which to plant seeds must

a) drain well.   c) be sterile.
b) hold moisture.   d) all of these

6) The first part of the new plant to emerge from the seed is the

a) stem.   c) root.
b) leaves.   d) endosperm.

7) Hybrid seeds are developed because they

a) provide greater vigor.   c) are more uniform.
b) produce better crops.   d) all of these

8) The best soil temperature for germinating most seeds is

a) 50° to 60°F.   c) 80° to 90°F.
b) 75° to 80°F.   d) 65° to 70°F.

9) Pansy and portulaca seeds require three days of darkness for germination to occur. This may be accomplished by

a) covering the flat or pot with newspaper.
b) burying the flat or pot in the ground.
c) planting the seeds at an unusually great depth.
d) none of the above

10) Which two of the following materials are mixed in the seeding medium to give good drainage and aeration?

a) peat moss and vermiculite   c) sand and perlite
b) sand and sphagnum moss   d) perlite and peat moss

11) Soil sterilization requires heating the soil at

a) 120°F for two hours.   c) 160°F for two hours.
b) 180°F for a half hour.   d) none of these

12) The date to start seeds is very important because

a) there must be greenhouse space made available for them.
b) seed houses have seeds available only at certain times.
c) the seeding medium must be ready.
d) the plants must often be ready for sale or planting at a certain time.

13) The label in a flat of seeds should include the

a) name or type of plant, variety, and date seeded.
b) date seeded and the selling date.
c) percent of germination listed on seed pack.
d) all of the above

14) Which of the following are perennials listed in the text as easily grown from seed?

    a) hardy aster, hollyhock, crab apple
    b) verbena, lupine, dianthus
    c) rhododendron, marigold, petunia
    d) petunia, tomato, sweet corn

15) The growth of seedlings is slowed down by withholding water and lowering the ground temperature. This process, called hardening off, is done to

    a) keep the seedlings from growing too quickly.
    b) prepare the seedlings for transplanting shock.
    c) hold the seedlings until they can be sold.
    d) none of the above

16) Seedlings are held by the first true leaves instead of the stem when transplanting because

    a) there are no cotyledons to grasp.
    b) the stems are too slippery and seedlings are dropped and lost.
    c) the stem may be bruised in handling and cause death of the plant.
    d) this bruises the leaves, which causes better growth.

17) The germinating medium that sticks to the seedling roots should not be knocked off because

    a) shaking the plant may bruise it.
    b) the roots will dry out more rapidly.
    c) it assures that the seedling will be planted at the right depth.
    d) it could spread disease to other seedlings.

18) Direct seeded plants are planted

    a) in the soil where they grow to saleable size.
    b) directly in flats in the greenhouse.
    c) in greenhouses only during winter months.
    d) all of the above

19) Which of the following would make a good seed germination medium?

    a) one-third sterilized soil, one-third sand, and one-third perlite
    b) one-third sterilized soil, one-third peat moss, and one-third sphagnum moss
    c) one-third sterilized soil, one-third perlite, and one-third peat moss
    d) one-third sterilized sand, one-third perlite, and one-third vermiculite

B) Provide a brief answer for each of the following.

    1) What is a primed seed?

    2) What is the advantage of seeds selected to be all the same weight?

    3) Why are primed or enhanced seeds best for direct field planting?

Softwood and Semi-hardwood Cuttings, and Micropropagation

OBJECTIVE

To collect softwood and semi-hardwood cutting wood, prepare cuttings, and place them in the propagating structure for rooting.

Competencies to Be Developed

After studying this unit, you should be able to

- ◆ select plants suitable for propagating by use of cuttings and determine if the maturity of wood is correct for optimum rooting.

- ◆ propagate at least ten softwood or semi-hardwood plants from cuttings following procedures outlined in the unit, including
  - — collecting wood from the parent plant,
  - — making the actual cutting,
  - — treating the cutting with the proper rooting hormone when necessary,
  - — preparing the rooting medium and placing the cutting, and
  - — caring for the cutting after it is placed in the medium.

  (A minimum of nine of the cuttings should root for development of competency.)

- ◆ write a brief description of the tissue culture procedure.

MATERIALS LIST

- ✔ parent plant material from which to make cuttings
- ✔ plastic bags for collection
- ✔ labels and label marking pen
- ✔ propagating knife
- ✔ propagating container (flat, pot, etc.)
- ✔ rooting medium (sphagnum moss, perlite, or sand)
- ✔ rooting hormone
- ✔ watering can
- ✔ polyethylene plastic cover or sprinkling system

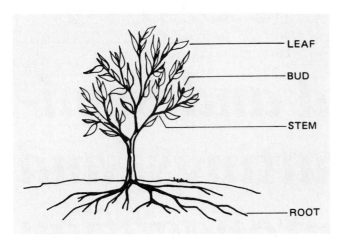

- LEAF
- BUD
- STEM
- ROOT

FIGURE 7-1 Cuttings may be taken from the bud, stem, leaf, or root.

Asexual reproduction is required for plants that are sterile (do not produce seeds), for hybrids as mentioned in Unit 6, and for plants difficult to grow from seed.

The most commonly used and commercially important method of asexual plant reproduction is propagation by the use of cuttings. *Cuttings* are leaves or pieces of stems or roots used for propagating plants. Most of the economically important perennials and many annuals are reproduced in this way. (*Perennials*, such as chrysanthemums, are plants that grow year after year without replanting. *Annuals*, such as petunias, normally complete their growth cycle in one year.)

There are various kinds of cuttings which are taken from different parts of the plant and at different stages of plant maturity (age). Cuttings may be taken from the leaf, bud, stem, or root, figure 7-1. They may be taken when the plant tissue is soft and tender (*softwood cuttings*), or when the plant tissue is hardened and more woody (*hardwood cuttings*). Some plants root more easily from softwood, others more easily from hardwood.

To root, cuttings require essentially the same conditions that seeds need to germinate: moisture, oxygen, and warmth (40°F [4°C] or above). In addition, cuttings need light—something that most seeds do not need to germinate. Light is essential in allowing the cuttings to produce food through the process of photosynthesis. Cuttings may begin to grow roots without light, especially in the first stages of *callus* formation (growth of tissue over a wounded plant stem), but generally, light is needed for food production. Food production supplies energy to the rapidly developing roots. Stored hardwood cuttings are an exception since they are able to develop roots without light.

Root formation on cuttings is stimulated because of the interruption of the downward movement of carbohydrates, hormones, and other materials from the leaves and growing tips. Since these materials cannot travel beyond the point at which the stem has been cut, they accumulate and stimulate the healing and rooting process.

Additional rooting hormones are sometimes used to aid in root formation. Some plants root more easily than others because they produce a higher level of natural rooting hormones. These plants need less synthetic or man-made rooting hormones to root satisfactorily.

SOFTWOOD AND SEMI-HARDWOOD STEM CUTTINGS

One method of asexual reproduction by cuttings is softwood and semi-hardwood cuttings of the stem. These cuttings are taken after the current or present season's growth has at least partially matured or hardened. The wood for semi-hardwood cuttings should be able to bend, but generally snaps if bent at an angle greater than 70 to 80 degrees. The softwood cutting is taken while there is still active *terminal* (tip) growth. This type of wood is very soft and breaks with a snap. The horticulturist best learns about maturity of wood through experience; it is not necessary to be exact with most plants.

PROCEDURE

NINE STEPS IN TAKING SOFTWOOD AND SEMI-HARDWOOD CUTTINGS

1. Collect the cutting wood of proper maturity from a selected parent plant.
2. Label the cutting wood.
3. Fill the propagating container with rooting medium.
4. Make the cutting.
5. Treat the cutting with rooting hormone.
6. Insert the cutting in the rooting medium.
7. Water to settle the medium around the cutting.
8. Insert the label in the rooting container.
9. Control the atmosphere around the cuttings by covering with plastic or placing under a mist system.

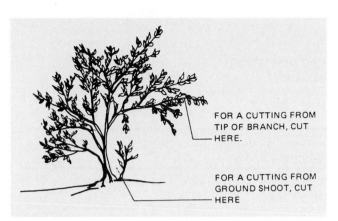

FOR A CUTTING FROM
TIP OF BRANCH, CUT
HERE.

FOR A CUTTING FROM
GROUND SHOOT, CUT
HERE

FIGURE 7-2 Softwood and semi-hardwood cuttings may be taken from the tip of a branch or from a shoot growing at the base of the plant.

COLLECTING CUTTING WOOD

The first step in making the cutting is to find a parent plant that has made at least 2 to 6 inches of new growth during the current year. This current season growth may be from the tip of a branch, or a shoot that is growing from the base of the plant, figure 7-2. Check to be sure that the wood is of proper maturity or hardness and that the plant is not wilted or dried out. If the plant appears to be suffering from lack of moisture, water it well and

wait until the shoots and leaves are *turgid* (swollen with moisture). This may require waiting overnight. Early morning is the best time to take cutting from the parent plant. The shoots and leaves contain more moisture at this time than later when the sun has had time to dry them out. Remember, the new cutting will have no roots with which to gather moisture. It should be started with as much moisture as possible.

Immediately after taking the cutting from the parent plant, place it in a bucket of water or a plastic bag containing water. This prevents drying of the wood after it is cut from the plant and before the actual cutting is made from it. At this time prepare a label with the variety name and the date on it. Place the label in the container with the cutting wood. Be sure the label is marked with waterproof pencil or waterproof felt tip pen.

Softwood and semi-hardwood cuttings are made when plants have leaves on them. There are various degrees of tissue hardness or maturity, depending on the plant being propagated. Figure 7-3 gives the best rooting date and desired maturity of various plants for the mid-Atlantic area.

Leaves help the propagator keep the cuttings right side up—they point upward. The buds in

PLANT NAME	ROOTING HORMONE	MONTH TO TAKE CUTTING IN MID-ATLANTIC STATES	CUTTING WOOD MATURITY	REMARKS
Arborvitae* (*Thuja occidentalis*) (*Thuja orientalis*)	Hormodin #2	Oct.–Dec.	hardwood	Wounding cutting helps.
Azalea (evergreen) (*Rhododendron*)	Rootone or Hormodin #1	early July	semi-hardwood	Use only terminal growth.
Barberry (*Berberis deciduous*)	Hormodin #2	July–Sept.	semi-hardwood	
Barberry (evergreen)	Hormodin #2	Nov.–Dec.	hardwood	
Boxwood* (*Buxus*)	Hormodin #2	July or Nov.–Dec.	semi-hardwood or hardwood	
Cotoneaster	Hormodin #2	July	semi-hardwood	
Dogwood (*Cornus*)	Hormodin #3	June–July	softwood	new growth after flowering
Deutzia	Rootone or Hormodin #1	June–July Nov.–Dec.	softwood hardwood	
English ivy* (*Hedera helix*)	Rootone or Hormodin #1	Nov.–Dec.	hardwood	
Euonymus*	Rootone or Hormodin #1	June–July–Aug. Dec.–March	evergreen-semi-hardwood deciduous-hardwood	

FIGURE 7-3 Rooting date and desired maturity of various plants

PLANT NAME	ROOTING HORMONE	MONTH TO TAKE CUTTING IN MID-ATLANTIC STATES	CUTTING WOOD MATURITY	REMARKS
Firethorn* (*Pyracantha*)	Rootone or Hormodin #1	July–Oct.	semi-hardwood	
Forsythia	Rootone or Hormodin #1	June–July Dec.–March	softwood in container hardwood, outside	
Holly (*Ilex*)* (evergreen)	Hormodin #3	July–Aug.	semi-hardwood	terminal growth
Hydrangea	Rootone or Hormodin #1	June–July Dec.–March	softwood hardwood	
Juniper (*Juniperus*)*	Hormodin #3	Nov.–Dec.	hardwood	
Lilac (*Syringa*)	Hormodin #2	directly after flowering	very soft wood	Use fungicide.
Mock orange (*Philadelphus*)	Rootone or Hormodin #1	July Dec.–March	softwood hardwood	
Pachysandra*	Rootone or Hormodin #1	Nov.–Dec.	hardwood	
Privet (*Liqustrum*)	Rootone or Hormodin #1	July Dec.–March	softwood hardwood	
Rhododendron	Hormodin #3	July and Sept.–Oct.	semi-hardwood	Wound cuttings; use only terminal growth.
Spirea (*Spiraea*)	Rootone or Hormodin #1	July Dec.–March	softwood hardwood	species: *vanhouttei*
Viburnum	Rootone or Hormodin #1	May–June July–Aug.	softwood semi-hardwood	fragrant types all others
Weigela	Rootone or Hormodin #1	July–Sept. Dec.–March	softwood hardwood	
Wisteria	Rootone or Hormodin #1	July	softwood	
Yew (*Taxus*)	Hormodin #3	Dec.–March	hardwood	

*May be taken in fall and midwinter.

Note: One of the new liquid dip rooting materials may be used in all cases above. Adjust the strength by diluting with water as recommended on the label. These materials penetrate plant stems better than powders. (Wear rubber gloves when working with all chemicals. Do not allow chemicals to get on your skin.)

FIGURE 7-3 Rooting date and desired maturity of various plants (continued)

the leaf axil are always on top on the leaf stem and toward the tip, figure 7-4. It is necessary to know the top of the cutting because the stem must be placed in the rooting medium with the bottom end down. Stem tissue is highly polarized and roots form much more easily on the bottom of the cutting. To avoid confusion, cut the branches from the parent plant at a 45-degree angle on the bottom end and straight across at a 90-degree angle on the top, figure 7-5. This is important only if the tip of the branch is cut off, a process which is usually necessary if the wood is too soft on the branch tip.

The cutting wood is now brought back to the bench or table where cuttings are to be made.

Tips: Make the cuttings as soon as possible; never allow the wood to dry out.

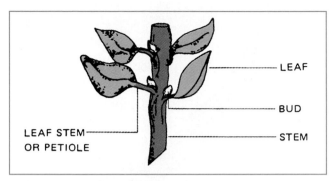

FIGURE 7-4 Parts of the cutting

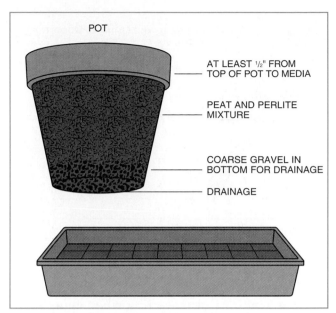

FIGURE 7-6 Any container at least 4 inches deep , such as a plant pot or flat, may be used to hold cuttings.

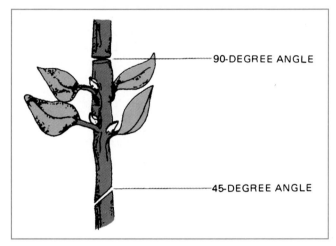

FIGURE 7-5 Cut branches from the parent plant at a 45-degree angle at the bottom and a 90-degree angle at the top.

PREPARING THE CONTAINER AND ROOTING MEDIUM

Before making the cuttings, a container for holding the medium must be prepared. A flat, pot, or any container at least 4 inches deep may be used, figure 7-6. The size of the container (other than its depth) is determined by how many cuttings are to be made. The container must have holes in the bottom to allow for drainage.

Tips: It is essential that the medium for rooting be loose to allow circulation of air, be able to hold moisture, and be sterile.

Remember that roots need oxygen and moisture to live. The cut end of the cutting makes it easy for organisms which cause rotting to enter the plant, so the medium must be sterile, or free from living organisms. A mixture of one-half horticultural perlite to help loosen the structure of the rooting medium and one-half coarse sphagnum moss provides a good medium. Washed con-struction grade sand may be used in place of the perlite and other types of peat moss may be used, however.

The medium may depend on the type of watering system that is to be used. Excellent success can be obtained with the peat-perlite mixture in containers that are to be covered with plastic until the cuttings root. When mist is used (a process in which fine droplets of water are sprayed on plants through nozzles), pure perlite or pure sand is satisfactory. If pure perlite or sand is not used, the percentage of peat moss should be reduced to 25 percent or less. The high percentage of peat moss recommended for pots and flats sealed with plastic would hold too much moisture under mist, thereby withholding sufficient oxygen and possibly causing the cuttings to rot. The rooting mixture should be mixed before it is wet so that a uniform mixture results.

FILLING THE CONTAINER

After the rooting medium is thoroughly mixed, it is placed in the container, leveled, and firmed. There should be at least one-quarter inch between the top of the container and the level of the medium surface so that water does not run off the surface as the medium is watered. The surface is leveled with a flat board, skew, or brick. The leveling is important so that water spreads evenly over the surface and all cuttings are watered equally. The medium is watered until water runs through the bottom of the container. If the peat moss used

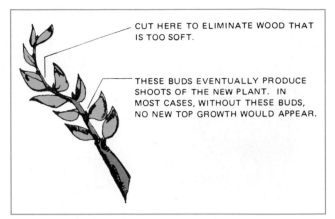

FIGURE 7-7   If the top of the branch or shoot is too soft, the more mature wood closer to the base should be used.

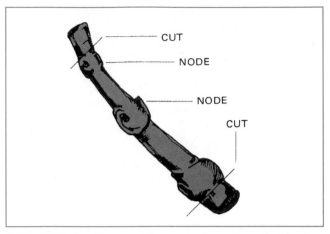

FIGURE 7-8   Stem cutting on hollow stem. Cutting is made below a node at the bottom and above a node at the top. (Leaves have been omitted from this illustration to allow clear view of the nodes.)

is dry, the moss may require an overnight soaking in water, since it resists wetting when completely dry. After soaking the peat moss in water, squeeze out the excess water before mixing and placing in the container. Hot water is recommended to thoroughly moisten peat moss and soilless planting media.

MAKING THE CUTTINGS

Now that the cutting wood is selected and the container is prepared, the cuttings are made. Inspect the cutting wood for proper maturity. If the tips of branches or shoots are too soft for the particular plant being propagated, cut them off and use more mature wood nearer the base, figure 7-7. This does not apply to plants such as the rhododendron, a plant in which the terminal end must be used as the cutting. This is because buds appear only near the terminal end on the rhododendron.

Make as many 3-inch or 4-inch cuttings as possible from each stem or shoot. A very sharp knife or pruning shears must be used so that a clean cut is made. Remember to cut the bottom on a slant and the top of the cutting straight across to indicate which end should be placed in the rooting medium. The bottom of cuttings are cut slanted for another reason. The new cutting has no roots, and the slanted bottom allows more surface area to contact the rooting medium for moisture uptake or absorption. It is not important to cut at or near a node unless the plant stem is hollow. On hollow stems, cut just below the node on the bottom and just above a node at the top, figure 7-8.

It is important to include two or three buds on each cutting. New shoots can grow only from leaf buds, which appear in the axil of the leaf petiole or leaf stem.

TREATING WITH HORMONE AND INSERTING IN MEDIUM

After the cutting is made, it is treated with the proper concentration of rooting hormone. Follow directions on the label, and be sure the substance is recommended for plants being rooted. Rooting hormones are chemicals which help cuttings grow roots more quickly and grow a larger number of roots. Hormones used for various plants are shown in figure 7-3. If the hormone does not contain a fungicide, the cutting may be sprayed, dipped, or watered with a fungicide such as captan to help prevent rotting. Captan has been reported to stimulate rooting in addition to controlling fungus infection.

The cutting is immediately placed in the rooting medium to one-half its length but not more than 2 inches deep. Cuttings should be spaced so that no leaves overlap. Large leaves may be trimmed. Do not press the medium around the cutting very firmly. Add water to settle the mixture around the cutting.

LABELING

The cuttings must now be labeled, using either the same label that was prepared when the wood was cut from the parent plant, or by preparing new labels. The name of the rooting hormone should be added to the information already on the label

> *Caution: Knives used in propagation are sharp. Never cut toward yourself or another person.*

(variety name and date, including the day, month, and year). Remember to use a waterproof pencil or pen. The label is placed in the rooting container.

CONTROLLING THE ATMOSPHERE

The new cutting has leaves, but no roots. Because of this, these leaves lose moisture through *transpiration* (loss of water through the leaves or stems of plants) more quickly than it can be drawn up through the slanted cut end. Therefore, it is essential to keep the relative humidity around the cutting close to 100 percent to reduce moisture loss.

If a greenhouse is available, one way in which high humidity can be maintained is to place the cuttings in a misting propagation bench. This system uses a greenhouse bench equipped with misting nozzles which spray a fine mist on the cuttings from a piped water system. The system may be operated manually or set up to water the cuttings automatically at specific intervals. An alternative method for use without a greenhouse is to enclose the containers with a film of polyethylene plastic (a clear plastic that holds in moisture but lets in light and a small amount of air). Very little water is lost from the container, but there is a need to water the cuttings every three or four weeks until they have rooted.

When a pot is used as a container, a plastic food bag may be slipped over it and tied in place or tucked under the pot as shown in figure 7-9. A flat may be covered by drilling three holes in each side and placing stiff wire over the flat. This structure is

FIGURE 7-10A   Flat with rooting medium being enclosed in a watertight framework. The plastic is pulled over the wire hoops and tied around the bottom or tucked under to form a watertight hut for propagating.

FIGURE 7-10B   A commercially available propagating flat similar to the framework hut. (Courtesy of Gardener's Supply Company)

covered with polyethylene film which is tied or tucked under the bottom of the flat, figure 7-10. Containers covered with plastic must be placed out of direct sunlight by locating them against a north wall, in the shade of a tree, or under a bench. Light aids formation of carbohydrates—and thus rooting—and should be available. However, the cuttings should not receive direct sunlight since this raises the temperature inside the container too much, thereby killing the cuttings. If they must be placed in the sun, the plastic may be covered with cheesecloth or burlap to give shade and hold down the temperature. When the misting system is used and cuttings are not in a tight container, direct sunlight is beneficial.

FIGURE 7-9   Rhododendron cutting ready to be inserted in a plastic bag. The bag is pulled over the pot and tied to seal in moisture. The pot is then placed under bright light, but not direct sunlight. (Ed Reiley, Photographer)

FIGURE 7-11 (A) shows rhododendron cutting as taken from parent plant; (B) shows cutting after trimming, with stem and leaves trimmed and the stem slices on each side at the base; (C) illustrates a well-rooted rhododendron cutting ready to be planted in the cold frame. (Ed Reiley, Photographer)

ROOTING

As time nears for the cuttings to be rooted, open the plastic-covered container or go to the misting propagation bench and check the cuttings by holding each cutting and tugging gently. If the cutting does not easily slip out, roots are developing. If there are no roots, merely close the container or leave the cuttings under mist, depending upon the system used, and wait another week to ten days. As long as the cutting has not rotted or turned dark and is holding its leaves, it will probably root if it is given more time. Growth on the tip or sides of the cutting is normal, but may not indicate that the cutting has rooted.

When the cuttings have a root ball measuring 2 or 3 inches across (see C of figure 7-11), it is time to harden them off in preparation for transplanting (removing and planting in a new place). The hardening off must be done very slowly over a period of seven to fourteen days so that the cuttings are not exposed to a drying atmosphere too suddenly. They are accustomed to high humidity and must adapt to new, drier conditions. Open the plastic a little more each day over the fourteen-day period until it is totally removed. Once the plastic is completely opened, care must be taken not to allow the rooting medium to dry out. The cuttings must now be watered to keep the medium moist but not wet. If a misting propagation bench is used, the misting cycle is changed so that the misting time becomes shorter and shorter until no mist is used and the cuttings are watered normally.

After the hardening-off process, the cuttings are ready to be *lined out* (planted outside in the nursery row under shade or sprinklers) or transplanted to other containers for further growth.

HERBACEOUS CUTTINGS (SOFTWOOD)

Herbaceous cuttings are made from succulent greenhouse plants such as the geranium, chrysanthemum, coleus, and carnation. Most florist crops are propagated in this way. Cuttings are made 3 to 6 inches long with leaves left on the upper, or terminal, end. Several cuttings may be cut from a plant stem. This allows the horticulturist to obtain many more cuttings than if only leaf or stem tips are used. Leaves are trimmed from the basal, or lower, end of the cutting, with three or four leaves left remaining on the top end. A fungicide should be used on all cuttings to prevent rotting. A sterile rooting medium is also essential.

Herbaceous cuttings are rooted under the same high humidity conditions required for softwood cuttings. Bottom heat helps to speed rooting. Under proper conditions, rooting is rapid. For this reason, this type of cutting is highly recommended for the beginner. A mild concentration of rooting hormone such as Rootone #1 may be used, but is not necessary. A high degree of success may be expected, with more than 90 percent of cuttings rooting. Using a flat, flowerpot, or mist system in the greenhouse, try propagating at least one of the herbaceous plants listed. Care for them in the same manner as noted for the softwood cuttings.

Additional plants easily propagated as herbaceous cuttings include the lantana, Swedish ivy, wandering Jew, and begonia.

MICROPROPAGATION— TISSUE CULTURE

One of the newest approaches to plant propagation is a method called *tissue culture* (also called *micropropagation* or *meristem culture*). This is an asexual method using sterilized terminal shoots or leaf buds placed on a sterile (disease-free) agar gel or other nutrient-growing medium, see figure 7-12. One part bleach to 10 parts medium is used

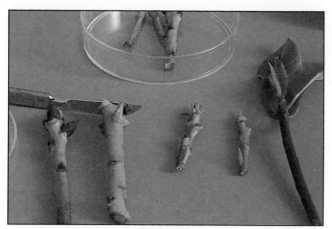

FIGURE 7-12 Small stems being trimmed and sterilized. They will be placed in a small jar or test tube on nutrient agar gel to start sending out shoots. Shoots will emerge from the buds or growing tips of the stem. (Courtesy of Briggs Nurseries)

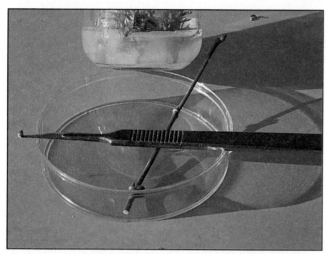

FIGURE 7-13 Small shoots (micro-cuttings) growing from stem in a jar. (Courtesy of Briggs Nurseries)

FIGURE 7-14 Close-up of tissue proliferation and new shoots sprouting. (Courtesy of Briggs Nurseries)

FIGURE 7-15 Shoots (micro-cuttings) being separated and placed in pots to root. They will root like any other softwood cutting taken from a full-sized plant. Since they are very tender, extreme care must be taken that they do not dry out. (Courtesy of Briggs Nurseries)

for sterilizing, and the proper growth-regulating chemicals and nutrients are mixed in the growing medium. The tips and medium are placed inside a test tube or small jar and kept sealed to keep out mold and other disease organisms.

Soon new tiny plants sprout out from the plant tissue placed in the sealed container, figures 7-13 and 7-14. As the tiny sprouts grow large enough to be moved, they are pulled off with sterile tweezers and placed in a new medium in another container to grow roots, see figure 7-15, or the new shoots may be placed back in a jar to produce more shoots. This process is repeated again and again as new sprouts appear. As soon as roots develop, the container may be opened, a little more each day for about a week, allowing the plants to harden off. Plants may then be transplanted to small containers and treated in the same manner as small seedling transplants.

> *Tips: It is very important that all containers, growing media, hands, etc. be sterile so that no plant disease organisms are present.*

Through the micropropagation method many thousands of new plants exactly like the parent can be produced very quickly. This is a good way to rapidly increase numbers of a new superior plant and get it on the market years earlier than could

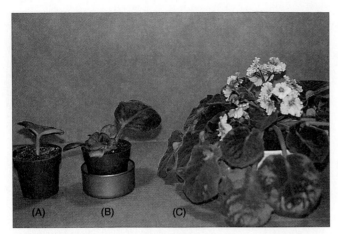

FIGURE 7-16 (A) African violet cutting as taken from the parent plant. (B) Rooted cutting showing the small plant growing from the base of the leaf stem. It is ready to be planted as a new plant. The old leaf is pruned off at this stage or soon after. (C) Adult African violet plant in bloom. (Ed Reiley, Photographer)

be done with cuttings or other propagation methods. It is estimated that in 28 weeks, one blackberry tip could produce 2.5 million tissue particles ready for rooting.

Micropropagation has been used for years on orchids, but is now being used for many kinds of plants. Different plant types and even different varieties of the same type of plant often need different amounts of chemical plant food, have different pH or acidity, and require different plant growth regulators for best results. This fact and the need for absolutely sterile conditions make it hard for the untrained person to propagate plants through micropropagation. For this reason no more detail will be given here. If you are interested in trying this newest propagating technique and have the laboratory equipment, get all the details from your local agricultural college or experiment station. Be prepared for almost daily

observation and a strict time schedule. Also, be sure you can keep sterile growing and transplanting conditions until the small plants are stuck for rooting.

An alternative way of obtaining tissue culture plants is to purchase them at the stages shown in figures 7-13 and 7-15 and then perform the rooting step yourself. This method does not require as much care in disease prevention.

Research done at McMaster University in Ontario, Canada has made it possible to produce small duplicate plants from the tip of a growing plant. A plant hormone is applied to the stem and a new plant starts to form right on the plant. No expensive sterile laboratory is needed in this process.

OTHER TYPES OF SOFTWOOD CUTTINGS

LEAF CUTTINGS

The African violet is a good example of a plant propagated by leaf cuttings. Examine figure 7-16 for details and try this easy-to-root plant. (See figure 7-18 for details on making mallet cuttings.)

It will be necessary to advance to another unit while the cuttings are rooting. A careful check should be made each day, however, to see if the cuttings are progressing and to be sure that no damage has been done to the containers or the cuttings.

LEAF-BUD CUTTINGS

Many plants normally propagated using softwood and semi-hardwood cuttings may also be propagated using leaf-bud cuttings. Rhododendrons root

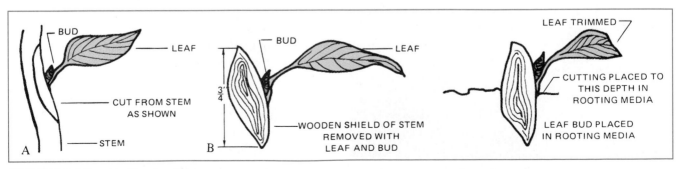

FIGURE 7-17 Leaf bud cutting

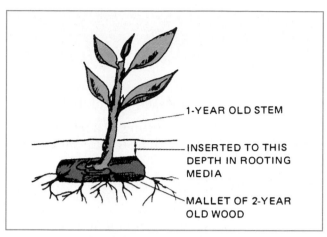

FIGURE 7-18 Mallet cutting

well this way and may be used as a project to test this method.

The cutting is taken by cutting a bud from the stem very much like the cut made for the bud in the T-budding procedure in Unit 11. The difference is that the leaf is left intact and thus the cutting must be protected from moisture loss the same as with a softwood or semi-hardwood cutting. The protection process was explained earlier in this unit.

The first step in the cutting procedure is to cut the leaf and bud from the plant as shown in figure 7-17(A). The wooden shield of the cutting should be about ¾ to 1 inch long and the leaf should cut off crossways to get it to that size. (See figure 7-11 for leaf trimming.) This smaller leaf will require less moisture to survive. Figure 7-17(B) shows complete leaf-bud cutting. Figure 7-17(C) shows the cutting inserted in the rooting medium. It must be firmed in, watered, and covered in a moisture-tight container. The advantage to leaf-bud cuttings is that many more cuttings can be made from the same number of parent plants than can be made using stem cuttings. This may be important in introducing new plants. Usually, fewer plants will root, and sometimes the bud is slow to break into new growth. Chemical rooting hormone strength should be weaker than with stem cuttings (about 50 percent or less in strength). Rooting hormones appear to slow initial bud growth and thus new shoot development.

MALLET CUTTING

The mallet cutting is made using a 2"-4" piece of 2-year-old wood with a current season or 1-year-old shoot on it, see figure 7-18. The mallet is placed horizontally and covered about 1 inch with rooting medium.

ROOT CUTTINGS

Root cuttings may be made from any plant that will sprout or sucker from the root. Roots from one-eighth to one-half inch in diameter cut into pieces 1 to 4 inches long are placed in sand in a flat or other container and watered well. Glass or clear plastic over the flat helps to hold moisture. When new shoots sprout from the flat, the new plants are moved to the nursery row. Plants commonly propagated this way include the Oriental poppy, wisteria, spirea, red raspberry, and trumpet creeper. Figures 7-19 and 7-20 illustrate the propagation of plants by use of root cuttings.

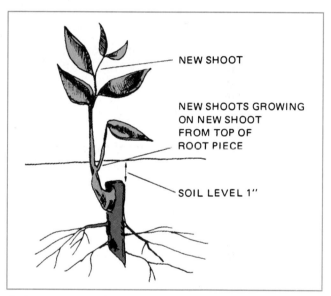

FIGURE 7-19 Fleshy root cutting. After cutting is placed in soil, it is given no special care.

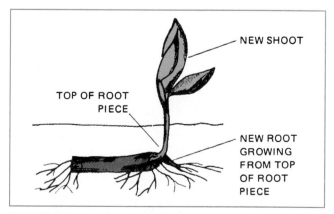

FIGURE 7-20 Small, nonfleshy root cutting. This cutting is placed horizontally in the soil.

Student Activities

1) Practice making softwood and semi-hardwood cuttings by completing each step of this exercise.

   a) Collect cutting wood for at least ten softwood or semi-hardwood cuttings. Label the cuttings and keep moist until placed in the rooting mix. Check with instructor on the proper maturity of cutting wood.

   b) Select a container and fill it with the proper rooting mix. (Provide for drainage.) Water properly as determined by the instructor.

   c) Using a sharp knife, make ten cuttings and insert in the rooting medium. treat with rooting hormone, label, and space properly. (The text may be used for reference.)

   d) Place the cuttings under a mist system or cover with plastic. Ninety percent of the cuttings should root. Check to be sure all of the previous steps in the exercise have been completed correctly. If nine out of ten cuttings root, you are a competent propagator. If less than 90 percent root, repeat the project if time allows. If not, review the steps in the procedure and trace your errors.

   Suggestion. Spray cuttings with a systemic fungicide or dip them in a fungicide when they are made. This is usually sufficient treatment to prevent rotting if a sterile medium is used.

2) Plants that are more difficult to root, such as evergreens, may be tried as an additional activity. A good choice might be the rhododendron. The following are necessary extra steps for this plant, which can be difficult to root.

   a) Take a cutting with three or four leaves from terminal current season's growth in July or from September 15 to October 30. (Trim leaves if necessary.)

   b) Cut to 2½ inches long.

   c) Trim a slice off each side of the cutting 1 inch from the base and extending to the base just deeply enough to cut into the wood.

   d) Treat with Hormodin #3, 0.8 percent indolebutyric acid, or a liquid like "Dip-N-Grow." (Liquid dip treatments are more effective.)

   e) Complete by following the same steps given for other cuttings.

3) Use one of the commercially available tissue culture kits to propagate.

Self-Evaluation

A) Complete each statement on the left regarding the propagation of plants from cuttings with the proper term on the right. Include both the letter and the word or phrase in your answer.

1) Cutting wood must be of the proper _____, or hardness.
2) A rooting medium that lets in _____ and holds _____ must be used.
3) The medium is placed in a container which measures at least _____ in depth.
4) The cutting wood must be kept _____ and identified with a (an) _____.
5) To aid in the rooting process, the cutting is treated with a (an) _____.
6) The cutting is inserted in the medium no deeper than _____.
7) The rooting medium is not pressed around the cutting to settle the medium; rather, it is _____.
8) The container holding the cutting is labeled according to _____, _____, and _____.
9) If the cuttings are placed under plastic, they should be given plenty of _____, but no direct sunlight.
10) If the cuttings are placed under _____, direct sunlight is preferred.
11) As the rooting time nears, the horticulturist checks for roots by gently _____ on the cuttings.
12) The cuttings are hardened off when they have sufficiently _____.
13) After hardening off, the cuttings are ready to be _____.
14) Tissue cultured plants are grown in _____ to prevent attack by disease.
15) The tissue culture method allows propagation of _____ plants than other methods.

a) moisture
b) 6 inches
c) light
d) pulling
e) rooting hormone
f) maturity
g) sterile
h) oxygen
i) 4 inches
j) variety
k) fungicide
l) moist
m) 2 inches
n) mist
o) rooted
p) label
q) watered
r) misting
s) date
t) hormone
u) transplanted
v) sterile container
w) more

B) Provide a brief answer for each of the following.

1) What is asexual plant propagation?

2) What type of plants are usually propagated by asexual propagation?

3) Why is it necessary to reproduce plants asexually?

4) List the nine key steps in taking softwood or semi-hardwood cuttings.

5) What part of the plant is used to propagate a plant by tissue culture?

Hardwood Cuttings

O BJECTIVE

To perform the six steps in taking hardwood cuttings so that success in rooting 60 to 80 percent of the cuttings is achieved.

Competencies to Be Developed

After studying this unit, you should be able to

- ◆ list five plants commonly propagated by hardwood cuttings.
- ◆ select, collect, and label mature cutting wood from one of the five identified parent plants.
- ◆ make and store ten hardwood cuttings.
- ◆ line out cuttings in the field at the proper planting time.
- ◆ keep accurate records of the percentage of cuttings that root.

M ATERIALS LIST

- ✔ cutting wood
- ✔ string with which to tie bundles
- ✔ storage area
- ✔ propagating knife
- ✔ labels and waterproof pen or pencil with which to mark labels

Propagation by hardwood cuttings is one of the easiest and least expensive methods of vegetative or asexual propagation. Cuttings are prepared in the dormant winter season when time is usually more available to the propagator. Hardwood cuttings may be shipped long distances or stored for long periods of time. Expensive misting equipment or rooting benches are not needed and the necessary cold storage may be done outdoors in low-cost facilities if temperatures are

SOME DECIDUOUS PLANTS GENERALLY PROPAGATED BY HARDWOOD CUTTINGS	
blueberry	mock orange
currant (ribes)	mulberry
deutzia	Persian lilac
dogwood	plum
euonymous	pomegranate
fig	privet
forsythia	rosa multiflora
gooseberry	spirea
grape	weigela
honeysuckle	wisteria
kerria	

SOME NARROWLEAF EVERGREENS PROPAGATED BY HARDWOOD CUTTINGS	
chamaecyparis	spruce
hemlock	thuja
juniper	yew
pine	

FIGURE 8-1 Deciduous and narrowleaf evergreens propagated by hardwood cuttings

FIGURE 8-2 Strong, reddish-brown young shoots growing from the base of a blueberry plant. These shoots (located to the left of the parent plant) make excellent hardwood cuttings. Notice that the leaves have all dropped off the plant; this is because the plant is dormant. (Ed Reiley, Photographer)

50°F (10°C) or lower. Many deciduous woody plants are easily propagated by hardwood cuttings. Figure 8-1 lists deciduous and narrowleaf evergreens propagated by hardwood cuttings.

Propagation by hardwood cuttings differs from softwood and semi-hardwood cuttings in several ways:

- the time of year in which the cuttings are taken,
- the hardness, or maturity, of wood used for the cuttings,
- the usual absence of leaves on the cuttings, and
- the storage of cuttings instead of immediate planting.

PROCEDURE

SIX STEPS IN TAKING HARDWOOD CUTTINGS

1. Identify the plant to be propagated and select proper cutting wood.
2. Collect wood for cuttings.
3. Make the cuttings.
4. Store the cuttings.
5. Line out or plant the cuttings.
6. Determine rooting percentage of the cuttings.

SELECTING THE CUTTING WOOD

Wood for hardwood cuttings is generally taken from current season's growth, as is done with softwood cuttings. The wood is cut from the ends of branches or from long shoots that grow from the base of the plant, figure 8-2. This material may be gathered soon after the plants become dormant. Plants that are *dormant* have lost their leaves and are preparing for the winter rest cycle. Cuttings may be taken throughout the winter months. The wood taken is current year's growth, but it is now mature, or hardwood. Plants that are healthy, vigorous, and grown in full sunlight yield cutting wood with more stored carbohydrates and more vigor in rooting.

COLLECTING THE CUTTING WOOD

Using a sharp knife, cut the selected wood from the parent plants. Since the wood has no leaves, drying out is not as much of a problem as it is with softwood cuttings. Label the cuttings according to variety name and date collected. The wood may be made into 6- to 8-inch cuttings for immediate use, or stored in a cool moist place (below 50°F or 10°C) for later use, figure 8-3. If stored, the branches should never be allowed to dry out

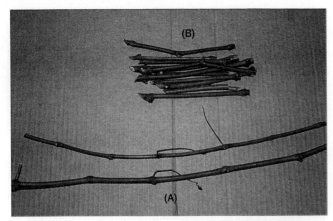

FIGURE 8-3 (A) illustrates grape hardwood cutting wood as collected from the plant. (Stems are longer than shown in the photograph.) (B) shows hardwood cuttings cut to proper length and ready to be tied in bundles and stored. (Ed Reiley, Photographer)

or become too wet. A covering of moist (but not wet) sawdust, sand, or peat moss works well to maintain moisture. There should be little variation in temperature during storage since temperature variation tends to hasten drying.

TAKING CUTTINGS

Cuttings, usually 6 to 8 inches long, are made with a sharp knife or pruning shears. For hollow-stemmed plants, such as the grape, the bottom cut is made just below a node and the top cut about 1 inch above a node or bud (see B of figure 8-3). Make the bottom cut (end toward the base of the plant) at a 45° angle and the top cut at a 90° angle so that cuttings can be planted bottom down. The diameter of cutting wood is not extremely important. Medium-sized wood, however, tends to survive storage and root better than wood that is very thin.

Cuttings should be made early enough so that they may be stored six to eight weeks prior to planting.

STORAGE OF CUTTINGS

The cuttings are treated with a rooting hormone for better growth. The finished cuttings are then tied in bundles for storage. They should be placed so that all the tops are on the same end of the bundle and buried in moist sawdust or similar material. The bundles should not be tightly sealed; the cuttings are alive and need oxygen to remain healthy. A tight seal could also cause excess moisture to collect, resulting in rotting.

The storage period allows the cut ends of the cuttings to callus and the rooting process to begin. Callus formation which occurs during storage is an advantage when cuttings are placed outdoors for rooting under drying conditions, since callused cuttings root more quickly. New roots begin to grow during storage and may be visible at this point.

The cuttings may be buried outside in sand-filled containers in a well-drained area. Outside temperatures must be cold enough to prevent growth beginning at the tops of the cuttings during the storage period. During the first four weeks of storage, the temperature should be 50° to 55°F (10° to 13°C). This relatively high temperature is favorable for callus formation. After callus has formed, the temperature may be lowered to below 40°F (4°C), but never below 32°F (0°C). This lower temperature prevents growth on the tops before the cuttings are lined out (planted).

LINING OUT CUTTINGS

As soon as the soil is ready in the spring, the cuttings are planted outside. This process is known to nursery workers as *lining out*. Before lining out the cuttings, the soil must be prepared. A well-drained, sunny site with some wind protection is best.

Rows are dug deeply enough so that the cutting can be covered with soil with only the top bud remaining above the soil level, figures 8-4 and 8-5. If they are to be left for only one year, the cuttings are placed 6 inches apart in rows 1 foot apart. When the cuttings are to be left more than one year, or when especially fast growing plants are involved, the cuttings are spaced 9 to 10 inches apart and in rows 2 to 3 feet apart.

A small shovel or other tool is used to break the soil. The cuttings are placed in the soil at the depth described above. The cutting should not be pushed into the soil under pressure; this could damage the tissue or break the cutting. After the cutting is placed in the row, the soil is pressed firmly

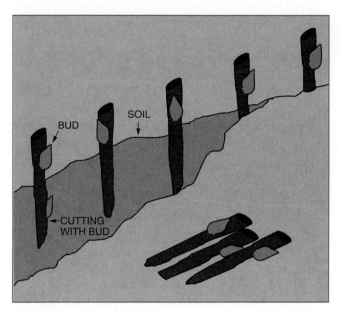

FIGURE 8-4 Hardwood cuttings lined out in the field. Only the top bud is left above the soil level.

Narrowleaf evergreens propagated by hardwood cuttings are often very difficult and slow to root. In general, the slow-growing *Juniperus* species root easily, the yews fairly easily, and the upright junipers, spruces, and hemlocks with difficulty. Pines are extremely difficult to root. Rooting hormones with high concentrations of indolebutyric acid (0.8 to 1.0 percent actual chemical solution or powder) are usually beneficial. These cuttings are taken from late fall to late winter and are best rooted in a greenhouse under high light intensity and in high humidity. This is achieved by placing the cuttings under plastic or using a misting system as specified for softwood and semi-hardwood cuttings. Cuttings should also be dipped in a fungicide such as captan to prevent rotting. Rooting media and other environmental conditions are essentially the same as those for softwood and semi-hardwood cuttings.

Hardwood cuttings of deciduous plants do not require the extensive care given to softwood and semi-hardwood cuttings. Since no leaves are present in the initial rooting stages, the demand for moisture is not high, and these cuttings need not be placed under plastic or in a mist system. The evergreen hardwood cutting, which does have needles at the time of rooting, must be handled as softwood cuttings are.

Rooted cuttings of deciduous plants are usually dug for transplanting when they are dormant and without leaves. They may be dug bare root. Evergreens, however, which have needles, should be dug with a ball of soil attached to the roots. Care must be taken to keep the root ball intact.

around the cutting. The soil surface may then be mulched to help hold moisture and control weeds or cultivated for weed control. The soil should never be allowed to dry out around the cuttings. Provision should be made for watering as needed.

The number of plants which will eventually root depends upon the variety of the plant. However, between 60 and 80 percent of the cuttings can be expected to root and grow.

As the weather becomes warm, the cuttings develop leaves and shoots although they may not yet be rooted. However, if growth continues into the summer months, it is almost certain that the cuttings have rooted. The care required beyond this stage varies with the plant being propagated.

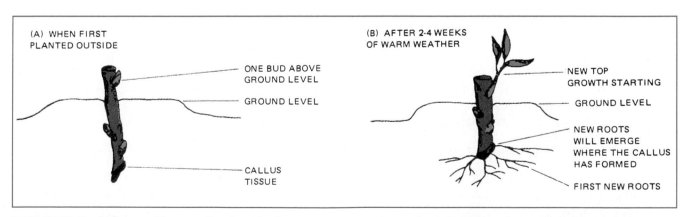

FIGURE 8-5 When cuttings are lined out in the nursery row, only one bud should be above ground level. At least one (preferably two) buds should be below ground level. (A) shows cutting directly after planting; (B) shows root development and initial top growth.

Student Activities

1) Visit a local nursery to observe a demonstration of hardwood cutting techniques.

2) a) Select at least one plant commonly propagated by hardwood cuttings. Perform the six basic steps to produce at least ten cuttings.

   b) Set up a calendar to determine

      1. the storage temperature and dates to change temperature.
      2. the approximate date for lining out certain cuttings.
      3. inspection dates so that such things as moisture conditions, temperature, and rooting date can be noted.

   c) Line out the cuttings in a nursery row in the spring and follow up to determine rooting success. Keep a record of observations, including dates when shoots and roots first develop.

3) Start a small nursery on the school grounds or at home in which to propagate and grow plants for future use. This is an inexpensive way to keep a supply of laboratory materials on hand.

Self-Evaluation

A) List three differences between hardwood and softwood cuttings.

B) Complete each statement on the left with the proper term on the right. Include both the letter and the word in your answer.

1) Ideally, hardwood cuttings are _____ long.
2) Hardwood cuttings are stored for the first four weeks at a temperature of _____.
3) Wood of _____ thickness makes the best cuttings.
4) Cuttings should not be sealed airtight since _____ would be excluded causing rotting or even death.
5) After the first four weeks, cuttings are stored at a temperture of between _____.
6) A storage temperature of between 32° and 40°F prevents cuttings from growing _____.
7) A soft wound-healing growth called _____ forms on the basal end of the hardwood cutting.
8) It is best if cuttings are stored a minimum of _____ weeks prior to planting or lining out.
9) Cuttings are planted so that only the top _____ is above the soil level.
10) In any group of cuttings which are properly cared for, between _____ percent of the cuttings should root.

a) oxygen
b) 6 to 8 inches
c) medium
d) 2
e) 32°–40°F
f) bud
g) 6
h) 60–80
i) tops
j) 50°–55°F
k) callus
l) 4 to 8 inches
m) moisture
n) 80–100

Separation and Division

OBJECTIVE

To propagate one plant by separation and one by division.

Competencies to Be Developed

After studying this unit, you should be able to

◆ describe the processes of separation and division and explain the major difference between the two.

◆ identify five specialized plant structures used in propagation and explain how each is used.

◆ list five plants propagated by separation or division and the type of specialized structure used in each case.

◆ write a brief definition of *bulb, corm, tuber, tuberous root,* and *rhizome.*

MATERIALS LIST

✔ bulbs, corms, tubers, tuberous roots, rhizomes
✔ sharp knife
✔ storage container and packing material

Many plants are propagated by one of two processes usually associated with one another—the process of separation, and the process of division. In both cases, plants are propagated by use of underground plant parts. The primary function of these plant parts is to act as an organ for food storage. These plant parts are found on herbaceous perennials having shoots which die back at the end of the growing season; the underground part lives through the winter (dormant season) and sends forth a new top the following growing season. The second function of these plant parts is that of vegetative or asexual reproduction.

Separation is a method of propagation in which natural structures produced by certain plants are removed from the parent plant to become new plants. Bulbs and corms are examples of removable structures. These parts are usually removed while the parent plant is in a dormant or resting stage. Some bulb-producing plants propagated by separation are the tulip, amaryllis, lily, narcissus, daffodil, hyacinth, and grape hyacinth. Examples of corm-producing plants are the crocus, gladiolus, and colchicum.

Division is a method of propagation in which parts of the plant are cut into sections, each capable of developing a new plant. Division is possible with plants which grow rhizomes, stem tubers, and tuberous roots. Examples of plants with rhizomes are the iris, canna, and calla lily. Those with stem tubers include the caladium, peony, and Irish potato. Tuberous roots are found on the dahlia, gloxinia, and bleeding heart.

PROPAGATION BY SEPARATION

As mentioned, plants that are propagated by separation are those which produce bulbs or corms as specialized underground plant parts which are responsible for food storage and propagation of the plant. Separation is a natural process.

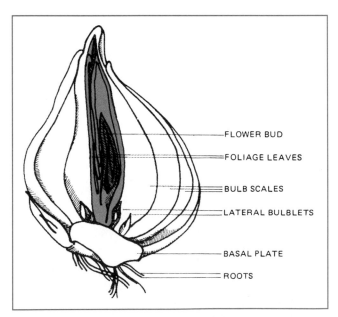

FLOWER BUD

FOLIAGE LEAVES

BULB SCALES

LATERAL BULBLETS

BASAL PLATE

ROOTS

FIGURE 9-1   Parts of the tulip bulb. The bulb scales grow small bulblets at their base. The bulblets eventually grow larger and may be removed and grown to flowering size.

BULBS

PARTS OF THE BULB   A *bulb*, such as the tulip bulb, is a plant structure containing many parts but primarily composed of leaf scales, figure 9-1. Immediately next to and outside of the foliage leaves are bulb scales which produce small *bulblets* (tiny bulbs) at their base. These tiny bulblets grow larger and become small bulbs which may be separated and planted as individual plants. After several growing seasons, the small bulbs become large enough to bloom. This produces *offset bulbs*. The process of developing offset bulbs is a simple commercial method for propagating the tulip, daffodil, bulbous iris, and hyacinth, figure 9-2.

Offset daffodil bulbs are called *splits* or *slabs* when first separated from the mother bulb. After one year of growth, the slab or split is called a *round bulb*, and is capable of flowering the following season. Within the second year, a second flower bud forms. At this time, the bulb is known as a *double nose* and is capable of producing two flower stalks. The round and double-nose sizes are sold commercially. Large mother bulbs are kept to produce more splits. Splits or slabs are planted to grow for another year to reach flowering size.

TYPES OF BULBS   The tulip bulb is a fairly solid structure and therefore does not require spe-

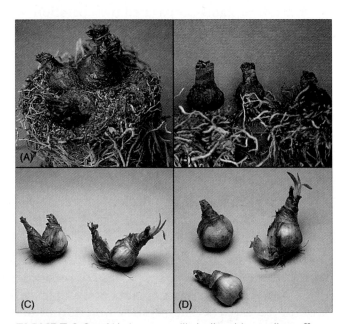

FIGURE 9-2   (A) An amaryllis bulb with smaller offset bulbs attached. (B) The same bulbs, after the offsets have been separated from the **mother**, or parent, bulb. (C) Two daffodil bulbs with smaller offset bulbs attached. (D) The same bulbs, after the offset bulbs have been separated from the parent plant. The two offset daffodils may require another year to grow to blooming size.

cial care in handling. The lily, however, is a loosely scaled bulb which cannot withstand rough handling or drying. The tulip is a *laminate* or *tunicate* bulb; these types of bulbs have dry membranous outer scales that protect them against drying and injury. The lily represents a *nontunicate* or *scaly* bulb, which has no tough outer cover.

PROPAGATION PROCEDURE   Bulbs may be dug and separated after the foliage of the plant dies back and the plant is in a dormant state. The new bulbs are then stored at 65° to 68°F (18° to 20°C) and planted at the proper time. For tulips, planting time is in the fall. The minimum size for flowering of tulip bulbs is 10 centimeters. Separated bulbs which are smaller than this must be grown for 1 to 3 years before they reach flowering size.

Dry bulbs are stored at 55° to 60°F (13° to 16°C). Bulbs should be washed clean of all soil before storage and treated for rot.

The tops of bulb-producing plants should not be cut off until they have turned brown. Early removal of tops greatly reduces the amount of food produced and stored, and thus reduces bulb size. This could result in the bulb being too small to flower the following season.

Potted plants, such as the amaryllis, should be allowed to continue to grow for 6 to 8 months after blooming before water is withheld to force them into dormancy. This allows time for the bulb to grow and store sufficient food reserves to bloom during the next growing period.

The other plants mentioned earlier such as narcissus, hyacinth, and grape hyacinth are propagated in a similar manner.

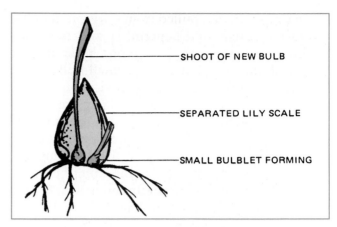

FIGURE 9-4   Bulb scale propagation of the lily

LILY PROPAGATION   The lily, figure 9-3, is also propagated by bulbs, but at a much slower rate. Mother bulbs are split at the base to force production of new bulblets. Some lilies naturally grow small bulblets on the stems and even in leaf axils. Commercially grown lily bulbs are separated and individual bulb scales are planted to produce new bulblets, figure 9-4. This is a more rapid method of increasing the number of bulbs since the lily bulb contains many loose, easily separated bulb scales.

The Easter lily is propagated commercially with underground stem bulblets, figure 9-5. The

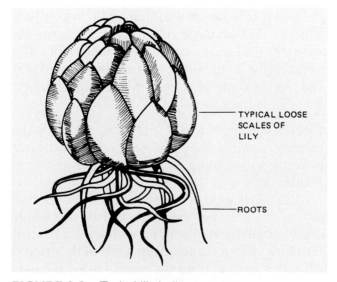

FIGURE 9-3   Typical lily bulb

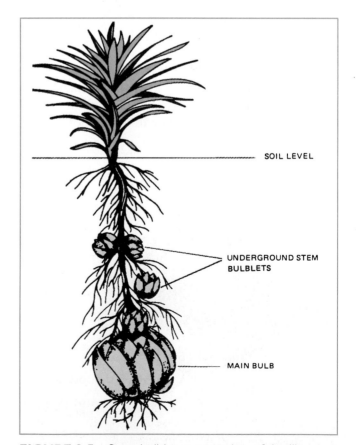

FIGURE 9-5   Stem bulblet propagation of the lily

flowering stems are pulled from the ground in late August through mid-September and the small bulblets and stems are kept moist by sprinkling them. About mid-October, the small bulblets are placed 4 inches deep and planted about 1 inch apart to grow the first season. They are moved again in September of the following season and planted about 6 inches deep, spaced 6 inches apart. By the end of the second year, they are dug and sold as flowering bulbs. At this time, bulb size should be 7 inches in circumference or larger. When dug, bulbs must be handled carefully to prevent bruising, and must not be allowed to dry out. The best storage is in moist, but not wet, sphagnum moss. A temperature of just below freezing holds lily bulbs in storage in an inactive state.

CORMS

Another natural plant structure which can be used in propagation is the corm. A *corm* is a very solid, compact stem with nodes and internodes. While the bulb is composed of leaf scales, the corm consists of a very short specialized stem for food storage. Corms have a dry covering which protects them from injury and drying, figure 9-6. In addition to being an organ for food storage, the corm is used as a reproductive structure. Development of small *cormels* is the principle means of reproducing (by separation) such plants as the gladiolus.

PROPAGATION PROCEDURE Cormels form naturally. If they are of flowering size, corms are planted shallower than normal, about 2 to 3 inches deep. More cormels will form at this depth

than if they are planted deeper. When the plant top dies back, the plant may be dug and the small cormels separated and grown to larger size. The plant top should be allowed to die back normally by the effects of frost, or grown at least three months after blooming so that the food supply manufactured in the top and stored in the specialized stem (the corm) is sufficient to develop good-sized cormels.

After frost or pulling of plants, the corms and cormels are separated from the rest of the plant and dried for storage. They should be treated with a fungicide and stored at 40°F (5°C) in a well-ventilated area with 80 percent humidity to prevent too much drying.

A hot-water treatment for disease control may also be necessary for cormels. Contact a commercial grower for information concerning the best treatment.

In the field, new cormels are planted in rows to grow larger, much as large seeds are. Two years are usually required for them to reach blooming size.

PROPAGATION BY DIVISION

Division differs from separation in that it is not a natural process; the parts used for propagation do not separate naturally from the mother plant. Parts of the plant which are to become new plants must be cut from the mother plant with pruners or a knife. Rhizomes, stem tubers, and tuberous roots are used in propagation by division.

RHIZOMES

Rhizomes are underground stems which grow horizontally and produce roots on the bottom and stems on the top. A rhizome may be thought of as a plant lying on its side with the stem covered with soil. An example of a plant which reproduces in this way is the iris.

PROPAGATION PROCEDURE When plants become crowded, they may be divided for the purpose of reproducing or increasing numbers of plants.

Rhizomes generally grow very near the surface of the soil. They are removed from the soil by digging underneath the plant with a garden (spading) fork or shovel and lifting it out of the ground. All soil is washed from the plant so that the parts are clearly visible for division. Division is done by

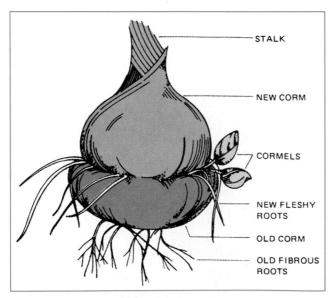

STALK

NEW CORM

CORMELS

NEW FLESHY ROOTS

OLD CORM

OLD FIBROUS ROOTS

FIGURE 9-6 Gladiolus corm

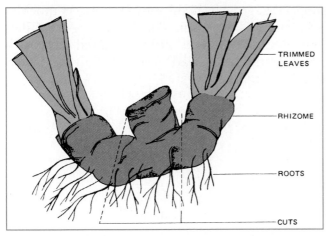

FIGURE 9-7 Division of the iris (rhizome structure)

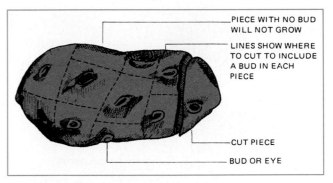

FIGURE 9-8 Division of the Irish potato (propagation by production of tubers)

cutting the rhizome into sections. Since a rhizome is an underground stem, the same care must be taken as with a stem cutting to make sure that each section has at least one bud (eye) and preferably several.

The iris is used here to illustrate division of a rhizome. Iris are divided in late summer after bloom. Each clump is washed clean and cut apart, figure 9-7. The tops should be cut back to about one-third of their original height to balance the root loss. After being cut apart, the new rhizomes are dusted with a fungicide, especially on the cut surfaces. They may then be planted in a new location.

TUBERS

A *tuber* is a swollen end of an underground side shoot or stem. Tubers are distinguished by their eyes. Each eye produces a separate plant as it sprouts, developing a shoot with roots at the base of the new shoot. The tuber contains stored food on which the new plant feeds until new leaves take over the job of food production. One common tuber is the Irish potato.

PROPAGATION PROCEDURE Propagation of plants by tubers is done by cutting the tubers into small pieces. Each piece must contain at least one bud or eye, figure 9-8. The pieces are planted in the same manner as larger seeds are. No storage is required; the cut surfaces are allowed to dry and the pieces are planted immediately.

TUBEROUS ROOTS

Tuberous roots are thickened roots which contain large amounts of stored food. They differ from tubers in that they are roots and have buds only at the stem end. Roots are produced at the opposite end, figure 9-9.

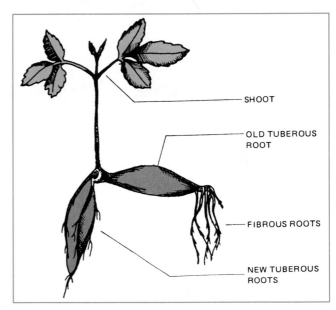

FIGURE 9-9 Tuberous root (dahlia)

PROPAGATION PROCEDURE Tuberous root crops are propagated by dividing the crown, or cluster of roots, when the plant is dormant. The plant is dug in the fall after frost has killed the top or it has died back for other reasons. The soil is washed away from the roots and they are allowed to dry. The roots are stored in dry sawdust, peat, or other materials at a temperature of 40° to 50°F (4° to 10°C) to prevent shriveling or complete drying out. Just before planting in early spring, the clumps or crowns are cut apart so that each piece has a bud. If sprouts are present, it is easier to see if a bud is part of the cut piece. The new pieces are planted in the garden or nursery row after the danger of frost has passed.

The sweet potato, a tuberous root, sends out many shoots from adventitious buds. (*Adventitious* buds are those which occur at sporadic and unexpected places on the vegetative structure.) The

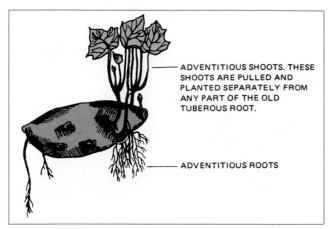

ADVENTITIOUS SHOOTS. THESE SHOOTS ARE PULLED AND PLANTED SEPARATELY FROM ANY PART OF THE OLD TUBEROUS ROOT.

ADVENTITIOUS ROOTS

FIGURE 9-10   Tuberous root (sweet potato)

shoots that grow from these buds root very quickly if the tuber is placed in a moist rooting medium. The shoots are pulled off and planted as rooted shoots without any part of the old tuberous root attached, figure 9-10.

Student Activities

1) Propagate at least one plant by separation and one by division. Plant on the school grounds or at home if possible.

2) Start a sweet potato in a glass of water or in moist sand. Observe the roots growing from one end of the tuberous root and the shoots from the other. This is known as *polarity*. The commercial grower increases the number of shoots for propagation by giving the sweet potato root a hot-water treatment to eliminate the polarity.

3) Cut a sweet potato root in half and heat the pieces at 110°F for 26 hours. Place in moist sand or water. Observe the greater number of adventitious shoots which grow when polarity has been eliminated. Cover the base of the shoots with moist sand as they grow and roots will develop.

4) If time permits, choose other plants which are propagated by separation and division and experiment with them. Some other plants that propagate by separation or division are day lily, hosta, orchids, and caladiums.

Self-Evaluation

A) Select the best answer from the choices offered to complete each statement.

    1) The main difference between separation and division is that

        a) separation is a natural method of propagation; in division, plants must be cut apart.
        b) division is a natural method of propagation; in separation, plants must be cut apart.
        c) separation is done in the spring of the year; division in the fall.
        d) There is no real difference.

2) The specialized plant parts used in division and separation have two functions. One is for propagation; the second is

a) to divide or multiply the plant.
b) to furnish food for human beings and wildlife.
c) to store food.
d) none of the above

3) A corm is a specialized

a) root.   c) leaf.
b) stem.   d) plant.

4) Daffodil bulbs are stored at

a) 55° to 60°F.   c) 80°F.
b) 60° to 70°F.   d) 40°F.

5) The tops of bulb-producing plants should be allowed to continue growing until they die naturally because

a) they add beauty to the flower garden.
b) they continue to manufacture food for storage in the bulb.
c) they grow larger and provide good mulch for the bulbs.
d) they shade the bulbs and keep them cool.

6) Potted plants such as a potted amaryllis can be forced into dormancy by withholding

a) fertilizer.   c) warmth.
b) light.   d) water.

7) Lily bulbs are loosely sealed bulbs and must be stored in moist sand, peat, or sawdust to prevent

a) drying out.   c) root growth.
b) forcing new tops.   d) none of these

8) The Easter lily is propagated commercially with underground

a) roots.   c) tubers.
b) stem bulblets.   d) corms.

9) The gladiolus is propagated through formation of small

a) seeds.   c) tubers.
b) roots.   d) cormels.

10) In propagation by division, the section of dahlia root which is cut from the mother clump must have a

a) bud.   c) number of small rootlets.
b) separation scar.   d) all of these

11) A rhizome is an underground

a) leaf.   c) stem.
b) root.   d) tuber.

12) An example of a plant propagated with rhizomes is the

a) tulip.   c) gladiolus.
b) potato.   d) iris.

13) Tuberous roots are thickened

    a) roots.   c) corms.

    b) stems.   d) bulbs.

14) Bulbs and corms reproduce by a process known as

    a) division.   c) offshoots.

    b) cuttage.   d) separation.

B) Match each plant in the right-hand column with the structure used in its propagation in the left-hand column.

    1) bulb   a) iris

    2) corm   b) tulip

    3) tuber   c) dahlia

    4) tuberous root   d) gladiolus

    5) rhizome   e) Irish potato

Grafting

OBJECTIVE

To perform three methods of grafting so that success in growing 60 percent of the grafted plants is achieved.

Competencies to Be Developed

After studying this unit, you should be able to

- ◆ identify three types of grafts from a sketch.
- ◆ list at least one plant that is propagated by each of the three grafting methods.
- ◆ list five reasons for grafting.

MATERIALS LIST

- ✔ rootstock and scion wood
- ✔ propagating knife
- ✔ tying material (plastic propagation tape or rubber ties)
- ✔ label and label-marking pen or pencil
- ✔ mulch
- ✔ polyethylene tent or other wilt-proofing material
- ✔ aluminum foil
- ✔ light hammer for driving knife
- ✔ cleft grafting tool or other wedge
- ✔ grafting wax

Grafting of plants is an ancient technique. It was practiced as early as 1000 B.C. Centuries later, it was employed widely by the Romans. *Grafting* is a process by which two different plants are united so that they grow as one. The *scion* is the newly installed shoot or top of the plant. The *rootstock* (or *stock*) is the seedling or plant used as the bottom half of the graft. The rootstock becomes the root system of the newly grafted plant. The union between scion and rootstock is a physical union (growing together of the tissue) which allows free movement of plant sap across the graft and from the rootstock to the new top and back again.

Many times, grafting is used as a method of rapidly increasing the number of a desirable plant. Grafting is also used to give plants stronger, more disease-resistant roots. In grafting, it is important that both the rootstock and scion be disease free. The two must also be compatible so that they can grow together to form a strong union.

If rootstock is carefully selected, the stock will have little effect on the newly attached top of the plant. The top will grow much the same as it grew on the parent plant before the graft and will generally duplicate all the characteristics of the parent plant. In some cases, rootstock reduces vigor and the overall size to which the new plant will grow. Dwarf fruit trees are a good example of this.

Grafting is not as widely used as budding in the commercial propagation of plants but is often used

- to topwork a large tree. This is done by grafting a different, usually improved, variety or varieties to many of the limbs of the tree,
- to insert a different variety on part of the limbs of a tree for cross-pollination, or
- to propagate plants that may be difficult to bud.

REQUIREMENTS FOR SUCCESSFUL GRAFTING

The following conditions must be met to assure a successful graft.

COMPATIBILITY   The two plants must be related to one another closely enough so that the stock and scion are able to grow together. As a general rule, it is best to graft a scion to a rootstock of the same type of plant, for example, apple is grafted to apple, pear is grafted to pear. Certain varieties are exceptions, however. Some Japanese-type plums, almonds, and apricots, for example, experience best growth on peach roots. The almond and apricot are both able to grow on peach roots, and cannot be grafted to one another.

It is extremely important to know which families and varieties of plants grow together best when grafted. When selecting rootstock for commercial production, also consider the climate and conditions of the local growing area.

SCION WOOD   The scion wood should be one year old and of vigorous growth.

TIMING   Grafting is generally done when the stock and scion are dormant or have no leaves. The rootstock may be actively growing, but the scion should not be. Of course, when evergreens are grafted, there are needles or leaves present on the shoot.

MATCHING OF TISSUE   The cambium layer of the two matched plant parts, the scion and the rootstock, must come in close contact and be held tightly together. The cambium must not be allowed to dry out.

WATERPROOFING   Immediately after the graft is made, all cut surfaces must be covered with a waterproofing material such as grafting wax, plastic ties, or rubber ties.

A discussion of three methods of grafting follows: the whip (tongue) graft, the side veneer graft, and the cleft graft. A brief discussion of stenting (a modification of whip or cleft grafting) also follows.

WHIP, OR TONGUE, GRAFT

The whip (or tongue) graft, accomplished during winter months, is used when small (1/4- to 1/2-inch) material is being grafted. Fruit trees are examples of plants propagated by the whip graft. If possible, the scion and root should be of the same diameter. The scion should contain three buds. The root piece should be at least 4 to 8 inches long and contain small fibrous roots. Figure 10-1 illustrates the steps in the whip graft.

The grafting cut is made below a bud on the stock. It should slant at such an angle that a smooth surface about 1 to 1½ inches long is produced. A very sharp knife allows the cut to be made with a single pass and usually gives a straight, smooth surface. This smooth surface is essential so that a close contact between the root and scion is made all along the cut surface. The cut made on the root piece should be exactly the same length and slope as that on the scion. This allows the two parts to fit together evenly. The cambium, the thin tissue just under the bark of the scion or stem,

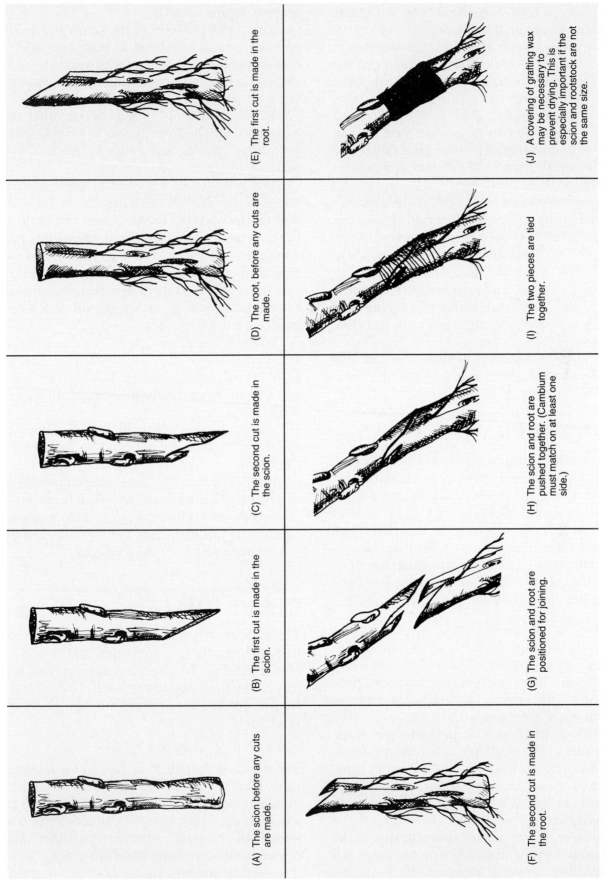

(A) The scion before any cuts are made.

(B) The first cut is made in the scion.

(C) The second cut is made in the scion.

(D) The root, before any cuts are made.

(E) The first cut is made in the root.

(F) The second cut is made in the root.

(G) The scion and root are positioned for joining.

(H) The scion and root are pushed together. (Cambium must match on at least one side.)

(I) The two pieces are tied together.

(J) A covering of grafting wax may be necessary to prevent drying. This is especially important if the scion and rootstock are not the same size.

FIGURE 10-1 Steps in the whip graft

must match the growing area at the edge of the root piece on at least one side. If the two pieces are the same diameter, it will match on both sides.

Before the two pieces are placed together, a second cut is made on the first cut surface in the reverse direction, starting about one-third of the distance from the tip and cutting nearly parallel to the first cut. This cut is made about half as long as the first cut. The two pieces are then slipped together with the tongues interlocking. Great care must be taken to match the cambium on at least one side.

The two pieces are then tied tightly together with plastic propagation tape or rubber bud ties and stored in moist sand or peat moss to heal. Storage for the first three to four weeks is at 50° to 55°F (10° to 13°C). After that, storage is at 32° to 40°F (0° to 4°C) to prevent top growth. The grafted plant may also be planted directly in the nursery row. In these cases, the graft union must be below ground level.

The new plant should send out shoots and begin to grow with warm spring weather.

SIDE VENEER GRAFT

The side veneer grafting method is a very effective way to graft evergreens such as the blue Colorado spruce. To obtain precisely that shade of blue which these trees possess, a small piece of stem from a selected blue spruce must be grafted to a small spruce seedling used as rootstock. The selection of the stem or scion is very important, since these seedlings vary in color and very few have the particular shade of blue that is so desirable.

GRAFTING PROCEDURE

The procedure for this graft is a simple one. Small seedlings may be grown in pots or in a nursery row. They graft more easily if they are about pencil size or a little larger. The grafts are made in early spring just before growth begins.

A shallow cut about 1½ inches long is made into one side of the seedling rootstock just above the soil level, figure 10-2(A). The cut is not complete at the bottom, but stops so that a second short downward cut leaves a ledge when the cut piece is removed, figure 10-2(B). The 4-inch scion, or new top piece, is cut on the opposite side at the bottom, making an end that fits on the ledge left on the seedling rootstock, figure 10-2(C). All cuts should be smooth and even so that the scion and

rootstock surfaces fit together along the entire cut surface, figure 10-2(D).

The cambium layer of the scion and seedling rootstock must match on at least one side. The two are placed with cut surfaces matching and are tied securely together with rubber bud ties or similar material, figure 10-2(E).

Placing mulching peat moss or other moist material over the graft union is very helpful in promoting healing and helps to insure a higher percentage of successful unions.

When grafts are exposed to sun and drying winds, a polyethylene tent may be tied to a stake and drawn over the seedling and the graft until the graft has healed and the new growth has started on the top. Aluminum foil is secured on the south side of the tent to keep down the temperature inside the tent. When the tent is removed, it should be opened gradually over a period of a week to harden off the new graft.

Spraying the stem or shoot grafted on the rootstock with one of the new plastic materials such as Wilt-pruf is also helpful in holding down moisture loss and drying of the new graft until the union heals.

As new growth starts, the top of the seedling rootstock is gradually cut away to force buds on the newly attached scion into active growth. The entire top of the seedling should be removed as soon as new growth is well established on the scion. The grafting process is now complete, but care must be taken in handling the new grafts so that the top is not broken loose from the rootstock.

CLEFT GRAFT

The cleft graft is used most often in topworking trees or grafting to a rootstock that is considerably larger than the scion. It is done in early spring just before the buds start to swell. The scion is cut from one-year-old wood and should include about three buds.

GRAFTING PROCEDURE

The selected rootstock or tree to be topworked must first be cut back to the point at which the graft is to occur. The limb or trunk is sawed off at a right angle, figure 10-3(A). A heavy knife is used to split the rootstock or understock through the center until a crack in which to place the scion opens wide enough, figure 10-3(B). A metal or wooden wedge is then placed in the center of the

(A) A long, shallow cut is made on one side of the rootstock.

(B) A second cut is made to remove the piece of wood.

(C) Two cuts are made on the scion to shape it so that it fits the cut made in the rootstock. The dark area is that piece which is cut out.

(D) The scion is inserted into the rootstock. The cambium layers must match on at least one side.

(E) The scion is tied tightly in place with string or rubber bud ties. A plastic tent may be used at this point. Mulch may be applied.

(F) The completed graft. A stake may be used to tie the new graft to prevent breakage.

FIGURE 10-2   Steps in the side veneer graft

stem to hold the split open, figure 10-3(C). A piece of scion wood is selected and two scions are cut, each containing three buds. The scion is cut with a long, smooth tapered cut on each side, making a wedge-shaped stick. The cut should be about 1 inch long and such that the wedge is slightly thicker on one side than the other, figure 10-3(D).

The thicker side is cut so that there is a bud directly above the top of the wedge. This thicker side is then placed against the outside of the rootstock where it is pressed firmly in contact with the cambium of the understock, figure 10-3(E). The scion is pushed down into the split in the rootstock until the entire cut area is inside the split.

Do not push the scion hard enough to cause tearing of the tissue. If necessary, the wedge is driven in farther to widen the split in the rootstock—this allows the scion to slip in more easily. Since the understock is much larger and has thicker bark than the scion, care must be taken to match the cambium layer of the scion with the cambium layer of the rootstock, not the bark of the rootstock. If properly placed, the scion will not be smooth with the outside of the understock, but will be set in, figure 10-3(F). The cambium layers must match for a successful graft.

After the two scions are in place, the wedge used to hold the split open is carefully removed.

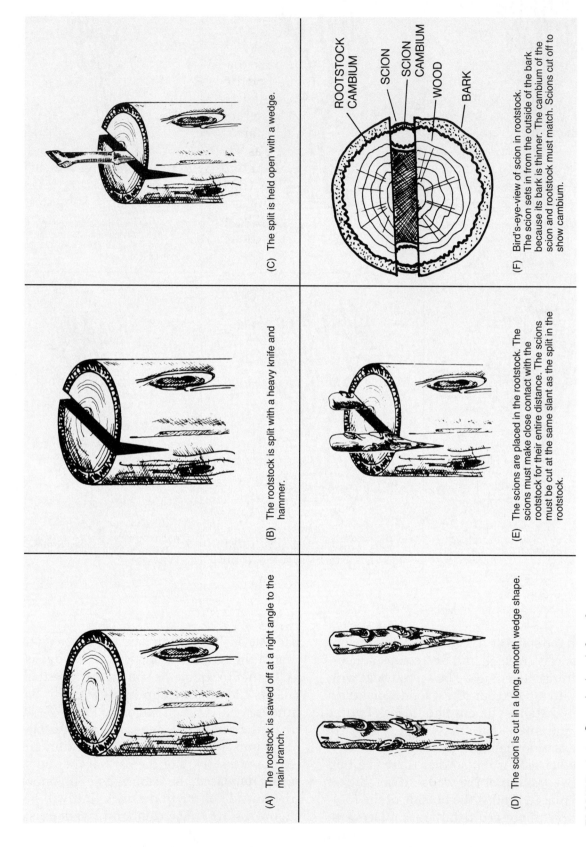

(A) The rootstock is sawed off at a right angle to the main branch.

(B) The rootstock is split with a heavy knife and hammer.

(C) The split is held open with a wedge.

(D) The scion is cut in a long, smooth wedge shape.

(E) The scions are placed in the rootstock. The scions must make close contact with the rootstock for their entire distance. The scions must be cut at the same slant as the split in the rootstock.

(F) Bird's-eye-view of scion in rootstock. The scion sets in from the outside of the bark because its bark is thinner. The cambium of the scion and rootstock must match. Scions cut off to show cambium.

ROOTSTOCK CAMBIUM
SCION
SCION CAMBIUM
WOOD
BARK

FIGURE 10-3 Steps in the cleft graft

The grafter should check to be sure that the cambium layers are in contact for the full length of the cut area on the scion. The pressure of the split closing is usually sufficient to hold the scions in place. No nailing or tying is necessary. The entire cut surface is then covered with grafting wax to prevent drying.

If both scions grow, one should be removed after the first season so that there is not a sharp crotch on the tree or limb at the graft union.

Where cleft grafting is used to topwork trees, some type of support may need to be tied to the tree branch and then to the scion to support the scion in extremely windy conditions. Wind could tear the newly attached scion loose during the first year of growth.

STENTING

A relatively new type of grafting called *stenting* is used on roses as well as other plants. With roses, a marketable plant can be produced one year sooner than with normally used budding. This method involves grafting a selected scion (top part of graft) of the desired plant onto a piece of *stem* which produces good roots. The grafted stems are then placed in a rooting chamber (the same as for semi-hardwood cuttings) for the graft to heal and roots to form on the bottom stem. Whip (or tongue) grafting or cleft (or saddle) grafting may be used to perform this grafting process. The graft is then treated exactly as a semi-hardwood cutting (see Unit 7).

Student Activities

1) Collect wood for propagation and accomplish at least one of the three grafts presented in this unit.

> *Caution: Grafting knives are sharp. Always cut away from yourself or other people in the area.*

Ask the instructor to check the graft for accuracy and for suggestions leading to a better graft.

2) Working either individually or with another student, prepare at least one graft, plant the resulting grafted plant, and allow it to grow. (Needed supplies are the scion wood, rootstock, a sharp knife, and waterproof material to tie the graft union.) After the plant is grafted, label it according to variety. If possible, invite an experienced propagator to assist in the actual grafting activities.

Self-Evaluation

A) Provide a brief answer for each of the following.

1) Define the term *grafting*.
2) List five reasons for grafting.

B) Select the best answer from the choices offered to complete the statement or answer the question.

1) Whip, or tongue, grafting is the best method to use when the rootstock and the scion are

a) the same size in diameter.
b) the same age.
c) of the same variety.
d) of different varieties.

2) The goal in grafting is to unite two plants

    a) so that they grow as one.
    b) in a physical union so that they grow together.
    c) that are difficult to root.
    d) all of the above

3) In grafting, great care must be used to match which layer of the scion and rootstock?

    a) woody   c) bark
    b) cambium   d) outside

4) The ideal time of the year to cleft graft is

    a) late summer, when buds are mature.
    b) early spring, just before the buds start to swell.
    c) in winter, when the plants are dormant.
    d) in spring, when new growth is young and tender.

5) Whip, or tongue, grafting is accomplished in

    a) winter, when plants are dormant.
    b) early spring.
    c) late summer, after buds mature.
    d) early summer, while there is still active growth.

6) The side veneer graft is the most effective way to propagate the

    a) apple.   c) tree peony.
    b) cherry.   d) blue Colorado spruce.

7) The top part of a grafted Colorado spruce should be protected very carefully from drying because

    a) spruce trees are naturally sensitive to drying.
    b) the graft has no roots to replenish moisture.
    c) it is an evergreen and dries quickly.
    d) none of the above

8) In cleft grafting, one side of the scion is cut thicker than the other

    a) so that there is more pressure on that side resulting in close contact with the cambium.
    b) so that the scion is handled more easily.
    c) so that the scion makes more rapid growth.
    d) all of the above

9) All cut areas that are left exposed in the grafting process should be covered with water-proof material

    a) to prevent bleeding of sap.
    b) to prevent drying of the plant tissue.
    c) to make the graft look better.
    d) to prevent insects form entering the graft.

10) A new technique modifying the usual whip or cleft graft method used to root roses more quickly is called

    a) rushing.   c) stem inserting.
    b) stenting.   d) all of the above

UNIT 11

Budding

OBJECTIVE

To propagate at least one plant by the budding process.

Competencies to Be Developed

After studying this unit, you should be able to

- ◆ name at least three plants commercially propagated by budding.
- ◆ list the seven steps in the budding process.
- ◆ identify on a diagram a piece of budwood, rootstock, a budding knife, and rubber bud ties.
- ◆ differentiate between a fruit bud and a leaf or vegetable bud by cutting each from a bud stick and labeling.
- ◆ propagate a plant by the T-budding process.
- ◆ propagate a plant by the chip budding process.

MATERIALS LIST

- ✔ budding knife
- ✔ rubber band ties
- ✔ plastic bag to collect budwood
- ✔ label and waterproof pen or pencil
- ✔ rootstock
- ✔ budwood
- ✔ plant for rootstock
- ✔ a sharp knife

METHODS OF BUDDING

Methods of budding include T-budding, patch budding, inverted T-budding, flute budding, I-budding, and chip budding, among others. The most commonly used methods in the United States are T-budding and chip budding. The T-budding operation is especially popular in the propagation of tree fruits and roses. It is a much quicker process than grafting; some rose budders insert up to several thousand buds a day if there are other people present to do the tying. Chip budding is used on grapes and is being used more on other plants because of the longer time period during which it can be done.

The process of budding is actually a form of grafting; in fact, budding is sometimes called *bud grafting*. Budding differs from grafting in several ways, however. In budding, a single bud is used instead of a scion. Because of this, many more plants can be reproduced from the same amount of parent wood. Budding is also accomplished more quickly. Another difference is the time at which the operation takes place—T-budding is accomplished when the rootstock is in active growth rather than in the dormant (resting) stage, either in spring (March or April), in June, or in late summer or fall (July through September).

T-BUDDING

T-budding is generally done on small, one- or two-year-old seedlings. Small branches in the tops of larger trees are also budded. Whichever type of rootstock is chosen, it is important that it be actively growing, disease resistant, and able to give the desired growth. If there is no sign of disease and the bark separates easily from the wood, the operation should be successful. It is also important that mature vegetative or leaf buds of the desired variety are available. As in other forms of grafting, the scion and rootstock must be compatible.

Some plants commonly propagated by budding include apples, pears, peaches, plums, cherries, roses, and the citrus group. Late summer is the best time to propagate fruit trees by budding. The rootstock generally grows from seed planted in fall to the size of a lead pencil by July or August of the following year. Leaf buds on the stock or parent plant are generally mature enough for use at that time.

PROCEDURE

STEPS IN THE BUDDING PROCESS

1. Plant seeds for seedling rootstock (the previous fall).
2. Select the variety of budwood to be propagated.
3. Determine the correct date to bud, as indicated by both the bud maturity of the desired plant and the active growth of the rootstock.
4. Cut the budwood, label it, and protect it in such a way that it does not dry out.
5. Perform the budding process.
6. Check to see if buds have taken.
7. Cut off the rootstock above the bud (the following spring).

COLLECTING BUDWOOD

Bud sticks, small shoots of the current season's growth, are collected on the same day the buds are to be inserted, figure 11-1(A). The budwood is kept wrapped in waterproof paper so that it does not dry out. The bundle of bud sticks is then labeled according to variety and date cut. Shoots

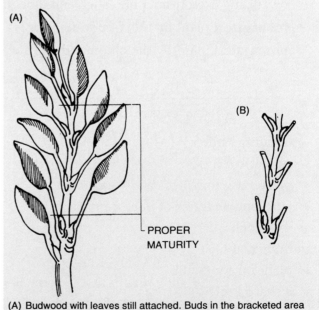

(A) Budwood with leaves still attached. Buds in the bracketed area are probably best for budding because of maturity.

(B) Budwood with top and bottom trimmed and leaves removed. The leaf petiole remains as a handle for holding the buds.

FIGURE 11-1 Preparing the budwood

that have a vigorous growth pattern usually have the best and most usable buds on them.

Remember that vegetative buds are necessary for propagation—fruit buds will not grow into new plants. Vegetative and fruit buds can be easily distinguished. The vegetative or leaf bud is slimmer and more pointed than the fruit bud. Fruit buds generally appear on the base or bottom end of the shoot.

Collect budwood of the proper variety. Carefully cut all the leaves from the stick, leaving a short piece of the leaf or petiole to act as a handle to hold the bud after it is cut from the stem, figure 11-1(B). Wrap in waterproof paper or place in a plastic bag.

ROOTSTOCK

The rootstock, which at this point has been developing from seed for one year or more, should now be about the size of a pencil for proper budding. Remember that the rootstock and budwood must be of related species, such as apple with apple, and cherry with cherry.

For T-budding, the seedling rootstock must be in active growth, indicated by soft, rapidly growing branch tips. The final test for determining if the seedling is receptive to the bud is to make a T-shaped cut and separate the bark from the wood of the stem. The bark should slip loose easily and the wood underneath appear moist and smooth, with no tearing or stringing of tissue. When this is established, the process may be continued.

CUTTING AND INSERTING THE BUD

MATERIALS LIST

- ✔ budding knife
- ✔ rubber bud ties
- ✔ plastic bag to collect budwood
- ✔ label and waterproof pen or pencil
- ✔ rootstock
- ✔ budwood

The next step is to prepare the seedlings being used for rootstock. The budding process is easier if a team of two people work together. One person makes the cut with a sharp budding knife and inserts the bud while the second person ties the bud securely in place.

Caution: **The person making the cut should exercise care when handling the knife.**

The person making the cut kneels beside the row of seedlings, bending them toward him or her. At a spot about 1 to 2 inches above ground level and where the stem is smooth, a 1-inch vertical cut is made through the bark. The cut should be made on the north side of the stem if possible. This protects the bud from the sun both in summer and winter. A horizontal cut is then made across the tip of the vertical cut to form the T shape for which this process is named. The knife is given a twist so that the bark and wood separate. See figure 11-2(A). If necessary the two sides of the vertical cut should be lifted at the top with the edge of the knife blade, opening the slit in the bark, figure 11-2(B). As mentioned, if the bark separates easily, the rootstock is growing rapidly enough to bud.

The next step is to cut the leaf bud from the bud stick. Generally, those buds in the middle portion of the bud stick are used for budding. Buds on the terminal end may not be mature enough, and buds on the basal end may be flower buds. The buds which are used may vary with the maturity of the bud stick. Whichever is used, the bud

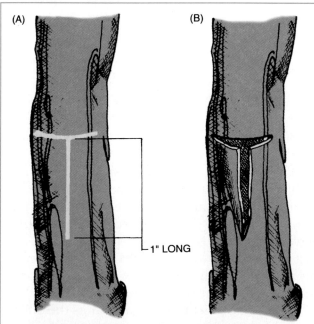

(A) A T-shaped cut is made in the rootstock. The horizontal cut is about one-third the diameter.

(B) The corners of the bark are lifted so that the bud can be inserted easily.

FIGURE 11-2 Preparing the rootstock

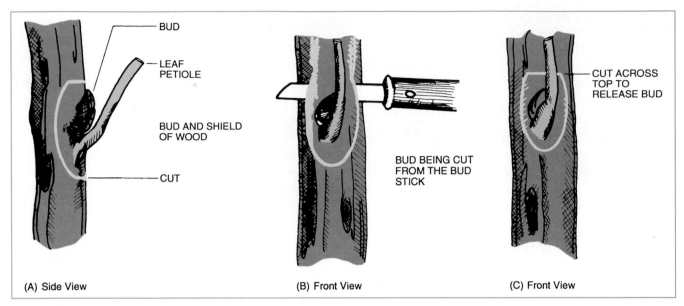

FIGURE 11-3 Cutting the bud

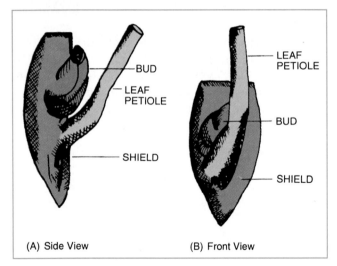

FIGURE 11-4 The leaf bud shield

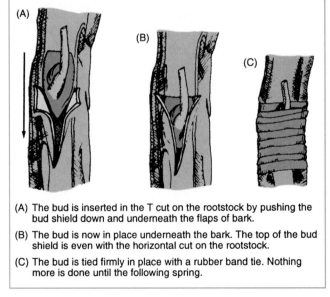

(A) The bud is inserted in the T cut on the rootstock by pushing the bud shield down and underneath the flaps of bark.

(B) The bud is now in place underneath the bark. The top of the bud shield is even with the horizontal cut on the rootstock.

(C) The bud is tied firmly in place with a rubber band tie. Nothing more is done until the following spring.

FIGURE 11-5 Insertion and tying of a bud

in the axil of the leaf stem must be visible and well developed.

The bud is cut with a small shield of bark and a sliver of wood attached, figure 11-3. The cut is made starting about 1/2 inch below the bud and cutting just deeply enough to include a small sliver of wood. The cut is continued under the bud and past it to about 1/4 inch above the bud. A second cut made at a right angle to the bud stick at the end of the first cut releases the bud from the bud stick. The bud shield should appear as the one in figure 11-4.

The bud is immediately inserted into the T-shaped cut made in the rootstock and pushed down into the open slit in the bark until the top of the bud shield is even with the top of the T cut

in the rootstock. This allows the bud shield to be completely inserted underneath the bark of the rootstock and to fit snugly against the wood, figures 11-5 and 11-6.

To hold the new bud firmly in place, the cut area is now tied with a rubber bud tie or a one-piece bud tie tape, which fits over the bud. The entire cut area is covered with the bud tie; only the bud itself is exposed. The bud tie is tucked under itself at each end to hold it in place. These ties, made of a special material, are designed to disintegrate or break in about three weeks. They do not require removal.

FIGURE 11-6 Budder inserting the shield of bark with a peach bud in the T cut. A small tree is being used as the rootstock. (Courtesy Stark Brothers Nurseries and Orchards Co.)

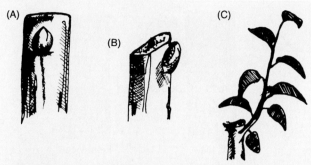

(A) Front view. The rootstock is cut directly above the bud in the spring before growth begins.

(B) Side view. The cut is made at a slant starting about ¼" above the bud.

(C) New shoot growing and healing of the wound.

FIGURE 11-7 Continuation of the budding process the following spring

In about three weeks, an inspection is made to see if the bud has *taken* (has attached to the stem and is still alive). If the leaf petiole has dropped off and the bud appears to be plump and of normal color, the bud has probably grown to the rootstock. Nothing else is done to the bud or

the rootstock until early the following spring. At that time, the top of the seedling rootstock is cut off directly above the bud as shown in figure 11-7(A) and (B). This forces the new bud into active growth, which then develops into the new top of the plant, figure 11-7(C). Any suckers or shoots that sprout out below the bud from the rootstock are pinched off. This process must be continued until no more new sprouts emerge. The budding

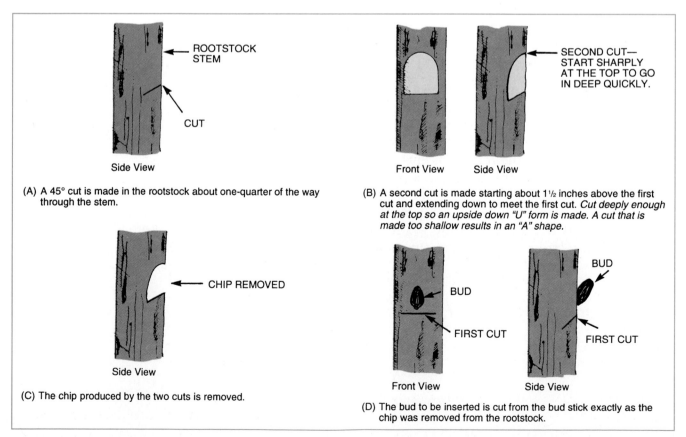

(A) A 45° cut is made in the rootstock about one-quarter of the way through the stem.

(B) A second cut is made starting about 1½ inches above the first cut and extending down to meet the first cut. *Cut deeply enough at the top so an upside down "U" form is made. A cut that is made too shallow results in an "A" shape.*

(C) The chip produced by the two cuts is removed.

(D) The bud to be inserted is cut from the bud stick exactly as the chip was removed from the rootstock.

FIGURE 11-8 Chip budding procedure

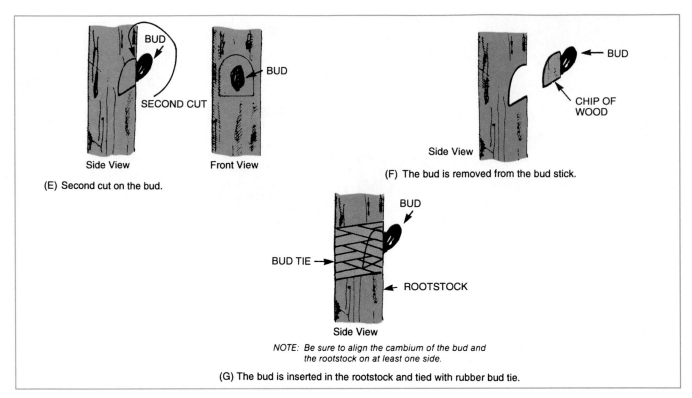

(E) Second cut on the bud.

(F) The bud is removed from the bud stick.

NOTE: Be sure to align the cambium of the bud and the rootstock on at least one side.

(G) The bud is inserted in the rootstock and tied with rubber bud tie.

FIGURE 11-8   Chip budding procedure (continued)

process is then completed. After one or two years' growth of the new bud, it is ready to be transplanted to a permanent site.

CHIP BUDDING

Chip budding is widely used for grapes and may be used for most other normally budded plants. The biggest advantage is that it can be done when the rootstock is not in active growth. You will recall that in T-budding, the rootstock must be actively growing so that the bark can slip loose without tearing, allowing the bud to be inserted beneath it. This is not necessary for chip budding, since the bark is not lifted from the wood. This means that chip budding can be done in summer or fall.

Selection of rootstock and budwood is the same as for T-budding.

The chip budding procedure is shown in figure 11-8.

MATERIALS LIST

- ✔ plant for rootstock
- ✔ budwood
- ✔ a sharp knife
- ✔ rubber band ties
- ✔ label to identify cultivar of inserted bud

AFTERCARE FOR CHIP BUDDING

For *spring budding*, cut the rootstock off just above the bud in 10 to 15 days. For *fall budding*, cut the rootstock off just above the bud the following spring, just as growth starts.

Student Activities

1) Take a field trip to a nursery to watch a budding operation.

2) Plant seeds to grow rootstock.

3) Practice the budding process on plant parts provided.

4) Bud plants in a nursery row.

5) Write a description of the physical condition of a bud that has grown quickly and of one that is dead. Pinpoint the differences in appearance.

Self-Evaluation

Select the best answer from the choices offered to complete the statement or answer the question.

1) When budding

   a) the bud stick must have mature buds.
   b) a single bud is used instead of a scion.
   c) only vegetative buds are used.
   d) all of these

2) T-budding is done when the rootstock

   a) is in active growth and about the size of a pencil.
   b) has matured for the season.
   c) is available.
   d) is about the same size as the scion.

3) Vegetative or leaf buds may be differentiated from flower buds because they

   a) are plumper.
   b) are more slender or pointed.
   c) are generally located at the base of the bud stick.
   d) are not yet mature.

4) Bud sticks are cut from

   a) actively growing wood.
   b) healthy wood.
   c) the current season's growth.
   d) all of these

5) Rootstock for budding is

   a) grown from seed planted in previous years.
   b) generally about the size of a pencil.
   c) of a closely related plant species.
   d) all of these

6) The bud is inserted into the rootstock

    a) on the south side of the rootstock 2 inches above the ground.
    b) on the north side of the rootstock 2 inches above the ground.
    c) on the north side of the rootstock 6 inches above the ground.
    d) none of the above

7) From which portion of the bud stick are the best quality buds obtained?

    a) end
    b) base
    c) middle
    d) none of these

8) Three plants commercially propagated by budding are the

    a) apple, peach, and rose.
    b) apple, rhododendron, and cherry.
    c) azalea, rhododendron, and viburnum.
    d) rose, peach, and chestnut.

9) The bud is cut with a small shield of

    a) bark.
    b) wood.
    c) bark and wood.
    d) none of these

10) The seedling is the proper state of growth for budding if upon opening the T-shaped cut in the seedling, the bark separates from the wood

    a) with difficulty.
    b) with a small piece of wood attached.
    c) enough so that the bud can be inserted.
    d) easily and cleanly.

11) The leaf stem or petiole is left on the bud because

    a) it acts as a handle for holding the bud.
    b) it helps the bud to heal.
    c) it simplifies tying the bud in place.
    d) the bud would be damaged if it were removed.

12) The bud is tied tightly in place with a bud tie so that

    a) it is held firmly in place against the rootstock.
    b) animals are prevented from loosening it.
    c) the flow of sap is stopped at that point.
    d) the rootstock is gradually girdled.

13) An inspection to see if the bud has grown and is alive should be made in about

    a) 6 weeks.
    b) 3 weeks.
    c) 1 year.
    d) 6 months.

14) The spring after the bud is attached, the seedling rootstock top is cut off

    a) just below the bud.
    b) even with the bud.
    c) directly above the bud.
    d) none of these

15) In the development of a new bud, what must be pinched off as they appear on the rootstock below the bud?

    a) roots
    b) suckers
    c) scars
    d) knots

16) The new plant may be transplanted to a permanent site in

a) 2 or 3 years.
b) 6 months.
c) 1 or 2 years.
d) 3 or 4 years.

17) Which of the following occurs in budding but not in grafting?

a) A single bud is used and the rootstock is in active growth.
b) A single bud is used and the rootstock is larger than that used in grafting.
c) A vegetative leaf bud is used and it is placed closer to the ground on the rootstock than in grafting.
d) A single bud is used and the rootstock is dormant.

18) Chip budding is widely used to propagate _____ as well as other plants.

a) apples
b) grapes
c) roses
d) azaleas

19) What is the biggest advantage in using chip budding?

a) Plants are more compatible.
b) Plants grow faster.
c) It is cheaper.
d) It can be done when the root system is not in active growth.

Layering

OBJECTIVE

To propagate one plant by one of the layering processes described in the unit.

Competencies to Be Developed

After studying this unit, you should be able to

◆ describe two situations in which layering is used to propagate plants.

◆ list five plants commonly propagated by layering and two plants which propagate naturally by layering.

◆ list the eight steps in the layering process.

◆ perform simple layering as described in the text.

MATERIALS LIST

✔ parent plant	✔ string
✔ shovel	✔ brick or stone
✔ peat moss or sawdust	✔ propagating knife
✔ two stakes, one with hooks	✔ hammer

Layering is a method of asexual propagation in which roots are formed on a stem or root while it is still attached to the parent plant. The stem or root which is rooted is called a *layer*. The layer is cut free from the parent only after rooting has taken place. This is the major difference between layering and other methods of asexual reproduction.

As a method of propagation, layering has advantages and disadvantages. It is a relatively simple process, but requires more time than many other methods of asexual propagation, since it requires a great deal of work by hand. Also, fewer plants can be started from each parent plant than with the use of cuttings. The major

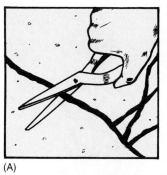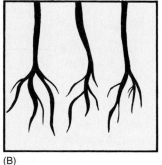

(A)   (B)

FIGURE 12-1   Natural layering of spirea. In (A), the plant has been dug from the ground. The three stems shown at the right of the pruner's hand in (A) are shown cut free in (B).

advantage of layering is the degree of success with which some plants will root when layered as compared with results achieved with other methods. (Some plants that are difficult to root by other methods may be rooted by layering.) In fact, many plants root naturally by layering, figure 12-1.

Cane fruits such as the black raspberry and the boysenberry root easily and economically by *tip layering*, one form of natural layering. In this method of natural layering, a shoot with current season's growth bends to the ground and then turns sharply upward. Roots develop from the point at which the shoot meets with the soil, figure 12-2. Other plants which root naturally by layering are strawberries (by runners), red raspberries and spirea (by root suckers), and African violets (by crown division).

Some plants do not reproduce naturally by layering, but can be propagated in this manner with assistance from the horticulturist. Two methods used for this purpose are simple layering and air layering.

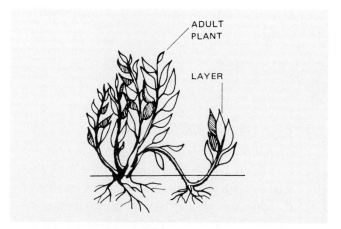

ADULT
PLANT

LAYER

FIGURE 12-2   Tip layering

SIMPLE LAYERING

In *simple layering*, a branch from a parent plant is bent to the ground where it is partially covered at one point with soil. The terminal end remains exposed. (See figure 12-3.) This process is usually accomplished in early spring.

PROCEDURE

STEPS IN SIMPLE LAYERING

1. Select a stem with one-year-old wood that can be bent to soil level.

2. Make a cut about halfway through the top of the stem at the point at which the stem is to be inserted in the ground, figure 12-3. Use a sharp knife for the best cut or girdle the stem completely of all bark and cambium.

> *Caution: Do not cut toward yourself or another person.*

3. Twist the stem 90 degrees and raise the stem to an upright position. (It is not necessary to twist girdled stem.) Some stems may break if twisted, as stated above. For stems that break easily, do not cut and twist; rather girdle the stem as shown in air layering (see discussion later in this unit).

4. Dig a hole or trench 3 to 6 inches deep at the point at which the twist in the stem touches the ground.

5. Dust the stem wound with rooting hormone and push the stem into the opening in the ground.

6. Drive a peg with hooks on it next to the stem to hold it upright, figure 12-4. Another peg holds the layer in the ground.

7. Fill the trench with soil, and mulch the top to help hold in moisture. A brick or stone may be placed on top to keep the layer in position. Keep the soil moist (by watering, if necessary). The layer should be rooted by the next spring (about one year).

8. Cut the layer free of the parent plant when it is well rooted and transplant to a new location, figure 12-5.

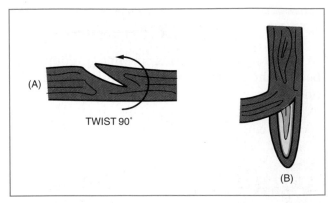

FIGURE 12-3 The stem is cut halfway through where it is to be inserted in the soil (A) and then twisted 90 degrees (B).

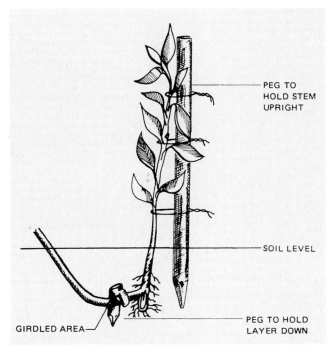

FIGURE 12-4 Pegs hold the layer upright and in the ground.

Since the root system of the new plant is so limited, it is necessary to water it liberally. Some shade may also be needed until the root system is well established.

AIR LAYERING

The ancient Chinese practiced a technique called *air layering*, a process which eliminates burying part of the parent plant in the soil. Instead, a part of the plant stem is slit, or girdled (to *girdle* is to completely remove the bark and cambium around a plant), and then surrounded by a moist growing medium in some type of enclosure. Roots form where the plant has been wounded. The

FIGURE 12-5 Rooted layer cut from parent plant

materials available in earlier years made the task much more time-consuming and difficult for the Chinese than it is today. The major problem in the past was keeping the layered area moist. The development of polyethylene film has made air layering a highly successful procedure. It is used commercially to propagate certain tropical and subtropical trees and shrubs which are difficult to root by other procedures. The Persian lime (*Citrus aurantifolia*) is commercially propagated in this way. The magnolia and the rubber plant (*Ficus elastica*) may also be propagated by air layering. In fact, most plants may be propagated using this method, but usually it is not economically feasible.

Air layers are generally made in the spring on wood of the previous year's growth or in late summer on partially hardened growth of the current season. Older wood is much more difficult to root.

PROCEDURE

STEPS IN AIR LAYERING

1. Girdle the selected stem in a band ¹/₂ to 1 inch wide. Girdle at a spot that is 6 to 12 inches from the top of the branch so that 6 to 10 leaves are left between the girdled spot and the branch tip. These leaves are

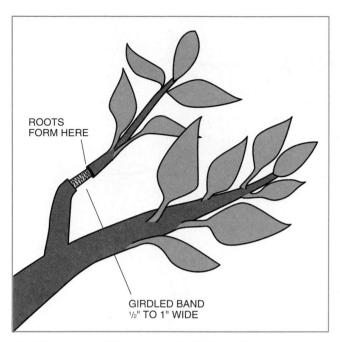

FIGURE 12-6 The stem is girdled to stimulate root formation.

FIGURE 12-7 Sphagnum moss is packed around the girdled area.

needed to produce carbohydrates and hormones needed for good root development. Since the stem is girdled, these materials, produced in the leaf, accumulate at the girdled point on the stem of the plant, greatly stimulating root formation. Scrape the girdled area to ensure that all bark and cambium are removed, thus slowing the healing process, figure 12-6.

2. Treat the girdled area with rooting hormone (Hormodine #3, Rootone #10, or a concentration of 0.5 to 0.8 percent indolebutyric acid in water). Treat the bark edges on the branch tip side of the girdled area especially well.

3. Pack about two handfuls of moist (but not wet) sphagnum moss around the girdled area, figure 12-7. Squeeze all the moisture out of the moss—the cut area will rot if the moss is wet.

4. Wrap a 10-inch-square piece of polyethylene plastic around the sphagnum moss so that the moss is completely covered. No moss should protrude from underneath the polyethylene wrap. Fold together both ends of the polyethylene film at the same time. Wrap the film until a snug-fitting tube is formed. Twist the two ends of the tube around the branch and fasten securely with plastic electrician's tape, figure 12-8. Extend the tape far enough onto the plant stem so that there is a watertight seal. The polyethylene allows the passage of air with no moisture loss. Additional watering is not needed.

5. Rooting may be observed through the polyethylene plastic. When strong roots have developed, the layer may be removed for transplanting. If the plant being layered is deciduous, wait until the leaves fall from the plant and it is dormant. If the plant is an evergreen, remove the layer when no new active growth is occurring. Simply cut the layer free below the rooted area.

If the top of the new plant has a great deal of growth in relation to the root system, plant it in a pot or flat and place it in an area of high humidity for a short time. After additional root development has occurred, the plant may be gradually hardened off by exposure to a drier atmosphere in preparation for the permanent planting site.

OTHER METHODS OF LAYERING

Three other methods by which plants are layered by the horticulturist are trench layering, stool layering, and compound layering.

FIGURE 12-8 The two ends of the polyethylene tube are twisted around the branch and fastened securely with electrician's tape.

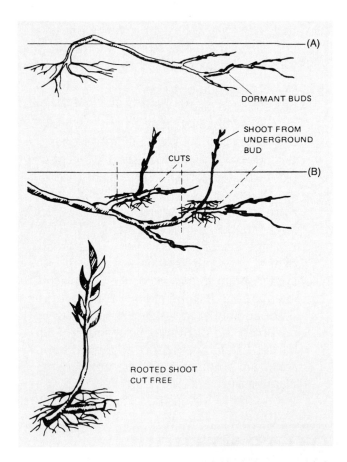

FIGURE 12-9 Trench layering. (A) Shows the mother plant in the trench exhibiting dormant buds. Soil is added to the trench periodically to cover the developing shoots. At the end of the growing season, the layers are separated from the mother plant, as shown in (B).

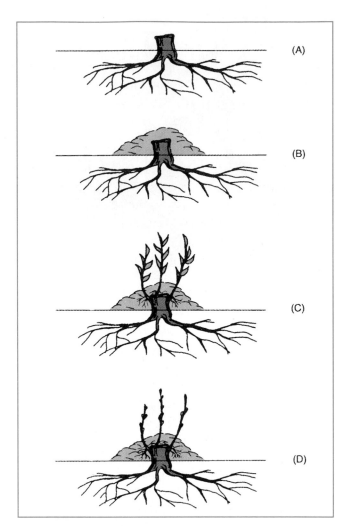

FIGURE 12-10 Stool (or mound) layering

TRENCH LAYERING

In *trench layering*, the mother plant is bent to the ground and buried in a trench, figure 12-9(A). As shoots arise from the buried buds, roots form on the covered portion of the plant. The shoots may then be separated from the mother plant, figure 12-9(B).

STOOL LAYERING

Stool (or *mound*) *layering* begins with the planting of a rooted layer in soil. After one season's growth, the parent plant is cut back to soil level, figure 12-10(A). The stem is then covered with a mound of soil, figure 12-10(B). Soil is added to the mound periodically as the shoots grow, figure 12-10(C). At the end of the season, the new shoots are rooted and dormant, figure 12-10(D). The shoots are cut free and planted in early spring. One year has passed from the beginning of the layering process to the planting of the new shoots.

COMPOUND LAYERING

Compound layering, accomplished in springtime, is very similar to simple layering except that a stem is covered by soil at two or more points along its length. Long-stemmed, vinelike plants such as wisteria, clematis, and grape work well with this system, figure 12-11.

The stem is girdled at a point below ground near where new roots are expected to form. The portion of the stem above soil level must have at least one bud to form a new shoot which will grow into the top of the new plant.

After rooting takes place, usually by the end of one growing season, the stem is cut to include roots and a new shoot on each new plant. Several plants may be produced from a single stem by compound layering.

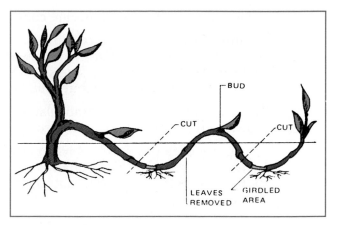

FIGURE 12-11 Compound layering

Student Activities

1) Visit a horticulturist and observe a layering process. Record the steps in the process in a notebook.

2) With a group of students, perform a layering process on a parent plant. Keep records on the experiment, including the date on which the plant is layered and the date on which it roots. Plants suggested for layering: azalea, rhododendron, spirea, forsythia, wisteria, grape.

Self-Evaluation

Select the best answer from the choices offered to complete the statement or answer the question.

1) Layering is used to propagate plants which are

    a) best grown from seed.
    b) difficult to root or root naturally by layering.
    c) easily cut apart.
    d) too thick to propagate by other methods.

2) The main difference in propagating by layering as opposed to cuttings is that in layering, the newly propagated plant part is

    a) cut free from the parent plant.   c) wounded.
    b) older.   d) still attached to the parent plant.

3) Which of the following plants are commonly propagated by layering?

   a) apple, pear, peach, cherry, plum
   b) black raspberry, boysenberry, strawberry, red raspberry, African violet
   c) bean, tomato, pea, corn, squash
   d) rhododendron, azalea, viburnum, holly, lilac

4) When a rooted layer being transplanted has a small root system and a large top, the plant's moisture requirements are balanced by

   a) cutting back the top to a smaller size.
   b) watering the plant more often.
   c) using a mulch to conserve moisture.
   d) none of the above

5) The age of the wood affects the speed at which layers root. Younger (one-year-old) wood generally roots

   a) more slowly than older wood.
   b) at the same speed as older wood.
   c) more rapidly than older wood.
   d) more heavily than older wood.

6) In air layering, cutting the bark off the stem, known as girdling,

   a) stimulates root formation just above the point of girdling.
   b) helps in breaking off the stem when it is rooted.
   c) gradually starves the stem.
   d) all of the above

7) The cut area on the layered stem is dusted with a rooting agent to

   a) prevent fungus diseases from killing the stem.
   b) cause roots to form faster.
   d) kill any insects that may be present.
   d) aid in attracting moisture.

8) The main reason for covering the sphagnum moss which is packed around the air layers with polyethylene plastic is that

   a) it holds moisture in and allows the roots to breathe.
   b) the roots can be observed growing inside.
   c) it is the least expensive material available.
   d) it prevents circulation of air.

Greenhouse Management and Crops

Poinsettias

OBJECTIVE

To produce a saleable poinsettia crop.

Competencies to Be Developed

After studying this unit, you should be able to

- ◆ prepare a soil mixture for growing poinsettias.
- ◆ transplant ten rooted poinsettia cuttings.
- ◆ identify one growth regulator and use it correctly.
- ◆ outline a growing schedule for poinsettias.
- ◆ identify four insects that affect poinsettias and the control measures for each.
- ◆ list the common names of three diseases that affect poinsettias and the control measures for each.

MATERIALS LIST

- ✔ 2¼-inch rooted poinsettias
- ✔ loamy soil
- ✔ sand
- ✔ horticultural grade perlite
- ✔ Pro-Mix B
- ✔ 4-, 5-, and 6-inch pots
- ✔ Cycocel or some other growth regulator
- ✔ assorted insecticides (Resmethrin, orthene, Sulfotepp)

The poinsettia has long been a traditional flower used during the winter holiday season. Poinsettias were first introduced into the United States from Mexico by J.R. Poinsett, the first ambassador to Mexico. Poinsettias were used as exotic plants in conservatories and

FIGURE 13-1 Poinsettia varieties are grown for the winter holiday season. In warmer climates poinsettias are grown outdoors. (Courtesy Eckespoint™ Lilo-Commercial Display, Paul Ecke Poinsettias)

botanical gardens until 1900. Through the work of Albert Ecke, poinsettias were then grown in southern California as cut flowers. Today, the growing of poinsettias is a big part of many horticultural operations, figure 13-1.

GREENHOUSE MANAGEMENT

The greenhouse is the structure used to start and grow plants throughout the year. It must be located to receive the maximum available sunlight. Other factors to consider in greenhouse production are temperature, moisture, ventilation, and climate.

Temperature in the greenhouse is an important key to growing a successful crop. To obtain optimum growth of plants, it is critical that the correct temperature be maintained throughout the greenhouse. Great variation in the greenhouse temperature can cause fast vegetative growth or the lack of plant growth. The general rule of thumb for greenhouses is the day temperature should be 5 to 10 degrees higher than the night temperature. It is important to monitor the greenhouse temperature with thermometers at the plant level to ensure optimum growth. Thermometers should be kept out of the direct sunlight, facing a northern exposure with a wooden hood on each side. A maximum/minimum thermometer works well in a school greenhouse because it helps students to realize the high temperature and the low temperature of the day.

Plant growth is dependent on moisture. It helps to hold the cell shape, to nourish, and to carry dissolved nutrients present in media into the plant for optimum growth. The amount of moisture in the greenhouse will also influence the relative humidity. Watering plants in the greenhouse is an important task since moisture supply controls the growth of plants. The amount of water needed depends on the type of plant and the outside conditions on a particular day (i.e., bright sunshine or clouds). The appearance or feel of the medium is an indication for watering plants. Good judgment must be used when watering.

Ventilation is the movement and exchange of air in the greenhouse. This air movement is important for optimum plant growth. The individual greenhouse grower is able to feel the air movement in the greenhouse. This ensures the correct temperature and relative humidity and helps control plant diseases.

Climate directly influences the type of greenhouse structure, heating, and cooling systems. It would be best to talk to a state land grant college, local extension service, or grower in the area.

THE POINSETTIA FLOWER

The poinsettia flower is a small, yellow flower that grows at the terminal end of the plant. Just below the flower are the *bracts*, the leaves that give poinsettias their color. The most common bract color is red, but through years of research, other colors have been developed. Some of these colors are white, pink, white variegated, and red and white variegated.

Poinsettias require a short day for the production of the flower, which in turn is necessary to produce the color in the bracts. Poinsettias should be grown in full sunlight. During months with long days, the plants develop leaves and increase stem length. The plant forms the flower buds naturally during late September and early October. During November and December, color develops in the bracts.

GROWING POINSETTIAS

PLANTING AND SPACING

Wholesale growers can supply 2¼-inch rooted poinsettia cuttings for planting, or producers with a smaller greenhouse can buy prefinished poinsettias from wholesalers and grow them to finish just prior to the Christmas holiday. Before pur-

SQUARE FOOT			2 FLOWERS PER	
POT SIZE	PLANTS PER POT	TREATMENT	Spacing Inches	Square Ft/Pot
4 inches	1	none	9 x 9 in.	0.56
4 inches	1	pinched	9 x 9 in.	0.56
5 inches	3	none	15 x 15 in.	1.56
5 inches	1	pinched	15 x 15 in.	1.56
6 inches	3	none	15 x 15 in.	1.56
6 inches	1	pinched	15 x 15 in.	1.56

FIGURE 13-2 Poinsettia planting guide (From Ball Red Book, 15th edition, 1991. Courtesy George J. Ball, Inc.)

chasing the cuttings or prefinished plants, check to be sure they are healthy.

When planting the cuttings, several things must be considered. The size of the pot is important, since it partially determines the size of the full-grown plant. Poinsettias are grown in a variety of pot sizes—4-inch, 5-inch, 6-inch, 7-inch, 8-inch, and 10-inch sizes, either hanging baskets or azalea pots. The size of the plant also depends upon the number of plants per pot and greenhouse space allowed for each potted plant. Each of the above factors also determines the price of the saleable plant.

Figure 13-2 gives the correct pot size and number of rooted cuttings per pot, treatment, and spacing for each pot in the greenhouse. Do not crowd poinsettias in the greenhouse. They need space to produce quality bracts and foliage.

TEMPERATURE

The greenhouse temperature is critical for growing top-quality poinsettias, figure 13-3. The temperature at night should be 65°F and 70° to 75°F during the day. It is important to maintain these temperatures during September and October while flower bud initiation is taking place. After the bracts start to show development, the nighttime temperature can be reduced to 60° to 62°F. There are newer cultivars on the market today that require a cooler finishing temperature after the bracts have developed. It is best to consult the whole-

DATE	TEMPERATURE (°F)		GREENHOUSE PROCEDURES
	Night	Day	
September 10	65	70–75	Arrival of 2¼-inch rooted cuttings from a reliable greenhouse grower. Transplant three rooted cuttings into a 6-inch azalea pot. Treat with a fungicide to control disease. Mix with water: Dexon—35% WP, 1¼ teaspoons per 5 gallons water.
September 11	65	70–75	Start slow release fertilizer program. Apply 1 level tablespoon Osmocote 18-9-9 per 6-inch azalea pot. Start a constant liquid feeding program.
September 23	60	75	Space each 6-inch pot in an area of 15 inches by 15 inches. Spray with growth regulator such as Cycocel 3000 ppm (3½ oz. per 1 gallon water). Lower temperature to 65°F. It is during this time that the flower buds form. Formation of flower buds is necessary for the production of color in the bracts of the poinsettia.
October 10	68	75	Increase night temperature to force color formation in bracts.
October 15	68	75	Drench soil with Dexon—35% WP, 1¼ teaspoons per 5 gallons water.
November 15	68	75	Color should be showing in upper bracts. Use a fan to provide continuous air movement in the greenhouse. Drench soil with Dexon—35% WP, 1¼ teaspoons per 5 gallons of water.
December 5	62	75	Lower night temperature to develop a bright color in the bracts. By this time the bracts should have achieved maximum size.
December 10	57–62*	75	Poinsettias are ready for sale.

WP—Wettable Powder
ppm—parts per million

*Lower nighttime temperature improves bract color, maintains plants at an optimal stage of development, and improves plant performance for the consumer.

FIGURE 13-3 Growing schedule for poinsettias

FIGURE 13-4 Growing poinsettias under optimum conditions will produce quality bracts and foliage. (Courtesy Paul Ecke Poinsettias)

saler of your geographic area for specific recommendations, figure 13-4.

PINCHING

Pinching poinsettias is a process in which the grower removes the terminal end of the plant. Pinching is accomplished by taking the top of the plant between the thumb and index finger and carefully breaking the top of the poinsettia completely off. Four or five nodes should remain above the soil level. Nodes occur at each point at which a leaf is attached to the stem. Therefore, four or five leaves should remain on the plant after it is pinched. After pinching, new shoots called *breaks* will develop from the bud in the axis of the node. As each shoot grows, it will produce a flower. After the flower develops, color will form on the bracts of the poinsettia. The major reason for pinching poinsettias is to increase the number of bracts per plant. It also causes the plant to appear fuller.

SOIL MIXES

Poinsettias may be grown in many different media. Whatever type of medium is used, it must be porous and well drained. This allows for maximum root development. Poinsettias also require a generous supply of fertilizer and moisture.

One suitable soil mixture consists of one-quarter soil (loam), one-quarter construction grade sand, one-quarter perlite (horticultural grade), and one-quarter peat moss. Another mix that works well is Pro-Mix B (in bags containing 6 cubic feet) and horticultural grade perlite (in bags containing 2 cubic feet). Perlite aids in drainage of the mixture. The Pro-Mix B and perlite should be mixed to evenly distribute the perlite.

It is important that all soil mixes be free of any insects and disease organisms, and possess good nutrient-holding capabilities. If soil is used, it should be sterilized at 180°F for thirty minutes.

FERTILITY

Poinsettias require a regular fertility program. The fertilizer used should be a complete fertilizer, which means that it contains nitrogen, phosphorus, and potassium. Fertilizers used to produce quality poinsettias are in a liquid or slow release form. The liquid fertilizer program may be constant with each watering or intermittent with every third watering. A prepared commercial soluble fertilizer such as 20-10-20 is an excellent choice. When using this fertilizer follow the manufacturer's recommendations for application. If the slow release fertilizer is used as a part of the fertility program, it is applied in a dry granular form. It has been prepared to dissolve slowly over a selected period of time (i.e., three months).

USING GROWTH REGULATORS

Growth regulators were developed to produce a better quality plant. Growth regulators have two main purposes: to control the height of the plant, therefore causing the plant to be saleable, and to improve the quality of color in the poinsettia bracts by darkening them.

Some poinsettia growers use a soil drench growth regulator. This type is mixed with water and applied to the soil mixture. It is then absorbed by the poinsettia plant, which in turn controls the height of the plant. Examples of this type of growth regulator are Cycocel (CCC), Ancymidol (A-Rest), and Bonzi.

Another type of growth regulator is applied as a foliar spray. The poinsettia foliage is sprayed until droplets form on the leaves. The leaves absorb the regulator, thus controlling their height. One example of this type of regulator is B-Nine. The rate of application is determined by the manufacturer's directions.

MAINTAINING POINSETTIAS

TRANSPORTING THE PLANT

When poinsettias are purchased from the local florist and transported to a new location, they should be

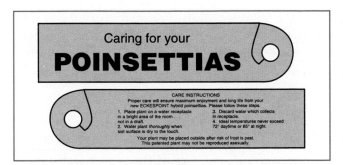

FIGURE 13-5 Instructions for care of poinsettias

Figure 13-5 illustrates a sample care tag accompanying poinsettias which are purchased from florists. By following these instructions, poinsettias can be maintained throughout the entire holiday season.

CONTROLLING INSECTS

For optimum growth, poinsettias must be protected from insects. For best protection, be sure the greenhouse is kept clean. Figure 13-6 lists pests that attack poinsettias and control methods for each.

CONTROLLING DISEASE

Just as poinsettias must be protected from attack by insects, so must they also be protected from attacks by disease organisms. One of the more common problems of poinsettias growers is root rot. To help prevent this problem, purchase rooted poinsettias from wholesalers who supply disease-free stock. When poinsettias are transplanted, treat them with a fungicide for further control.

handled very carefully. The plants should be protected from cold temperatures, and special care should be taken that the branches are not broken. This can be accomplished by carrying the plant in a sleeve.

Sleeves are made for the purpose of protecting plants. They are constructed in several materials such as paper or plastic. Sleeves are tapered in a funnel shape. The base of the sleeve is the same size as the pot, allowing the plant to be completely covered by the sleeve.

When removing the sleeve from the plant, cut it off with scissors. Lift the sleeve upward and off the plant. Be very careful not to break any of the plant's branches or bracts.

Although it may seem that poinsettias have more than their share of attacks by disease organisms, keep in mind that fungicides provide good control. Figure 13-7 lists common poinsettia diseases and methods of control.

PEST	MATERIALS	FORMULATION	DOSAGE	GENERAL REMARKS
Mealybugs	Sulfotepp (Dithio)	5% aerosol 15% smoke	Follow label directions.	Treat at 7-day intervals; foliage dry, vents closed
	Acephate (Orthene PT 1300)	75% SP 3% aerosol	Follow label directions.	Use 2 to 3 times at 7- to 10-day intervals
Whiteflies	Resmethrin (SBP 1382) (PT 1200)	2% EC 1% aerosol	Follow label directions.	Good late season control
	Sulfotepp (Dithio)	5% aerosol 15% smoke	Follow label directions.	Treat at 7-day intervals; foliage dry, vents open
	Acephate (Orthene PT 1300)	75% SP 3% aerosol	Follow label directions.	Use 2 or 3 times at 7- to 10-day intervals
Spider Mites	Dienochlor (Pentac)	50% WP	Follow label directions.	Treat every 7 days until controlled
	Acephate (Orthene PT 1300)	75% SP 3% aerosol	Follow label directions.	Use 2 to 3 times at 7- to 10-day intervals
Fungus Gnats	Resmethrin	EC	Follow label directions.	Good late summer control
	Sulfotepp	5% aerosol	Follow label directions.	Treat at 7-day intervals; foliage dry, vents closed

SP—to apply material in a mist form, giving coverage from top to bottom of foliage area
EC—emulsifiable concentrate
WP—wettable powder
Special Note: It is also advisable to consult local university extension services for special recommendations.

FIGURE 13-6 Poinsettia pest control schedule (Courtesy Cornell Poinsettia Guidelines for New York State, 1987)

NAME	PLANT SYMPTOMS	ORGANISM CHARACTERISTICS	CONTROL
Rhizoctonia (Stem and Root Rot)	Plant has a brown rot on the stem at soil line; roots have brown lesions; plant appears stunted with yellowing leaves from the bottom. In sever cases, plant will fall over.	Fungus carried in the soil and by infected plants; spreads very easily by water. Grows best under conditions of high humidity and high temperatures.	Remove infected plants from the greenhouse; drench with fungicides. Terraclor—(75% WP); drench; Dexon—(35% WP); Chipco 26019 (50% WP); spray; keep soil somewhat dry.
Botrytis (Gray Mold)	Rotting of young and immature tissue on leaf edges.	Fungus with airborne spores. Occurs in very humid areas with moderately low temperatures.	Keep a clean greenhouse; provide good air circulation; remove all dead plant material. Fungicides to use: Exothermal, Termil—20% (Dust); Ornalin (50% WP) spray.
Pythium (Water Mold, Root Rot)	The root tips are rotted; rot moves up the stem; plants are stunted; lower leaves turn yellow and drop; soil remains wet.	Fungus is carried by infected plants and soil.	Reduce amount of moisture. Subdue 2E; drench; Truban (30% WP); drench; Banol 66.5%; drench.

FIGURE 13-7 Diseases of poinsettias

Student Activities

1) Demonstrate the proper techniques involved in mixing soil for growing poinsettias.

2) Propagate poinsettias from cuttings. Compare the full-grown plants with those propagated from 2¼-inch rooted cuttings.

3) At the proper stage of development, pinch the poinsettias grown in Activity 1 to increase the number of flower bracts per plant.

4) Contact a local university and inquire about field days on poinsettia production. If possible, become involved in some of the activities.

5) Tour a local greenhouse operation and observe the greenhouse management techniques.

6) Show a video on greenhouse management techniques and procedures for optimum plant production.

Self-Evaluation

A) Select the best answer from the choices offered to complete the statement or answer the question.

　1) A poinsettia flower is the

　　　a) large blue flower at the terminal end of the plant.
　　　b) large yellow flower at the terminal end of the plant.
　　　c) small red flower at the terminal end of the plant.
　　　d) small yellow flower at the terminal end of the plant.

2) Poinsettias produce their color in the leaves below the flower called the

a) branches.   c) canes.
b) bracts.   d) none of these

3) What do poinsettias require for flower production?

a) a long day   c) no light at all
b) a short day   d) none of these

4) Poinsettias form the flower buds naturally during

a) late June and early July.
b) late January and early February.
c) late September and early October.
d) late May and early June.

5) Pinching poinsettias is a process of removing

a) flower buds.
b) bracts.
c) terminal ends.
d) none of these

6) The major reason for pinching poinsettias is to

a) increase the number of flowers per stem.
b) decrease the number of flowers per stem.
c) separate cuttings for propagation.
d) none of the above

7) Which of the following are growth regulators that can be used on poinsettias?

a) 2, 4-D   c) Dexon and 2, 4-D
b) Cycocel and B-Nine   d) none of the above

8) Poinsettias are affected by

a) whiteflies.   c) mites.
b) mealybugs.   d) all of these

9) Pesticides applied to soil in which poinsettias are grown are

a) sprayed.   c) wettable powders.
b) drenched.   d) none of these

10) One of the most common diseases suffered by poinsettias is

a) flower drop.   c) blight.
b) root rot.   d) none of these

B) Briefly answer, or fill in the blanks, each of the following.

1) Write an essay describing the five main factors to be considered by a greenhouse grower.

2) The greenhouse should be kept _____ to _____ degrees (cooler or warmer) at night.

3) The air movement in a greenhouse ensures the control of _____ _____.

4) Plant moisture in the plant helps to _____ _____ _____ and dissolve the _____ in the growing medium.

Chrysanthemums

OBJECTIVE

To produce a commercially acceptable potted chrysanthemum crop, beginning with rooted cuttings.

Competencies to Be Developed

After studying this unit, you should be able to

- ◆ outline in writing a growing schedule for potted mums for any preselected market date.
- ◆ pot rooted cuttings.
- ◆ regulate formation of chrysanthemum flower buds by control of night length.
- ◆ determine the proper cultural requirements, such as fertilizing and watering, for production of a marketable chrysanthemum crop.

MATERIALS LIST

- ✔ rooted chrysanthemum cuttings
- ✔ growing medium and 5-inch or 6-inch pots
- ✔ lights or shade cloth as needed to control length of night
- ✔ chemical growth retardants such as B-Nine or A-Rest®
- ✔ fertilizer
- ✔ insecticide and fungicide if needed

According to sales figures, the chrysanthemum (mum) is the most popular cut flower sold in the United States. It also competes very well in sales as a potted plant. Because it is now possible to artificially control the length of day and night in the greenhouse, the mum can be grown throughout the entire year. The most popular markets for mums are Mother's Day, Easter, and Memorial Day.

The mum is considered an easy plant to grow. It offers a wide range of colors and keeps well in the home after being sold. In this unit, the mum is considered a winter crop, flowering for sale at Easter time. It will be prepared for sale as a potted plant.

DETERMINING MARKET AND VARIETY

If sale of mums at Easter is desired, a schedule which assumes that the plants will be blooming before that date must be established. This necessitates starting the plants a number of weeks before the Easter date.

The growing schedule is largely dependent upon the variety grown. Some varieties bloom in eight weeks after the start of short days; others bloom in ten, twelve, and fourteen weeks. This is known as the *response time* (response to short days). Varieties of chrysanthemums are grouped in terms of response time. For example, a catalog might specify an *8-week variety* or *10-week variety*.

REGULATING LIGHT

Mums are considered a short-season crop, that is, the plant sets flower buds and blooms only when nights are long and days are short. To grow mums satisfactorily, the length of day and night must be regulated at two times.

1. When rooted cuttings are first potted (for one to three weeks), days must be kept long and nights short. (Plants should be exposed to no more than seven hours of continuous darkness.) This allows the plant to establish good vegetative growth before the formation of flower buds.

2. When summer flowering is desired, days must be kept short and nights long to encourage the proper timing of flower bud formation and flowering.

Mums naturally set flower buds in the fall. This means that unless lighting is regulated, the newly started cuttings will set blossom buds immediately after planting in the fall and winter, when nights are long. This causes the plants to blossom when they have not attained enough growth to be attractive.

To prevent flower buds from forming too soon, artificial lighting must be used to lengthen the day or, more accurately, to shorten the night. To delay flower bud formation, the plants must not be exposed to an uninterrupted period of darkness over seven hours in length. The best way to accomplish this is to add a period of light in the middle of the night.

The intensity of light to which the plants should be exposed is about 7 foot candles. Bulbs with reflectors are usually needed for best results. Two 4-foot beds may be lighted with one row of 100-watt bulbs. Space the bulbs 6 feet apart and 5 feet above the bed's soil level.

EXAMPLE If sunset occurs at 5:00 P.M. and the sun rises at 7:20 A.M., lights could be turned on for four hours from 10:00 P.M. until 2:00 A.M. This means that darkness is provided from 5:00 P.M. until 10:00 P.M. (five hours) and from 2:00 A.M. until 7:20 A.M. (five hours and twenty minutes). In this way, both periods of darkness are shorter than seven hours. Under these conditions, the formation of flower buds is delayed and the plants are allowed to increase in plant size and stem length. If there are other plants growing nearby that would be affected by the extra light, shade curtains should be installed to shield them from the light. In warm, sunny climates where plants naturally grow taller, lighting is not needed.

For summer flowering, shading is needed since days are long. Mums set flower buds only when exposed to short days (twelve hours or less) and to night temperatures of 60°F (16°C) or lower. In summertime, black sateen shade cloths are used to shorten the days. Wires strung the length of the growing bed are used to support the cloth. The cloth is pulled over the bed in the evening and left there long enough to shorten the day to twelve hours or less. In very hot weather, the cloth should not be pulled over the plants until 6:00 or 7:00 P.M. If the cloth is pulled over the plants any earlier, heat buildup under the cloth could damage the plants.

OBTAINING STARTER PLANTS

Mums are propagated by softwood cuttings. They are relatively easy to root. There is, however, a variety of diseases that may be passed along from parent plants to the rooted cuttings. For this reason, it is very important to purchase rooted cuttings from

FIGURE 14-1 Typical rooted chrysanthemum cutting ready for planting in the pot

specialists who can guarantee that the cuttings are disease free. Contact a local greenhouse operator for a source. The operator may offer to order for you if you need only a few cuttings. Order at least several months in advance to be sure the selected varieties are available. Figure 14-1 shows a typical rooted cutting.

ESTABLISHING A GROWING SCHEDULE

Before ordering rooted cuttings, a growing schedule must be established. The schedule must be planned starting at the end of the schedule, the blooming date.

EXAMPLE Assume that Easter Sunday occurs on April 10. If the mums are being grown as a class project and the school closes for the holiday on Friday, April 8, the mums must bloom prior to that date. To be sure that the blooms are available for sale the week of April 4 through 7, the blooming date should be set for April 1. (Mums hold their blossoms for a fairly long time.) The schedule would appear as in figure 14-2.

According to variety, mums receive what is called *short treatment, medium treatment,* or *tall treatment.* These labels are necessary since some varieties naturally tend to grow short, some medium, and some tall in height. So that the plants grow to the same height and appear uniform, the length of the *vegetative period* (the long day in summer before shade is used, and the short night in fall and winter when artificial light is used) must be regulated differently for each of the three groups.

For example, the mum crop being prepared for the Easter market is grown in winter; therefore, the length of the vegetative period is controlled by using lights to shorten the number of hours of continuous darkness. After planting the rooted cuttings, the lights are left on for two-and-one-half weeks for the short treatment varieties, two weeks for the medium treatment varieties, and one week for the tall treatment varieties. Notice that the planting date changes in figure 14-2 according to variety. The schedule shown in 14-2 is for ten-week response varieties; the schedule for eight-week response varieties would be two weeks shorter. These varieties would simply be planted two weeks later.

			SCHEDULE			
TREATMENT	PLANT	SHADE	LIGHTS ON*	PINCH	B-NINE (2 Weeks After Pinching)	BLOOM DATE
Short (2½ weeks of lights)	Jan. 4	not needed	Jan. 4–Jan. 21	Jan. 22	Feb. 5	April 1
Medium (2 weeks of lights)	Jan. 7	not needed	Jan. 7–Jan. 21	Jan. 25	Feb. 8	April 1
Tall (1 week of lights)	Jan. 14	not needed	Jan. 14–Jan. 21	Jan. 29	Feb. 12	April 1

*From end of "lights on" to bloom date is 10 weeks (when long nights start).

FIGURE 14-2 Growing schedule for Easter flowering (ten-week response variety)

PLANTING ROOTED CUTTINGS

CONTAINER AND MEDIUM (FOR POTTED MUMS)

A 6-inch plastic pot is a good choice for planting rooted mum cuttings. This size pot requires five cuttings per pot. Varieties which send out more side shoots may be planted with only four per pot. If a 5-inch pot is used, plant four cuttings to each pot. Some markets use a 4-inch pot and place two cuttings to each pot.

The medium should be one that is well drained and yet holds moisture well. If natural soil is used, it must be sterilized first. Peat moss for moisture retention and sand or perlite for drainage should be added to soil in this ratio: one-third soil, one-third sand or perlite, and one-third sphagnum peat moss. There are other growing mixes on the market which contain no soil and are very effective. The medium should be such that water added to the surface of the pot drains quickly through the pot and out the bottom. If the soil mix has the proper texture, it will not crack when dry. Be especially careful that the potting medium does not dry out after the mums are potted. Mums should never be allowed to wilt from dryness. To prevent this, provide a constantly moist, but not wet, soil. Add water when the top of the potting medium first appears to be drying out.

FIGURE 14-3   Mum cuttings planted in a 6-inch pot. Notice that the cuttings are slanted toward the outside of the pot at an angle of about 45 degrees. The roots of the cuttings are barely covered.

PROCEDURE

STEPS IN PLANTING

As mentioned, mum cuttings should be planted in 6-inch shallow pots (azalea pots), with four or five rooted cuttings per pot. Select cuttings for each pot that are all approximately the same size. This gives more uniformity to the finished product.

1. Fill the pot with planting medium to within 1 inch of the desired level after planting. **Press the medium firmly in the pot so that it will not settle later.**

2. Space the cuttings evenly in the pot. Slant the cuttings at approximately a 45-degree angle so that the top of the cuttings protrudes over the edge of the pot, figure 14-3. This angle encourages the growth of more shoots and, thus, better, more compact plants.

3. Place just enough soil in the pot to cover the roots (about 1 inch or less). When the pot is planted and filled, the medium level should be about 1 inch below the top of the pot. This makes watering the cuttings an easier job.

4. Water the cuttings until the water runs through the bottom of the pot.

5. Place the pot in the growing area, figure 14-4. Later, these pots will require more space. Pots should have at least 1 inch of space between them for the first three to four weeks, and later about 8 inches between pot rims. This wider spacing gives the plant room to spread out, resulting in a better shaped plant with more blossoms. Watch for signs of overcrowding within the pots. One of the first signs may be drying of lower leaves. (However, this may also signal underfeeding, disease, or poor drainage.)

CARING FOR POTTED CUTTINGS

TEMPERATURE CONTROL

When first placed in the growing area, the greenhouse temperature should be kept at 63°F (17°C)

FIGURE 14-4 Pots of mum cuttings placed in the growing area of the greenhouse. It will be necessary to provide more space between the pots in about three weeks. (Ed Reiley, Photographer)

TIME PERIOD	TEMPERATURE		
	Night	Cloudy Day	Sunny Day
From planting through the first 2 weeks of short days*	63°F	63°F	70°F
Up to last 3 weeks of growing cycle	62°F	62°F to 65°F	70°F
Last 3 weeks of growing cycle	55°F	55°F	65°F

*If temperatures are cool in the early stages of short days, an uneven bud set may occur.

FIGURE 14-5 Temperature control schedule

at night, figure 14-5. The temperature should be held to 63°F (17°C) on cloudy days and 70°F (21°C) on sunny days. This temperature range is maintained through the first two or three weeks of short days. The higher temperatures promote faster vegetative growth. If the temperature drops below 60°F (16°C) during this period, flower buds may not set evenly. Uneven bud set may be one of the first signs of improperly regulated temperature.

At the end of this period, the temperature should be dropped to 62°F (16°C) at night; to 62°–65°F (16°–18°C) on cloudy days; and to 70°F (21°C) on sunny days. For the last three weeks of growth, the temperatures should be dropped to 55°F (13°C) at night and on cloudy days, and to no higher than 65°F (18°C) on sunny days. This cooler temperature at the end of the growing period tends to harden off the plants.

FERTILIZING CUTTINGS

Mums may be fertilized using one method or a combination of several different methods.

FERTILIZING BY CONSTANT APPLICATION A very diluted fertilizer may be added to the irrigation water, supplying the mums with constant fertilizing. Wait a few days after planting to begin this program. If this form of fertilizing is chosen, a water-soluble fertilizer prepared specifically for mums should be used according to directions. The fertilizer should be high in nitrogen and potash. For example, if a 15-10-30 fertilizer is chosen, it should be mixed in a stock solution of 18 ounces per gallon of water and diluted again to 1 part solution to 100 parts water before application or as it is being applied through a siphoning device.

Caution: Follow directions on the container label.

Note: During cloudy weather, less fertilizer and water are needed. This fertilizing system works well to reduce both at the same time when necessary.

FERTILIZING BY SLOW RELEASE A slow release fertilizer may be mixed in the potting medium at the time of planting. Use 8 ounces of fertilizer for every bushel of potting medium. This fertilizer will furnish all or most of the fertilizer needs for a period of three to four months. If plants appear to need more nitrogen, additional fertilizer may be added in the irrigation water. *Do not sterilize a soil mix after a slow release fertilizer has been added.*

FERTILIZING BY TIMED APPLICATIONS Fertilizer may be applied once every two weeks in the irrigation water or to each pot as needed. This method is not as effective as either of the other methods listed, and the crop produced may not be as good.

WATERING

How often mums are watered varies with the temperature and humidity in the room and water-holding capacity of the medium. In hot, dry weather, the plants may require watering twice a day. In cool, cloudy weather, once a day or every other day may be sufficient. Check the soil mix frequently for dryness and add water whenever the top of the medium is beginning to appear dry. Do not keep the soil saturated, however; the plants' roots may rot. At the same time, allowing the plants to wilt will cause damage to the crop. Add enough water at each watering so that water runs through the pot and out the drainage holes.

> *Caution:* Newly planted cuttings require special care and attention the first week. During this time, be especially watchful for signs of dryness.

PINCHING AND DISBUDDING

Pinching means simply to break or pinch off tips from plant stems. Hold the tip of the stem and pinch off ¼ to ½ inch of growth, figure 14-6. This causes side shoots to develop, thereby resulting in more branches to set flowers. The final result is a more compact, flowerful plant.

FIGURE 14-6   A plant tip is held directly above the mum from which it was pinched. The tip measures about ½ inch in height. Because of the pinching, many more side shoots will form, resulting in a bushier, more flowerful plant. (Ed Reiley, Photographer)

The ideal potted mum should have about ten to fifteen flowers per pot. If more buds are present, remove some or all of the smaller side buds. This will result in larger, more attractive individual flowers, figures 14-7A and B.

GROWTH RETARDANTS

The chemicals B-Nine and A-Rest®, growth retardants, tend to shorten stem growth. This causes

FIGURE 14-7A   One stem of a mum plant showing many buds at tip and along the stem.

FIGURE 14-7B   The same stem after all buds except the terminal bud are removed. The removal of the buds results in larger, single flowers on each stem. (Ed Reiley, Photographer)

the blossoms to grow closer together, resulting in a more compact plant. B-Nine is sprayed on the foliage about two weeks after pinching, or when side shoots are about 2 inches long. Do not wet or water the plant foliage for twenty-four hours after applying B-Nine to prevent the chemical from being washed off. The treatment may be repeated if plants appear to be growing too tall. Follow label directions for second treatment.

Caution: Read label directions carefully before using these chemicals.

CONTROLLING INSECTS AND DISEASE

Preventing attacks on plants by insects and disease organisms is a far better method of plant protection than fighting a serious infestation. Good sanitation and cultural practices are often all the control that is needed. If insects do become a problem, spray with an emulsifiable concentrate spray. Use a recommended fungicide if necessary for fungus disease.

Note: Wettable powders leave a residue on plants and should not be used within six weeks of sale date. Emulsifiable concentrates do not leave visible residues. (An *emulsifiable concentrate* is a chemical in liquid form that will mix with water.) Emulsifiable concentrates and wettable powders may not mix well if applied in the same solution.

Caution: Unless insect or disease problems arise, do not spray. Never use highly toxic substances. When any pesticide is used, read directions carefully and follow all safety precautions.

SPACE REQUIREMENTS FOR YEAR-ROUND PRODUCTION

To produce fifty pots of mums every two weeks year-round, about 380 square feet of bench space is needed. Provision must be made for lighting on at least part of the space, figures 14-8 and 14-9.

FIGURE 14-8 A potted mum ready for market. (Ed Reiley, Photographer)

FIGURE 14-9 The Yellow Diamond, a relatively new mum variety, which has open, showy petals. (Ed Reiley, Photographer)

GARDEN MUMS

Many horticulturists prefer to cultivate plants that can be set outdoors after blooming in the pot. This may be done by using a special variety, the garden mum, for the spring crop. After planting outdoors in the spring, the garden mum can be expected to bloom again the following fall.

Plant cuttings of garden mums in 3- to 5-inch pots and grow according to the schedule for that particular variety. Most garden mums are the short response variety and flower in eight to nine weeks of short days. This gives a quick return on a very popular crop.

Student Activities

1) Develop a potting soil for chrysanthemum cuttings. Determine by feel and appearance if the potting medium has the proper characteristics.

2) Plant at least one pot of four cuttings of mums and care for it throughout the growing period.

3) Develop a complete growing schedule for an Easter mum crop. Compare it with your instructor's schedule for accuracy.

4) Visit a commercial greenhouse operator who produces mums. Observe the operation and discuss a prepared growing schedule with the grower.

Self-Evaluation

Select the best answer from the choices offered to complete the statement or answer the question.

1) The mum is a short-season crop. This means that it flowers

    a) in a very short period of time.
    b) when days are short and nights are long.
    c) in areas that have a shorter than average growing season.
    d) only on chrysanthemum plants that are very short.

2) Mum cuttings planted in fall and winter require special lighting to lengthen days for the first one to three weeks so that

    a) the plants grow larger.
    b) the plants are more saleable.
    c) the plants do not set flower buds too quickly.
    d) all of the above

3) The amount of or intensity of light needed to prevent flower bud formation in mums is

    a) about 5 foot candles.
    b) about 15 food candles.
    c) about 7 foot candles.
    d) about 70 foot candles.

4) Shading of mums in summer is necessary because

    a) days are too long.
    b) temperatures are too high.
    c) heat tends to fade the flowers.
    d) the sunlight is too strong, providing too many foot candles of light.

5) When a mum is labeled an *8-week variety* (response time), this means that

    a) eight weeks is the entire growing time for the crop.
    b) eight weeks is the length of time lighting must be provided for the crop.
    c) eight weeks is the period of time from the start of short days until bloom.
    d) all of the above

6) Mums that are labeled for short treatment are

    a) plants that naturally grow to a shorter than average height.
    b) plants that require more pinching than the average plant.
    c) plants that bloom in a short period of time.
    d) plants that require stronger light, but for a shorter period of time.

7) The suggested potting arrangement for mums is

    a) six cuttings to each 4-inch pot.
    b) one cutting to each 4-inch pot.
    c) three cuttings to each 5-inch pot.
    d) four to five cuttings to each 6-inch pot.

8) When potting mums, the most important characteristic of the potting medium is that it

    a) be inexpensive.
    b) drain and hold moisture well.
    c) be locally available.
    d) contain good, loamy soil.

9) Mum cuttings are equally spaced around the pot and slanted toward the outside of the pot at and angle of

    a) 45 degrees.
    b) 35 degrees.
    c) 60 degrees.
    d) 25 degrees.

10) After mum cuttings are placed in the pot, how much potting medium should be placed in the pot?

    a) 2 inches
    b) ½ inch
    c) 3 inches
    d) 1 inch or less

11) After the first three to four weeks, how far apart should potted mums be spaced (measuring from pot rim to pot rim)?

    a) 1 inch apart
    b) 2 inches apart
    c) 8 inches apart
    d) 12 inches apart

12) For the first two weeks, the temperature of the greenhouse in which potted mum cuttings are kept should be _____ at night and on cloudy days and _____ on sunny days.

    a) 55°F; 65°F
    c) 45°F; 55°F
    c) 63°F; 70°F
    d) 70°F; 75°F

13) From two weeks after planting until the last three weeks of the growing period, the temperature of the greenhouse in which mums are kept should be _____ at night and on cloudy days and _____ on sunny days.

    a) 62°F; 70°F
    b) 55°F; 65°F
    c) 63°F; 75°F
    d) 70°F; 75°F

14) During the last three weeks of the growing cycle, the temperature of the greenhouse in which mums are kept should be _____ at night and on cloudy days and _____ on sunny days.

    a) 62°F; 70°F
    b) 63°F; 75°F
    c) 70°F; 75°F
    d) 55°F; 65°F

15) Mums should be watered when

    a) the top of the medium is beginning to appear dry.
    b) medium begins to run through the bottom of the pot.
    c) the plants have wilted.
    d) it is convenient.

16) An uneven bud set on mums is usually caused by

    a) improper lighting control.
    b) improper temperature control.
    c) crowding pots together.
    d) insect damage to buds.

Easter Lilies

OBJECTIVE

To produce a marketable crop of Easter lilies.

Competencies to Be Developed

After studying this unit, you should be able to

- ◆ properly sterilize containers, medium, and a planting area, demonstrated by either actually doing it or writing up the procedure.
- ◆ prepare or select the proper medium for growing lilies and explain the percentage of each material in it.
- ◆ pot plant Easter lily bulbs to the proper 2-inch depth, with properly filled pots.
- ◆ calculate the planting date for lilies to flower by Easter Sunday.
- ◆ outline in writing a fertilizer schedule for the crop.
- ◆ describe in writing which growth regulators are used and why.
- ◆ describe in writing the disease control methods needed to produce a clean crop.

MATERIALS LIST

- ✔ chemicals or steam to sterilize the working area (Captan, Terraclor, and Ferbam)
- ✔ 6-inch, three-quarter-size pots
- ✔ well-drained planting medium
- ✔ lily bulbs
- ✔ fungicide for drenching newly planted bulbs (Lesan 25% WP and Terraclor 75% WP)
- ✔ ground limestone
- ✔ dolomitic limestone
- ✔ greenhouse growing area with good temperature control
- ✔ the fertilizers potassium nitrate and calcium nitrate
- ✔ superphosphate
- ✔ Ancymidol (A-Rest®) for height control, if necessary

The Easter lily is an important potted plant as well as a popular cut flower. We will deal with it here as a potted plant being grown for the Easter holiday.

VARIETIES OR CULTIVARS

The two most popular varieties of lilies on the market are called "Ace" and "Nellie White." "Ace" produces more flowers and grows slightly taller than "Nellie White." If you prefer not to use or to limit the use of growth regulators to dwarf the plants, "Nellie White" would be the best choice. "Nellie White" is an excellent plant for potted culture and may be a slight favorite for this purpose.

BULBS

Most lily bulbs are purchased from Washington State, Oregon, and northern California. They are shipped in crates packed in slightly moist peat moss, figure 15-1. The lily bulb is a soft structure with no heavy protective coating and loses moisture rapidly; bulbs should never be allowed to dry out. To simplify the growing procedure it is wise to purchase precooled bulbs. Bulbs from 7 to 9 inches in circumference are large enough. Large bulbs may produce more flowers, but a controlled lower temperature can offset this difference.

ESTABLISHING A GROWING SCHEDULE

It is very important to plant the Easter lily bulbs on the proper date so that the plants will flower

FIGURE 15-1 A crate of Easter lily bulbs as shipped from California to an Eastern grower. Note that the bulbs are packed in peat moss to prevent drying out. (Courtesy of Michael Dzaman)

1996—April 7	2001—April 15
1997—March 30	2002—March 31
1998—April 12	2003—April 20
1999—April 4	2004—April 11
2000—April 23	2005—March 27

FIGURE 15-2 Dates for Easter Sunday, 1996–2005

DATE BULBS ARE RECEIVED	PROBABLE WEEKS OF PRECOOLING	APPROXIMATE DAYS TO BLOOM (60°F at night, 70°F during the day)
November 25	5	125
December 2*	6	120
December 9	7	114
December 16	8	108
December 23	9	103
December 30	10	100

FIGURE 15-3 Flowering times (Adapted from "Cornell Recommendations for Commercial Horticulture Crops")

just before the Easter holiday, figure 15-2. The demand for potted Easter lilies drops off after Easter. How soon the bulbs will flower after planting is determined by two things: the length of the cold treatment given; and, after bud set, the temperature at which they are grown.

According to work done at Cornell University, the number of days to flowering time is influenced by how much cold treatment the bulbs have had. The approximate days to flowering may be predicted using the table in figure 15-3.

*Based on the information in figure 15-3, if Easter Sunday falls on April 19 and precooled bulbs were received on December 2, flowering date would be approximately 120 days. The potting date would then be calculated at December 21. It may be a good idea to plant a week earlier than this date to be sure. Plant growth can be slowed down later, if necessary, by lowering the greenhouse temperature.

If the bulbs need to be held a week or two before planting, store them in the shipping crate at 40°F. Do not allow the bulbs to dry out, but do not soak then either; misting the crate may help.

PLANTING BULBS

GROWING MEDIUM

Lilies should be planted in a very well-drained potting mix. The mix should be high in perlite, sand, or coarse vermiculite to improve drainage.

Many commercial mixes are available, but if you mix your own use at least one-third perlite, sand, or vermiculite. If any soil is used in the mix, it should be sterilized or pasteurized at 180°F for at least 30 minutes. The mix should be one part loam, one part peat moss, and one part perlite. The best policy is to use a soilless mix of peat and perlite or sand. A good soilless mix is one-half peat moss, one-quarter perlite, and one-quarter vermiculite.

The pH of the medium should be adjusted to 6.8–7.2 with limestone. To adjust the pH of most mixes requires about 7 or 8 pounds of limestone per cubic yard of mix. About 3 pounds of this should be dolomitic limestone and the other 4 or 5 pounds regular ground limestone. About 12 to 16 ounces of regular superphosphate is also helpful.

POTTING

Bulbs are planted in 6-inch, three-quarter-size pots, figure 15-4. Each bulb should be placed so the tip is covered with 2 inches of potting medium. Remember that the lily bulb is very tender and must be handled carefully. Do not push the medium around the bulb hard enough to bruise it, but do pack it firmly. *After the medium is settled*, the pot should be filled to within 1 inch of the top and the bulb tip should be 2 inches below the surface of the medium.

FIGURE 15-4   A bulb is placed in the pot so that it will be covered with about 2 inches of potting medium when the pot is filled. (Ed Reiley, Photographer)

The pot should be watered well to thoroughly soak the medium. Watering after planting should be done often enough to keep the medium *moist, not wet.*

Immediately after potting, drench the medium in the pot with a combination of Lesan and Terraclor to help control root rot. Mix according to directions on the label and apply one-half pint per pot.

Caution: Use nitrile *gloves when handling chemicals. Nitrile or Neophene gloves are best for chemical protection.*

CARING FOR POTTED LILIES

FERTILIZING

When shoots emerge from the soil, the pots should be watered with a mixed solution of ¾ pound of potassium nitrate per 100 gallons of water (.12 ounces per gallon) and 1½ pounds of calcium nitrate per 100 gallons of water (.24 ounces per gallon) every 2 weeks, figure 15-5.

When flower buds are ½ inch long, use only calcium nitrate at 2 pounds per 100 gallons of water (.32 ounce per gallon) weekly until the plants are sold, figure 15-6. Do not use an acid-forming nitrogen source such as ammonium sulfate.

WATERING

Soak the containers when the bulbs are first planted. After that, keep the medium moist but

FIGURE 15-5   When plants reach the stage of growth shown in this picture, it is time for the first fertilizer application. (Ed Reiley, Photographer)

FIGURE 15-6   A plant in full flower, ready for market. (Ed Reiley, Photographer)

never wet. In other words, wait to water until the medium is barely moist on top. If the plants are allowed to dry out, plant quality and the number of flowers is decreased. If they are overwatered, root rot is a strong possibility. Good air circulation around the plants will help to prevent very wet conditions by drying the soil quicker after each watering.

LIGHT

The potted lilies should be placed in full light. If the greenhouse glass has shade fabric, dirt, or shade spray on it, these should be removed.

GROWING TEMPERATURES

Day and night temperatures for Easter lilies should be closely controlled. A nighttime temperature of 60°F and a daytime temperature of 70°F should be carefully maintained. This is especially important up to the time flower buds form, usually about the middle of February. If temperatures go above 70°F at night flower bud formation may be delayed.

After flower buds have formed, temperature control is not as important. Temperature may now be used to control growth and the time of flowering. Warmer night temperatures especially will speed up crop progress and cause earlier flowering. Cooler temperatures both day and night will slow development and result in a later flowering date. Under warmer temperatures, plants will need to be watered more often; under cooler temperatures, less often.

GROWTH REGULATORS

Extending the day length to eighteen hours will cause lilies to grow taller. Long days are started when plants are 4 to 6 inches tall and continued for five to six weeks. Tall lilies are needed for the cut flower trade but are not desirable, however, in plants sold in pots. Shorter-than-normal day length will shorten stem growth, but chemical growth retardants are more effective.

The most commonly used growth retardant is Ancymidol (A-Rest®) and is used as a soil drench. For lilies which normally are short growers, such as "Nellie White" and "Ace," the amount of chemical used is small. *Read the label for proper use and cautions.* Plants are treated when they are from 4 to 7 inches tall. Chemical retardants do not affect the number of leaves or flowers; their purpose is to shorten the stem internodes or distance between leaves.

Keep a record of the chemical treatment used and its effect on the plants. Adjustments can be made in the amount of chemical used for the following crop based on this record. More chemical use results in shorter plants. *Follow label directions.*

Cultural practices also affect plant height. A proper fertilizer program results in shorter plants than would otherwise develop without proper fertilization.

CONTROLLING INSECTS AND DISEASES

ROOT ROT   The best method of disease prevention is to be sure the entire growing environment is sterilized. This includes the pots, medium, potting table, greenhouse bench, workers' hands, and so forth. The bulbs should not come into contact with anything that might be contaminated with the fungi *Rhizoctonia* or *Pythium*. These are the two fungus disease organisms responsible for root rot in lilies. Growing benches may be drenched with a Captan-Terraclor-Ferbam mixture. Steam is also effective to sterilize the growing environment.

If proper sanitation is provided and if good cultural practices are maintained, such as starting with a well-drained medium and *not* overwatering, root rot will probably not be a problem.

Immediately after planting, water the pot with a good preventive fungicide. Cornell University recommends the following: Lesan (8 ounces of 35% WP per 100 gallons of water) + Terraclor (4 ounces of 75% WP per 100 gallons of water). Apply ½ pint of drench per 6-inch pot. After this initial soil drenching, a drench using ½ pint Lesan (4 ounces of 35% WP per 100 gallons of water) + Banrot used according to label directions at monthly intervals. *Wear protective gloves.*

INSECTS   The most common insect pest is the aphid, or plant louse. If this pest is present, use any good systemic insecticide labeled for lilies and indoor use.

> *Caution: Use extreme caution with all chemicals.*

STORAGE

According to researchers at Cornell University, if plants flower too early for Easter, they can be stored for up to two weeks in the dark at 31°F. Storage should start when the first bud begins to open. Plants should be sprayed with Zineb (½ pound per 100 gallons of water) before storage to protect against botrytis. Any open flowers should be removed before storage, since they are more susceptible to botrytis.

Removal from storage should be done one full day before full bloom is desired. Move plants to a shaded area to warm up before placing them in full sunlight in the greenhouse.

Student Activities

1) Visit a greenhouse where Easter lilies are grown and talk with the manager.

2) Have your local greenhouse grower come to class and speak with you about the crop and demonstrate his methods.

3) Plant and grow a crop of Easter lilies for sale.

Self-Evaluation

Select the best answer from the choices offered to complete the statement or answer the question.

1) The potting area and containers may be sterilized using the chemicals _____ , _____ , and _____ , mixed together according to direction.

   a) Captan, Lesan, Ferbam
   b) Terraclor, Ferbam, Lesan
   c) Captan, Terraclor, Ferbam
   d) none of these

2) Most Easter lily bulbs used in the United States are purchased from

   a) Holland.
   b) the West Coast.
   c) Florida.
   d) Colorado.

3) Bulbs from _____ to _____ inches in size are large enough for pot culture.

   a) 6, 8
   b) 6, 9
   c) 9, 11
   d) 7, 9

4) Precooled lily bulbs received around the first of December

   a) take about 120 days to flower after planting.
   b) are usually not yet mature.
   c) need cold storage treatment before they will flower.
   d) all of the above

5) Lily bulbs may be stored in the shipping case at a temperature of

a) 30°F.   c) 50°F.
b) 40°F.   d) 70°F.

6) How soon bulbs will flower after planting is determined by two things: _____ and _____ .

a) length of cold treatment, temperature at which they are grown
b) length of cold treatment, how often they are watered
c) temperature at which they are grown, pot size
d) pot size, bulb size

7) The planting medium must drain very well and have a pH of

a) 6.0 to 7.0.   c) 4.5 to 6.0.
b) 7.0 to 8.0.   d) 6.8 to 7.2.

8) Soil pH is adjusted using

a) sulfur.
b) ground limestone and dolomitic limestone.
c) gypsum.
d) superphosphate.

9) After planting, the first fertilizer application is made

a) in one week.
b) when shoots are 2 inches tall.
c) when shoots first emerge from the soil.
d) when flower buds first appear.

10) Bulbs are planted so that the tip of the bulb is _____ inches below the medium surface.

a) 3   c) 4
b) 2   d) 1

11) Potted lilies placed in the greenhouse should get _____ light.

a) full   c) low
b) filtered   d) any of these

12) Growing temperatures must be carefully controlled through mid-February and kept at

a) 60°F at night and 70°F in the daytime.
b) 40°F at night and 70°F in the daytime.
c) 70°F at night and 80°F in the daytime.
d) 60°F at night and 80°F in the daytime.

13) Plants are often treated with a growth regulator to

a) cause production of more flowers.
b) lengthen stem growth.
c) slow down flowering time.
d) shorten stem length.

14) The most common disease of Easter lilies is

a) root rot.   c) aphids.
b) leaf rust.   d) root grubs.

15) Plants that flower too early for Easter may be stored in the refrigerator

a) for up to 4 weeks at 40°F.
b) for up to 2 weeks at 31°F.
c) no more than 1 week at 28°F.
d) as long as necessary if held at 31°F.

Integrated Pest Management

Integrated Pest Management and the Biological Control of Pests and Diseases

OBJECTIVE

To understand the control of plant pests through natural, biological means.

Competencies to Be Developed

After studying this unit, you should be able to

- explain orally or in writing what biological control means.

- explain orally or in writing what integrated pest management means.

- outline a pest control program, explaining when biological control should be used and at what point chemicals must be used.

- list at least three insects that have been effectively controlled without man-made chemicals.

- list at least one plant disease controlled by biological means.

- list at least one weed and the biological method used to control it.

MATERIALS LIST

✔ *1986 Yearbook of Agriculture—'Research for Tomorrow'* for research

INTEGRATED PEST MANAGEMENT

A new approach to pest control of plants has been tested and found to be very effective in not only controlling pests, but also greatly reducing the need for the application of chemicals.

The key to an effective integrated pest management (IPM) program is regular monitoring to detect the pest when it first becomes active. This is done through regular inspection of plants by specialists and by using traps or other collection devices to spot the first few pests on the scene. Action is taken immediately, before the number of pests has a chance to increase.

The first control used should be nonchemical such as hand collecting of pests and pruning out infected or diseased limbs or plants. Next, a biological control system is considered and used if deemed adequate. Lastly, chemicals are used. Treatment should be early when only a small area is affected, spraying only infected plants, thus reducing the need to spray entire fields or other large areas. Often, one well-timed spray application on a small area is all that is needed for control the entire season. The least toxic chemicals available should be used first. Spray oils, which are now highly refined and effective, are very safe for the plant. If used early in the season, spray oils may eliminate the need to spray with more toxic chemicals later, thus protecting the environment. IPM is working well and for many pests has eliminated the need for costly regular applications of chemicals once thought necessary.

Plant diseases are not quite as easy to control with IPM as insects, but good sanitation to prevent carry-over of fungus spores and early identification and prompt control can greatly reduce the need for regular preventive sprays. Disease resistant varieties can also reduce or eliminate the need for chemical control of some diseases.

IPM can work. A report in Maryland stated that nurseries using the IPM program used 98 percent fewer pesticides than reported for other nurseries. This reflects tremendous savings in money spent for chemicals as well as reducing pollution of the environment.

A HISTORY OF BIOLOGICAL CONTROL

The biological control of plant pests is the oldest control method available. All creatures in nature have natural enemies, such as diseases, predators, or parasites which repel, kill, and consume them. Plant pests are controlled by naturally occurring bacteria, fungi, insects, nematodes, birds, toads and frogs, lizards, snakes, voles (micelike creatures), and others. See figure 16-1. If these natural enemies were not here, the world would be overrun with insects. If natural enemies did not

FIGURE 16-1 The green lacewing, one of the most effective biological controls of aphids and many other soft-bodied insects. (Courtesy USDA)

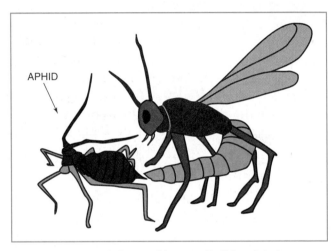

APHID

FIGURE 16-2 A Braconid wasp (1½" long) laying an egg in an aphid. The wasp larva will kill the aphid.

exist, a single aphid could be responsible for the production of more than 318 million aphids in one month. The fact that aphids do not increase in numbers to this extent in nature (where no spraying is done) is proof that biological control works, figures 16-2 and 16-3.

In the first half of the twentieth century and in earlier years, when farms were smaller and crops were grown on smaller acreage and in rotation, the natural predators did a good job of controlling, not eliminating, many of these pests. Then came the large, specialized grower with hundreds or thousands of acres of one crop. This was a paradise for plant pests. Whichever direction a fungus disease spore blew in the wind, it hit a plant it could grow on. Soil-borne diseases multiplied as the same crop was planted in the same field year after year. Pests survived in large

FIGURE 16-3 A ladybug larvae preparing to eat an aphid. (Courtesy USDA)

numbers under the same conditions. Why didn't the plant pest-control organisms also build up and control these pests? There are three main reasons the pests' natural enemies did not keep up with and do their control job.

1. Many of these organisms needed a different environment as adults in order to survive. They may have needed nectar from wildflowers or some cover in which to lay their eggs. Large acres of the same crop destroyed this natural habitat and the adult pest-control organism died out.

2. The American consumer wanted a perfect food product with no blemishes from insect bites. To do this, intense chemical control was used, killing good organisms as well as pests.

3. The farmer felt that the only good insect was a dead insect, and started using chemicals to kill them all. For biological control organisms to survive, there must be some pests on which they can feed.

In order for biological control to work, man must rethink his strategy on pest control, and the methods of *integrated* pest management may provide the answer. We must be willing to live with a few pests when no real economic damage is being done, and we must learn what environment the natural biological control agents need in order to survive.

It is now known that many adult parasitic pest control insects need wild flowering plants such as mustard, wild carrot, yarrow, dill, anise, black-eyed susan, catnip, and milkweed, to name a few, as food sources of nectar and pollen. Some areas set aside for these wild flowers would help to ensure survival of the adult insects and guarantee a new generation of natural pest-control insects.

EXAMPLES OF SUCCESS

Brief mention will be made here of the successful biological control of some insects that cause serious damage. More detail will be given later in this unit.

In the 1940s and 1950s, the Japanese beetle was a serious pest in the eastern United States. Millions of dollars were spent on chemical controls, which had to be repeated year after year. When two natural enemies of the beetle were brought into the United States and placed in the environment, the beetle was controlled to the point where it is not as damaging and chemical sprays are less often needed. The biological controls live from year to year and need not be reapplied.

Researchers at Cornell University recently released and established the European sevenspotted lady beetle (*Coccinella septempunctata*) as a predator of aphid pests. This species is highly adaptable to many environments and should survive to provide continued control of aphids and greatly reduce the need for chemical control. See figure 16-4.

The U.S. Department of Agriculture has had great success against isolated pockets of gypsy moth infestations. Through the use of *Bacillus thuringiensis*, a bacterial disease, and traps baited with sex attractants, the Department has had consistent success in eradicating the moth. In thirty out of thirty-five infested sites, the moths were eliminated with no chemical pesticides.

FIGURE 16-4 A sevenspotted ladybug adult. (Courtesy USDA)

Other exciting success stories using biological control are being reported for the Colorado potato beetle, which is resistant to nearly all chemical pesticides, the alfalfa weevil, the Mexican bean beetle, the musk thistle, and many others.

The Mexican bean beetle can be controlled on a crop if a special management system is set up whereby some beans planted early are used to initiate the bean beetle population and start the parasitic control agent on these beans. When the commercial crop is planted later, enough parasites have built up to maintain control on the larger planting. It is necessary to build up the pest control agents in this way or grow them in a laboratory for release in the field. Otherwise, there is diminished effectiveness until the predators have time to increase in numbers large enough for satisfactory control.

Some predators remain in the environment and give control in the future; others must be released from the laboratory each year. Through selection and genetic changes, researchers are developing control agents which can live longer in the environment.

THE NEED FOR BIOLOGICAL CONTROL

The most exciting thing about this success in biological pest control is that we are finally looking to nature for control of plant pests, rather than using highly poisonous chemical materials. Many of these chemicals are no longer effective anyway, because insects become resistant to them. The Chinese use biological control to a very large extent and we can learn much from them in this respect.

In spite of close monitoring, new plant pests and weeds are still being introduced into the United States. In the article "Biological Control" by Dr. Lloyd A. Andres in the *1986 Yearbook of Agriculture—'Research for Tomorrow'* it is estimated that eleven new immigrant species make the United States their home each year and that about seven of these will become "pests of some importance." Bringing in the natural enemies of these new pests will be important in controlling them. Nature normally has a system of balances where each creature, large and small, has some natural enemy to help control the population. It is easy to understand how a new insect, weed, or disease, introduced in an area where there are no natural enemies, would multiply rapidly.

CHEMICAL-FREE CONTROL

Insects are controlled to a great extent by natural forces. Natural enemies of pests, such as birds, insects like the ladybug and lacewing, and diseases, help to keep the pest population under control. For example, tiny wasps lay eggs on the tomato hornworm, figure 16-5; the eggs hatch into worms that eat the tomato worm from the inside out. Other parasitic wasps kill insects such as the alfalfa weevil, figure 16-6, and the Mexican bean beetle, figure 16-7. If natural enemies are not present in a certain area, they may be introduced to provide the necessary control. Figure 16-8

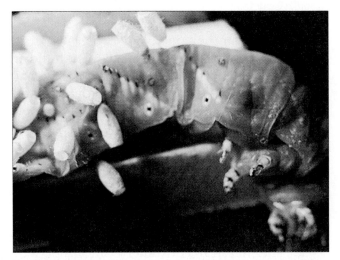

FIGURE 16-5 A tomato hornworm with tiny, white eggs of the Braconid wasp attached to its back. The eggs will hatch and the tiny larvae will enter the tomato worm and kill it. (Courtesy USDA)

FIGURE 16-6 The tiny stingless wasp, a parasite of adult and larvae stage alfalfa weevils, depositing an egg in a weevil larvae. When the egg hatches, the wasp larvae eats its way out, killing the weevil. (Courtesy USDA)

FIGURE 16-7   A tiny stingless wasp (Pediobius faviolatus) lays eggs inside a Mexican bean beetle larva. (Courtesy Dr. Gary Cunningham, Animal and Plant Health Inspection Service, Hyattsville, Maryland)

FIGURE 16-8   A digger wasp parasite of the Japanese beetle in its larval form feeding on the larva of a beetle. The U.S. Department of Agriculture introduced the parasite from Japan to the United States, where it prevents the Japanese beetle from becoming a serious economic threat. (Courtesy NYS Experimental Station)

shows a small, wormlike larva of a wasp which was introduced to the United States to aid in the control of the Japanese beetle.

Commercial preparations containing pest-control organisms that attack insects can also be used for chemical-free control. Naturally occurring disease organisms which attack insects, such as fungi, viruses, and bacteria, are grown on host insects in the laboratory and then prepared for use as pesticides. In this way, the natural diseases of the insects are multiplied to such a great extent that the insect population is greatly decreased or eliminated.

Before chemical pesticides are used the insect or disease build-up is carefully monitored by university extension specialists and growers. If a problem is found to exist, natural enemies may be used, such as parasitic wasps for bean beetle, figure 16-7, or Thuricide (*Bacillus thuringiensis*), which is a bacterium that kills many worms or caterpillars. The bacillus kills by causing high gut pH, thus upsetting the digestive tract and causing fatal starvation. Such caterpillars as the spruce bud worm, cabbage looper, imported cabbage worm, and gypsy moth are killed.

Another example of new experimental biological control is the use of the insects' own uric acid as poison. There is also a control for Verticillium wilt: a fungus *Talaromyces flavius*, which fights other harmful fungi such as Verticillium. If established naturally, these new controls could save millions of dollars each year in crop damage and chemical pesticide cost. These biological controls are not harmful to other living things and do not harm the environment.

BIOLOGICAL CONTROL SYSTEMS

As mentioned, biological control or natural enemies of plant pests should be considered first in any pest-control program. After being mass produced, these biological organisms are added to the soil or sprayed on plants. In most cases, these controls are very effective and do not poison the environment.

The first thing that must be understood if biological control is to be used effectively is that we cannot kill every pest in one or two days. This is not a problem if the pest does not enter the part of a plant to be eaten by humans or marketed. In other words, a potato beetle can eat some plant leaves without damaging the crop, which grows underground. If large numbers of beetles eat lots of leaves and the crop is reduced, this *is* a problem. Biological control may keep the number of beetles down to where crop production is not reduced. This is successful control, even if a few beetles do survive.

As discussed, biological control organisms are living things, such as viruses, bacteria, fungi, or insects. They may also be naturally occurring chemicals which repel insects, sterilize them so eggs do not hatch, or stop them from maturing to adulthood. Most of these living organisms or chemicals are very specific; in other words, they affect only one insect or disease.

SPECIFIC EXAMPLES OF BIOLOGICAL CONTROL

Biological control agents work in a number of ways.

1. The plant itself is resistant to the disease or insect.
2. Chemicals in the plant are toxic or repellant, or there are sticky hairs on the leaves and stems that repel insects.
3. Biological control agents produce chemicals that injure or kill the plant pest.
4. Parasitizing—they grow on or in the pest and kill it.
5. They eat the pest or suck out its body fluids.
6. They compete for food with plant pests in the soil, or eliminate the food supply.

DISEASE- AND INSECT-RESISTANT PLANTS

Development of insect- and disease-resistant varieties of plants is a most important step in biological control, and there are many such plants available which should be selected and used, figure 16-9. Some of these varieties are so resistant to pests that no chemical sprays are needed. For example, the tomato variety Better Boy is listed as VFN resistant. The *V* stands for Verticillium wilt, the *F* for Fusarium wilt, and the *N* for nematodes. Therefore, this tomato variety needs no help in controlling all three of these pests.

Insect pests never bother the neem tree or the Ginkgo tree. We know that the neem tree has a chemical which deters or kills insects. But there are other reasons insects do not attack certain varieties of plants while they seriously damage other varieties of the same type of plant. For example:

- *Rhododendron 'Mist Maiden'* never has lace bug damage because of a thick layer of hairs (indumentum) on the underside of its leaves where the lace bug feeds.
- *Tomato variety 'Brimmer'* deters the tomato fruit worm because of its thick skin.
- *Bush lime 'Henderson'* deters all small insects because of leaf hair.

ENDOPHYTES

Microorganisms called endophytes are fungus and bacterium which live in other plants without harming them. The endophytes protect plants against insect attack. Some lawn grasses naturally infected with fungal endophytes have excellent resistance to lawn insects. Researchers are now inserting genes into the endophytes from other bacteria such as *Bacillus thuringiensis* (Bt) to provide specific insect control in other plants. One example is control of the corn borer in corn. The treated endophytes are forced under pressure into the corn seed through tiny naturally occurring cracks in the corn seed. When the plant grows the endophyte grows with it and controls the corn borer without harming the plant.

NATURAL PESTICIDES

For years we have used rotenone and pyrethrums as natural insecticides. Some of the newest and most effective insecticides are the synthetic pyrethroids made by copying the chemical found in the chrysanthemum. In addition, six plants show real promise as natural pesticides. They are:

- **Basil**—an effective repellant and larvacide for mites, aphids, and mosquitos.
- **Big Sagebrush**—Extract of branch ends is very effective in preventing the Colorado potato beetle from feeding.
- **Chilcuan**—Root extracts are effective against many agricultural pests.
- **Mamey**—flowers, fruit, and leaves are effective against a large number of insect pests.
- **Calamuse**—an insect repellant especially effective on stored grain insects.
- **The Neem Tree**—by far the most promising of these six. Extracts of the seed are outstanding repellants to insects.

THE NEEM TREE   The chemical azadirachtin is a natural pesticide taken from the neem tree. The chemical is most concentrated in the seeds of the neem tree. Azadirachtin acts very effectively as an insect repellant and as a growth regulator, interrupting the growth cycle of insects. It does not kill adult insects, but prevents them from eating. It does kill the larva stage as it stops the growth cycle of the insect.

The neem extract is not toxic to warm-blooded animals and is easy on beneficial insects since they do not eat the plant and swallow the azadirachtin. It gives effective control of more than 170 different plant insects. It is a systemic chemical; it gets

VEGETABLE	PLANT VARIETIES	PLANT PEST RESISTED
Asparagus	Jersey Giant, Jersey Knight	Asparagus rust, Fusarium wilt, crown and root rot
Bean, bush or snap	Tendercrop	Bean mosaic virus and pod mottle virus
	Top Crop	Bean mosaic virus and pod mottle virus
Bean, pole	Kentucky Wonder	Rust
Bean, lima	Eastland	Downy mildew
Broccoli	Emperor Hybrid	Black rot, Downy mildew, Hollow stem
	Green Dwarf	Downy mildew
Corn	Burpee's Honeycross, Ambrosia	Wilt
Cucumber	Saladin Hybrid	Cucumber mosaic virus, Powdery mildew, Bacterial wilt
	Sweet Success	Cucumber mosaic virus, Scab, Target leaf spot
	Amira Hybrid	Cucumber mosaic virus, Downy mildew, Powdery mildew
Melon	Bush Charlston Gray	Fusarium wilt, Anthracnose
	Ediato Muskmelon	Downy mildew, Powdery mildew
	Dixie Queen	Wilt
	Dixie Queen	Wilt
	Sweet Favorite, Crimson Sweet	Anthracnose, Fusarium wilt
Pea	Sugar Bon snap pea	Powdery mildew
	Green Arrow garden pea	Downy mildew, Fusarium wilt
Pepper	Golden Summer hybrid	Tobacco mosaic virus
	Bell Boy	Tobacco mosaic virus
Potato	Kennebec	Late blight
Tomato	Super Beefsteak	VFN*
	Roma	VF*
	Better Boy	VFN*
	Park's OG 50 Whopper Improved	VFFNT*
	Better Bush	VFN*
	Celebrity	VFFNT*
	Big Pick	VFFNTA*

TREES

Chinese Chestnut American Chestnut	Dunstan Hybrids	Chestnut blight Chestnut blight
Ginkgo		All
Apple	Liberty	Scab, Cedar rust, Fire blight, mildew
	Priscilla	Scab, Fire blight, Cedar apple rust
	Prima	Scab, Fire blight, mildew

*V   Verticillium wilt
 F   Fusarium wilt
 N   Nematodes
 T   Tobacco mosaic
FF   Race 1 and 2 Fusarium wilt
 A   Crown wilt

FIGURE 16-9   Some plants that resist insects and diseases

into the plant sap and offers longer protection since rain cannot wash it off.

Azadirachtin is effective as an insect feeding deterrent at concentrations as low as one tenth of a part per million (ppm) parts of water. It also breaks down rapidly in the environment. Recent research shows that Azadirachtin is also effective as a fungicide.

The naturally occurring chemical from the neem tree is now commercially available under trade names such as Bioneem, Azatin, and Margosan-O. It shows great promise as a naturally occurring, safe insecticide for the future.

PEST LIFE CYCLE

The life cycle of a pest to be controlled must be known for best control. This is true for chemical control as well. There are times in the life cycle of a pest when it is easier to control and times when control is not possible. Also, a pest that gets inside a plant part, such as a fruit or vegetable used for consumption, should be controlled in the egg or first hatch stage, if possible, before the plant is damaged, or the adult must be prevented from laying eggs in the first place. An excellent nonpolluting control for eggs is spray oil. There are now new, highly refined oils that can be used all year round instead of just when plants are dormant, as with the older types of oils. Many insects, if hit by oil spray during egg or early hatch stage, are effectively controlled. The oils are not toxic and do not harm the environment.

CROP ROTATION

Crop rotation is a method of controlling some types of insects that do not move rapidly from place to place. Crop rotation consists of planting a different crop in a field from year to year, rather than growing the same crop on the same land for several years. Many soil-borne diseases, such as tomato wilt, can live in the soil for only two or three years if the plants on which they feed (host plants) are not present. Thus, if crops are rotated so that the host plant or related plants are not grown in a field more than once in three years, the disease will die out. In the case of tomatoes, this means that tomatoes, eggplant, or potatoes (plants which also support tomato wilt) should be planted in a field only once every three or four years. A typical rotation might be: tomatoes the first year, beans the second year, corn the third year, and then tomatoes again the fourth year.

Rotations also help in the control of many insects which do not travel for long distances, such as weevils and nematodes.

HAND PICKING

Where only a few insects or diseased plants are present, it is effective to pick them off by hand and destroy them.

BIOLOGICAL CONTROL OF INSECTS

Figure 16-10 shows specific types of biological control agents and the various plant pests that these agents control.

BIOLOGICAL CONTROL OF PLANT DISEASES

Ninety-nine percent of soil organisms are beneficial; less than 1 percent cause diseases. Yet, soil fungi and bacteria cause many plant root diseases. Seed decay, blights, root rots, wilts, and damping-off of seedlings are some of them. These diseases cost an estimated $4 billion each year in the United States. The good soil microbes are now being genetically improved to live longer in the soil and to be resistant to chemicals used to fight diseases. If good soil microbes are resistant to fungicides or other chemicals used to fight disease, they will not be destroyed when these materials are used and thus will remain to aid in the fight against diseases. Figure 16-11 lists specific types of biological agents and the various plant diseases that these agents control.

NEW STRAINS OF BIOCONTROL MICROBES FOR PLANT DISEASES

By using chemicals and ultraviolet light, scientists at the plant research labs in Beltsville, Maryland, have been able, over the past ten years, to produce improved biological agents to fight soil-borne diseases. Some of the microbes were not killed by benomyl, a commonly used fungicide. Some of the new strains were stronger in their action against disease and some had longer shelf life in the laboratory and longer life in the soil. The fungi *Trichoderma spp.* and *Talaromyces flavius* attack soil-borne plant diseases. The above treatment has made them more effective control agents.

BIOLOGICAL CONTROL AGENT	PLANT PESTS CONTROLLED	ADDITIONAL INFORMATION
Parasitic wasp, Trichogramma wasp	Colorado potato beetle	The wasp lays its eggs on the beetles' eggs. Use of this wasp reduced need of chemical spray from 16 to 1 spray of Vydate in the early season. Eight thousand wasps per acre give good control.
Parasitic wasp, Trichogramma wasp	Tomato hornworm	The wasp lays its eggs on the hornworm. Wasp larvae enter the worm.
Parasitic wasp, Trichogramma wasp	Cabbage looper and other worms	The wasp lays its eggs on the eggs of moths and butterflies.
Bacillus thuringiensis (can be bought under the names Dipel, Biotrol, and Thuricide), Javelin	Colorado potato beetle; all caterpillars, including gypsy moth; cabbage looper and imported cabbage worm	A bacterium which is extremely effective on caterpillars, upsetting their digestive tract so they stop eating in about 2 hours and die in 2 days.
Predatory mite	Spider mite and whitefly	Some of these mites are resistant to chemicals.
Beauveria bassiana	Colorado potato beetle	This fungus gives 77% control by attacking the pupal stage in the soil, and also adults. Small fungus spores only need to touch the beetle to cause infection.
Orange peel	Many insects controlled, such as ants, fleas	Still experimental, but looks promising.
Ladybug	Aphids	Eats aphids at all stages of growth.
Nedalia beetle, Parasitic fly	Cottony cushion scale on citrus	Imported in 1888 to California and gave complete control of the scale insect.
Green lacewing	Aphids	Both nymphs and adults eat aphids.
Green lacewing biologicals.	Mealybugs	Operates in cooler temperatures than many
Bacillus popilliae	Japanese beetle	Known as milky spore disease; kills grubs in the soil.
Digger wasp	Japanese beetle	A tiny parasite kills the grubs in the soil. See figure 16-4.
Pediobius faviolatus	Mexican bean beetle	Eggs are laid on the bean beetle larvae. Wasp larvae destroy the bean beetle larvae.
Marigold and asparagus roots	Nematodes	Planting marigolds in an area greatly reduces the number of nematodes in the soil. Asparagus roots produce a chemical which causes a decline in the number of stubby root nematodes.
Eats over 100 kinds of harmful insects.	Spined Soldier bug *Podisus* spp.	One of the most effective insect predators.
Diatomaceous earth	Slugs, snails, mites, aphids	Finely ground sharp silica material kills soft-bodied insects, slugs, and snails.
Predatory nematode	Grubs of weevils, cutworms, fungus gnats, cabbage root weevil, wireworms	Controls many insects which spend some of their life cycle in the soil.

FIGURE 16-10   Biological control of insects

BIOLOGICAL CONTROL AGENT	PLANT DISEASE CONTROLLED	ADDITIONAL INFORMATION
Tilleteopars	Powdery mildew	A yeast which effectively controls powdery mildew in the greenhouse.
Talaromyces flavius	Potato wilt	A fungus is adapting well to the soils where potatoes are grown and giving reduction of the disease.
	Verticillium wilt	Reduced wilt on eggplant by 75% when added to pots containing small plants for later field growing.
Crop rotation and raised beds	*Phytopthora*, root rot, many wilts	Many soil-borne diseases cannot live in the soil for more than 2 or 3 years if a host plant is not available.
Plant disease resistance	Wilts, bacterial diseases, rust diseases	As new disease-resistant plants are produced, less and less spraying with chemicals will be needed. The Ginkgo tree is immune to all insects and diseases. There are hundreds of disease-resistant fruits, vegetables, grasses, and flowers, and they should be used whenever possible. Many more new ones are on the way as disease resistance is found in wild plants and bred into commercial varieties.

FIGURE 16-11 Biological control of plant diseases

In addition, a new strain of mite (*Melaseiulus occidentalis*) is resistant to several pesticides. This beneficial mite is used to control spider mites feeding on plants.

Newer methods are being developed to mass produce these good microbes so they are available in large quantities and at reasonable prices.

BIOLOGICAL CONTROL AGENT	WEED CONTROLLED	ADDITIONAL INFORMATION
Chrysolina quadrigemina	St. Johnswort weed	St. Johnswort was a serious weed on much grazing land in California. This shiny blue beetle is on the way to controlling the weed. Feeds only on this weed and its close relatives.
Fungus rust *Puccinia chondrillina*	skeleton weed	A rust disease destroys the seed pod of the plant, preventing it from reproducing, and stunts its growth. This is still experimental.

FIGURE 16-12 Biological weed control

BIOLOGICAL CONTROL OF WEEDS

Plant diseases which attack only a specific weed are being developed. An example is a type of rust that affects only the skeleton weed. Another fungus gives good control of the weed jointvetch.

Insects also can be used in the same way. An imported fly has been released in Maryland recently to control a noxious weed, the musk thistle. Figure 16-12 contains just two examples of what has been done.

ALLELOPATHY

Some plants secrete chemicals that inhibit root growth or seed germination of other plants. For example, sweet potatoes reduce growth of the weed yellow nutsedge. As little as ten parts per million of sweet potato root extract inhibits nutsedge, velvetleaf, and pigweed seed germination.

WHY USE BIOLOGICAL PEST CONTROL?

The following are some reasons:

1. In many cases, the dollar cost is much less in the long run.

2. These controls do not poison the environment. Most poisonous chemicals from plants are more easily broken down and made harmless in the environment.

3. Insects are becoming resistant to many chemical pesticides.

4. Many man-made chemicals are suspected of or proven to cause cancer or pose other health hazards to man and other mammals.

5. Some chemical pesticides are getting into our drinking water.

6. The Environmental Protection Agency is restricting the use of many chemical pesticides, so fewer chemicals will be available.

Thirty-one pesticides have been determined or preliminarily determined by the EPA as having some tumor-causing effects on animals. This does not mean that all these chemicals will be proven dangerous to humans when used for pest control. It does mean that they are suspect and will be monitored and tested more. It does make sense to use biological agents known not to be harmful to control plant pests whenever possible.

There is much yet to be done before we can put away all the chemicals, but progress is being made to reduce the need for them. Through use of insect- and disease-resistant plants and other forms of biological control, fewer chemicals may be needed in the future.

Microorganisms are also being used to clean up chemical spills in the soil and water by literally eating and breaking down the chemicals into harmless materials. One such organism was developed which could completely break down large vats of pesticide in forty-eight hours, turning the chemical into harmless materials.

Through genetic engineering (changing the heredity of an organism), selection, and understanding of habit, there are limitless possibilities for the future in biological control.

NEW BIOLOGICAL CONTROL METHODS UNDER STUDY

◆ *ClandoSan*—a by-product of crab and shellfish industries which controls nematodes. It works by stimulating the growth of the nematodes' natural enemies which then destroy the nematodes.

◆ *Vectobac*—a biological insecticide for fungus gnats applied as a soil drench.

Viable endospores and crystals kill in twenty-four to forty-eight hours.

◆ *Potato with a built-in insect repellant*—The new hybrid potato produces leptine, a chemical which repels the Colorado potato beetle.

◆ *Corn Vaccination*—A single gene from the bacterium *Bacillus thuringiensis* is forced into the corn seed and becomes a part of the seed passed on to the plant. The gene produces a toxin that is deadly to the corn borer.

◆ *Japanese Beetle Control*—nematodes Heterohabditis HP88 achieved 90 percent control of beetle larvae. This is as good or better than any chemical control. The nematodes over-winter.

INSECT GROWTH REGULATORS

Insect growth regulators are the newest class of biological insecticides. They work against the insects' own internal chemical system and change the insects' normal growth cycle.

Young immature insects go through a series of molts during which they shed their outer skin or exoskeleton (outside skeleton). As they grow larger they must shed the old skin and grow a new larger one. One type of growth regulator called *chitin inhibitors* prevents the proper formation of the exoskeleton or the shedding of the old skin. Another form of growth regulator called *Hormone analogs* prevents the insect from maturing into a normal adult.

Insect growth regulators have advantages and disadvantages.

Some advantages are:

1. IGRs are effective at very low concentrations, as low as 0.001 percent.

2. IGRs provide long-term control.

3. IGRs kill pesticide-resistant insects.

Some disadvantages are:

1. IGRs take longer to work than chemical pesticides.

2. IGRs are unstable in the environment and quickly break down, becoming ineffective.

3. IGRs must be carefully applied to be effective.

Student Activities

1) Write your state agricultural college for copies of research on biological control being done in your area.

2) While on a walking field trip, see how many beneficial (biological control) insects you can find.

Self-Evaluation

A) Select the best answer from the choices offered to complete the statement.

1) Biological control of plant pest is   (Select the best two answers.)

    a) the use of man-made chemicals to control pests.
    b) using living natural organisms or material for control of plant pests.
    c) the newest control method.
    d) the oldest pest-control method.

2) A good example of where biological control has worked with at least some success is

    a) the Japanese beetle.
    b) the Mexican bean beetle and the Colorado potato beetle.
    c) the gypsy moth.
    d) all of the above

3) Some plants produce natural pesticides. One of the most successful new plants producing a natural insecticide is

    a) the neem tree.   c) alfalfa.
    b) the onion plant.   d) none of the above

4) Crop rotation can help control some soil diseases and some insects. To be effective, rotations should be

    a) three or four years long.   c) at least five years long.
    b) one or two years long.   d) five to ten years long.

5) *Bacillus thuringiensis* is a biological control agent which will control

    a) aphids.   c) most caterpillars.
    b) the Mexican bean beetle.   d) Verticillium wilt.

6) Allelopathy is a natural phenomenon where

    a) bright light produces disease symptoms in plants.
    b) one plant secretes chemicals that inhibit root growth or seed germination of other seeds.
    c) two plants grow better when planted close to one another.
    d) plants grow toward sunlight.

7) Small microorganisms called endophytes protect plants from insects by

    a) growing inside the plant and protecting it from insects.
    b) forming a protective layer on the plant's surface to protect it from insects.
    c) filling the air with spores which infect insects.
    d) all of the above

8) The newest class of insecticides are insect growth regulators. They control insects by
    a) stomach poisoning.
    b) changing their normal growth cycle.
    c) killing their eggs.
    d) all of the above

B) Briefly answer each of the following.

  1) Why is it important to know the life cycle of a pest in order to effectively control it?

  2) List crops that might be grown in a three- or four-year rotation for soil disease control.

  3) Why is it so important to be looking to natural pest control at this time?

The Safe Use of Pesticides

OBJECTIVE

To use pesticides safely and competently.

Competencies to Be Developed

After studying this unit, you should be able to

- identify the three main routes by which pesticides enter the body.
- examine five pesticide labels and
  - identify the type of each and its degree of toxicity.
  - demonstrate the recommended precautions in the mixing and handling of each.
- list first aid steps to be taken in case of poisoning by one pesticide from each of the three families of pesticides.

MATERIALS LIST

- ✔ pesticide labels or containers with labels
- ✔ sprayer and duster for hand application of pesticide
- ✔ all articles of protective clothing specified on pesticide labels, including rubber gloves, respirator, rubber boots, waterproof coveralls, goggles, and rubber hat
- ✔ local pest control schedules and calendars from a local university or extension service
- ✔ pesticides for application

THE WORKER PROTECTION STANDARD (WPS)

Concern for worker protection has led to the passage of a new Worker Protection Standard (WPS). This new law presents a new set of rules to reduce pesticide-related illnesses for all who come in contact with agricultural chemicals. The WPS does not regulate or ban any chemical. It sets guidelines to follow during and after application, such as how soon workers can re-enter a treated area, warning signs put up to point out treated areas, information so applicators are aware of the toxic effect of chemicals and what protective clothing to wear. Facilities to wash up and change clothes after chemical use must also be provided within one-quarter mile of the area being treated with chemicals.

All pesticide labels must be changed to meet the new standard. Dealers will be allowed to sell pesticides with old labels until the new law takes effect. Exact dates for compliance can be found by contacting the EPA or your state pesticide regulation agency.

The WPS does not cover homeowner use of pesticides.

TYPES OF PESTICIDES

A *pesticide* is any material used to control *pests* (plants or animals that are harmful to human beings or to the crops that they cultivate). These materials may be natural or man-made. Biological controls may be living organisms. There are seven types of pesticides that concern horticulturists, each classified according to the type of pest it destroys.

- *Insecticides* are used to control insects. The insects are killed by body contact with the chemical or by swallowing the insecticide.
- *Miticides* are used to control *mites* (tiny spiderlike animals) and ticks. Mites and ticks are usually killed by coming in contact with the chemical or being attacked by a biological organism.
- *Fungicides* are used to control fungus disease. To be effective, they must come in contact with the fungus. Fungicides are usually used to prevent a plant from becoming diseased and are applied before the disease is present.
- *Herbicides* or *weed killers* are used to kill unwanted plants. *Nonselective* herbicides kill all plants, while *selective* herbicides kill only certain plants.
- *Rodenticides* kill rodents, such as rats and mice. These are chemicals that are usually applied as bait; the rodents are poisoned by eating the chemical.
- *Nematocides* kill nematodes (tiny hairlike worms that feed on the roots of plants). Nematocides are usually applied in the form of *fumigants*, substances which produce a smoke, vapor, or gas when applied, or as a biological agent.
- *Molluscicides* are used to kill slugs and snails (types of *mollusks*). These are chemicals that are usually applied as bait which attracts the slugs and snails and poisons them.

SELECTING THE PROPER PESTICIDE

The first step in selecting a pesticide is to identify the pest to be controlled and determine if control is needed. If the pest is doing very little damage, or if the crop is ready for harvest, control may not be needed.

If control is needed, methods which do not employ chemicals should first be considered. It is possible that the spread of disease can be halted simply by proper *sanitation*, removing and destroying any diseased plant material and disinfecting all tools and work areas. (This procedure should be followed while making cuttings even if there is no sign of disease.)

If the pest to be controlled is an insect, insects which eat other insects (such as the ladybug) may be effective and less expensive than chemicals. Such insects may be purchased and released on the infested plants. It may be unnecessary to introduce the insect to an area; natural parasites (such as the tiny wasp that kills tomato worms) may be present in large enough numbers to minimize attack by the pest. In the same way, mulches, rather than herbicides, may be used for weed control to minimize the use of harsh chemicals.

It is important to select the best pesticide for a particular problem. If correctly chosen, the pesticide will meet the following requirements.

- It kills or controls the pest.
- It does not injure the plant on which it is used.
- Of all the materials which could be used for the problem at hand, it is the least harmful to the environment.
- It is suitable for use with available equipment.
- The label recommends its use for the plant and pest for which it has been chosen.

TOXICITY

Because many pesticides can be poisonous to humans and the environment, the United States has established standards for their handling and use through the Environmental Protection Agency (EPA). Other countries have similar regulations. Because of their poisonous nature, pesticides must be used with a degree of caution, depending on how poisonous, or *toxic*, they are.

To poison or injure humans and other animals, pesticides must be present in or on the victim's body. There are three main routes by which poisons enter the body:

- ◆ oral contact (by swallowing)
- ◆ dermal contact (by contact with the skin)
- ◆ inhalation (by breathing)

When dealing with the danger of children coming in contact with poisons, the major concern is preventing the child from taking the poison orally. On the other hand, the individual who applies pesticides is more likely to be poisoned by contact dermally or through inhalation. Chemicals such as organophosphates are absorbed rapidly through the skin. Any chemical that vaporizes, has a strong odor, or is a fine dust or mist can be inhaled and absorbed through the lungs. Pesticides pass very rapidly through the skin on the back of the hands, wrists, armpits, back of the neck, groin, and feet. Cuts or scrapes also allow pesticides to enter the body more easily.

> *Caution: Never allow a pesticide to come in contact with the skin. If this does occur, wash the affected area immediately with soap and water.*

TYPES OF TOXICITY

Acute toxicity is a measure of how poisonous a pesticide is after a single exposure. Pesticides are generally rated according to acute toxicity. *Chronic toxicity* is a measure of how poisonous a pesticide is over a period of time and after repeated exposure. Chronic toxicity is a danger of chemicals that accumulate in the body, such as chlorinated hydrocarbons.

MEASURING ACUTE TOXICITY

ORAL AND DERMAL TOXICITY The method used to measure acute oral and dermal toxicity is LD_{50}. The *LD* stands for *lethal dose*

(amount necessary to cause death). The 50 means that 50 percent of test animals (generally rats and/or rabbits) are killed by this dose. The lower the LD_{50} number of a pesticide is, the more poisonous it is. LD_{50} values are given in milligrams of substance per kilogram of test animal body weight. This is a metric measurement which is the same as parts per million.

Figure 17-1 gives LD_{50} values for some important pesticides. The pesticides in the chart are listed with the most poisonous first, as shown by the smaller LD_{50} numbers of these substances. Please note that this figure contains chemicals that are too toxic for use by students or unlicensed applicators. They are listed only to point out how toxic they are.

INHALATION TOXICITY Acute inhalation toxicity is measured by LC_{50} values. *LC* stands for lethal (deadly) concentration. LC_{50} values are measured in milligrams per liter. A *liter* is a volume measurement (in the metric system of measurement) equal to about 1 quart. Again, the lower the LC_{50} number, the more poisonous the pesticide is.

There is no standard measure for chronic toxicity.

> *Caution: Before purchasing or using a pesticide, read and understand the label. Watch for signal words indicating toxicity of the chemical. Be prepared to take any precautions listed. If the pesticide is highly toxic, try to purchase a material that is as effective yet less toxic.*

STANDARD LABEL INFORMATION

Labels on pesticide containers always include mixing instructions. These must be followed carefully. Other standard points of information which appear on pesticide containers include:

1. name and address of manufacturer
2. trade name (may or may not be the same as the chemical name)
3. active ingredients, including the official common name of each ingredient; when an accepted common name is not available, the chemical name appears
4. type of pesticide such as insecticide or fungicide
5. form of substance such as dust, wettable powder, or emulsion
6. EPA registration number
7. storage and disposal precautions
8. hazard statement (read carefully)

PESTICIDE NAME*	ACUTE TOXICITY		TYPE OF PESTICIDE OR USE
	ORAL (by mouth) (rats)	DERMAL (on skin) (rabbits)	
✓ aldicarb (Temik)	0.9	5	insecticide
✓ demeton (Systox)	2	8	insecticide
✓ parathion	3	6.5	insecticide
✓ carbophenothion (Trithion)	6	22	insecticide
✓ azinphos-methyl (Guthion)	16	220	insecticide
zinc phosphide	45	—	mouse poison
nicotine	50	50	insecticide
warfarin	1 to 186	—	mouse and rat poison
bifenthrin (Capture)	50	> 2,000	insecticide
rotenone	60	1,000	insecticide
lindane	88	500	insecticide
PCP (penta)	100	100	wood preservative
dursban	135	200	insecticide
caffein	200	—	beverage
metaldehyde (Deadline)	250	—	snails and slugs
diazinon	300	379	insecticide
carbaryl (Sevin)	307	2,000	insecticide
2,4-D	375	1,500	herbicide
nabam (Dithane)	394	—	fungicide
dicofol (Kelthane)	575	4,000	miticide
metalaxyl (Subdue)	669	3,100	systemic fungicide
aspirin	750	—	pain killer
acephate (Orthene)	866	2,000	systemic insecticide
malathion (Cythion)	885	4,100	insecticide
EPTC	1,367	10,000	herbicide
atrazine (Aatrex)	1,869	>3,100	herbicide
resmethrin	2,000	2,500	insecticide
dichlobenil (Casaron)	3,160	1,350	herbicide
table salt	3,320	—	mineral food item
cyromazine (Larvadex)	3,387	3,387	insecticide
trifluralin (Treflan)	3,700	3,000	herbicide
ethyl alcohol	4,500	—	beverage
borax	4,980	—	cleaning compound insecticide
azadirachtin	5,000	—	insecticide from neem tree
simazine (Princep)	5,000	>3,100	herbicide
glyphosphate (Roundup)	5,600	5,600	herbicide
oxyfluorfen (Goal)	5,600	>10,000	herbicide
fosetyl (Aliette)	5,800	3,200	systemic fungicide
daminozide (Alar, B-Nine)	8,400	1,600	growth retardant
captan	9,000	—	fungicide
ferbam (Fermate)	17,000	—	fungicide
kerosene	28,000	—	fuel

*   Pesticides listed in order of oral acute toxicity.

— Where no dermal figure is given, the chemical is not absorbed through the skin in dangerous amounts.

✓   Use restricted to licensed or certified applicator.

FIGURE 17-1 LD_{50} (lethal dose) of common farm chemicals measured in milligrams of substance per kilogram of body weight. (Measurement is same as parts per million.)

CATEGORIES	SIGNAL WORD REQUIRED ON LABEL	LETHAL DOSE			PROBABLE ORAL LETHAL DOSE FOR A PERSON WEIGHING 150 LBS.
		LD_{50} mg/kg Oral	LD_{50} mg/kg Dermal	LC_{50} mg/l Inhalation 24 hr. Exposure	
I. Highly Toxic	DANGER skull and crossbones Poison	0–50	0–200	0–2	a few drops to a teaspoonful
II. Moderately Toxic	WARNING	over 50 to 500	over 200 to 2,000	over 2 to 20	over 1 teaspoonful to 1 ounce
III. Slightly Toxic	CAUTION	over 500	over 2,000 to 20,000	—	over 1 ounce to 1 pint or 1 pound
IV. Relatively Nontoxic	CAUTION (or none)	over 5,000	over 20,000	—	over 1 pint or 1 pound

FIGURE 17-2 Categories of acute toxicity (From Pesticide Applicator Training Manual, courtesy Northeastern Regional Pesticide Coordinators)

9. directions for use (read carefully and follow)
10. net contents
11. worker protection procedures under the new WPS regulations

SIGNAL WORDS

To alert the user to the toxicity of a pesticide, certain *signal words* appear on the container label. There are four categories of acute toxicity and signal words, figure 17-2.

It is essential to read labels to determine the toxicity of pesticides and the necessary precautions in their handling. The label should be examined before the product is purchased. In this way, it can be determined if the chemical is effective against the particular pest which is causing the problem, and if perhaps it is too toxic a substance with which to work.

Caution: These labels are approved by the EPA at the time of publication of this book. Pesticide product label changes are effected frequently so the labels in this text should not be used for field application instructions. Whenever a pesticide is purchased or used, follow strictly the recommendations and precautionary information as stated on the product package label. The labels published in this book may be changed at any time and the manufacturer will not be held responsible for the accuracy of the recommendations and precautions as stated here.

Examine the partial sample label in figure 17-3. The markings on the label of this pesticide indicate it is in the highly toxic category; therefore, it should

not be used around the school or home. Compare the information from this label with that found on a relatively nontoxic pesticide, figure 17-4.

Caution: When considering the use of any pesticide, be sure that the intended use appears on the label. It is violation of federal law to use any pesticide in a manner which is inconsistent with the labeling. A use that is **inconsistent** with a label includes use on any pest or on any site that is not specifically given on that label.

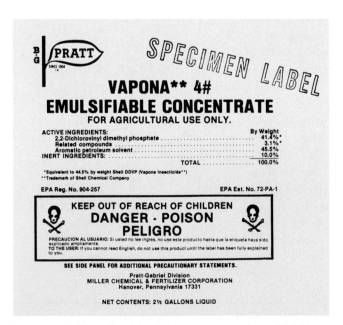

FIGURE 17-3 Partial label from a container of highly toxic pesticide. Note the signal word (DANGER) and the skull and crossbones, indicating the extremely high toxicity of this chemical. (Courtesy Miller Chemical & Fertilizer Corp., Richard Kreh, Photographer)

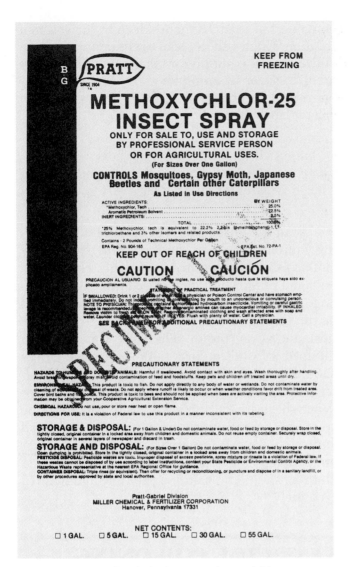

FIGURE 17-4 A relatively nontoxic pesticide. (Courtesy Miller Chemical & Fertilizer Corp., Richard Kreh, Photographer)

FIGURE 17-5 The chemical being applied here is relatively nontoxic. Note the lack of protective clothing.

FIGURE 17-6 To apply a highly toxic chemical requires complete protection. The applicator is wearing a respirator, protective gloves, waterproof clothing, goggles, and boots. This prevents the solution from coming into contact with any part of the applicator's body.

> *Caution: The proper protective clothing can be determined only by reading the label on the pesticide container.*

SAFETY PRECAUTIONS

Safety precautions must be considered before beginning to mix or apply a pesticide. Adequate safety precautions include:

Reading the label carefully. Prepare to follow all directions and precautions exactly.

Checking the recommended uses to be sure it is the correct substance for the intended use.

Having clean water and detergents available to wash spills.

Wearing protective clothing such as rubber gloves, a respirator, and any other protective gear called for on the label. Some pesticides require little protective clothing, figure 17-5. Other highly toxic materials require complete coverage of the applicator's entire body with waterproof material, figure 17-6.

Use extra caution with concentrated chemicals. (Chemicals are considered concentrated prior to mixing or diluting with water or dust.) Protective covering is especially important when handling the concentrated chemical in mixing and filling the spray tank. Always wear protective gloves and goggles. If a concentrate is spilled or splashed on clothing, wash and change clothes immediately. Always protect the eyes with goggles when handling concentrates; the eyes absorb pesticides rapidly.

Any pesticide spilled on the ground or floor should be removed immediately.

Apply the chemical with care. When applying pesticides, be sure that the chemical is directed only at the target area and that as little as possible is allowed to spread anywhere else. Apply only the amount needed to do the job.

Always mix just enough for the job at hand. The label lists the proper amount needed for various crops and pests. The sprayer may need *calibration* (adjustment of the nozzle to a certain size) to ensure that the proper volume of chemical is applied. Applying at the proper rate is especially important when working with herbicides (weed killers).

Guard against inhalation or ingestion (swallowing) of the chemical; do not inhale its fumes, dust, or mist. Never eat or smoke while handling any pesticide.

Consider weather conditions. Pesticides should be applied when weather conditions are favorable. Be aware of forecasted weather conditions before application of the chemical. If used outdoors, application during heavy rain will wash the chemical away. Strong winds prevent applications of sprays and dust from hitting and adhering to the target crop. Some chemicals must be applied only within certain temperature ranges. Read the label and use good judgment. When applying pesticides, avoid contact with the mist or dust that drifts through the air. Do not spray where the pesticide can drift into streams, lakes, or apiaries (bee yards), or near sensitive plants or animals that might be injured.

Store and dispose of chemicals properly. Store all concentrated pesticides in the original container. Keep in a locked area, away from children and animals. Do not store near foods, animal feed, fertilizer, or in a building which houses people or animals. Sprayers and dusters should be treated in the same manner for storage.

Dispose of all diluted chemical that is in the spray tank. If any is left after spraying or dusting, try to use it on another area or crop where it will not contaminate or damage the environment, livestock, or people. However, the best way to eliminate problems associated with leftover chemicals is to mix only the amount needed for the job at hand. Empty containers must be disposed of out of the reach of all persons and animals, and in such a way that they will not contaminate the environment. Never reuse a pesticide container. Bury containers at least 18 inches deep in the soil in a location which will not lead to contamination of a stream or water supply, or take them to a sanitary landfill. Never set empty or partly empty containers out to be collected as trash unless they are clearly labeled.

> *Caution: Before disposing of any pesticides or empty containers, check local regulations, since the procedures for disposal may vary greatly.*

Clean up thoroughly. The sprayer, all measuring equipment, gloves, and any other protective clothing should be washed thoroughly. The sprayer tank and pump should be rinsed three times with large amounts of water. Some of the water should be run through the sprayer each time to flush the hose and spray pump. The rinse water is then dumped into the mixing tank. Protective gloves should be washed thoroughly with soap and water while still on the hands and hung up to dry. When the pesticide application is finished and the sprayer or other application equipment is cleaned and safely stored, the applicator should shower and change clothes immediately. Contaminated clothing should be handled carefully and washed separately.

PESTICIDE POISONING

If the warnings mentioned are not heeded, or if accidental spillage results in poisoning, it is important that any symptoms of poisoning be recognized quickly. In cases of spillage or other types of exposure to a toxic chemical, the most important consideration is to remove the contaminated clothing and wash yourself thoroughly with a good soap or detergent and plenty of water. A shower should be used rather than a tub bath because fresh, clean water is continually supplied.

> *Caution: Dilution of the poison is the most important first aid practice.*

FIRST AID FOR PESTICIDE POISONING

First aid is just what the term indicates—the first help given to a victim until medical professionals arrive. If poisoning has occurred, telephone a doctor immediately and read the pesticide label to him or her. Before assistance arrives, you can do the following.

- For any pesticide spilled on the skin, wash with plenty of soap and water to dilute the chemical.

- Remove contaminated clothing. Shower, dry, and wrap or dress the victim in warm blankets or clothing. Cover any chemical burns with loose, clean, soft cloth.

Caution: *Be careful that you do not come in contact with the pesticide while coming to someone's aid.*

- For eye poisoning, hold the eye open and flush with clean water for at least five minutes.

- For inhaled poisons, carry the victim immediately to fresh air. Do not allow the victim to walk. Loosen clothing and apply artificial respiration if breathing has stopped. Keep the patient quiet and warm. Do not give the victim alcohol.

The label on the pesticide container may give additional first aid information. Use any information available. After notifying a doctor, follow his or her advice. Provide expert medical attention as soon as possible. Do not panic and frighten the victim; instead, remain calm and reassure the victim that help is on the way.

Caution: Always choose the least toxic chemical that will do the job.

EPA RULES FOR THE USE OF CHEMICALS

The Environmental Protection Agency published guidelines for the handling of chemical pesticides. The agency established four toxicity categories and regulations stipulating how soon it is safe for a person to re-enter an area after it has been treated with chemical pesticides. The categories are:

- *Toxicity I chemicals*—areas treated with chemicals carrying a "Danger—Poison" signal word. Warning signs must be posted and no one may re-enter the treated area for forty-eight hours after application.

- *Toxicity II chemicals*—areas treated with chemicals carrying a "Warning" label. Restriction from the treated area is for twenty-four hours.

- *Toxicity III chemicals*—areas treated with chemicals carrying a "Caution" label. People may re-enter the area as soon as the dust or spray mist settles.

- *Toxicity IV chemicals*—areas treated with pesticides carrying no warning on the label may be re-entered immediately.

This is one more attempt at reducing the health risk of using pesticides.

SAFE PESTICIDES

A new approach in the development of pesticides is being used. The goal is to produce chemicals which are toxic to insects but not to warm-blooded organisms such as human beings. Most pesticides now in use affect the nervous system of insects. The problem is that they also affect any other organism with a nervous system, including humans.

Some progress has been made in the development and use of pesticides which affect only insects. A good example is chemicals which prevent the molting or shedding and hardening of the insects' skin. If insects cannot shed and then harden a new skin as they grow they die. Some new biologicals act as repellants or interrupt digestion of food.

Future research will zero in on more man-made and natural pesticides which kill only insects and if possible only the bad insects.

 # Student Activities

1) Obtain labels from containers of commonly used pesticides. Draw up a list of pesticides, including the following:

   a) name of pesticide

b) crops on which pesticide is commonly used

c) pests controlled by the pesticide

d) toxicity level of the pesticide and necessary precautionary measures, including protective clothing

2) Demonstrate the proper procedure for mixing and applying a pesticide.

3) Try on and properly adjust a respirator, rubber gloves, waterproof hat, and goggles.

 # Self-Evaluation

A) Briefly answer each of the following.

1) List the seven categories of pesticides and define each briefly.

2) What is the first step in selecting a pesticide?

3) List the four signal words found on pesticide labels. Also list the category of pesticide that each word signals.

4) What is a toxic chemical?

5) List the three routes by which pesticides can enter the body.

B) Complete each of the following statements with the proper word or words.

1) The toxicity of a pesticide is determined by reading the _____.

2) Acute toxicity is a measure of how poisonous a pesticide is after _____ exposure.

3) Chronic toxicity is a measure of how poisonous a pesticide is after _____ exposure.

4) The more poisonous a pesticide is, the _____ (higher, lower) the LD_{50} number is.

5) If a pesticide is spilled on the skin, it should be washed off immediately with _____ and _____.

6) If a pesticide is splashed in the eyes, they should be flushed with clean water for _____ minutes.

7) When handling concentrated pesticides, always protect the eyes with _____ and wear _____ on the hands.

8) Store all pesticides in a _____ storage area.

9) Leftover spray that has been mixed for use should be disposed of by _____ it on a crop that the pesticide will not damage.

10) Empty containers should be _____ where they will not contaminate the environment.

11) If a person is poisoned by pesticides, the most important first aid practice is to _____ the poison.

12) In cases of inhaled poisons, the first step in first aid is to carry the victim to _____ _____.

Insecticides

OBJECTIVE

To identify insect pests and select and apply the appropriate insecticides.

Competencies to Be Developed

After studying this unit, you should be able to

- identify common insect pests and select an effective control method for each.
- describe the six ways in which insects are killed by insecticides and the type of insect against which each is most effective.
- list the names and characteristics of the three major groups of insecticides (according to their chemical makeup).
- compare the six ways in which insecticides are applied.
- explain the relationship between the life cycle of insects and timing of insecticide application.

MATERIALS LIST

- ✔ illustrations of insect pests
- ✔ live insects
- ✔ recommendations for spraying from local extension service
- ✔ insecticides with labels (none should be highly toxic)
- ✔ spraying and dusting equipment
- ✔ safety gear, including elbow-length protective gloves and respirator

Insecticides, one of the seven types of pesticides mentioned in the previous unit, are used by most people at one time or another in the control of flies, mosquitoes, and other insect pests. They are used widely by horticulturists to protect plants from insect damage. Although insecticides can effectively and safely control insects, they can be dangerous and even deadly to human beings when used improperly. For this reason, it is extremely important that they be handled with care and only according to directions on the label.

This unit deals with chemical insecticides, not the biological insecticides discussed in Unit 16.

WHAT IS AN INSECT?

An *insect* is a small animal with three clearly defined body regions and three pairs of legs. The three body regions are the head, thorax, and abdomen, figure 18-1. The proper identification of insects is an important step in their control. Only after the insect is identified can the best method of control be selected. Guides to insect identification are available from the U.S. Department of Agriculture.

HOW INSECTICIDES KILL

To be considered good insecticides, chemicals must kill a specific type of insect while doing little or no damage to the plant. They must also be as safe for the handler and environment as possible.

STOMACH POISONS   These poisons work against insects that actually eat a part of the plant.

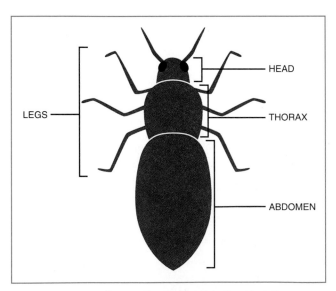

FIGURE 18-1   Parts of an insect. Notice that there are three definite body regions (head, thorax, and abdomen), with the legs attached to the thorax.

The chemical is sprayed or dusted on the plant; as the insect eats the plant, it is poisoned through the stomach. Insects that chew their food, such as caterpillars, grasshoppers, and beetles, are controlled by stomach poisons. Rotenone is considered to be an effective stomach poison.

CONTACT POISONS   These insecticides kill insects when they are hit by or come in contact with the poison; the insect does not have to consume it. Any type of insect can be controlled by contact poisons, including insects that suck plants, such as aphids and leaf hoppers. Contact poisons kill by upsetting the insect's nervous system or breathing system. An example of a contact insecticide is malathion.

New, highly refined oil sprays are excellent contact poisons against insect eggs and soft-bodied insects, such as aphids, mites, scale, and whiteflies. Oils do not contaminate the environment and so are safe to use.

SYSTEMIC POISONS   These are chemicals that enter the plant sap and move throughout the entire plant. When insects eat parts of plants or suck juice from plants, the chemical is swallowed. Systemic poisons are effective in the control of insects with either chewing or sucking mouth parts. They are especially effective in controlling insects that hide under leaves or underground and that therefore are not affected by contact sprays. If used on plants that are used as food, the chemical must be given time to break down within the plant and become nonpoisonous before the harvest date for the crop. These poisons are sprayed on the plant foliage or mixed with water and applied to the roots. Both methods result in the chemical spreading throughout the entire plant. One systemic insecticide is Orthene.

FUMIGANTS   Fumigants are actually contact poisons applied in gaseous form. The gases or fumes kill the insect after entering its system through breathing pores. The insect must actually absorb the poison for this insecticide to be effective. Fumigants are used to control soil-borne insects that damage roots, such as weevils and June beetle grubs. One example of a fumigant is methyl bromide.

Fumigants are also used in greenhouses. When applied in the greenhouse, the fumigant, consisting of tiny particles of insecticide, settles on the insect and is breathed into its system.

FIGURE 18-2 Aluminum foil strips around cucumber plants protect the plants from flying aphids by reflecting ultraviolet rays from the sky, making the area undesirable to the aphids. (Ed Reiley, Photographer)

REPELLANTS Repellants generally do not kill insects, but, instead, drive them away before they attack the plant. One example of a repellant is aluminum foil which, when placed around plants, repels flying insects, figure 18-2.

ATTRACTANTS AND PHEROMONES

Attractants and pheromones work in the opposite way from repellants; they lure insects to their death. One example is the Japanese beetle bait used in traps to catch the beetles. Another example is the sex lure used to trap the gypsy moth. The sex lure used in these traps is made from naturally occurring or synthetically produced hormones which attract the adult insects. The male gypsy moth is irresistibly attracted to the sex lure; this attraction is stronger than the normal attraction of the female moth. Since most of the male moths are caught in the trap, the females remain unfertilized and lay sterile, unfertilized eggs which do not hatch. The sex lure acts in the same manner on the Japanese beetle.

THE CHEMICAL MAKEUP OF INSECTICIDES

Insecticides are classified in three major groups according to their chemical makeup.

INORGANIC COMPOUNDS

Inorganic compounds are of mineral origin. That is, a mineral is used as the basis for the poison. They usually work in the form of stomach poisons. Sulfur is one type of inorganic insecticide and is also widely used as a fungicide. Lead arsenate is an inorganic insecticide but is not legal in many areas.

ORGANIC COMPOUNDS

Organic compounds are those derived from plants. They usually work in the form of stomach poisons or contact poisons. Two examples of organic compound poisons are rotenone and pyrethrum, which are very safe when used according to the directions on the label.

SYNTHETIC ORGANIC COMPOUNDS

These chemicals are of relatively recent origin, many of them having come into use in the last thirty to thirty-five years. They are very effective against insects, but many of them are also toxic to human beings. This large group of insecticides is subdivided into three smaller groups: chlorinated hydrocarbons, organophosphates, and carbamates. They are not found naturally as are the organic compounds listed above, but are produced in the laboratory.

CHLORINATED HYDROCARBONS These insecticides contain chemicals that have long *residual* control, that is, they continue to kill long after the initial application. DDT is one of the best-known chemicals in this group. It is no longer sold in the United States, however, because of concern about the buildup of the poison in the environment and the bodies of warm-blooded animals. One chemical from this group which is still being used is methoxychlor.

ORGANOPHOSPHATES This group of insecticides is very effective in controlling insects. It also contains some of the chemicals most toxic to warm-blooded animals. Parathion is so toxic that one drop of the concentrated chemical in the eye can kill a human being. They are also absorbed rapidly through the skin. Malathion, another chemical in this group, is a relatively safe chemical and one that can be used to control many insects.

> Caution: *Read the insecticide label for caution in use of these chemicals.*

Organophosphates break down quickly in the environment (within fifteen to thirty days) and do not build up in the bodies of warm-blooded animals.

CARBAMATES This group of synthetic organics contains some of the safest insecticides on the market. Carbamates such as carbaryl (Sevin) are very effective in killing some insects although not effective in killing others. They are slightly toxic to warm-blooded animals. Carbamates break down rapidly (in two to seven days) and leave no residue to contaminate the environment. These chemicals do not build up in the bodies of warm-blooded animals.

CHEMICAL STERILANTS

The most successful use of sterilants involves the use of gamma radiation treatments which are used to sterilize insects so that they cannot produce offspring. One method is to sterilize male insects and release them to mate with females. Since the eggs are not fertilized, they will not hatch when deposited by the female. This is a very effective method in the control of certain insects, such as the screw worm, which infests cattle. This process is also being tested for use on plant insect pests such as the gypsy moth.

APPLICATION OF INSECTICIDES

DUSTS

Insecticides are applied with dusters if purchased as dusts. The chemical is already diluted when bought; it requires no mixing before application.

Dusts are easy to apply with inexpensive equipment, but tend to blow or drift from the target, figure 18-3. This drifting can damage other crops or contaminate the environment.

WETTABLE POWDERS

Wettable powders (WP) resemble dusts in appearance, but are concentrated and must be diluted with water before application. Read the directions for the amount of insecticide to mix per gallon of water. Wettable powders tend to settle while in solution and must be stirred while being sprayed. Some powders are pressed into granules which release less dust into the air during mixing.

With a water-mixed spray better coverage of plants is usually possible and less drift occurs, figure 18-4.

FIGURE 18-3 Dust being applied to potato plants. Note that the dust drifts away from the area.

FIGURE 18-4 Sprayer being used to spray potatoes. There is very little drift with spray, a water solution. (Ed Reiley, Photographer)

EMULSIFIABLE CONCENTRATES

Emulsifiable concentrates (EC) are liquids which are mixed with water in the same manner as wettable powders. The concentrate is safer to handle since there is no powder or dust to blow during mixing. Emulsifiable concentrates do not settle and separate from the solution, as wettable powders do, and give the same good coverage of plants. Shelf life of these chemicals is generally longer than dusts or wettable powders if the container is kept closed and stored in a cool, dark place.

GRANULES

Granules are insecticides in the form of small pellets. They are spread on the soil surface where they either penetrate the soil after the application of water, or turn into gases which fumigate the insects. Some of the systemic insecticides mentioned in this unit are sold in the form of granules. They may be applied to the soil in granulated form and watered in, after which the chemical is absorbed by the roots of the plants.

BAITS

Baits are poisonous materials which attract insects. The pest eats the bait and is killed by the poison. Slugs, snails, cutworms, grasshoppers, and weevils are examples of pests that are controlled in this manner.

AEROSOLS

Aerosol insecticides are contained in pressurized cans. They are generally used for small insect control jobs around the house or in the greenhouse. The chemical is already diluted and ready for use when purchased, figure 18-5. Many aerosols in the form of bombs are available for use in the greenhouse. The bombs are punctured and the chemical is slowly released as a fine mist.

SPIKES

Spikes of various sizes and shapes containing systemic pesticides are available. Small spikes are pushed into the soil in pots and larger ones driven in around plants in the landscape. Plant roots pick up the chemical and it spreads throughout the plant. The pesticide is released over a long period of time, giving extended protection.

EQUIPMENT

HAND-OPERATED DUSTERS AND SPRAYERS

Application of insecticides on school grounds or in greenhouses can be done safely using inexpensive hand-operated sprayers and dusters. An explanation of the use of each of these pieces of equipment will be given by the instructor.

LARGER DUSTERS, SPRAYERS, AND GRANULAR APPLICATION

Large-scale commercial application of sprays, dusts, or granules is accomplished by the use of airplanes, helicopters, tractor sprayers, or trucks with special granular spreaders or sprayers. Figure 18-6 illustrates the use of a large fan jet orchard sprayer.

Airplanes and helicopters are used to spray and dust large acreage of crops, and crops which cannot be treated by ground-operated equipment, figure 18-7. Commercial applicators apply the specified chemical and charge the grower at a cost

FIGURE 18-5 Sprays commonly used by the homeowner. The can on the left is an aerosol. The others use manual power to spray or dispense the chemical. (Ed Reiley, Photographer)

FIGURE 18-6 A large commercial sprayer applying pesticide to field-grown rhododendrons (Courtesy Chesapeake Nursery)

FIGURE 18-7 Airplanes are used to apply pesticides on a large scale. Many acres can be covered rapidly by this method. (Courtesy USDA)

per acre. This is an economical way to apply insecticides when large areas must be covered quickly.

APPLYING THE PROPER AMOUNT

Whatever method of application is used, it is important that the proper amount of the chemical is applied to a given area. This requires calibration of sprayers and dusters. Calibration involves measuring the amount of chemical applied during a specified period of time or over a measured area and then adjusting the applicator or speed of application to obtain the proper amount of chemical required.

Dusts and sprays must be applied so that there is total coverage of all leaf surfaces. When dusts are applied to plants, they can be seen settling on the surface of the plants, thereby making it easy to see the extent of the coverage. When sprays are used, leaves of the treated plants should be wet just to the point of dripping.

Tips: Safe, efficient control of insects requires that the proper concentration of the chemical be mixed in sprays. Read the label for proper mixing instructions.

IDENTIFYING INSECT DAMAGE TO PLANTS

To identify the insect causing damage to plants, the damage itself must be examined. Chewing in-

sects actually eat away part of the plant. This results in holes in the leaves or missing pieces of bark. Chewing insects are controlled by a stomach poison or a contact spray, figure 18-8.

Sucking insects suck the juices from the leaves of plants. The injury caused by sucking insects is not as easily detected as that which is caused by chewing insects. Some signals of damage by sucking insects include twisted plant tips and rolled leaves. Plants that are infested by sucking insects

FIGURE 18-8 Damage to leaves by Japanese beetles. These chewing insects are actually eating parts of the leaves. Notice that the tough leaf veins are left uneaten. (Courtesy USDA)

FIGURE 18-9 Aphids attacking a plant. These insects suck juices from the plant, thereby greatly reducing plant vigor and yield. Notice the ants also present. Ants tend aphids like farmers tend cows. They collect honeydew (a sweet liquid) from the aphids as food and protect the aphids. (Ed Reiley, Photographer)

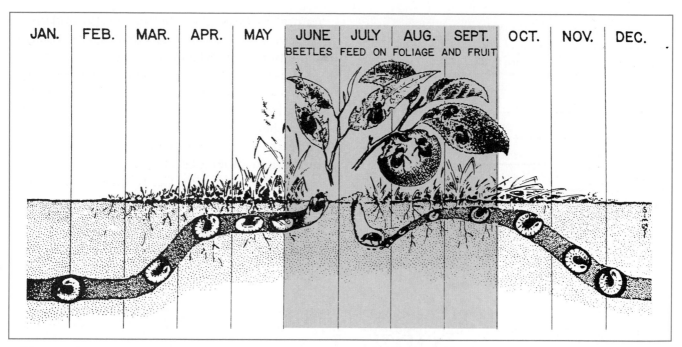

FIGURE 18-10 The adult Japanese beetle is killed with an insecticide applied as a spray only during June, July, August, and September. The rest of its life cycle is spent underground where it does a great deal of damage to grass roots. Insecticides, in the form of soil chemicals, or the introduction of natural enemies, such as "milky spore," are used during the larval stage, which is October or in early spring. (Adapted from USDA)

may show no signs of infestation at all, but simply appear to be less healthy. Figure 18-9 shows damage to a plant by a sucking insect, the aphid. Sucking insects must be controlled with a contact chemical, a fumigant, or a systemic.

TIMING IN INSECT CONTROL

Insects must be killed when they are actively feeding or moving about on the plant. When in the resting (*pupa stage*) or egg stage, very few chemicals can affect insects; however, the new spray oils give control in the egg stage. It is important to determine what stage of growth the insect is in before spraying. This is done by examining the insect on the plant.

Insects go through one of the following basic life cycles. The stages during which the insects are controllable are boxed.

- ◆ Complete metamorphosis:
  egg — larva — pupa — adult
- ◆ Incomplete metamorphosis:
  egg — nymph — adult

If the entire insect population of a plant is in an inactive stage, insecticides must not be applied until the eggs hatch or the pupa emerges as an adult. If insects in both active and inactive stages are present, an immediate application of insecticide should be made, with a second application made in seven to ten days when the eggs have hatched and before the insect has a chance to mature and lay more eggs. Figure 18-10 illustrates the life cycle of the Japanese beetle.

The integrated pest management system relies on exact timing and less use of chemicals to control pests.

PLANTS AS INSECT REPELLANTS

According to organic gardening experts, certain plants may be used to prevent attacks by insects on other plants. Figure 18-11 lists some of those plants, along with the insects they repel.

THE ORGANIC GARDENER

According to *Webster's Dictionary*, *organic* is defined as relating to, produced with, or based on the use of fertilizer of plant or animal origin without employment of chemically formulated fertilizers or pesticides. These gardens rely on insect repellants that do not include chemicals in any form—

PLANT	REPELS
mint	flea beetles, cabbage butterflies
onions, garlic*, chives, leeks	aphids
marigolds	root nematodes
nasturtiums	aphids, cucumber beetles
sage	carrot flies, cabbage pests
horseradish	potato bugs, other flying insects

*Garlic also inhibits the growth of asparagus.

FIGURE 18-11   Plants reputed to repel insects

natural enemies, plants, extracts of ground insects used as sprays, natural insecticides, crop rotation, sanitation, and resistant varieties.

It is a generally accepted fact that the incorporation of organic matter into the soil and the use of mulches result in better plant growth. This is due to better moisture retention in the soil and to an increased availability of minor plant food elements and other plant growth stimulants. This leads to the production of healthier, faster growing plants which are better able to resist attack by insects and disease.

Student Activities

1) Observe insect damage to plants. Identify the insect involved and select the proper pesticide for effective control.

2) With guidance from your instructor, demonstrate the proper mixing and application procedure of an insecticide to a live insect pest discussed in the unit and evaluate the results.

3) Use each piece of equipment described in this unit to properly apply an insecticide. The instructor will demonstrate the proper filling and spraying technique.

4) Visit a local greenhouse operation to observe insect control techniques.

Self-Evaluation

A) Please list

1) six ways in which insecticides work to control insects. Include one insecticide in each category and one insect it controls.

2) the three major groups or families of insecticides and one chemical belonging to each.

B) Select the best answer from the choices offered to complete the statement or answer the question.

1) Stomach poison insecticides work best

a) on sucking insects.
b) as a contact insecticide.
c) on chewing insects.
d) on nematodes.

2) An example of an inorganic insecticide is

a) sulfur.
b) Sevin.
c) malathion.
d) rotenone.

3) An example of a natural organic insecticide is

a) lead arsenate.   c) malathion.
b) Sevin.   d) rotenone.

4) An example of a synthetic organic insecticide is

a) lead arsenate.   c) malathion.
b) diazinon.   d) rotenone.

5) Systemic insecticides are most effective in killing insects that feed by

a) sucking.   c) night.
b) chewing.   d) none of these

6) An example of an insecticide that is used as a fumigant is

a) malathion.   c) DDT.
b) Sevin.   d) methyl bromide.

7) Insecticides that remain active in the environment for the longest period of time are

a) organophosphates.   c) chlorinated hydrocarbons.
b) carbaryls.   d) none of these

8) The insecticides that are most toxic to human beings are the

a) carbamates.   c) carbaryls.
b) chlorinated hydrocarbons. d) organophosphates.

9) Application of dust insecticides requires

a) a duster.
b) no dilution.
c) a quiet day to reduce drift of the insecticide.
d) all of the above

10) Wettable powders (WP) are powdered insecticides that

a) must be diluted with water.
b) are ready to apply with a duster when purchased.
c) are used only on chewing insects.
d) all of the above

11) Insecticides that come in the form of small pellets and are scattered on the soil surface are called

a) pellets.   c) granules.
b) baits.   d) none of these

12) Emulsifiable concentrates (EC) are liquid insecticides that must be

a) diluted with water.   c) used for chewing insects only.
b) used as contact sprays. d) all of these

13) An insecticide packaged in a pressurized can is called a (an)

a) homeowner's spray.   c) home and garden spray.
b) convenience package.   d) aerosol.

14) The proper amount of spray insecticide has been applied to a plant when

a) the bugs begin to die.
b) the leaves are damp.
c) the liquid just begins to drip from the leaves.
d) none of the above

15) When buying an insecticide to eliminate a specific insect, one should purchase a chemical

a) that is recommended for that insect.
b) only after reading the label.
c) that is safe to use.
d) all of these

16) How many pairs of legs do insects have?

a) six
b) three

c) two
d) four

17) The legs of every insect are fastened to the body part known as the

a) thorax.
b) head.

c) abdomen.
d) shank.

Fungicides, Rodenticides, Molluscicides, and Nematocides

OBJECTIVE

To identify fungus diseases, rodents, mollusks, and nematodes and prescribe a chemical control for each.

Competencies to Be Developed

After studying this unit, you should be able to

- identify fungus diseases on plants and set up a spray or dusting schedule for effective control.
- recognize at least two types of rodents and apply a rodenticide in a bait station.
- recognize slugs and snails on sight and develop a slug control program, including chemicals used in the bait station or as contact poisons.
- identify two nematocides and explain in writing their application to specific situations.

MATERIALS LIST

- ✔ plants infected with fungus disease
- ✔ one live snail and slug
- ✔ one plant with nematode damage to roots and one healthy plant of the same variety
- ✔ sample fungicides, rodenticides, molluscicides, and nematocides

FUNGICIDES

Fungicides are pesticides used to control plant diseases caused by fungi. *Fungi* are tiny nongreen plants such as rusts, mildew, mold, and smut that lack chlorophyll and live as parasites on green plants. Fungus diseases are spread by *spores* that resemble tiny seeds floating through the air. As fungi grow on the green plants, the plant tissue is killed or the plant is weakened. These diseases must be controlled if attractive, highly productive crops are to be grown.

CHEMICAL MAKEUP

Fungicides are manufactured from a variety of chemicals. Generally, compounds of copper, sulfur, mercury, and dithiocarbamates make effective fungicides. The following are some of the fungicides in use today.

BORDEAUX MIXTURE The bordeaux mixture is a combination of copper sulfate (blue vitriol) and hydrated lime in water. It is the copper in the bordeaux mixture that kills.

LIME SULFUR AND WETTABLE SULFUR Lime sulfur, a mixture of lime and sulfur, has been used as a fungicide for many years. Sulfur, in the form of finely ground wettable powder, is a very effective fungicide.

FERBAM, MANEB, AND ZINEB Classified as dithiocarbamates, these are relatively new fungicides that have been used successfully.

MERCURY COMPOUNDS Mercury compounds have proven to be good fungicides, *but most of these are off the market because of heavy metal contamination of the environment.*

HOW FUNGICIDES WORK

Fungicides are most effective when used to prevent germination and/or growth of the fungus spore, thus preventing the attack to the plant. There are, however, some new fungicides which can control a fungus disease even after the disease is present in the plant. Systemics, such as Boyleton and Aliette, work by entering the plant and circulating in the sap. However, they work primarily as

---

SPECIMEN LABEL
NOT FOR RESALE

MILLER
Potato Seed-Piece Fungicide Dust
(CONTAINS DITHANE M-45)***

ACTIVE INGREDIENT:
  Zinc ion and manganese ethylene bisdithiocarbamate 80%, a coordination product of manganese 16%, zinc 2%, ethylene bisdithiocarbamate 62% .. 8%
INERT INGREDIENTS: _____ 92%

GENERAL INFORMATION

MILLER POTATO SEED-PIECE FUNGICIDE DUST is a special fungicide dust formulation for application on whole or cut potato tubers to protect them from seed-piece decay caused by **Fusarium sambucinum.**

DIRECTIONS FOR USE

Use 1 lb. of **MILLER POTATO SEED-PIECE FUNGICIDE DUST** per 100 lbs. of whole or cut potato seed-pieces. Apply the dust in a manner such that the entire surface of the seed-piece becomes thoroughly covered with the fungicide dust. Plant seed as soon after treatment as possible. Do not use treated seed potatoes for food or feed purposes.

CAUTIONS

MAY CAUSE IRRITATION OF NOSE, THROAT, EYES AND SKIN. Do not breathe dust. In case of skin contact, wash immediately with soap and water; for eyes, flush with plenty of water, then get medical attention.

Rinse equipment and containers and dispose of wastes by burying in non-crop lands away from water supplies. Containers should be disposed of by punching holes in them and burying with wastes, or by burning (Keep out of smoke).

This product is toxic to fish. Keep out of lakes, ponds and streams.

CAUTION NOTICE: Keep away from fire and sparks. Store in a cool, dry place. Do not allow to become wet or overheated in storage. This may bring on chemical changes which will impair the fungicidal effectiveness of Dithane M-45 and may also generate flammable vapors. Keep container closed when not in use.

The use of this material being beyond our control and involving elements of risk to human beings, animals or vegetation, we do not make any warranty, express or implied, as to the effects of such use, when this product is not used in accordance with the directions as stated on this label.

CAUTION: KEEP OUT OF REACH OF CHILDREN

NET WEIGHT — 50 POUNDS

Manufactured By
MILLER CHEMICAL & FERTILIZER CORPORATION
Hanover, Pennsylvania 17331, U.S.A. · An Alco Standard Company

104M2NLS   EPA Est. No. 72-MD-2
            EPA Reg. No. 72-544   ***Reg. T.M. Rohm and Haas Company   BATCH NO.

FIGURE 19-1 Typical fungicide label. Notice the different types of information that appear on the label. (Courtesy Miller Chemical Company)

other fungicides do, that is, to prevent the disease rather than cure it after it is established in the plant.

Fungicides are generally not as toxic to humans and other warm-blooded animals as insecticides are. Review figure 17-1 in Unit 17, which lists the LD_{50} for some of the most frequently used fungicides. Compare the toxicity level of fungicides listed in the chart with those of other pesticides. Figure 19-1 illustrates a typical fungicide label.

EFFECTS OF FUNGUS

Fungus diseases attack all parts of the plant, the leaves, stem, fruit, seed pods, and roots.

LEAVES Plant leaves are attacked by many so-called *leaf spot diseases*. The term *leaf spot* describes the appearance of the diseased leaf. The leaf is marked by round or irregular spots which are beginning to turn a color different from the rest of the plant, figure 19-2. The spot or spots on the leaf grow larger and larger as the fungus disease spreads. The entire leaf may be killed, but this is unusual.

Leaf spot diseases damage the plant by destroying the cells that manufacture food. As more cells are killed, the plant is less able to make food and becomes weaker. The appearance of the plant is also affected. A plant sold for its beauty may be damaged so much that it is unsaleable. A plant which produces a crop of fruit or seeds does not produce as large a crop when affected by leaf spot disease because the leaves cannot manufacture enough food for best growth of the fruit or seeds.

Leaves are also attacked by another fungus disease, known as *rust disease*, figure 19-3. Again, the name describes the appearance of the disease; the leaf which takes on a rusty color in the beginning is gradually killed by the disease. Mildews and molds are other fungus diseases. Their names describe their appearance on the plant.

Fungicides are sprayed on leaves to prevent the entry of the fungus disease into the plant.

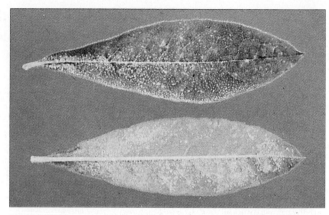

FIGURE 19-3 Rust disease on rhododendron leaves

When the fungus spore sprouts on the plant leaf, the fungicide kills it. A spray or dusting calendar complete with dates must be established so that the fungicide can be applied on a regular schedule. Systemic fungicides such as Subdue, which enter the plant sap, are more effective than nonsystemic ones in preventing leaf diseases.

Tips: Most fungicides are used to prevent disease and must therefore be sprayed or dusted on the plant before the disease is present to prevent spread of disease. Some fungicides do cure or eliminate disease already present on the plant, but even these chemicals work best as preventors of disease, and not as a cure.

Figure 19-4 is a list of fungus diseases and the fungicide used to prevent them.

STEMS, FRUIT, AND SEED PODS These parts of plants are also attacked by fungus diseases. The effect and control are the same as those listed for leaves. Anthracnose, for example, is a fungus disease that attacks the stems of many plants.

High humidity and rain help to spread fungus diseases of stems and leaves. Most fungus diseases grow best in high moisture conditions.

ROOTS Plant roots are attacked by many types of root rot or root mold. *Root rot* and *root mold* describe the appearance of the root after attack by the fungus disease. Root diseases are controlled by drenching the soil with fungicides or by spraying.

Fungus root diseases are most often found in soils that do not drain well and are wet too much of the time. Improved soil drainage generally helps to control fungus diseases of roots.

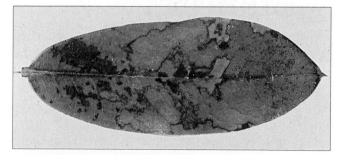

FIGURE 19-2 The effects of leaf spot disease on a rhododendron leaf

PLANT DISEASE	SYMPTOMS	FUNGICIDE TO USE	HOW TO APPLY	REMARKS
Black spot	Circular black spots up to ½ inch across, usually on the upper leaf surface.	Phaltan or captan	spray or dust	A serious disease of roses. Difficult to control.
Bean rust	Rusty growth on underside of leaves; may cause leaves to fall off.	sulfur or maneb	dust	No bean variety is resistant.
Anthracnose (Attacks the black raspberry.)	Circular reddish brown sunken spots with purple margins and gray centers.	lime sulfur in dormant stage; ferbam	spray	Grows on young shoots; becomes large cankers on older canes.
Powdery mildew (Attacks the lilac.)	Covers leaves with a thick white coating.	Phaltan or Karathane (best)	spray	Worse in dry season.
Apple scab	Appears on leaves as small smoky area, moldly ¼ inch across fruit, raised spots; scablike.	ferbam, captan, or Cyprex	spray	Worst apple disease. Attacks leaves and fruit.
Damping-off	Seeds either fail to appear or small seedlings rot off at soil level.	captan-Arasan seed treatment	seed treatment or soil drench	Spreads rapidly; prevention is better than cure.
Rhododendron root rot	Top of plant starts to wilt on hot afternoons; watering does not help.	As a soil drench Aliette and Subdue will work. Excellent as preventatives but will not cure infected plants.	soil drench	Destroy plant and sterilize soil.
Brown rot (of peach)	Starts as brown spot; spreads rapidly until entire fruit turns brown and dries up.	wettable sulfur	spray	Attacks blossom, stem, and fruit. Easiest to identify on fruit.
Botrytis (Attacks the peony.)	Buds blast (wither) and turn black; leaves get gray mold and rot off.	zineb, ferbam	spray	Attacks many bulb crops.

Note: Most fungus diseases are at least partly controlled by sanitation. If leaves, fruits, stems, or any part infected with the disease are collected and burned at the end of the season, further control measures may not be needed.

FIGURE 19-4 Some common fungus diseases and their control

Small seedlings are often attacked by fungus diseases such as *damping-off*. The use of sphagnum moss to grow seedlings often prevents the start of the disease. Dusting or drenching the soil mix with captan often gives excellent control of seedling root and stem rot.

APPLICATION

Fungicides are sprayed or dusted on plants or poured around the roots as drenches. The methods of application are the same as those for insecticides.

One difference between the application of fungicides and insecticides is in the timing. Insecticides are usually applied after it is determined that an insect is on the plant and is causing damage. Fungicides are applied to prevent attack. The chemical acts as a protective film to kill the fungus spore as it sprouts and before it can penetrate the surface of the plant.

RODENTICIDES

Rodenticides are chemicals used to control rats, mice, other rodents, and bats. Most commonly, rodenticides kill as stomach poisons. They are applied as bait which is attractive to the rodent and is eaten. In limited cases, rodenticides are spread or sprayed on the rodent or in an area through which the rodent travels. This technique is usually avoided, however, because other wildlife or domesticated animals could be poisoned accidentally.

A common chemical used for rats and mice is warfarin. This chemical is applied to grain in *bait stations* where the rodent feeds. Field mice in orchards are often poisoned by a coating of deadly poison on pieces of apple. The apple is placed in bait stations for the mice to eat.

Caution: Always place poisoned baits in bait stations where other animals cannot reach the poison. Place the bait under a wooden or metal box with a small hole cut in it which is just large enough for mice to enter.

MOLLUSCICIDES

Molluscicides are chemicals used to control snails and slugs. They are applied as poison baits or contact poisons. These chemicals kill as stomach poisons when eaten or as contact poisons when the slug or snail is covered by the poison or crawls through an area to which the poison has been applied. Slugs and snails usually hide during the daytime; therefore, it is difficult to hit them while spraying or dusting.

Snails and slugs look much alike, except that slugs do not have a shell and snails do, figure 19-5. Both travel on the soil surface leaving a slimy trail behind, and feed on plant leaves and tender stems at night. Because they feed at night, slugs and snails actually doing damage to plants may be difficult to locate, thereby causing other insects to be blamed for the damage. To pinpoint snails and slugs as the pest, look for the slimy trail.

To find slugs and snails lift stones and mulch and look underneath. Any cool, moist place is a good hiding place for them.

FIGURE 19-5 Snail (A), Slug (B) (Courtesy NYS Experimental Station)

CONTROL

Rough sand, ashes, gravel, or cinders act as barriers to slugs and snails; the rough surface provided by these materials is difficult for them to crawl on. Another chemical-free control method is to set a shallow pan filled with beer in the soil, with the top of the pan at the soil level. The slugs follow the smell of the beer into the pan and drown. Yeast mixed with water is also effective.

Various slug killers are sold by pesticide dealers. Some effective chemicals include Slugit, which is a stomach and contact killer; Slug-kill (15 percent methaldehyde dust) which is a contact poison; and Zectran.

Caution: Keep these and all pesticides out of the reach of children.

NEMATOCIDES

Nematocides are chemicals used to control nematodes. *Nematodes* are small hairlike worms that feed on plant roots. These tiny worms bore into the root, usually causing knots or bumps to appear on the attacked root, figure 19-6. Nematodes live in the soil and swim in soil water to move from plant to plant. Some nematodes also live inside plant leaves.

CONTROL

Soil fumigants are the most effective treatment for nematodes. The nematocide must contact the

FIGURE 19-6 Nematode damage to plant roots. The plant on the left is badly infested with nematodes and the root system is severely damaged. The plant on the right has a normal root system.

nematodes as a gas or liquid passing through the soil. Two chemicals meant specifically for nematodes are Dasanit and Nemagon. Systemic insecticides are effective on leaf nematodes.

Student Activities

1) Under the instructor's supervision, apply a fungicide to a plant on the school grounds infected with fungus disease. Use a small aerosol. As the spraying progresses, check for disease control and record the results. If two plants are available, spray one and keep one as a control.

2) Put together two bait stations, one for rodents and one for slugs and snails. Actually use the stations under the supervision of your instructor if circumstances permit.

> *Caution: These activities should be accomplished only under the supervision of the instructor. When applying any pesticide, carefully read and follow all instructions and precautions.*

Self-Evaluation

Select the best answer from the choices offered to complete each statement.

1) Fungicides are chemicals used to control

    a) plant rust diseases.
    b) fungus diseases.
    c) plant mildew.
    d) all of these

2) Fungus diseases are spread by small seedlike structures called

    a) spores.
    b) roots.
    c) pollen.
    d) nematodes.

3) Fungicides cause a protective film to coat plant leaves and

    a) keep fungus spores from sticking to the leaves.
    b) kill the fungus spore before it can enter the plant cell.
    c) smother fungus diseases on the leaves.
    d) prevent oxygen from reaching the fungus spores so that they cannot grow.

4) Rodenticides generally kill rats and mice

    a) as stomach poisons, so they must be eaten by the rodents.
    b) while they crawl on the poison.
    c) by fumigating them.
    d) all of the above

5) Snails and slugs are not often seen eating plants because

    a) they only eat roots and underground stems.
    b) they feed at night.
    c) they drop to the ground and hide if any vibration occurs.
    d) they are the same color as most plants and therefore difficult to see.

6) Snails and slugs belong to the same family, but have one main difference. This difference is

    a) slugs are larger than snails.
    b) slugs do not have shells; snails do.
    c) slugs leave a slimy trail when they move; snails do not.
    d) slugs have longer antennae than snails.

7) Contact sprays are not completely effective for slugs because

    a) their slimy bodies resist the chemicals.
    b) they form a path with slime on which to walk.
    c) their bodies are slippery so chemicals do not stick to them.
    d) they hide under objects such as stones and mulch and are therefore difficult to hit.

8) Nematodes are tiny hairlike worms that

    a) belong to the insect family.
    b) feed on plant roots.
    c) attack animals, including human beings.
    d) all of these

9) Nematodes are controlled with fumigants or soil drenches because

    a) they live in the soil where these chemicals are able to reach them.
    b) they do not eat plants, so stomach poisons do not work.
    c) no other chemicals are made for use on them.
    d) all of the above

Herbicides

OBJECTIVE

To select and apply herbicides so that weed control is accomplished without damage to the crop being cultivated or contamination of the environment.

Competencies to Be Developed

After studying this unit, you should be able to

- ◆ identify a weed problem and select a herbicide to control the problem.
- ◆ define and differentiate between selective and nonselective herbicides.
- ◆ describe three ways in which herbicides destroy weeds.
- ◆ outline in writing how a sprayer is calibrated.
- ◆ list three possible reasons for failure of a herbicide to work properly.
- ◆ properly apply a herbicide.

MATERIALS LIST

- ✔ weed killers complete with label and recommendations
- ✔ weed charts for weed identification
- ✔ sprayer for use in calibration and application of herbicides
- ✔ an area of lawn or crop land for use in locating and identifying weeds and applying herbicides

A *herbicide* is a chemical that kills weeds. Herbicides comprise a large percentage of the total pesticides used each year. In the past twenty years, many new weed killers have been introduced on the market, and each year new uses are found for traditional weed killers. It is estimated that in the United States, weeds cause damage and loss of production to crops in excess of $4.5 billion each year. Herbicides are used to kill weeds in crops, on lawns, along roadways, and in many other places.

A *weed* is considered to be any plant growing where it is not wanted. Grass which is growing in a lawn is not a weed, but is considered a weed if it is growing in the garden.

Weed control can be accomplished without the use of chemicals. In some cases, mulching and the cultivation of crops may be more effective and just as economical to control weed growth. For example, it would be very difficult to spray weed killers in a small garden without damaging other plants. It is in large acreages of single crops that selective weed killers can be used to greatest advantage.

There are many different types of chemicals used as weed killers. A partial list of herbicides and their toxicity levels is given in figure 20-1.

Notice that some of these pesticides may be very poisonous if swallowed (oral). They are generally not as toxic dermally (on the skin).

TYPES OF HERBICIDES

NONSELECTIVE HERBICIDES

Nonselective herbicides kill all plants to which they are applied. These herbicides are used in places where no plant growth is wanted. Lumber yards, railroad tracks, fence lines, driveways, and parking areas are a few examples of areas where these chemicals are used. Examples of some weed killers in this category are Atratol, Aatrex, and Roundup.

Nonselective weed killers sometimes may also be used as selective weed killers. Examples of this

TOXICITY CLASS	COMMON NAME OR DESIGNATION	SOME COMMON TRADE NAMES	ACUTE ORAL LD$_{50}$ mg/kg (rats)	PROBABLE LETHAL DOSE OF UNDILUTED CHEMICAL FOR 150 LB. PERSON
Very Toxic	✔ sodium arsenite	Atlas A	10	2 drops to 1 teaspoon (less than 1/6 oz.)
Moderately Toxic	paraquat	Weedol	150	1 teaspoon to 1 ounce
	diquat	Aquacipe	400	
Slightly Toxic	2,4-D	various brands	500	1 ounce to 1 pint (or up to 1 pound)
	MSMA	Ansar, Daconate	700	
	cacodylic acid	Phytar 560	830	
	aspirin	(for comparison)	1,240	
	linuron	Lorox	1,500	
	DSMA	Ansar, Sodar	1,800	
	norea	Herban	2,500	
	✔ amitrole	Weedazol	2,500	
	borate	Borax, Borascu	2,500	
	dicamba	Banvel	2,900	
	prometon	Pramitol	2,980	
	DCPA	Dacthal	3,000	
	atrazine	Aatrex	3,080	
	table salt	(for comparison)	3,320	
	diuron	Karmex, Krovar 1	3,400	
	monuron	Telvar	3,600	
	chloroxuron	Tenoran	3,700	
	prometryn	Caparol	3,750	
	AMS	Ammate	3,900	
Almost Nontoxic	siduron	Tupersan	5,000	1 pint to 1 quart or 1 to 2 pounds
	propazine	Miloguard	5,000	
	bromacil	Hyvar X, X-L, X-P Krovar 1	5,200	
	✔ picloram	Tordon	8,200	
	dalapon	dowpon	9,300	
	benefin	Balan	10,000	

✔ Use restricted to licensed or certified applicator.

FIGURE 20-1 Relative toxicity of some herbicides to mammals

type of substance include Atratol and atrazine (Aatrex). Used at a rate of 2 to 4 pounds of active chemical per acre, it kills only small weeds or selected weeds; if application is increased to from 5 to 10 pounds per acre, it kills all plants.

SELECTIVE HERBICIDES

Selective herbicides are just what the name implies—chemicals which kill some plants, but not others. This is by far the largest group of weed killers. Weed control in a particular crop can be accomplished safely and effectively with selective weed killers. These chemicals are designed to kill the weed but not the crop.

In some cases, a selective weed killer which kills broadleaf plants will not kill grasses which have narrow, bladelike leaves. This makes it possible to spray a lawn with 2,4-D to kill dandelions which have a broad leaf and not kill the grasses which have narrow leaves. Since corn is a grass, broadleaf weeds in a corn field can be killed with a selective herbicide without damaging the corn. The herbicide 2,4-D is one of the best known selective weed killers used to kill broadleaf plants without killing grasses.

Herbicides that have the potential to kill a particular crop at high concentrations can be used at low concentrations to kill young weeds that are growing in the crop and not damage the crop. By applying the weed killer at a low concentration, young tender weeds are killed and the older, more mature crop plant is not harmed. Strawberries are protected from weeds in this way.

THE EFFECT OF HERBICIDES

Herbicides work by upsetting the *metabolism*, or life functions, of the plant. The plant either starves to death or wears itself out due to the increased rate of activity caused within its system by the chemical. Herbicides are designed to affect the metabolism of some plants and not others.

The following are some specific examples of the effect of herbicides.

ATRAZINE

Atrazine is widely used to kill both broadleaf weeds and grassy weeds in corn and turf grass sod as well as those in tree and shrubbery plantings.

APPLICATION   Atrazine may be applied *preplanting* (before planting has taken place); *preemergence* (before emergence of seedlings aboveground); or *postemergence* (after seedlings have grown above ground but before the crop matures). However, it must be applied before the weeds reach a height of more than 1½ inches. The chemical is usually applied with an ordinary sprayer.

BEHAVIOR IN THE PLANT   The chemical is absorbed through both roots and foliage. It moves through the plant in the xylem or wood and accumulates in the stem tips and leaves of plants. Atrazine kills the plant by preventing photosynthesis from occurring. Since the plant cannot manufacture food, it starves to death. Plants such as corn are not killed by atrazine when applied in usual dosages because they are able to break down the chemical before any harm is done to them. Cyanazine (Bladex) also works as a photosynthesis inhibitor.

2,4-D

2,4-D is a systemic herbicide and is widely used to control broadleaf weeds. Most dicot (broadleaf) plants are susceptible to the chemical at normal application rates.

APPLICATION   Application is usually by spray directed on the plant foliage. Both plant leaves and roots absorb 2,4-D.

BEHAVIOR IN THE PLANT   The chemical moves through the plant phloem or bark. Movement is more rapid in fast-growing plants and therefore plants in active growth are killed more easily. The chemical 2,4-D causes abnormal growth in the weed and affects respiration, food reserves, and cell division, figure 20-2.

FIGURE 20-2   The weed killer "Roundup" was used to kill this weed. The roots are killed and the top dries out and turns brown. (Ed Reiley, Photographer)

DACTHAL

Dacthal is used to control annual grasses and certain annual broadleaf weeds.

APPLICATION Application is generally at planting time before the weeds come up. The chemical is usually diluted with water and applied as a spray.

BEHAVIOR IN THE PLANT Dacthal is not absorbed by leaves and it does not move through the plant. Rather, it kills sprouting seeds. The exact way it acts is not yet known.

The weed killer Treflan kills weeds in a similar manner.

SUMMARY

The three herbicides just discussed kill weeds in three different ways:

- ◆ atrazine ➞ prevents photosynthesis
- ◆ 2,4-D ➞ upsets cell division, respiration, and food reserves
- ◆ Dacthal ➞ destroys sprouting seeds (specific action not yet known)

All weed killers do not fall into these three groups, but they are representative of a great percentage of herbicides.

APPLICATION OF HERBICIDES

Herbicides are either sprayed with ordinary sprayers or applied already mixed in fertilizers or in granules. The same type of equipment is used as with insecticides. However, a sprayer that is used to apply herbicides should not be used to apply insecticides or fungicides unless it is thoroughly cleaned. This does not mean simply rinsing out the sprayer; it must be washed with a substance that neutralizes the herbicides, such as ammonia. Read and follow the directions carefully; otherwise, enough herbicide may remain in the sprayer to damage desirable plants sprayed for insects or diseases. When using small hand sprayers, a separate sprayer should be used for the application of herbicides.

TIME OF APPLICATION

Application of herbicides must be made at one of three times to avoid injuring desirable plants.

PREPLANTING Herbicides are mixed into or sprayed onto the soil or seed bed. Some herbicides are effective only when incorporated with or mixed into the soil; others need no mixing. Read the label before buying.

PREEMERGENCE Treatment is made before any plant growth appears or after the crop comes up but before the weeds appear. Read the label to see if the chemical should be applied before the crop emerges or after the crop emerges and before the weeds emerge. When used at the recommended concentration, these chemicals usually prevent seeds from germinating or kill only small tender weeds without damage to desirable plants.

POSTEMERGENCE Applications must be made very selectively in terms of the chemical used. These chemicals are applied after the crop plant has come up.

> *Tips: The horticulturist must be very careful to choose a herbicide that kills the weeds but not the crop since at the time of application, both weeds and crop are aboveground.*

CALIBRATION OF SPRAYERS

It is extremely important that the proper amount of actual chemical be applied to a particular area when spraying herbicides. Even if the right mixture or concentration of spray is in the spray tank, it is possible to apply the wrong amount of weed killer. To avoid this problem, the sprayer should be properly *calibrated*, or adjusted, before use. The correct amount of active chemical must be applied on a measured area for the most effective control.

LARGE SPRAYERS The first step in the calibration of large sprayers is to measure and mark off an acre. Spray the acre with water and refill the spray tank, measuring exactly how many gallons of water are required to refill the tank, the amount of water applied to the acre. This is the rate of application per acre. If 10 gallons of water are applied per acre and 2 pints of chemical are to be applied per acre, then 2 pints of chemical are added to each 10 gallons of water. For example, for a 100-gallon tank, add 20 pints, or 2.5 gallons, of chemical.

> *Caution: The rate of travel over the area and the sprayer pressure must be the same as in the trial run.*

SMALL HAND SPRAYERS Follow these steps to calibrate a small hand sprayer.

1. Lay out an area measuring 10 feet x 100 feet (1,000 square feet).
2. Fill the sprayer with water.
3. Spray the 1,000-square-foot area.
4. Measure the number of pints of water required to refill the sprayer to the same level as before. This is the number of pints applied per 1,000 square feet.

If 8 pints (1 gallon) is sprayed on the 1,000 square feet of land, and 2¼ teaspoons of chemical are required per 1,000 square feet, add 2¼ teaspoons of chemical to each 1 gallon of water. This is the same rate as 1 pint per acre. (One acre is 43,560 square feet.)

> *Caution: Spray with the same pressure and speed of application as in the trial run.*

Another way to determine the amount of water that was applied to the 1,000-square-foot area is to measure the length of time required for the application. Spray into a container for the same length of time and measure the amount of water collected.

> *Caution: Use the same spray pressure as when spraying the 1,000-square-foot area.*

DRIFT

Any substance that is sprayed can blow or drift in the wind. If a herbicide drifts onto a nontarget plant, there is a chance that this plant will also be damaged or killed. When spraying herbicides, use spray nozzles that apply large droplets of spray. These larger droplets do not blow or drift as much as a fine mist and greatly reduce the danger of damage to other crops nearby. Tomatoes and tobacco are injured especially easily if 2,4-D drifts over them. Injury of other crops by drifting chemicals is a greater problem with herbicides than with other types of pesticides.

BREAKDOWN OF HERBICIDES

Most herbicides are broken down in the soil by soil microorganisms and eventually become harmless materials. The length of time required for this breakdown varies from several weeks to several years. Read the label for recommendations. If a herbicide persists or stays in the soil for a year or more, it may kill or damage the crop planted in that soil the following year or later in the same season. Crops which are not damaged by the chemical should be planted there until the soil is clean enough for crops which could be damaged by the herbicide.

Since preemergent herbicides are only effective in controlling germinating weed seeds, they must be applied on clean, cultivated soils before weed seeds germinate. Of all the preemergent herbicides, only Chloro-IPC, Casoron, and Preemerge will control existing chickweed as well as germinating chickweed seeds.

These recommendations should be used only as a guide in selecting the proper test herbicide. Under no circumstances should a herbicide be applied over the entire nursery without first spending one growing season testing it in a limited area under the same soil conditions and the same growing program. Herbicides are noted for not being equally dependable under all growing conditions.

Always read herbicide labels for instructions, including the species that have been cleared for use for that herbicide. Even when materials are used as recommended, the manufacturer cannot be held responsible for injury when it is applied to ornamentals not listed on the label.

WHY HERBICIDES SOMETIMES FAIL

Why herbicides are sometimes ineffective is a very complex question. The following are some of the more common reasons for failure.

INCORRECT SELECTION OF THE HERBICIDE

Unsatisfactory weed control may result from use of the wrong herbicide. It has been noted that many weed killers are selective, that is, they kill only certain weeds. Therefore, it is most important that the right chemical be selected to control a specific weed. This requires accurate identification of the weed to be controlled. To use herbicides effectively, the applicator must be able to accurately identify weeds.

IMPROPER APPLICATION OF THE HERBICIDE

Herbicides are designed for application in a variety of ways and at various stages of the growing process. Poor application and poor timing of application are reasons for herbicide failure. For example, preemergence herbicides must be spread evenly on the surface of the soil. To make contact with sprouting weed seeds, the entire soil surface must be covered with the chemical; if not, the herbicide could be rendered ineffective.

Some chemicals also must be washed into the germination zone by rainfall or irrigation within seven to ten days after application. If no rain occurs or irrigation is not possible, poor weed control will result. Preplant weed killers such as Eptam must be mixed into the soil immediately after application; if not, they evaporate into the air. Much of the active chemical is thus lost and control is poor. Postemergence weed killers such as 2,4-D must be sprayed onto the plant to be killed or over the roots of the plant so that the chemical can be absorbed into the plant. The chemical 2,4-D also works most effectively if the weeds are in active growth.

Sprayer calibration is also very important for proper application of the herbicides. This may involve tests to determine the proper amount of chemicals to apply per acre of land. Too low a rate results in poor weed control; too high a rate could kill the crop as well as the weeds.

ENVIRONMENTAL VARIABLES

THE MATURITY OF THE WEEDS Weed maturity is an important variable; most weeds are killed more easily when they are young and in active growth. In fact, this in the only time many weed killers work effectively. An old, hardened plant requires much more chemical, or may even require a different type of weed killer. An exception to this is the herbicide glyphosphate (Roundup) which works better on mature plants.

RAIN Rain which falls just after application of the herbicide can be an advantage with some preemergence chemicals. However, excessive rain can leach or wash the chemical through the soil or off foliage and dilute it so much that weeds are not controlled. Roundup is washed off easily. It requires six to seven hours to move into the plant.

SOIL TYPE AND ORGANIC MATTER CONTENT The type of soil and its organic content affect the amount of active chemical that must be applied. Some chemicals are absorbed or held on the surface by clay particles and organic matter. There are two dosages for some chemicals. For example, Dacthal is recommended in doses of 10-15 pounds of active ingredients per acre. Ten pounds is recommended for sandy soils, and 15 pounds for silt and clay soils. A soil high in humus absorbs and makes some of the weed killer inactive, requiring a higher rate of chemical application.

Not all chemicals are affected by organic matter in the soil to the same degree. For example, dalapon is not absorbed at all, while Dacthal is easily absorbed and would be greatly affected by organic matter level of the soil.

Weed control with chemicals is not a simple job; many variables must be considered to achieve effective control. Herbicides have limitations and are not a magic cure-all for weed problems, but if used carefully and accurately, good results can be expected.

Caution should be exercised in the use of any of these chemicals.

Student Activities

1) Demonstrate the proper calculation and mixing of at least one herbicide, working in groups of three to five students.

2) If land is available, mark off plots of 100 square feet. Apply the herbicide mixed in Activity 1 to an assigned area. Work in the same groups.

    a) Identify weeds present and estimate number of each.

    b) Select the proper weed killer.

    c) Mix and apply the chemical.

    d) Record the results.

3) a) Calculate how much chemical to use in 100 gallons of water if the weed killer is to be applied at the rate of 1 pint per acre and the sprayer, at the speed traveled, sprays 20 gallons of water per acre.

    b) If the speed the sprayer is pulled changes from 2 miles per hour to 4 miles per hour, how will this affect the amount of spray applied per acre?

> *Tips:* *Complete these projects only under the supervision of your instructor.*

Self-Evaluation

Select the best answer from the choices offered to complete the statement or answer the question.

1) A weed is a plant

    a) growing in the lawn.
    b) growing where it is not wanted.
    c) growing in the wild.
    d) that is not a food item.

2) Weeds can be controlled without the use of herbicides by

    a) hand weeding and cultivation.
    b) mulches and cultivation of crops.
    c) hoeing and weeding.
    d) all of these

3) Selective weed killers can be used to greatest advantage

    a) on large acreages of a single crop.
    b) in the home vegetable garden.
    c) where there is a mixture of weeds.
    d) where the land is level.

4) Nonselective weed killers

    a) should not be used to kill weeds.
    b) can be used in any weed problem situation.
    c) kill all of the plants to which they are applied.
    d) are the best weed killers for the home garden.

5) The herbicide 2,4-D controls the serious weed problem caused by

    a) crabgrass.
    b) bluegrass.
    c) orchard grass.
    d) dandelions.

6) The weed killer atrazine kills plants by

    a) starving them to death.
    b) preventing photosynthesis.
    c) accumulating in the leaves and buds and causing damage therein.
    d) all of the above

7) The herbicide 2,4-D, a synthetic herbicide, is widely used to control

   a) grasses.
   b) broadleaf weeds.
   c) weeds in the garden.
   d) weeds in commercial tomato plantings.

8) Dacthal is a weed killer that is generally applied

   a) at planting time.
   b) at the time weeds are 1½ to 2 inches tall.
   c) at postemergence.
   d) to control perennials.

9) Postemergence application of an herbicide means that the chemical is applied

   a) before the crop is up.   c) after the crop is up.
   b) at planting time.   d) after the crop is harvested.

10) To be sure a herbicide is recommended for use on a specific weed and crop

   a) test the chemical on a small sample area first.
   b) read the label for recommendations.
   c) ask your neighbors if they have used it.
   d) ask the salesperson for advice.

11) Preemergence application of herbicides means that the chemical is applied

   a) before the crop is up.   c) after the crop is up.
   b) at planting time.   d) after the crop is harvested.

12) Herbicides kill plants by all the methods listed below *except* for

   a) preventing photosynthesis.
   b) upsetting cell division and respiration.
   c) preventing seed germination.
   d) drying them up.

13) A sprayer used to apply herbicides should not be used to spray insecticides on crop plants because

   a) it is difficult to clean all of the weed killer out of the sprayer.
   b) insecticides and weed killers do not mix.
   c) weed killers may damage the sprayer.
   d) you may forget to clean the sprayer.

14) The three different times for applying weed killers are

   a) spring, summer, and fall.
   b) preplant, preemergence, and postemergence.
   c) preplant, postplant, and after harvest.
   d) preplant, postplant, and postemergence.

15) If a herbicide is to be applied at 2 pints to the acre and the sprayer applies 10 gallons of water to the acre, how much chemical should be mixed in each gallon of water?

   a) 2 pints   c) 0.2 pint
   b) 20 pints   d) 5 pints

16) Even if the proper amount of chemical is mixed per gallon of water, the wrong amount of chemical can be applied per acre

   a) if the wrong amount of spray solution is applied per acre.
   b) if the wrong amount of chemical is added to each gallon of water.
   c) if the wrong amount of chemical is put in the spray tank.
   d) if the weeds are too old and tough.

17) Drift of weed killers to nontarget plants can be reduced by

    a) using a nozzle that sprays larger droplets.
    b) spraying downward.
    c) using the proper chemical.
    d) mixing the proper amount of chemical per gallon of water.

18) Most weed killers are broken down into harmless materials in the soil

    a) by freezing and thawing.
    b) by rainwater.
    c) by soil microorganisms.
    d) by plant uptake and detoxification.

19) A soil high in organic matter tends to

    a) hold a great deal of water, thereby diluting herbicides.
    b) absorb herbicides and cause them to be ineffective.
    c) grow weeds so quickly that a higher dosage of herbicide is needed.
    d) hold herbicides in the soil, thereby making them more effective.

20) An example of a nonselective weed killer is

    a) Roundup.   c) Dacthal.
    b) 2,4-D.   d) atrazine.

SECTION 6

Container-Grown Plants

◆

Terrariums

OBJECTIVE

To construct and care for terrariums.

Competencies to Be Developed

After studying this unit, you should be able to

◆ prepare a terrarium for planting by mixing a suitable medium and placing it with the proper drainage material in the container.

◆ identify ten plants suitable for use in terrariums.

◆ plant a terrarium.

◆ identify five cultural conditions that affect the growth of plants in terrariums.

◆ list four disorder symptoms of terrarium plants.

◆ prune two terrarium plants to control growth.

◆ list and identify five insects that affect terrarium plants.

MATERIALS LIST

✔ 2½-inch tall potted foliage plants
✔ plant mister
✔ single-edge razor blades
✔ bamboo stakes
✔ 6-inch wooden pot labels
✔ containers (brandy snifter, bottle, Plexiglas box, etc.)
✔ sphagnum moss or peat moss
✔ small pebbles or crushed rock
✔ activated charcoal

✔ teaspoon
✔ number 4 wooden pencil (very sharp)
✔ 16-gauge florist wire
✔ small funnel
✔ small paint brushes in assorted sizes
✔ lazy Susan
✔ sand in assorted colors
✔ construction grade sand

The original *terrarium* (container, usually covered, in which plants are grown) was developed by Dr. Nathaniel Ward, a nineteenth-century English surgeon. It was quite by accident, and in an experiment unrelated to horticulture, that Dr. Ward made his terrarium discovery. He had been attempting to grow bog ferns, but found that the polluted air from London factories near his home inhibited their growth. During this same period, Dr. Ward investigated many aspects of nature, including the life cycle of butterflies. As part of this work, he placed the pupa (cocoon) of a butterfly in a glass jar to keep it alive, along with some soil from the same area. After several weeks, he noticed a seedling growing inside. He found that when the seedling lost moisture from its leaves (transpiration), it collected on the walls of the terrarium. As the water formed and moved down the sides of the container, it was reabsorbed by the plant and soil. This established a rain cycle for the plant inside the container, making it an ideal place to grow healthy plants with a minimum of attention. Despite the rain cycle that developed, Dr. Ward found it necessary to occasionally add a few drops of water to prevent the plant from drying out. After several weeks, the plant continued to grow without any particular attention.

Dr. Ward went on to experiment with growing plants in containers of various sizes. As a result of this work, the *Wardian case* was developed. This enclosed case containing medium and moisture in which plants could be started and grown evolved into the modern-day terrarium. It was originally used to transport sensitive tropical plants which needed protection from sea air and the rough weather of sea voyages. Over the past 150 years, the case has continued to be used for both growing and transporting plants. Today, many nurserymen, horticulturists, and plant enthusiasts use the case for starting and displaying plants.

THE TERRARIUM CONTAINER

SELECTING THE CONTAINER

There are various containers suitable for the construction of terrariums, figure 21-1. When choosing a container, several things should be considered.

FIGURE 21-1 Various terrarium containers

LOCATION OF TERRARIUM The terrarium should harmonize with, as well as accent, the furnishings and materials already present in the surroundings. For example, a modernistic terrarium would not be a good choice for a room furnished with antiques.

SIZE OF TERRARIUM Terrarium size should be chosen in relation to the size of the area in which it is to be located. A terrarium which is too large or too small is less effective than one which is in proportion with the rest of the room.

COLOR OF GLASS The chosen container must allow sufficient light to reach the plants so that photosynthesis can occur. Dark tinted glass should be avoided since it filters out too much sunlight; plants are unable to grow in this environment.

SIZE OF CONTAINER OPENING The size of the opening in the container limits the type of plant which may be used. For example, if a one-gallon bottle is selected, small plants which will fit through the narrow neck of the container must be chosen. A wider variety of plants is possible when a container with a large opening, such as a brandy snifter, is used.

CONDITION OF CONTAINER Check to be sure the container is free from cracks or scratches. It should also be clean and dry. Water spots on the inside or outside should be removed so that the finished terrarium appears clean and attractive. If a closed container is chosen, it must be watertight. This helps to maintain the necessary humidity in the terrarium to allow best growth of plants.

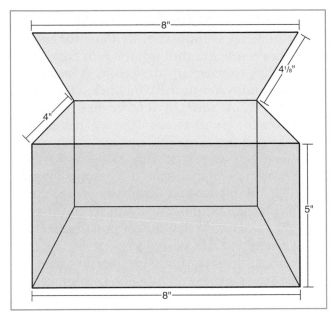

FIGURE 21-2 Plexiglas terrarium assembly

ᴘROCEDURE

CONSTRUCTING A PLEXIGLAS TERRARIUM

Many people enjoy constructing their own terrarium containers. One way in which this can be done is by using Plexiglas and special Plexiglas glue, both of which can be purchased at most hardware stores. To make a 2¼-gallon terrarium, follow the directions given below. The diagram for assembly is shown in figure 21-2.

1. Cut pieces the following sizes from a sheet of ⅛ inch Plexiglas, leaving its protective paper covering attached:

    —two pieces, 4 inches by 5 inches, for end walls

    —two pieces, 8 inches by 5 inches, for sides

    —two pieces, 8 inches by 4⅛ inches, for top and base

2. Glue the end pieces to the base of the terrarium. Set one end in place and carefully brush the glue onto the seam. Allow a few minutes for drying; the glue seals immediately.

3. Repeat this procedure for the other end and sides.

4. After allowing the sides to dry for a few minutes, place the top on the terrarium. Do not glue the top to the rest of the structure.

5. After planting, seal the top with masking tape.

PREPARING THE CONTAINER FOR PLANTING

Several layers of material are placed in the terrarium in preparation for planting, figure 21-3. These include a layer of drainage material, sphagnum moss, charcoal, and soil mix.

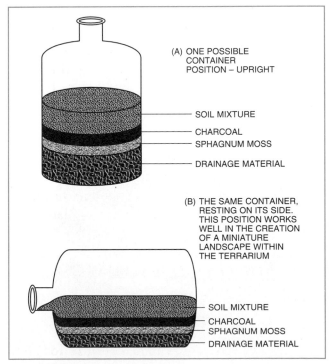

FIGURE 21-3 Two terrarium containers prepared for planting. Container (B) illustrates an interesting variation in container position.

DRAINAGE MATERIAL

Proper drainage is important to allow the movement of excess water through the root zone area. It also helps to maintain the proper balance between air and water in the container. Excess water in the root zone of terrariums makes it impossible for plants to absorb oxygen from the air in the soil mixture. The average soil mixture contains about 25 percent air, 25 percent water, 45 percent mineral matter, and 5 percent organic matter. If the plants in a terrarium are not provided with the best possible drainage in the root zone, they will turn yellow and eventually die from excess water and lack of oxygen.

Drainage in terrariums can be provided by crushed stone, pea gravel, washed river gravel, or pieces of broken clay pots. These materials are available at florists, nurseries, and garden centers. The material selected is the first to be placed in the container. It is spread directly on the bottom of the terrarium.

SPHAGNUM MOSS AND CHARCOAL

Sphagnum moss, composed of sterile leaf and stem tissue from acid bog plants, is spread in a layer over the drainage material in the container. The main purpose of the moss layer is to prevent the soil mixture from seeping into the drainage layer, figure 21-4.

FIGURE 21-5   A layer of charcoal helps in the absorption of bacterial odors in the terrarium

A thin layer of activated charcoal is spread over the sphagnum moss to absorb any odors in the terrarium, figure 21-5. These odors usually develop from bacteria which result from breakdown of soil particles in the terrarium.

SOIL MIXTURE

A good soil mixture placed over the charcoal layer is important since a wide variety of plant materials are usually used in terrariums. A good general potting mixture consists of two parts loamy soil, one part organic matter (peat moss), and one part horticultural grade perlite. If this mixture is used, it must be sterilized to kill all weed seeds and harmful disease organisms which might be present. To sterilize the soil, spread a layer of the mixture in a shallow pan, place in an oven, and bake at 180°F for thirty minutes.

Commercial media also work very well in terrariums. If only a small amount of medium is required for a terrarium, it is most practical to purchase commercial terrarium soil from a florist, nursery, or garden center. Fertilizer should not be added to closed terrariums. When the fertilizer breaks down, ammonia is released and causes the plants to burn and eventually die.

FIGURE 21-4   A ¼-inch layer of sphagnum moss is added to prevent the potting soil from seeping into the drainage material (in this case, a sand design)

SELECTING PLANTS FOR THE TERRARIUM

The process of selecting plants for a terrarium can be a very interesting and enjoyable experience. Usually, the first consideration in plant selection is the appearance of the plant and how it will fit into the overall design of the terrarium. One very important plant characteristic is leaf color. Some plants have solid foliage colors, while others are *variegated* (have more than one color), for example, a leaf may have a dark green center and a white outer edge. A combination of solid-colored and variegated plants creates an attractive contrast in a terrarium.

The texture, shape, and finish of the leaves and the shape and size of the plant are also important factors. Plants selected for terrariums are usually small, relatively inexpensive, and easy to work with when planting. Small, newly rooted cuttings or small potted plants produce the best results. However, larger plants work well if a large container with a good-sized opening is selected.

Terrariums vary in their environmental requirements, depending on the type of plants that are used. Since the entire terrarium provides the same environment, it is wise to choose plants that have similar basic needs in terms of light, soil, and moisture. Terrariums are often constructed by selecting one of the groups described below and using a variety of plants from that group.

TYPES OF PLANTS

CACTI AND SUCCULENTS Cacti and succulents require natural desert conditions and full sun. They do not require a great deal of humidity, and therefore work well in open containers. During the spring and summer, cacti and succulents are watered to promote active growth. They enter a period of rest in winter, during which watering is reduced. The medium previously recommended (one part loamy soil, one part organic matter, and one part horticultural perlite) with the addition of one part coarse sand is recommended for these plants.

WOODLAND PLANTS Woodland plants are native to forests or woods. When grown away from their natural environment, they require similar conditions to those present in a typical woodland setting. The woodland terrarium requires dense shade, soil with a high organic content, and a good supply of moisture. The soil mix previously mentioned (one part loamy soil, one part organic matter, and one part horticultural perlite) usually works well in woodland terrariums.

Familiar woodland plants include ferns, mosses, yellow root, ground ivy, jack-in-the-pulpit, lichens, oxalis, partridgeberry, trailing ar-

butus, and violets (Johnny-jump-ups). These plants grow well in closed terrariums because of the humidity created by the accumulation of moisture in the closed container. Generally, woodland terrariums are easy to manage and keep clean.

TROPICAL PLANTS Tropical plants respond favorably to a more humid type of terrarium than either cacti or woodland plants since their natural environment is the moist tropical regions. Light requirements for this group vary with different plants. It is advisable to use those plants which require low light. Although there is a great variety of large tropical plants, the smaller varieties are best for terrariums since they are easier to work with and less expensive.

PLANTING THE TERRARIUM

PLANTING EQUIPMENT

There are a few simple tools which simplify planting terrariums. A funnel is used to work the soil mixture into the container. In this way, soil can be evenly distributed in the terrarium. When performing this operation, it is best to avoid getting soil on the sides of the container, since it is difficult to remove after planting. A wooden dowel is used to form planting holes in the soil mixture. One end is sharpened for digging while the other end is used to tamp the soil around the plant roots. When planting in a narrownecked bottle, a bamboo stake slit about 2 inches on one end can be used to hold plants. The plant is slipped into the slit in the stick to hold it securely while planting. A bent spoon is helpful in adding soil around the plants. After the terrarium is planted, a mister is used to spray the foliage.

PLANTING TECHNIQUE

The following steps comprise the planting process in the construction of terrariums.

Procedure

PLANTING A TERRARIUM

1. Push the wooden dowel into the soil mixture to form holes in which to place the plants.

2. With the plant held by the split in the bamboo stake, transplant the plant into any container with a small opening, such as a bottle. (Gently bending the leaves does not harm the plant.)

3. After the plant is in position, carefully remove the plant from the bamboo stake with the dowel. Lightly tamp the soil around the plant roots with the dowel to help hold the plant straight and firm.

4. Add plants until the terrarium design is complete.

5. Water the completed terrarium with a mister. Spray the foliage until water droplets form on the leaves. Finally, seal the terrarium and enjoy watching it grow.

FIGURE 21-7 A terrarium with a plastic lid after fogging has cleared. It contains common house plants growing in a mixture of equal parts of garden soil, sand, and peat moss. A 1-inch layer of charcoal has been placed in the bottom of the container for drainage. The sides of the terrarium below the soil level have been lined with moss.

Plants give off moisture. During this process (transpiration), moisture is given off from the small openings on the underside of the leaves known as *stomata* (plural of *stoma*). When plants are in a closed terrarium, the moisture collects on the sides of the container and drops back into the soil so that the plants may use it again, figure 21-6. If the moisture is concentrated in the terrarium, the container may fog up. In this case, remove the top and allow the moisture to escape. This establishes the balance of air and moisture in the terrarium. After the fogging has cleared, replace and seal the container top.

If a terrarium lacks moisture, the leaves of the plants may droop, and the soil may appear lighter than normal. When this occurs, spray the terrarium with a mist of water until beads of water form on the leaves of the plants. Seal the terrarium

and allow the rain cycle to become reestablished. The plants can then continue their natural growth process, figure 21-7.

SAND DESIGN

Unique effects can be developed in terrariums by creating designs in the sand which is used as the drainage material. This process, known as *sand art*, is not new but has recently gained popularity in terrarium construction.

MATERIALS AND EQUIPMENT

Most of the materials and equipment for sand design are easy to obtain. The most important item is colored sand. It is best to buy it already colored rather than attempt to color it. The amount of colored sand used will vary with the size of container selected. A construction grade sand is also needed. Its use as filler material reduces the amount of the more expensive colored sands which are needed.

Basic equipment for sand design consists of an old metal spoon, a piece of 16-gauge florist wire, a 4H wooden pencil, a funnel with a small piece of plastic hose, and several small paint brushes. The metal spoon is bent at a right angle and is used to place small quantities of sand in the container. Larger amounts of sand are added to narrownecked containers by using the funnel and piece of plastic hose. Various designs are created with the aid of the pencil and florist wire. The paint brushes are used to replace sand which falls outside the design. If a lazy Susan is available, place

FIGURE 21-6 Terrarium with established rain cycle. Notice the condensation forming on the glass.

the terrarium on it while the design is being created. This permits turning of the terrarium without disturbing the sand.

DESIGN TECHNIQUE

The first step in sand design is to decide upon the type of design which is to be created. This may range from simple layering of colored sand to the creation of a landscape or picture in the terrarium. A layered design is the easiest type to construct and the best choice for the beginner in sand design.

Almost any clear container can be used for sand design, but the wider the container opening, the easier it is to create the design. A brandy snifter is an ideal container.

When using colored sand, small quantities can be placed in plastic drinking glasses or similar containers. Taking the sand from these smaller containers when the designer is ready to use it in the terrarium helps to avoid wasting sand.

CREATING A LAYERED DESIGN

If a layered design is desired, it is best to select three to five colors of sand. The first layer is placed as a base in the bottom of the container. It should completely cover the bottom. To place this sand, use the bent metal spoon described in the section on equipment.

After the base color is in place, use the spoon to add construction grade sand in the center of the colored sand. The construction sand should come within a ¼ inch of touching the glass sides of the container. In other words, the colored sand need only be a ¼-inch layer at the outside of the design. It is held in place against the glass by the construction sand and appears to be the only material used.

The layers may consist of a variety of colors and may be placed in any sequence which gives a pleasing appearance. Interest may be added by varying the thickness of the different colored layers, figure 21-8.

> *Caution: Do not shake or tip the terrarium while working on the design.*

If the terrarium is shaken but the colors are not mixed, more colored sand can be added to fill in the spots where the sand has fallen to the center. Additional filler sand is then added to hold the colored sand in place.

FIGURE 21-8 Creating a pattern with layers of colored sand is probably the best method of sand design for the beginner.

FIGURE 21-9 A 2- to 3-inch layer of moist soil is placed over the sand design.

When the colored sand is built up to the desired level, add the prepared soil mixture in a 2- to 3-inch layer over the sand design, figure 21-9. Be sure to keep the glass clean as the soil mixture is added.

CREATING LANDSCAPE SCENES

To create a landscape in sand design, first imagine how the scene will appear in the terrarium. A good practice is to draw a picture to follow when placing the sand. Land forms such as snowcapped mountains, valleys, rivers, seashores, and deserts can be created.

Snowcapped mountains are made by forming the base of the mountains in black or dark brown sand and topping them with white sand to make the snowcaps. Once the base of the mountain has been formed, carefully push a sharpened wooden pencil or florist wire down the side of the glass to make an indentation in the black sand. The white sand falls down into the indentation as the pencil or wire is slowly withdrawn. By repeating this procedure several times, a series of peaks and valleys is created. This pattern may be used on all sides of the container.

In one valley, create a sunset or sunrise by filling in the valley with red or red softened with red-yellow sand. The appearance of the sunset is up to the individual designer. To form the sunset, carefully place the empty spoon in the area where the sunset is to be located. Press firmly to prevent the sand from sliding out of place. Take a small sample of the sand to be used for the sunset and add it to the terrarium. If grains of red sand drop out of place, brush them back to the center with a small paint brush. Continue adding sand until the desired sunset is created.

A river or ocean can be created in another valley by using a deep blue sand and following the same procedure as was used to form the sunset.

The sky is made by mixing equal amounts of blue and white sand. Using the spoon, spread this mixture over the entire landscape, making it thicker over the water and sunset. This gives depth and contrast to the mountains.

Birds in flight are designed by taking the spoon and firming the sand in the place where the bird is to be located. Place about ½ teaspoon of black sand in the sky. Then push a sharp wooden pencil down through the center of the black sand about ¼ inch into the sky. The black sand flows down into the indentation made by the pencil to form the shape of the bird. Any color sand may be used for birds. To show birds which appear to be in the distance, make them smaller. This technique adds depth to the design.

SEALING THE SAND DESIGN

If it is important that the colored sand be prevented from fading, the sand design can be covered with a sealer. Sealer can be applied in several ways. A commercially prepared sealer, available at most florists, garden centers, and large department stores, should be applied according to manufacturer's direction. A homemade sealer consists of melted paraffin (hot, liquid wax) which is poured in a thin layer over the sand. However, this method is difficult and hazardous and should be avoided by the novice sand designer. If it is used, extreme caution should be exercised to avoid scalding spills.

If colored sand is sealed, it does not serve as drainage material, but strictly as a decorative feature. To prepare a terrarium for planting in which a sand design has been created, follow the same procedure as that for preparing a terrarium without a sand design. Place the drainage material, sphagnum moss, and soil mixture on top of the sealed design. Refer to figure 21-3 for proper placement of these materials.

CARING FOR THE TERRARIUM

After the terrarium has been planted, it should be checked periodically. It is especially important to see that it continues to have the proper light, moisture, and temperature.

DISORDER SYMPTOMS

MUSTY ODOR This can be corrected by the addition of activated charcoal, mixed into the medium, to the terrarium. The charcoal absorbs the bacterial odor and aids in drainage.

CHLOROTIC (YELLOWISH) AND STUNTED GROWTH OF PLANTS This condition could result from improper drainage, improper ventilation, or improper soil mixture.

DECAYING FOLIAGE This may be the result of several things. Improper ventilation, one possible cause, may be corrected by removing the sealed lid to ventilate the terrarium. High nitrogen fertilizers may be present in the soil mixture, giving off ammonia and thus burning the plant leaves. In this case, it is best to replace the soil mixture.

BROWN FOLIAGE This indicates that the terrarium has not received enough water.

PRUNING

In some cases, the growth of plants in the terrarium is so great that they become too big for the container. This may be due to the plant type or very favorable cultural conditions. Whatever the reason, the plants must be pruned to remove excess foliage. Pruning plants causes them to branch out and become fuller.

Use a single-edge razor blade for pruning. If the terrarium container is a long-necked bottle, insert a single-edge razor blade into a bamboo stake.

> *Tips: Remember to remove any material that is pruned from the terrarium to prevent it from decaying inside the container. Some areas in the terrarium may be difficult to reach. Pruning can be accomplished using long scissors.*

INSECT PESTS

Insect pests are controlled most efficiently by planting the terrarium with clean, insect-free plants. At times, however, pests may be present in the soil and not noticed when the terrarium is planted. Some insecticides may be used for insect control in terrariums; consult the local extension service for recommendations.

Specific insects that plague terrarium plants include the whitefly, mite, aphid, mealybug, scale, and ant.

WHITEFLIES Whiteflies have very small, white, wedge-shaped wings. When the plant is disturbed, these insects fly around the plant's leaves where they can be seen. Whiteflies feed on the underside of the leaf. Control: Spray with orthene or insecticidal soap.

MITES These are very small insects and difficult to see. If it is suspected that a plant has mites, take a sheet of white paper, insert it under the leaf, and tap the leaf lightly. If mites are on the plant, they will fall onto the paper. Check the paper for small moving insects. Draw a finger across the insect and the paper; if mites are present, there will be a red mark on the paper. Control: Apply a systemic insecticide to the soil. (*Systemic* insecticides—used to control sucking insects—are absorbed into the plant's system through its leaves and roots. As the insect sucks the plant juices containing the insecticide, the body system of the insect is affected, thus causing death.)

MEALYBUGS Mealybugs are small, flat, white insects that resemble a small spot of cotton on the plant. Mealybugs suck the plant juices until the plant dies. Control: Swab the areas of the leaves where the bugs are located with alcohol.

SCALE INSECTS These pests harm plants by sucking the juices from them. Scale insects are small, gray to brown in color, and flat, with a soft or hard shell. They can be found sticking to the stems of plants. After scale attacks, the plant becomes yellow, then turns brown, and finally dies. Control: Spray with orthene.

ANTS Tiny black ants are drawn to some plants by the presence of honeydew on their leaves. *Honeydew* is sap which is secreted from the leaves of plants after an attack by aphids, scale insects, or, sometimes, a fungus. These ants usually attack the root system of the plant. Control: Spray with orthene.

WATERING

If moisture beads are present on the sides of the terrarium container, there is a good chance that the rain cycle has been established. Be cautious of overwatering. If the terrarium is overwatered, the plants may die from poor drainage and a lack of air circulation in the soil. In the case of overwatering, it is best to open the top of the terrarium and allow it to dry out somewhat. When enough water has evaporated, reseal the terrarium to reestablish the rain cycle.

TEMPERATURE

The ideal temperature of terrariums depends to some extent upon the type of plants they contain, but a temperature of 65°F to 75°F is suitable in most cases. Remember that the temperature is greatly influenced by the location of the terrarium. For example, if it is located in or very close to a window, it will experience greater changes in temperature than a terrarium which receives indirect light.

LIGHT

Different types of plants require different amounts of light. When selecting plants for terrariums, refer to Unit 23, Section 6 of this text for correct light exposures. Remember that plants in terrariums located near windows grow toward the light source (phototropism).

ARTIFICIAL LIGHT Terrariums are excellent decorative additions to the home, office, or school. The sunlight in these areas may be very limited, however, making artificial light necessary. There are special lights on the market today that are especially good for this purpose. A fluorescent light is a better choice than an incandescent light (standard light bulb).

Various manufacturers of fluorescent lights publish information concerning the placement of the lights for best results in the terrarium. On the average, it is recommended that they be located about 15 to 18 inches away from the terrarium. The light should be provided from twelve to fourteen hours per day. Flowering plants require more light than others.

FERTILIZERS

If plants are grown in a good medium, additional fertilizer is not needed. Fertilizer tends to stimulate excessive growth of terrarium plants, causing the terrarium to appear unkempt and cramped.

If the plants must be fertilized, use an organic fertilizer such as bone meal. Simply sprinkle a small amount of the fertilizer in the terrarium. Organic fertilizers release slowly and do not burn plants, as inorganic fertilizers tend to do.

PREVENTIVE DISEASE CONTROL

When the terrarium has been completed, preventive measures should be taken to control disease organisms that may be present in the terrarium. One chemical that is commonly used is Benlate. Follow manufacturer's directions for its use. Use a spray bottle to ensure even distribution of the chemical on the foliage and soil.

 Student Activities

1) Ask a local florist to visit the class and demonstrate the various techniques used in terrarium construction and planting.

2) Organize a field trip to a woods area and collect native plant materials for terrarium plantings.

3) Propagate plants for the terrarium. For woodland plants, students might visit a nearby forest or woods and collect seeds, sow them, and select the terrarium plants from the results.

4) As a class project, construct and plant three terrariums, using the following media: (1) soil; (2) two parts loamy soil, one part organic matter, and one part horticultural perlite; and (3) Pro-Mix. Compare the growth in each terrarium. (Students might consider constructing and selling other terrariums as a fund-raising project.)

5) As the plantings in the three terrariums grow, observe and record the results. Care for the plants, employing the techniques found in the text. Prune when necessary.

6) Construct stands with artificial lighting for the terrariums.

 Self-Evaluation

Select the best answer from the choices offered to complete the statement.

1) When selecting a container for terrariums, one must consider

  a) size.
  b) color of glass.
  c) desired environmental conditions.
  d) all of these

2) Charcoal is used in a terrarium for its

  a) color.
  b) drainage ability.
  c) ability to absorb bacterial odors.
  d) none of these

3) Crushed rock or pebbles are used in terrariums for

   a) soil mixture.
   b) added weight.
   c) drainage material.
   d) none of these

4) A musty odor in a terrarium can be eliminated by adding

   a) stone.
   b) peat moss.
   c) more soil.
   d) activated charcoal.

5) Terrarium plants which are chlorotic (yellow) and are stunted in growth may have this condition because of

   a) improper drainage.
   b) improper ventilation.
   c) improper soil mixture.
   d) all of these

6) The use of high nitrogen fertilizers in terrarium soil results in the release of

   a) moisture.
   b) new plants.
   c) ammonia gas.
   d) none of these

7) The pruning of terrarium plants is needed to

   a) remove excess growth.
   b) remove dead leaves.
   c) remove broken plants.
   d) all of these

8) Terrarium plants that have small, white, wedge-shape-winged insects on the underside of the leaves are probably infested with

   a) whiteflies.
   b) horseflies.
   c) ants.
   d) none of these

9) The ideal temperature for terrariums depends upon the

   a) amount of soil used.
   b) size of the terrarium.
   c) number of plants in the terrarium.
   d) type of plants in the terrarium.

10) Plants tend to grow toward light sources. This is known as

   a) photosynthesis.
   b) transpiration.
   c) the rain cycle.
   d) phototropism.

11) Flowering plants in terrariums usually require

   a) more light than foliage plants.
   b) less light than foliage plants.
   c) the same amount of light as foliage plants.
   d) none of the above

12) Plants requiring a natural desert condition and full sun are

   a) woodland.
   b) tropical.
   c) cacti and succulents.
   d) all of these

13) Plants which are native to the forest or woods are

   a) woodland.
   b) tropical.
   c) cacti and succulents.
   d) none of these

14) Ferns, mosses, ground ivy, and lichens are classified as

   a) tropical.
   b) woodland.
   c) cacti and succulents.
   d) none of these

15) The loss of moisture from plant leaves is known as

   a) condensation.
   b) evaporation.
   c) tropism.
   d) transpiration.

16) Moisture in plants is lost through small openings on the underside of the leaves known as

a) air holes.

b) stomata.

c) stems.

d) none of these

17) Once a terrarium has the correct balance of air and moisture, it has established its

a) air cycle.

b) rain cycle.

c) transpiration.

d) life cycle.

18) If there is too much moisture in a terrarium, it will

a) dry out naturally.

b) fog up.

c) pop its top.

d) none of these

19) A sealer over a sand design in a terrarium will prevent _____ of the colored sand.

a) fading.

b) erosion.

c) transpiration.

d) none of these

20) Ideally, a woodland terrarium should have

a) dense shade.

b) soil which is high in organic matter.

c) a good supply of moisture.

d) all of the above

The Art of Bonsai

OBJECTIVE

To construct and maintain an attractive, healthy bonsai planting.

Competencies to Be Developed

After studying this unit, you should be able to

- ◆ explain the purpose and general technique of bonsai.
- ◆ list the steps in the construction of a bonsai planting. This may be done orally or in writing, followed by a demonstration of the technique.
- ◆ describe the aftercare of bonsai plantings.
- ◆ explain how to repot bonsai plantings.

MATERIALS LIST

- ✔ bonsai containers
- ✔ 9- to 20-gauge copper or painted aluminum wire (for shaping)
- ✔ screen wire (to cover drain holes)
- ✔ potting medium and gravel
- ✔ chopsticks or similar sharp sticks
- ✔ pruners (some should be for fine work)
- ✔ potted plants (to be trained to bonsai)

Bonsai, a word of Japanese origin meaning "tray planting," is the art of dwarfing plants by growing them in shallow pots and trays and by judicious trimming of roots and tops. The end result is a plant which resembles a large tree in miniature. By applying the techniques of bonsai, the horticulturist is able to produce an 18-inch tree which resembles an 80-foot tree in all aspects except size. Part of bonsai tradition

FIGURE 22-1 Japanese White Pine (Pinus parviflora) 150 years old. From the collection of His Majesty Hassan II, King of Morocco. Now at the National Arboretum. (Ed Reiley, Photographer)

FIGURE 22-2 Folmina juniper (Juniperus chinensis). A group planting 52 inches high and 49 inches wide. Trained since 1953 from nursery stock and donated to the National Arboretum collection by John Y. Nako of Los Angeles, California, and named "Goshin" meaning "protector of the spirit." (Peter L. Bloomer, Photographer)

is to make these plants appear old and rugged or graceful and wispy, just as they appear in nature after the passing of many generations, figure 22-1. Bonsai designs may be simple or complex, ranging from a simple one-tree planting to a miniature forest, figure 22-2.

PLANTING THE BONSAI

CHOOSING THE PLANTS

Plants used in bonsai may be evergreen or deciduous. The plants that are selected should have small leaves. This quality helps them to appear more as reduced size models of larger trees or plants. Plants

selected should also be able to grow in crowded conditions. Flowers, berries, and fruit can add to the attractive quality of bonsai specimens. The following species are some of the best choices for bonsai plantings.

Evergreens
Japanese white pine (*Pinus parviflora*)
Creeping juniper (*Juniperus procumbens "Nana"*)
Hinoki cypress (*Chamaecyparis obtusa*)
Azaleas (Satsuki and Kurume)
Japanese yew

Deciduous plants
Japanese maple (*Acer palmatum*)
American hornbeam (*Carpinus caroliniana*)
Japanese flowering apricot (*Prunus mume*)
Japanese flowering quince (*Chaenomeles lagenaria*)

SOURCE OF PLANTS

There are two sources of bonsai: naturally dwarfed and artificially dwarfed plants.

NATURALLY DWARFED PLANTS It is sometimes possible to find old plants growing in the wild that have been twisted and dwarfed by cold or dry conditions. Collecting such plants is discouraged due to depletion of wild plants. It is best to grow seedlings for bonsai use.

ARTIFICIALLY DWARFED PLANTS The best source of plants is a nursery, where potted plants that have the potential to be pruned to the desired shape may be purchased. Plants with small leaves or needles and which have an interesting trunk line should be chosen. They should also be

FIGURE 22-3 A 250-year-old Sargent juniper trained to resemble an old, twisted tree. The juniper is 29 inches tall. (Ed Reiley, Photographer)

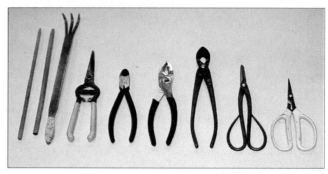

FIGURE 22-6   Tools used in preparation of plants for bonsai. (Ed Reiley, Photographer)

FIGURE 22-4   An evergreen plant (Chamaecyparis obtusa "Nana Gracilis") chosen for the bonsai done in this unit. When selecting container-grown plants for bonsai, the desired shape must be visualized and one branch selected for the main trunk.

selected according to their ability to grow well in the area, figure 22-3.

Plants for bonsai may also be grown from seed or cuttings. These plants are trained from the very beginning by restricting their growth and by pruning to obtain the desired shape. The first step is selection of the plant. Figure 22-4 shows a container-grown plant selected for a bonsai.

CONTAINERS

The bonsai container should

- ◆ have one or more drain holes.

- ◆ be shallow.

- ◆ have a length two-thirds the height of the plant placed in it.

- ◆ not detract from the beauty of the plant placed in it. Subdued earthy colors are best; browns, greens, and grays are most natural.

- ◆ have a simple form (round, square, or rectangular).

- ◆ be made of pottery and unglazed on the inside.

Figure 22-5 illustrates one type of bonsai container.

TOOLS

Various tools used to cut and prune plants in preparation for bonsai are shown in figure 22-6. All of these tools were used to prepare the bonsai shown in this unit.

DWARFING PLANTS

DETERMINING PLANT SHAPE

The first step is to visualize the bonsai design in its final form. Will it have one trunk or several? Will there be more than one plant in the container? Is the plant to be set straight or slanted in the

FIGURE 22-7 The first cut in starting the bonsai is to shorten the plant. (Ed Reiley, Photographer)

FIGURE 22-8 Other branches to be saved are selected and all other branches are removed. (Ed Reiley, Photographer)

container? Will roots show above the soil level or be covered? These are some of the questions that must be answered before pruning is begun. When pruning, remember that the trunk, the rootage, apex, and branches of the plant are the focal points of the design.

Pruning the top of the tree is essential in developing the rugged but miniature shape which is characteristic of bonsai. Any active shoots must be shortened and undesirable buds or shoots removed. There should be some space between limbs or side branches to give the effect of a tall tree. See figures 22-7 and 22-8.

INITIAL PRUNING

Remove all of the unwanted branches. This step must be done very carefully to ensure that enough branches remain to form the desired shape for the entire plant. Figure 22-7 shows the same plant as the one in figure 22-4 after the first pruning is done.

WIRING

The remaining branches are wired with copper or aluminum wire, figure 22-9. Branches are wired to force growth in a direction other than the direction in which they naturally grow. Copper or painted aluminum wire is desirable because it does not rust and is not highly visible. Heavy wire is used for the larger branches and trunk and smaller wire for small branches. The size of the wire used varies from 9 to 20 gauge. The wire must be pliable enough to bend and yet rigid enough to hold the plants in shape. Cut the wire one-third longer than the combined length of the trunk and branch to be wired. Stick one end of the wire into the soil mix to hold it firmly and wire the trunk. When wiring branches, do two wires starting at different levels on the trunk. The two wires overlap and stay in place better. Wrap the branches with spirals about 1 inch apart, being careful not to tear the bark on the twigs. Avoid wrapping plant foliage under the wire since this foliage may die and disfigure the plant. When the wrapping is finished, cut the end of the wire off close to the branch so that the ends do not show.

When two branches are to be pushed apart, one end of the wire is used on each branch. The first branch is wired and pushed down, and then the second branch is wired, figure 22-10. Notice that the two pieces of wire are twisted in opposite directions. After the wire is twisted on both branches, the branches are pushed in the desired direction. The wire holds them in place. The ends

FIGURE 22-9 Painted aluminum wire to twist around and hold branches in place. (Ed Reiley, Photographer)

FIGURE 22-10 Selected branches wired and twisted into place to form the shape of the new bonsai. (Ed Reiley, Photographer)

FIGURE 22-11 The root ball is dug down from the top and in from the sides to make it shallower and smaller in diameter. (Ed Reiley, Photographer)

of the wire are cut off close to the branches so that pieces of wire do not protrude from the plant. The wire should not be left on long enough to girdle the branches as they grow.

> *Tips: Be careful not to twist the wire too far and break the branches or to twist the branches from the trunk.*

ADDITIONAL PRUNING

The top of the plant is now pruned to remove some of the small branches and shorten others. The branches on top are pruned to form the desired shape. The needles of the bonsai shown in this unit were pulled in half to shorten them.

ROOT PRUNING

After the top is pruned, attention is focused on the roots. Dig the soil ball apart by gently pulling down with a chopstick or other tool through roots which make up the root ball, figure 22-11. Continue until most of the soil is removed, and the roots are exposed. Expect many roots to be torn off in this process. The large cutters are used to cut off the center tap root, figure 22-12. Many of the roots must be cut back so that the new plant fits in the bonsai container, figure 22-13.

PLACING THE PLANT IN THE CONTAINER

The plant is now ready to be placed in the bonsai container. The container is prepared by placing a screen wire over the drainage holes in the bottom. A thin layer of gravel is then spread over the screen

FIGURE 22-12 The bottom of the root ball is cut off. (Ed Reiley, Photographer)

FIGURE 22-13 Root ball cut to final size to fit the container. Note that many small fibrous roots remain. (Ed Reiley, Photographer)

FIGURE 22-14 Small gravel in the bottom of the pot, ready to plant. (Ed Reiley, Photographer)

wire to ensure proper water drainage, figure 22-14. The plant is set on top of the gavel and pushed down so that the root level is higher than the edge of the container. It may be necessary to cut off the bottom of the root ball, but save enough fibrous roots. Notice that the plant is placed in the center of the round container. Bonsai are generally not placed in the exact center of their containers except for hexagonal, round, or square pots. Before adding soil, tie the plant to the container by guiding a piece of copper or aluminum wire up through a drain hole and over part of the roots or main stem.

ADDING POTTING MEDIUM

A potting medium with good drainage is now placed on top of the root system and gently worked through the roots with a pointed stick to fill any air pockets, figure 22-15.

FIGURE 22-15 The growing medium is worked down around the roots with a small stick. No large air spaces should remain around the roots. (Ed Reiley, Photographer)

The potting medium should be of such structure that it is well drained and yet holds moisture. The Japanese use fired silt clay called "Akadama." The mix used by Dr. Daniel Chiplis, Assistant Curator at the U.S. National Arboretums "National Bonsai and Penjing Museum" to plant the bonsai shown in this unit is as follows:

> 1 part rock (which is 2 parts crushed granite and 1 part lava stone)
>
> 1 part organic (pine or fir bark and a nursery mix of decomposed peat moss, bark, sand, and vermiculite or perlite)
>
> 1 part fired clay particles (turface)

The mix is screened to take out all fine particles which might plug up pore spaces and reduce drainage.

The medium should cover all of the roots except for large roots near the trunk, which are sometimes left exposed for effect. Exposed roots cause the plant to look like an older tree whose large roots have gradually been exposed.

Fill the container to a point within ½ inch of the top of the edge and form a gradual mound toward the center of the container to cover the plant roots. Work the medium down among the roots so that there are no large air pockets.

WATERING

Soak the newly potted plant in water until the roots and medium are wet.

FIGURE 22-16 The finished bonsai. Note the taper of the trunk and the spacing of branches causing the miniature plant, about 12 inches tall, to closely resemble an old tree. Moss placed on top of the soil breaks up the smooth surface adding appeal to the finished planting. (Ed Reiley, Photographer)

FIGURE 22-17 Kyushu azalea (Rhododendron indicum "Chinzan"). Trained to a miniature informal upright style, this plant is 5 inches tall. It was created from nursery stock about 1985 and donated by Thomas G. Wright of Atlanta, Georgia, to the National Collection of North American Bonsai, located at the National Arboretum. (Peter L. Bloomer, Photographer)

Figures 22-16 and 22-17 show the finished product. The new plants should be protected from wind and sun for at least two weeks until they recover from the severe pruning.

CARE AND MANAGEMENT OF BONSAI

Aftercare of the plant itself consists of pruning the roots every one to three years so that they fit the container and shaping the top to obtain the desired effect.

WATERING

Keep the soil moist; the bonsai has a small root system and must not be allowed to dry out. In summer this means watering every day. Add water until it runs through the bottom of the container. Do not keep the roots wet. Indoor bonsai may need watering only two or three times a week.

LOCATION

During summer months, keep the bonsai outdoors in light shade or sun depending on the type of plant. Protect it in a cold frame or greenhouse in winter months to prevent the roots from freezing to temperatures below 23°F (−3°C) and winds from drying the plant out. Storage in a cool basement in winter is adequate. The plant must be watered and cared for while it is inside. A bonsai should not be kept inside except when protection from heat or cold is needed.

FERTILIZING

Organic materials such as cottonseed meal, well-rotted animal or vegetable manure, or the newer liquid fertilizers work well. Sprinkle on the surface (2 teaspoons per plant of organic fertilizer every six weeks from midspring to August 1). Liquid fertilizers, at one-quarter strength, may be applied with watering. Apply them about three times a year, beginning when active growth starts, and repeat every four weeks until August 1.

REPOTTING AND TRAINING PROCEDURE

As the new plant grows, additional pruning is required. Repotting should be done whenever the plant becomes root-bound or roots have curled around the root ball. It is very important to repot bonsai to keep the soil fresh and the roots actively growing. Repotting should generally be done every one to two years for deciduous trees and every two to three years for evergreens. Fast growers need repotting sooner than slower growers. The best time to repot is in early spring, just before new growth begins. When repotting bonsai plants, root pruning is the most important consideration. Roots must be pruned to bring about the growth of new roots. Many small feeding roots will emerge from the cut ends of large roots.

PROCEDURE

REPOTTING

1. Remove the plant from the container.
2. Straighten out roots if curled in a ball.
3. Remove one-third of the soil from the outside of the root ball, being careful not to break up the rest of the root ball. This process is much easier if the soil is dry.
4. Trim the roots back to the soil left in the root ball with sharp scissors or shears. (One-third of the root length is cut off.)
5. Repot the tree, using the same procedure as was used when the bonsai was originally potted. The new root ball should have 2 inches of space between it and the outside edge of the pot. New soil is worked firmly into this area with a small stick. The 2 inches of

new soil provide room for new active feeding roots to develop and keep the root system active. Soil is packed tightly against all roots so that no air space is left. Take the time to do this job thoroughly. The soil added should be dry so that it sifts down around the roots.

6. If support is needed, tie the plant to the container by guiding a piece of copper or aluminum wire, threaded through a small plastic tube where it contacts the root system. The wire will not show, since it is covered with soil. Do not wrap the wire completely around the roots, or they will be girdled.

Bonsai can be both a rewarding hobby and a business, and can be accomplished by anyone with the patience to work with the plant and wait for its beauty and character to develop. It is not necessary to wait fifty to one hundred years for an attractive bonsai. The small plant used for demonstration purposes in this unit will develop shape, character, and beauty in just a few years.

Student Activity

Select a container and suitable plant and prepare a bonsai planting following the steps in the text. This may be done in groups of two to four students or individually.

Self-Evaluation

A) Select the best answer from the choices offered to complete each statement.

1) Plants selected for bonsai should have small leaves because

   a) large leaves hide too much of the trunk.
   b) large leaves get in the way when wiring and shaping.
   c) they make the plants more closely resemble large trees, and the scale of all parts appears more nearly correct.
   d) large leaves lose too much moisture.

2) The bonsai container should

   a) be shallow and well drained.
   b) have a simple form and not detract from the plant placed in it.
   c) be made of pottery and glazed on the inside.
   d) all of the above

3) Copper or painted aluminum wire is used for shaping plants because

   a) it is inexpensive.
   b) it is pliable but strong and does not rust.
   c) it is the only wire available in the sizes needed.
   d) it does not girdle plants.

4) The large center root of plants used in bonsai is cut off

   a) to stimulate growth of small feeding roots.
   b) to cause root branching.
   c) to make the root system shallow enough to fit in the container.
   d) all of the above

5) The new bonsai plant is wired in the container because

    a) the shallow root system may not be able to anchor the plant and hold it up.
    b) it helps to restrict root growth.
    c) the potting medium is too light for good anchorage.
    d) all of the above

6) The new bonsai plant must be watered regularly because

    a) it has a small, shallow root system.
    b) the small container cannot hold much moisture.
    c) any drying out would damage a valuable plant.
    d) all of the above

7) The bonsai should be kept

    a) outside most of the time in light shade.
    b) inside year-round in a protected area.
    c) in a cool basement.
    d) either inside or outside in a very dark place.

8) Bonsai plants generally must be repotted after one to three years

    a) so that new soil can be added, encouraging new root growth.
    b) to invigorate the root system.
    c) because the roots become pot-bound in this length of time.
    d) all of the above

B) Briefly list the eight steps in the construction of a bonsai planting.

The Interior Landscape: Houseplants and Plantscaping

OBJECTIVE

To grow and maintain healthy interior landscape plants.

Competencies to Be Developed

After studying this unit, you should be able to

- ◆ list the four major concerns in caring for interior plants.
- ◆ mix a medium to use for growing various interior plants.
- ◆ describe the best cultural conditions (water, light, and soil requirements) for ten selected interior plants of your choice.
- ◆ list the four methods of watering interior plants.

MATERIALS LIST

- ✔ fifteen interior plants from the plant list in the text
- ✔ one bag of soluble fertilizer
- ✔ one bag of slow-release fertilizer
- ✔ peat moss, soil, construction grade sand, clay pots, and pieces of clay pots

Plants have been a special part of people's lives and environments throughout time. Of course, plants that bear food and materials for making articles have always been essential to life, but the plant that simply adds beauty to its surroundings serves a special purpose. Since the initiation of the environmental protection and "green survival" movements, the popularity of interior plants has grown rapidly. In 1971, $38 million worth of interior plants was sold. In 1976, the figure jumped to $480 million.

This increased interest in live indoor plants is reflected in both public places and private homes. Shopping malls, office buildings, and motel complexes are now using live plants to landscape indoor

FIGURE 23-1 Interior plantscaping has expanded throughout the country. It enhances shopping malls and office complexes by providing a natural garden atmosphere. (Courtesy Creative Plantings, Inc.)

FIGURE 23-2 Interior plantscaping creates attractive natural barriers to direct the flow of pedestrian traffic. It also provides an area where shoppers can relax in a peaceful environment. (Courtesy Creative Plantings, Inc.)

areas. This method of interior plant use is called *plantscaping*. The ancient practice of building homes around an *atrium* (indoor garden accessible from all main rooms of a house) is receiving renewed attention. People who live in crowded urban areas often find the idea of a private garden of interior plants particularly appealing. Interior plants can be used solely for the interest they create or to serve a specific function. Plants act functionally when used as room dividers, screens, and to accent pieces of furniture. See figures 23-1 and 23-2.

CARING FOR INTERIOR PLANTS

To care for interior plants properly, the following questions must be answered for each type of plant being grown.

- ◆ What is the proper way to water the plant?
- ◆ How much light does it need?
- ◆ What is the best temperature for optimum growth?
- ◆ What type of soil mixture is needed?

WATERING

To be sure that houseplants are watered properly, there must be careful observation of the plant's behavior. There are several signs of improper watering. Drooping leaves or leaves that do not seem as full as they should may indicate a lack of water. Too much water can cause leaves to wilt or turn yellow and finally drop off the plant. Other factors besides the amount of water given a plant are important to consider in moisture control, such as using the proper medium and providing good drainage.

There are several methods of watering interior plants.

- *Drench/let dry.* Enough water is added to moisten the medium which is then permitted to dry before watering again. In the drenching method, the desired amount of water is added all at one time.

- *Drench/let dry slightly.* Enough water is added to moisten the medium. It is then permitted to dry slightly so that it is damp to the touch.

- *Keep medium constantly moist.* Enough water is added so that the medium is damp to the touch at all times.

LIGHTING

As we landscape interiors with tropical and semi-tropical plants, the interior environments cause stress because of low available light. Therefore, to create a long-term plantscape, it is necessary to create appropriate artificial environments to grow certain plant cultivars. Interior plants can be grown under natural, filtered sunlight, and incandescent, fluorescent, and high-intensity discharge lights. All of these light sources allow the plants to receive enough energy for photosynthesis (process of manufacturing food).

Light meters (small machines that measure illumination, or light) can be used to determine if the location selected for the plant is suitable. Light meters are sold at florists, garden centers, and department stores with plant shops. Light intensity requirements, which vary from plant to plant, are classified into the following groups.

- *Direct sun.* Plant should receive full intensity of natural sunlight.

- *Partial sun or weak sun.* Plant should receive less than 50 percent of natural sunlight.

- *Indirect or filtered light.* Plant should receive no direct sunlight at all.

In general, the more light that is available, the greater the variety of plants which may be grown. In most cases, foliage plants require less light than flowering plants.

ARTIFICIAL LIGHTING

The advent of the fluorescent light greatly influenced the art of growing interior plants. The fluorescent light enables the horticulturist to:

- root cuttings of interior plants more easily.

- start seeds of new and different interior plants.

- grow tropical plants, cacti, and succulents with greater ease.

TEMPERATURE

Many times, interior plants are exposed to a great range of temperatures within one twenty-four-hour day. For this reason, plants must sometimes be given special protection. For example, plants located in windows during winter months may receive full sunlight during the day. During the night, however, the temperature next to the window will be much cooler. If the temperature is below 32°F (0°C) and the plants are very close to the window, the plants may freeze. The temperature recommended for interior plants (50°F to 85°F) is the daytime temperature; the night temperature can generally drop 10 to 15 degrees below this temperature without harming the plants.

The common use of air conditioners in homes and offices can be a source of trouble for plants. The plants can be protected, however, by locating them away from drafts and direct air flows.

HUMIDITY

Humidity is the percentage of moisture in the air—an important factor to consider when growing interior plants. Humidity varies with seasons; it is lower than usual in the winter; therefore, additional moisture is needed in the atmosphere. During the summer the humidity is usually higher than in the winter.

Humidity around interior plants can be increased by:

- spraying a fine mist of water on the foliage of the plant with an inexpensive plastic spray bottle or any type of mister.

- using a watertight tray filled with small pea gravel to hold the potted plants. Water is added to the tray to come to the top of the gravel and plants are set on the gravel, making certain that the pots are not actually in the water.

- using a humidifier to provide moisture in the air.

- placing the plants in that part of the house that has the highest humidity. For example, areas of highest humidity are probably the bathroom or the kitchen just above the sink.

SOIL MIXTURE

The composition of the soil mixture varies with the requirements of the particular interior plant. A good general potting mixture includes two parts loamy soil; one part organic matter such as peat moss, leaf mold, or well-rotted compost; and one part coarse sand or a horticultural grade perlite. When using sterilized soil, the addition of trace elements is usually not necessary.

A good loamy soil consists of equal parts of sand, silt, and clay. Organic matter, rich in carbohydrates, helps to improve the water-holding capacity of the mixture. It makes the mixture easy to work with and helps to keep the soil crumbly. Peat moss, leaf mold, and well-rotted compost act as fertilizers and soil conditioners. The use of coarse sand or horticultural grade perlite helps to keep the mixture loose and provides good drainage. This allows better movement of air and nutrients carried by the water, and is called *permeability*.

FERTILIZERS

There are two types of interior plant fertilizers on the market today—slow release and soluble. *Slow-release fertilizer* is in the form of small beads, which are coated with plastic and dissolve over a period of time, and is applied to the soil mixture. As it comes in contact with normal soil moisture, plant nutrients are steadily released. The plant absorbs the released nutrients as needed, thus eliminating the danger of overfeeding or fertilizer burn.

Soluble fertilizer is available in liquid or solid form. Solid forms of soluble fertilizer must be dissolved in water before being applied to interior houseplants. Applying this fertilizer dry may burn the roots of the plant. Soluble fertilizers in liquid form are concentrated, requiring that water be added to the mixture before application.

> *Tips: Always follow the manufacturer's recommendations when using any fertilizer.*

CONTAINERS

There are many beautiful ceramic and pottery plant containers on the market today. Plants grow in almost every kind of container, provided that drainage is available. However, if there are no drainage holes in the container, the soil mixture will become waterlogged. Plants which take up the oxygen in the medium cannot survive under these conditions. In addition to drainage, there must be some air spaces in the mixture so that the plant roots breathe properly. Proper air circulation is an important ingredient of the soil mixture.

Clay has long been the standard type of container for growing plants, but many greenhouse growers have now changed to other types of containers such as plastic, glass, fiberglass, and metal. All plant containers should be set in some type of saucer to catch excess drainage water. Since clay is porous, water can escape through the clay saucers to table surfaces. This problem can be eliminated by coating the saucer with plastic or paraffin wax.

POTTING PLANTS

After the proper container and soil mixture have been selected, potting or repotting interior plants is a simple process, figure 23-3. First, some type of drainage must be provided in the bottom of the pot. For clay pots, a crock may be used. A *crock* is a broken piece of clay pot. Place the crock so that it is curved upward over the drainage hole in the bottom of the pot. If it is placed with the curve down, it may seal the drainage hole and cause problems by not allowing excess water to drain out of the pot. Stones may also be used in containers. If they are used, be sure that the stones do not completely close the drainage holes in the bottom of the pot.

Add the selected potting mixture to the container so that it covers the drainage stone or crock. Place the plant in the container, making sure to check the depth of the plant. If the plant is below the rim of the pot, adjust it so it is 1/2 inch below the top of the pot. This allows for an area at the top of the pot to hold additional water while the plant is being watered. If no space is allowed, the water will run off over the edge of the pot. Not having received much of the water added, the plant will dry out much more quickly, thereby inhibiting growth.

REPOTTING

Interior plants should be repotted when they become pot-bound. To determine if the plant is pot-bound, remove the plant from the original pot. If the roots of the plant are growing around the root

(A) Select a pole 3 to 5 times taller than the container in which the totem is to be used.

(B) Wrap the pole with a layer of sphagnum moss 2 to 3 inches deep and bind the moss with string.

(C) Push the pole into the soil.

(D) Wind the vine around the totem.

(E) Use hairpins to fasten the vine to the totem.

(F) Keep the moss damp; roots will grow into the moss and the leaves will form a solid mass.

FIGURE 23-3   Making a totem

ball of the plant, and are firm to the touch, the plant should be repotted. If the plant is pot-bound, the roots must be separated; this allows the roots to develop and make new growth. Repotting should be done before new growth starts. If the plant is flowering, wait until after flowering is completed.

INTERIOR PLANTSCAPE PLANTERS

As the interior plantscaping business continues to grow, more and more businesses are using interior plants for the aesthetic improvement of their offices.

There are various types of planters used today. The moveable, decorative planter is one type, which includes floor planters, hanging baskets, and file-top planters. Another type of planter is the fixed-bed planter. This one is constructed in place and has a drainage system that is linked to the office plumbing system.

When using either of these types of planters, there are two planting procedures that may be employed: direct planting or double potting. *Direct planting* is a process whereby the plant is removed from the nursery container and planted directly into the medium in the container bed. In *double potting,* the plant is *not* removed from the nursery container; both are set into the medium. This allows for easier seasonal changes and replacement of plants that have been damaged or experienced physical changes, such as yellowing leaves.

GENERAL CARE

The leaves of interior plants should be cleaned occasionally with a fine spray of water at room temperature. Wipe the leaves on both the top and underside. This is a good way to remove dust and insects, help to maintain humidity, and keep the plant fresh-looking and attractive, figure 23-4.

CONTROLLING INDOOR PLANT DISEASES AND PESTS

Interior landscape plants have few problems with diseases and insects. The secret to controlling these pests is to be able to recognize them and control them on the plant.

FIGURE 23-4   Cleaning the leaves of a philodendron to remove dust. (Courtesy Creative Plantings, Inc.)

Some of the most common insects affecting interior plants are aphids, whiteflies, mealybugs, mites, and scale.

Aphids are small green insects with piercing mouth parts. These insects are called plant lice because they suck the plant juices. Aphids work on the stem of plants and the underside of the leaves. The leaves then become sticky with honeydew causing the growth of a black, sooty mold. Heavy infestation causes the leaves to curl, disfigure, and turn yellow. Aphids can be controlled with orthene.

The *whitefly* is a small white, winged, sucking insect that works on the underside of the plant leaves. Whiteflies are heavy producers of eggs; therefore, infestation can occur fast. These egg masses are laid on the underside of the plant leaves. They cause the plant to turn yellow and have a mottled appearance. A whitefly infestation is easy to recognize because the adult fly will leave the plant with movement of the leaves. They can be controlled with orthene insecticide.

Mealybugs are white, wooly masses like a ball of cotton. They will concentrate at the nodes of plants and the underside of leaves. They suck the plant juices to give the plant a faded yellow appearance. These bugs are controlled in an organic fashion by dipping Q-tips in alcohol and swabbing the insect thus removing it from the plant. Orthene can also be used.

Mites are small spiderlike insects on the underside of the leaf. They suck the plant juices and cause the plant to turn a faded yellow. Mites are so small that it is impossible to see them with the naked eye. An easy way to determine mite infestation is to take a piece of white paper and hold it under the infested leaf; then tap the leaf, causing the insect to fall onto the paper. Look for spiderlike insects moving on the paper. If you suspect they are there, take your finger and strike it over the paper. If there are spider mites, there will be a streak of red across the paper. Control of mites can be effective with orthene or kelthane.

Scale in its mature stage attacks the stems of indoor plants in a cluster. They look like small white, gray, or brown lumps. Scale is very noticeable on the infected plant as the leaves turn yellow and the plant itself loses vigor. Scale can be controlled by maintaining a clean environment. It can be removed by rubbing with a soap solution or spraying with malathion or orthene.

Botrytis and mildew are two common diseases of indoor plants. Botrytis causes indoor plant leaves to turn black and then develop a gray mold. They will rot off if the disease is allowed to advance. Environmental controls include providing good air circulation and sanitation. A fungicide that can be used is Benlate. Mildew causes the indoor plant leaves to get a heavy white coating. A good control is to provide the plants with proper sanitation, air circulation, and a fungicide of Benlate.

THE INDOOR PLANT CHART

Following is an indoor plant chart which gives the essential information for growing forty-two indoor plants. Each plant is illustrated with a picture to aid in identification.

The common name of the plant is given first and the scientific name given second.

◆ *Watering* tells how to water each plant.

◆ *Sunshine* gives the amount of light exposure each plant needs, ranging from full sun to indirect light.

◆ *Temperature* indicates the optimum temperature at which to grow each individual plant.

◆ *Soil mixture* gives a general recommendation for the plant using a mixture of sand, peat moss, and soil.

◆ *Ease of growth* is a rating given for each plant on a scale of 1 to 4.

Number 1 means that the plant is very easy to grow and requires little care.

Number 2 means that the plant is easy to grow but needs some special attention.

Number 3 means that the plant is very difficult to grow.

Number 4 means that the plant is very difficult to grow and must have very special care.

This rating may be helpful when selecting plants to grow.

 INDICATES PLANT WILL GROW IN A CLOSED TERRARIUM

 INDICATES PLANT WILL GROW IN A HANGING BASKET

 INDICATES PLANT WILL GROW IN A DISH GARDEN OR POT

 INDICATES PLANT WILL GROW IN AN INDOOR PLANTER

INDOOR PLANT CHART

BOUGAINVILLEA
(*Bougainvillea spectablis 'Barlara Karst'*)

WATERING:	Moist in summer/dry in winter
SUNSHINE:	Full sun
TEMPERATURE:	50° to 60°F
SOIL MIXTURE:	2 peat, 1 sand, 1 soil
EASE OF GROWTH:	2

INDOOR PLANT CHART

ASPARAGUS FERN
(Asparagus densi-florus 'Sprengeri')

WATERING:	Drench/let dry slightly
SUNSHINE:	Weak (partial) sun
TEMPERATURE:	50° to 70°F
SOIL MIXTURE:	2 peat, 1 sand, 2 soil
EASE OF GROWTH:	1

(Courtesy George C. Ball, Inc.)

GARDENIA
(Gardenia jasminoidies)

WATERING:	Keep moist
SUNSHINE:	Direct sun
TEMPERATURE:	60° to 75°F
SOIL MIXTURE:	1 peat, 1 sand, 1 soil
EASE OF GROWTH:	4

POINSETTIA
(Euphorbia pulcherrima)

WATERING:	Keep moist
SUNSHINE:	Direct sun
TEMPERATURE:	70° to 80°F
SOIL MIXTURE:	1 peat, 1 perlite, 1 soil
EASE OF GROWTH:	2

(Courtesy Interior Plantscape, division of ALCA)

BEGONIA
(Begonia)

WATERING:	Drench/let dry slightly
SUNSHINE:	Direct sun, weak (partial) sun, no direct sun
TEMPERATURE:	50° to 70°F
SOIL MIXTURE:	1 peat, 2 sand, 1 soil
EASE OF GROWTH:	1

INDOOR PLANT CHART

GERANIUM
(*Pelargonium x hortorum*)

WATERING:	Drench/let dry slightly, Mist often
SUNSHINE:	Direct sun
TEMPERATURE:	50° to 60°F
SOIL MIXTURE:	1 peat, 1 sand, 2 soil
EASE OF GROWTH:	1

IMPATIENS
(*Impatiens walleriana*)

WATERING:	Keep soil moist
SUNSHINE:	Weak (partial) sun
TEMPERATURE:	50° to 60°F
SOIL MIXTURE:	1 peat, 2 sand, 1 soil
EASE OF GROWTH:	1

GRAPE IVY
(*Cissus rhombifolia 'mandaiana'*)

WATERING:	Drench/let dry slightly
SUNSHINE:	Weak (partial) sun to filtered light
TEMPERATURE:	50° to 70°F
SOIL MIXTURE:	1 peat, 1 soil, 1 sand
EASE OF GROWTH:	1

(Courtesy Interior Plantscape, division of ALCA)

CHINESE FAN PALM
(*Livistona chinensis*)

WATERING:	Moist
SUNSHINE:	Filtered
TEMPERATURE:	65° to 80°F
SOIL MIXTURE:	1 peat, 1 sand, 1 soil
EASE OF GROWTH:	2

(Courtesy Interior Plantscape, division ALCA)

INDOOR PLANT CHART

SOUTHERN YEW
(*Podocarpus macrophyllus*)

WATERING:	Evenly moist/not constantly wet
SUNSHINE:	Bright to filtered
TEMPERATURE:	62° to 85°F
MIXTURE:	1 peat, 1 sand, 1 soil
EASE OF GROWTH:	2

KALANCHOE
(*Kalanchoe pinnata*)

WATERING:	Drench/let dry slightly
SUNSHINE:	Direct sun to partial sun
TEMPERATURE:	55° to 65°F
SOIL MIXTURE:	1 peat, 2 sand, 1 soil
EASE OF GROWTH:	2

(Courtesy George C. Ball, Inc.)

CORN PLANT
(*Dracaena fragrans massangeana*)

WATERING:	Moist
SUNSHINE:	Weak (partial) sun
TEMPERATURE:	60° to 80°F
SOIL MIXTURE:	3 peat, 2 sand, 1 soil
EASE OF GROWTH:	2

(Courtesy Interior Plantscape, division of ALCA)

AZALEA
(*Rhododendron*)

WATERING:	Drench/let dry slightly
SUNSHINE:	Weak (partial) sun
TEMPERATURE:	50° to 65°F
SOIL MIXTURE:	1 sphagnum peat, 1 perlite
EASE OF GROWTH:	3

(Courtesy Interior Plantscape, division of ALCA)

INDOOR PLANT CHART

BOSTON FERN
(Nephrolepis exaltata bostoniensis)

WATERING:	Keep mixture moist, Mist often
SUNSHINE:	Partial sun
TEMPERATURE:	50° to 70°F
SOIL MIXTURE:	1 peat, 2 sand, 1 soil
EASE OF GROWTH:	1

NORFOLK ISLAND PINE
(Araucaria heterophylla)

WATERING:	Keep moist
SUNSHINE:	Weak (partial) sun
TEMPERATURE:	50° to 65°F
SOIL MIXTURE:	3 peat, 2 sand, 1 soil
EASE OF GROWTH:	1

DRAGON TREE
(Dracaeana marginata)

WATERING:	Moist
SUNSHINE:	Weak (partial) sun
TEMPERATURE:	60° to 80°F
SOIL MIXTURE:	3 peat, 2 sand, 1 soil
EASE OF GROWTH:	2

(Courtesy Interior Plantscape, division of ALCA)

BROMELIAD
(Aechmea fastciata)

WATERING:	Even moist (fill vase of plant with water)
SUNSHINE:	Partial sun to filtered light
TEMPERATURE:	62° to 65°F Night, 80° to 85°F Day
SOIL MIXTURE:	1 peat, 1 osmunda fiber, 1 sand
EASE OF GROWTH:	2

(Courtesy Interior Plantscape, division of ALCA)

INDOOR PLANT CHART

DUMBCANE
(Diffenbachia amoena 'Tropic Snow')

WATERING:	Drench/let dry slightly
SUNSHINE:	Filtered light
TEMPERATURE:	60° to 75°F
SOIL MIXTURE:	3 peat, 2 sand, 1 soil
EASE OF GROWTH:	1

(Courtesy Interior Plantscape, division of ALCA)

FIDDLELEAF FIG
(Fiscus lyrata)

WATERING:	Evenly moist, but not constantly wet
SUNSHINE:	Filtered light
TEMPERATURE:	62° to 65°F Night, 80° to 85°F Day, Keep out of draft
SOIL MIXTURE:	1 peat, 1 perlite, 1 soil
EASE OF GROWTH:	1

(Courtesy Interior Plantscape, division of ALCA)

CHRYSANTHEMUM
(Chrysanthemum sp.)

WATERING:	Moist to touch
SUNSHINE:	Direct sun
TEMPERATURE:	60° to 75°F
SOIL MIXTURE:	1 peat, 1 perlite, 1 soil
EASE OF GROWTH:	2

(Courtesy Interior Plantscape, division of ALCA)

WEEPING FIG
(Fiscus benjamia)

WATERING:	Evenly moist, but not constantly wet
SUNSHINE:	Filtered light
TEMPERATURE:	62° to 65°F Night, 80° to 85°F Day, Keep out of draft
SOIL MIXTURE:	1 peat, 1 perlite, 1 soil
EASE OF GROWTH:	1

(Courtesy Interior Plantscape, division of ALCA)

INDOOR PLANT CHART

SPIDER PLANT
(*Chlorophytum comosum*)

WATERING:	Keep moist
SUNSHINE:	Weak (partial) sun
TEMPERATURE:	50° to 70°F
SOIL MIXTURE:	1 peat, 1 sand, 2 soil
EASE OF GROWTH:	1

UMBRELLA TREE
(*Schefflera digitata*)

WATERING:	Drench/let dry slightly
SUNSHINE:	Weak (partial) sun
TEMPERATURE:	40° to 60°F
SOIL MIXTURE:	1 peat, 2 sand, 1 soil
EASE OF GROWTH:	1

(Courtesy Interior Plantscape, division of ALCA)

PRAYER PLANT
(*Maranta leuconeura*)

WATERING:	Keep moist
SUNSHINE:	Weak (partial) sun
TEMPERATURE:	60° to 70°F
SOIL MIXTURE:	1 peat, 2 sand, 1 soil
EASE OF GROWTH:	3

(Courtesy Interior Plantscape, division of ALCA)

JANET CRAIG DRACAENA
(*Dracaena deremensis 'Janet Craig'*)

WATERING:	Moist
SUNSHINE:	Weak (partial) sun
TEMPERATURE:	60° to 80°F
SOIL MIXTURE:	3 peat, 2 sand, 1 soil
EASE OF GROWTH:	2

(Courtesy Interior Plantscape, division of ALCA)

INDOOR PLANT CHART

SNAKE PLANT
(Sansevieria trifasciata)

WATERING:	Drench/let dry
SUNSHINE:	Direct to indirect
TEMPERATURE:	50° to 75°F
SOIL MIXTURE:	1 peat, 2 sand, 1 soil
EASE OF GROWTH:	1

(Courtesy Interior Plantscape, division of ALCA)

EXACUM
(Exacum offine)

WATERING:	Quite moist
SUNSHINE:	Filtered
TEMPERATURE:	62° to 65°F Night, 80° to 85°F Day
SOIL MIXTURE:	2 peat, 1 sand, 1 soil
EASE OF GROWTH:	3

(Courtesy Interior Plantscape, division of ALCA)

DWARF DATE PALM
(Phoenix roebelenii)

WATERING:	Quite moist
SUNSHINE:	Filtered
TEMPERATURE:	62° to 65°F Night, 80° to 85°F Day
SOIL MIXTURE:	1 peat, 1 sand, 1 soil
EASE OF GROWTH:	3

(Courtesy Interior Plantscape, division of ALCA)

SPINESLESS YUCCA
(Yucca elephantipis)

WATERING:	Drench/let dry
SUNSHINE:	Direct
TEMPERATURE:	55° to 70°F
SOIL MIXTURE:	1 peat, 1 sand, 1 soil
EASE OF GROWTH:	1

(Courtesy Interior Plantscape, division of ALCA)

INDOOR PLANT CHART

SPATHIPHYLLUM
(Spathiphyllum wallisii)

WATERING:	Quite moist
SUNSHINE:	Filtered to low light
TEMPERATURE:	62° to 65°F Night, 80° to 85°F Day
SOIL MIXTURE:	1 peat, 1 perlite, 1 soil
EASE OF GROWTH:	1

(Courtesy Interior Plantscape, division of ALCA)

SWEDISH IVY
(Plectranthus australis)

WATERING:	Keep soil moist
SUNSHINE:	Indirect sun
TEMPERATURE:	50° to 65°F
SOIL MIXTURE:	3 peat, 2 sand, 1 soil
EASE OF GROWTH:	2

ENGLISH IVY
(Hedera helix)

WATERING:	Drench/let dry slightly
SUNSHINE:	Partial sun
TEMPERATURE:	50° to 65°F
SOIL MIXTURE:	1 peat, 2 sand, 1 soil
EASE OF GROWTH:	1

(Courtesy Interior Plantscape, division of ALCA)

CLADIUM
(Claldium sp)

WATERING:	Evenly moist
SUNSHINE:	Filtered light
TEMPERATURE:	62° to 65°F Night, 80° to 85°F Day
SOIL MIXTURE:	3 peat, 1 perlite, 1 soil
EASE OF GROWTH:	1

(Courtesy Interior Plantscape, division of ALCA)

INDOOR PLANT CHART

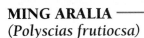

MING ARALIA
(Polyscias frutiocsa)

WATERING:	Evenly moist
SUNSHINE:	Direct light
TEMPERATURE:	62° to 65°F Night, 80° to 85°F Day
SOIL MIXTURE:	3 peat, 1 perlite, 1 soil
EASE OF GROWTH:	1

(Courtesy Interior Plantscape, division of ALCA)

NEPHTHYTIS OR ARROWHEAD PLANT
(Syngonium podophyllum 'White Butterfly')

WATERING:	Quite moist
SUNSHINE:	Filtered to low light
TEMPERATURE:	62° to 65°F Night, 80° to 85°F Day
SOIL MIXTURE:	1 peat, 1 perlite, 1 soil
EASE OF GROWTH:	1

(Courtesy Interior Plantscape, division of ALCA)

IVY-LEAVED GERANIUM
(Pelargonium peltatum)

WATERING:	Drench/let dry
SUNSHINE:	Direct sun
TEMPERATURE:	50° to 60°F
SOIL MIXTURE:	1 peat, 1 sand, 1 soil
EASE OF GROWTH:	2

BANANA
(Musa maurelii)

WATERING:	Keep soil moist
SUNSHINE:	Partial sun
TEMPERATURE:	60° to 85°F
SOIL MIXTURE:	1 peat, 1 sand, 1 soil
EASE OF GROWTH:	3

INDOOR PLANT CHART

BIRD OF PARADISE
(Strelitzia reginae)

WATERING:	Keep soil moist—Summer
	Drench/let dry—Winter
SUNSHINE:	Direct sun
TEMPERATURE:	50° to 75°F
SOIL MIXTURE:	1 peat, 1 sand, 1 soil
EASE OF GROWTH:	3

(Courtesy Interior Plantscape, division of ACLA)

MEDICINE PLANT
(Aloe Vera)

WATERING:	Drench/let dry
SUNSHINE:	Direct sun
TEMPERATURE:	55° to 70°F
SOIL MIXTURE:	1 peat, 2 sand, 1 soil
EASE OF GROWTH:	1

HEART LEAF PHILODENDRON
(Philodendron)

WATERING:	Drench/let dry slightly, mist leaves often
SUNSHINE:	Indirect
TEMPERATURE:	50° to 75°F
SOIL MIXTURE:	3 peat, 2 sand, 1 soil
EASE OF GROWTH:	1

(Courtesy Interior Plantscape, division of ACLA)

AMARYLLIS
(Hippeastrum sp.)

WATERING:	Keep soil moist
SUNSHINE:	Direct sun
TEMPERATURE:	55° to 65°F
SOIL MIXTURE:	1 peat, 1 sand, 1 soil
EASE OF GROWTH:	3

(Courtesy Michael Dzaman)

INDOOR PLANT CHART

JADE PLANT
(Crassula argentea)

WATERING:	Drench/let dry
SUNSHINE:	Partial sun
TEMPERATURE:	60° to 70°F
SOIL MIXTURE:	1 peat, 1 sand, 1 soil
EASE OF GROWTH:	1

POTHOS
(Scindapsus aureus or Epipremnum aureus)

WATERING:	Drench/let dry
SUNSHINE:	Indirect/filtered
TEMPERATURE:	50° to 75°F
SOIL MIXTURE:	3 peat, 2 sand, 1 soil
EASE OF GROWTH:	1

(Courtesy Interior Plantscape, division of ACLA)

KING SAGO
(Cycas revoluta)

WATERING:	Drench/let dry between waterings
SUNSHINE:	Filtered light
TEMPERATURE:	62° to 65°F Night, 80° to 85° Day
SOIL MIXTURE:	1 peat, 1 perlite, 1 soil
EASE OF GROWTH:	1

(Courtesy Interior Plantscape, division of ALCA)

CROTON
(Codiaeum variegatum pictum)

WATERING:	Keep soil moist
SUNSHINE:	Partial sun
TEMPERATURE:	60° to 80°F
SOIL MIXTURE:	3 peat, 2 sand, 1 soil
EASE OF GROWTH:	2

(Courtesy Interior Plantscape, division of ACLA)

INDOOR PLANT CHART

COLUMN CACTUS
(*Cereus peruvianus*)

WATERING:	Drench/let dry between waterings
SUNSHINE:	Direct
TEMPERATURE:	62° to 65°F Night, 80° to 85°F Day
SOIL MIXTURE:	1 peat, 1 perlite, 1 soil
EASE OF GROWTH:	1

(Courtesy Interior Plantscape, division of ALCA)

CHINESE EVERGREEN
(*Aglaonema 'Silver Queen'*)

WATERING:	Evenly moist
SUNSHINE:	Indirect/filtered
TEMPERATURE:	50° to 75°F
SOIL MIXTURE:	3 peat, 2 sand, 1 soil
EASE OF GROWTH:	1

(Courtesy Interior Plantscape, division of ALCA)

STAGHORN FERN
(*Platcerium bifurcatum*)

WATERING:	Moist, Mist daily
SUNSHINE:	Partial shade
TEMPERATURE:	60° to 80°F
SOIL MIXTURE:	Osmunda Fiber Mix
EASE OF GROWTH:	3

PONYTAIL PALM
(*Beaucarnea recurvata*)

WATERING:	Drench/let dry
SUNSHINE:	Full sun/partial shade
TEMPERATURE:	60° to 80°F
SOIL MIXTURE:	2 peat, 1 sand, 1 soil
EASE OF GROWTH:	2

Student Activities

1) Visit a local retail florist to see the available indoor landscape plants.

2) Attend a home and garden show to observe the use of these plants.

3) Participate in the FFA State and National Floriculture Contest.

4) Review interior decorating magazines to see how interior plants are used in decorating.

5) Propagate a large variety of indoor plants and use them for identification.

6) Read trade magazines, for example, *Florist Review* and *Interior Plantscaping*.

7) Design and install an interior plantscape garden.

8) Have an interior plantscape professional speak to the class.

Self-Evaluation

Select the best answer from the choices offered to complete each statement.

1) The major concerns in caring for interior plants are

   a) watering, light, disease control, and growth.
   b) watering, light, temperature, and soil mixture.
   c) mature height, light, pest control, and soil mixture.
   d) none of the above

2) A general potting mixture includes

   a) two parts loamy soil, one part peat, and one part sand.
   b) two parts coarse stone, one part soil, and one part perlite.
   c) three parts peat, one part sand, and one part loamy soil.
   d) none of the above

3) When indoor plants are overwatered,

   a) they droop their leaves.   c) their leaves turn yellow.
   b) their leaves drop off.   d) all of the above

4) The nighttime temperature of indoor plants should be _____ degrees cooler than that of the daytime temperature.

   a) 5 to 10   c) 3 to 4
   b) 20 to 30   d) 10 to 15

5) Humidity around indoor plants can be increased by

   a) spraying a fine mist of water on the leaves.
   b) using a watertight tray with pea gravel and water around the gravel.
   c) using a humidifier.
   d) all of the above

6) A white, wooly insect that resembles a ball of cotton and usually infests at the nodes of plants is the

a) mite.

b) mealy bug.

c) aphid.

d) whitefly.

7) Azaleas need a soil mixture of

a) one part peat, one part sand, and one part soil.

b) three parts peat and one part gravel.

c) all peat.

d) none of the above

8) Norfolk Island pine grows best in a temperature of

a) 40°F to 50°F.

b) 100°F to 110°F.

c) 50°F to 65°F.

d) 32°F to 45°F.

9) The genus name for croton is

a) *Sansevieria.*

b) *Codiaeum.*

c) *Ficus.*

d) *Araucaria.*

10) The common name for Dieffenbachia is

a) Umbrella tree.

b) Snake plant.

c) Bougainvillea.

d) Dumb cane.

11) Spathiphyllum grows best in a

a) bright area.

b) low light.

c) full sun.

d) none of the above

12) Flowering plants create an attractive seasonal display of color. The _____ is used for its flowering quality.

a) Boston fern

b) Chinese evergreen

c) azalea

d) pothos

13) Cereus is a member of the _____ family.

a) legume

b) rose

c) cactus

d) none of the above

14) Caladium is used in the interior landscape for its

a) flower.

b) leaf color.

c) root size.

d) none of the above

15) The common name of Hedera helix is

a) Swedish ivy.

b) Boston ivy.

c) English ivy.

d) none of the above

UNIT 24

Shrubs and Trees

OBJECTIVE

To provide enough basic facts so that a container crop could be grown to marketable quality.

Competencies to Be Developed

After studying this unit, you should be able to

- ◆ list and explain at least four advantages and three disadvantages of container growing.
- ◆ list at least five types of plants that are commonly grown in containers.
- ◆ list or explain orally the characteristics of a good growing site.
- ◆ list the factors to consider in selecting a container.
- ◆ prepare a soil mix, fill a container, and successfully place a plant in a container.
- ◆ set up a growth control schedule.
- ◆ discuss marketing of the finished product.
- ◆ outline at least one winter storage method where temperatures fall to below freezing.
- ◆ list at least one reason for shading container plants.

MATERIALS LIST

- ✔ containers
- ✔ growing medium
- ✔ plants to grow
- ✔ fertilizers
- ✔ growing area
- ✔ watering system
- ✔ pesticides, if needed
- ✔ hand pruners
- ✔ weed killers, if needed

Container growing of shrubs and trees is becoming more important and is gradually replacing field growing of many kinds of nursery plants. Container growing in the nursery trade means growing the plant to marketable size in a container. Many plants are grown in pots for short periods of time as seedlings, grafts, or rooted cuttings and later transplanted to a field or to a large container for container growing. This short-term pot culture is not considered container growing.

ADVANTAGES OF CONTAINER GROWING

There are a number of advantages of container growing:

- Plants may be sold and transplanted year-round. Since the roots are in a container they are not cut off and damaged as may be the case with field dug plants. This also allows for more year-round use of labor. If plants are pot-bound and roots circling the inside of the pot must be cut, this advantage is lost.

- More plants can be grown in the same area, thus reducing cost.

- Uniform growing medium (mix used instead of soil) is readily available and plant nutrients and water may be more accurately controlled. This results in more uniform plant growth.

- Plants require no digging or balling (wrapping the root ball in burlap) of roots to hold the root ball together and prevent drying out. Labor is saved because of this.

- Growing medium is usually lighter than soil, making it easier to haul larger loads of plants.

- The container, if sturdy, protects the root ball from being crushed when plants are stacked on top of each other in transporting to the retailer or planting site.

- The growing medium is generally sterilized, thus reducing damage to plant roots from insect and disease organisms. This results in faster, healthier plant growth.

- The cost to produce a plant to marketable size is lower because many of the items listed previously result in less labor, fewer insect and disease problems, lower shipping costs, and if managed properly, more uniform plant growth.

DISADVANTAGES OF CONTAINER GROWING

The major disadvantages of container growing are:

- A higher initial cost in getting established. The growing area should be graded flat with a slight slope, treated for weed control, and covered with plastic or ground-cover fabric.

- Personnel must be trained in potting technique and watering becomes a more precise growth factor in container growing.

- More water is required since the plants cannot tap soil reserves. A good supply of irrigation water is a must.

- If plants are overwintered, some protection must be provided where temperatures fall much below freezing. Roots are damaged at higher temperatures than the top part of the plant. As the roots are not down in the soil, they get colder in containers than in field grown situations.

- Transplanting to the landscape is more difficult for some plants because the soilless mixes generally used are too different in structure from the soil in which the plant will be transplanted. (Adjustment may be made by adding soil to the container mix or by digging pine bark and sphagnum peat moss into the soil at the planting site to make the difference less.) Plant roots are thus encouraged to leave the container ball and grow out into the soil.

- If left in containers too long plants develop a circular, pot-bound root system, figure 24-1. This system may never grow well in the soil, and may not provide adequate anchorage or spread out to absorb water and nutrients needed for normal growth.

TYPES OF PLANTS GROWN IN CONTAINERS

No attempt will be made to list all the plants grown in containers in the nursery trade. It can be generally said, however, that nearly all shrubs and trees are container grown. Some plants such as yews are difficult and not as often grown in containers. The variety of plants ranges from garden flowers such as herbaceous perennials, ground cover,

FIGURE 24-1 Notice the circling roots on the plant on the left. This plant is badly root-bound and will be difficult to transplant. The plant on the right was grown in a pot treated with a copper solution, Spin Out, and is not pot-bound. (Courtesy Griffin Corporation)

vines, broadleaf evergreen shrubs, and narrowleaf evergreen shrubs, to evergreen and deciduous trees. (Refer to the list in figure 24-8.) Indoor tender foliage plants are also container grown.

Deciduous trees and shrubs that are usually planted during the dormant season only (such as fruit trees) are generally still field grown and dug and shipped bare root (no soil on the roots). This is a cheaper method of handling and shipping large quantities of this type of plant. Flower trees, however, are often grown in large containers such as the bushel baskets shown in figure 24-2.

SITE

Site (the location where plants are to be grown) selection for container plant growing is very important. A flat, gently sloping area is necessary. Containers will fall over on too steep a slope but enough slope should be provided to drain water off the surface so that no water collects around the containers, figure 24-3. It is generally necessary to use a road grader to grade and slope the land.

Where plants must have individual attention, such as pruning, beds should be made no more than 5 feet wide with an 8-foot service road on at least one side of each bed for loading and unloading from a truck. Beds should be planned to make the best use of the irrigation (watering) system. If underground water lines are used, they should be installed at the time that the beds are graded and laid out.

FIGURE 24-2 Flowering trees grown in bushel baskets have greater success with year-round transplanting since the container remains on the root ball. The basket will rot away and need not be removed unless it was rot-proofed with a copper chemical compound. (Ed Reiley, Photographer)

FIGURE 24-3 Potted plants placed on gravel, under irrigation, to grow on. Notice the wide roadway between rows of pots for truck or wagon traffic to load and unload plants. (Ed Reiley, Photographer)

The growing area should also be located to be convenient to the water supply and irrigation system. Since the plant roots must depend entirely on the medium within the pot for water and nutrients, a dependable supply of irrigation water is critical. On hot, windy days plants may need to be watered daily or even twice daily.

WINDBREAKS

Windbreaks are also very important in site selection. A strong wind can blow containers over. These containers do not get watered and soon dry out, becoming damaged to the point where the plants are not marketable. Winds also interfere with sprinkler irrigation systems. A row of evergreen trees, a snow fence, or other windbreaks should be put in place. The height of the windbreaks will determine how far apart they should be placed. A good rule of thumb is that a windbreak is effective for a distance of about seven times its height. A 4-foot snow fence would thus be spread 28 feet apart whereas 45-foot trees could be spread about 300 feet. Figure 24-4 shows a natural windbreak of evergreen trees surrounding a growing area.

WEED CONTROL

Weed control is best done through the use of black plastic groundcover fabric or stones on the soil surface before the pots are put in place. A weed killer may be placed on the soil before the material is put down. Black plastic lasts only one season unless it is covered with 4 inches of gravel or crushed stone to shield it from sunlight. If the crop is to remain on the same site for more than one year the gravel cover may be a good investment. The stones or gravel will also hold the plastic in place when plants are marketed and new containers may be placed in the same beds without reworking them.

FIGURE 24-4 An evergreen windbreak around a container growing area at a nursery. (Ed Reiley, Photographer)

CONTAINERS

Proper selection of the container that the plant will grow in is very important. The following considerations must be made.

Cost of the container must be low since container cost is one of the major expenses in the production of container-grown nursery stock.

Durability should be matched with the length of time the crop will grow in the container prior to marketing. An expensive container that will last three years would not be a good business investment if the crop is marketed at the end of one year. Containers are made of materials such as wood, plastic, metal, and pressed paper. Durability information should be obtained prior to purchase. Metal containers rust and wooden containers rot out in the bottom if drainage is not good.

Appearance of the container may be almost as important as plant appearance to some retail buyers. A container that remains attractive through the marketing cycle is important. Painted metal and plastic containers last longer and paper and peat have the shortest life.

Insulation value to the plant roots is also important. Where there are freezing temperatures during the growing cycle, the container must withstand freezing without becoming brittle and breaking. Plant roots grow close against the container wall and are affected by container temperature and insulating value. A metal container has poor insulating value and roots would be damaged from heat or cold sooner in them than in a wooden or fiber container.

Shipping durability is another consideration. If plants are to be stacked one on the other, fiber pots or wooden baskets may crush and damage roots. Metal or hard plastic would protect root balls better under these conditions.

The *shape* of the container is another factor. Tapered containers tend to store in less space when empty because they can be stacked one in the other. Too much taper, however, could result in easy tipping or blowing over if the bottom is too narrow compared with the top. Tall, slim containers also blow over easier than short, broad ones. The container should be shaped to permit normal plant growth. This varies with different plants.

Drainage and *soil aeration* are important to all plants and especially in the case of azaleas and rhododendrons. The potting medium must be coarse enough to allow good drainage and aera-

FIGURE 24-5 Different-sized containers used in growing plants. Pot sizes are 1, 2, 3, 5, 10, and 15 gallons. (Ed Reiley, Photographer)

FIGURE 24-6 Evergreens growing in a field potted to fabric root control bags. (Courtesy Root Control, Inc.)

FIGURE 24-7 Compact fibrous root system of a tree grown in a root control bag. There is no circling of roots. (Courtesy Root Control, Inc.)

tion. Wooden baskets also provide aeration to the roots. Without good drainage and aeration, roots will rot or at best be stunted in growth. See figure 24-5 for some sample containers of various sizes.

The *size* of the container affects plant growth. A plant in a container that is too small will become root-bound and not grow as well as if it were in a proper-sized container. Once roots penetrate to the walls of the container, plant growth starts slowing. Plants should be moved to larger containers before they become root-bound or they should be sold. Figure 24-8, page 244, gives some standards set forth by the American Association of Nurserymen for minimum container size.

The container must not contain any material that is toxic (poisonous) to the plant.

Selection of the proper container is a very important decision. The grower should get advice from an experienced grower or the State Extension Service. Purchase of the wrong container could be an expensive mistake.

NEW CONTAINER DESIGNS

After considerable research, Dr. Carl E. Whitcomb (retired from Oklahoma State University) developed the field-grow fabric container for trees. It was designed to modify the root growth and prevent circling of roots of the plant grown in it.

The *field-grow fabric container* is used for growing trees in the field. A tree is planted in a fabric container using the same soil as in the field, and the container is placed in the soil with just the top inch or two of the container above ground, figure 24-6. The pot restricts the tree's roots and when the tree is dug up, none of the roots are cut off. The container is made of a new fabric with small

holes which allows the root tip to grow through the pot for a short distance before it is stopped. As the root grows larger in diameter, the hole in the pot stops its growth and many small side roots form on the inside of the container. The result is a root system with many more small roots in a smaller root ball. This smaller ball is easier to move and it contains many more feeding roots than a field-grown tree without the pot, figure 24-7.

The growth of trees in the pots is not different from trees grown in the open field. The big difference is in digging and transplanting. Trees grown in the fabric container are much easier to dig and when transplanted survive much better. The major advantages are less labor cost in digging, far less loss from transplanting to the landscape, and a better growth rate after transplanting.

Another new and promising technique for reducing circling roots of pot-grown plants is to coat the inside of the pots with copper. The chemical compound Copper Hydroxide ($Cu(OH)_2$) is sprayed or rubbed onto the inside walls and bottom of the pot. The only currently EPA registered copper compound labeled for root control is Spin Out. This is a copper paint made by Griffin Corporation.

SHADE AND FLOWERING TREES, DECIDUOUS SHRUBS, CONIFEROUS EVERGREENS, BROADLEAF EVERGREENS

All container-grown plants shall be healthy, vigorous, well-rooted, and established in the container in which they are sold. They shall have tops which are of good quality and are in a healthy growing condition.

An established container-grown plant shall be a plant transplanted into a container and grown in that container sufficiently long for the new fibrous roots to have developed so that the root mass will retain its shape and hold together when removed from the container.

The container shall be sufficiently rigid to hold the ball shape protecting the root mass during shipping.

Dwarf- and light-growing varieties may be one or two sizes smaller than standard for a given size container.

The following tables give plant sizes and acceptable container sizes.

SHADE AND FLOWERING

Tree Height	Container Size
12 in. 18 in. 2 ft. 3 ft.	1 gal. (trade designation) Minimum of 5½ inches across top and height of 6 inches or equivalent volume
2 ft. 3 ft. 4 ft.	2 gal. (trade designation) Minimum of 7 inches across top and height of 7½ inches or equivalent volume
4 ft. 5 ft. 6 ft.	5 gal., egg can or square can (trade designation) Minimum of 9 inches across top and height of 10 inches or equivalent volume

CONIFEROUS EVERGREENS

Spread (Type 1, Spreading and Type 2, Semi-Spreading Conifers); Height (Type 3, Globe or Dwarf Conifers)

Height	Container Size
6 in. 9 in. 12 in.	1 gal. (trade designation) Minimum of 5½ inches across top and height of 6 inches or equivalent volume
12 in. 15 in.	2 gal. (trade designation) Minimum of 7 inches across top and height of 7½ inches or equivalent volume
18 in. 2 ft. 2½ ft.	5 gal., egg can or square can (trade designation) Minimum of 9 inches across top and height of 10 inches or equivalent volume

Types 4, 5, and 6*

Height	Container size
6 in. 9 in. 12 in. 15 in. 18 in.	1 gal. (trade designation) Minimum of 5½ inches across top and height of 6 inches or equivalent volume
12 in. 15 in. 18 in. 2 ft.	2 gal. (trade designation) Minimum of 7 inches across top and height of 7½ inches or equivalent volume
18 in. 2 ft. 2½ ft. 3 ft. 3½ ft.	5 gal., egg can or square can (trade designation) Minimum of 9 inches across top and height of 10 inches or equivalent volume

*Except for extreme columnar types as Cupressus sempervirens (Italian cypress) which is acceptable 1 or 2 sizes taller than standard for a given container.

DECIDUOUS SHRUBS

Height	Container Size
6 in. 9 in. 12 in. 15 in. 18 in. 2 ft.	1 gal. (trade designation) Minimum of 5½ inches across top and height of 6 inches or equivalent volume
12 in. 15 in. 18 in. 2 ft. 18 in.	2 gal. (trade designation) Minimum of 7 inches across top and height of 7½ inches or equivalent volume
2 ft. 3 ft. 4 ft.	5 gal., egg can or square can (trade designation) Minimum of 9 inches across top and height of 10 inches or equivalent volume

BROADLEAF EVERGREENS

Spread (Type 1, Spreading and Type 2, Semi-Spreading) Height (Type 3, Dwarf and Globe)

Height	Container Size
6 in. 9 in. 12 in.	1 gal. (trade designation) Minimum of 5½ inches across top and height of 6 inches or equivalent volume
12 in. 15 in.	2 gal. (trade designation) Minimum of 7 inches across top and height of 7½ inches or equivalent volume
12 in. 2 ft. 2½ ft.	5 gal., egg can or square can (trade designation) Minimum of 9 inches across top and height of 10 inches or equivalent volume

Type 4, Broad Upright Type and Type 5, Cone Type* Broadleaf Evergreens

Height	Container Size
6 in. 9 in. 12 in. 15 in. 18 in.	1 gal. (trade designation) Minimum of 5½ inches across top and height of 6 inches or equivalent volume
12 in. 15 in. 18 in. 2 ft. 18 in.	2 gal. (trade designation) Minimum of 7 inches across top and height of 7½ inches or equivalent volume
2 ft. 2½ ft. 3 ft. 3½ ft.	5 gal., egg can or square can (trade designation) Minimum of 9 inches across top and height of 10 inches or equivalent volume

*Except for extreme columnar types as Prunus laurocerasus (Cherry Laurel) and Ligustrum japonicum (Japanese Privet) which are acceptable 1 or 2 sizes taller than standard for a given container.

FIGURE 24-8 Container plant standards (From American Association of Nurserymen)

Researchers at Texas A & M University found the copper spray to be very effective in reducing root circling in pots, figure 24-1. It also resulted in a more compact root system with more fibrous roots in the interior of the root ball. In many cases no root pruning at transplanting time to correct and stop circling roots was necessary.

When root ends touch the copper coating of the container wall they stop growing. Instead of circling the pot roots send out side roots (laterals) back inside the root ball. This produces a superior root ball resulting in better plant growth and higher survival rate after transplanting.

When removed from the pots root ends again start extending out into the soil and grow normally.

Pot-bound root systems must have the circling roots cut or pulled out of the root ball to stop the circling growth pattern when transplanted to the garden. This is necessary to prevent girdling and to form a natural root system and to get roots out into the soil where moisture is available.

GROWING MIXES

After selecting the container for a particular crop the next important decision is the medium or soil mix the plant will grow in. Again, some plants require a more porous or open mix than others to allow for better aeration or oxygen supply to roots. For example, rhododendrons and azaleas require a more porous mix than does a maple tree. A tight container may also require a more open mix than does a basket. Container mixes must be more porous than mixes for the same plant grown in the field or open bench. Containers do not allow for as good drainage or air movement as do good field conditions.

A good growing medium must:

◆ Provide good drainage (sand, perlite, and bark are often used).

◆ Be porous for good aeration (sand, perlite, and bark are often used).

◆ Hold moisture well (peat moss is most often used).

◆ Hold plant nutrients (some soil may be added).

◆ Have proper pH range (different plants require different pH).

◆ Be sterile (free of insects, disease organisms, and weed seeds).

◆ Be economical and easily obtained.

◆ Be consistent in quality (always the same contents in the same percentages).

◆ Encourage good root development (be compatible to the plant).

◆ Transplant well to the soil in the plant's permanent location (a plant grown in sand and peat may not transplant well to a clay soil).

Growers prefer to stay with a mix that works best for them. This is probably a good idea since fertilizers and watering can be scheduled better if the mix remains the same. When growing many different kinds of plants at least two mixes will probably be required: one mix for plants requiring a pH of 6.0 to 7.0, and another for acid-loving plants with a pH of 5.0 to 5.5. Fertilizers used as additives to the growing medium may differ for different plants. Many commercial mixes are available already mixed for use.

In large businesses where growers mix their own planting media, sterilization may be done at the nursery. The mix should be steam heated to 180°F for thirty minutes. Most bagged commercial mixes are sold as sterilized and should be ready to use directly from the bag.

Many soilless mixes containing sphagnum peat moss and perlite or vermiculite are very light in weight. If wind tends to blow containers over, it may be necessary to add sand or gravel in the bottom of the containers to hold them down. Be sure this material is sterilized before placing it in the container. Fine pine bark (¼ inch and smaller) is being used extensively as a container growing mix now. It provides excellent drainage but has poor moisture holding capacity. A mix of one part sphagnum moss and three parts pine bark holds moisture better.

Dr. J.R. Gouin at the University of Maryland has found the following mixes to be popular with growers in Maryland who mix their own:

A) 1 part soil
1 part sand or perlite
1 part peat moss

B) 1 part soil
1 part peat moss

C) 1 part peat moss
1 part sharp sand

D) 3 parts peat moss
2 parts sharp sand

E) 1 part sand
1 part ground bark
1 part peat moss

F) 2 parts sand
1 part peat moss
1 part sawdust

G) 2 parts ground bark
1 part sand

Note: If soil is used in the mix, sterilization is necessary. After the mix is prepared it is either fumigated or steam sterilized at 180°F for thirty minutes. Keep the pile covered with plastic until used. (Steam sterilize before fertilizers are mixed in.)

When soilless mixes are used all the plant nutrients needed for growth, including minor elements, must be added. To adjust pH upward, and to add calcium and magnesium, use dolomitic limestone. The amount of lime to add will depend on the pH of the mix and the desired final pH. Sulfur or iron sulfate will lower pH. A test for pH should be made to insure accuracy. Superphosphate should be added at the rate of 6 pounds per cubic yard if tests indicate a need. Potash may also be added if tests indicate a need.

Since soilless mixes contain few trace elements such as manganese, iron, copper, or zinc, it is essential to add 1 or 2 ounces of slow-release trace elements per cubic yard.

Commercial slow-release fertilizers such as Magamp and Osmocote are often mixed into the medium to provide the first supply of plant food for the newly potted plants. (Add after steam sterilization.)

Note: It is essential that all of the above is mixed thoroughly.

Some of the materials in soilless mixes such as peat moss are difficult to wet once they dry out. A wetting agent may be added if this is a problem. Wetting agents also improve uptake and uniformity of wetting. They also shorten the time needed to wet the mix.

PLANTING

Placing plants in the containers is generally done by hand. Some large nurseries have machines, figure 24-9, that fill the containers with the growing medium and some plants are planted by machine. Even with the use of machines, actual placement of the plant in the container is a hand operation. Plants are usually seedlings, rooted cuttings, dor-

FIGURE 24-9 Potting machine (Photo courtesy of Dr. Allen Hammer)

mant grafts, or plants being transferred from smaller pots or flats to containers. The plants may have been purchased from another nursery or grown by the same nursery.

Planting is usually done in the spring. If storage space is available to protect plants from freezing, planting may be done all winter to make better use of labor. The plant is placed in the center of the container (if only one plant is used) at the same depth it was growing before. The growing medium is firmed around it and watered as soon as possible after planting until water runs through the bottom. After planting, the growing medium should be about ¾ inch down from the top of the container to allow space for water to collect when plants are watered. If more than one plant is placed in the container, the plants should be evenly spaced within the container.

Planting may be done in the field from wagons loaded with containers, pots, and plants, and the plants placed in the growing beds, or it may be done at a permanent spot. In permanent locations or on flatbed wagons or trucks, an assembly line is usually set up using a five-person work team. Each person has one job to do repeatedly, such as filling the container with growing medium, making planting holes, planting the plant, watering, and placing plants in the growing bed. A very efficient team can plant up to 8,000 plants per day. Notice that the machine in figure 24-9 fills the container and drills a hole for the plant. Different-sized holes can be made using various machine attachments. As the hole is drilled the medium is firmed in the container. The plant is placed in the hole by hand and firmed in. The next person levels the medium to the proper depth and moves the plant toward the wagon. Plants are hauled to the growing area, put in place, and watered immediately. If the plants being transplanted are in peat pots, the top of the pot must be torn off or covered by the growing medium in the container. Peat pot tops, if allowed to stick out above the medium, tend to dry out and pull moisture away from the roots of the young plant being transplanted.

The depth of planting is also important. Some plants prefer shallow planting and if planted too deeply will get stem rot, or roots will rot for lack of oxygen. For most plants, it is better to plant shallow so that roots are just covered.

CONTAINER SPACING

Spacing of plants in the bed is very important. When plants are small the containers may be placed so that they touch one another. As the plants grow it may become necessary to remove every other container to allow space for the plant's branches to spread and grow normally. Ideal spacing is when the ends of the branches of plants almost touch the plant next to them, figure 24-10. This allows maximum use of space and results in compact plants. Too close spacing results in shading out and browning or killing of lower branches or foliage. Too wide spacing is wasteful of space and irrigation water if sprinklers are used. Plants are often pruned to shape them and cause compact growth. This also helps to prevent crowding and may allow closer spacing. Chemicals can be used in place of hand pruning to cause plants to grow more compactly. An example is the use of a

FIGURE 24-10 Plants spaced with room to grow for the current season. (Ed Reiley, Photographer)

chemical which kills all the tender tips on azaleas to cause branching and compact growth. Without the chemical each tip would need to be pinched off by hand.

A definite growing schedule should be drawn up for each plant, including the space needed for the first year and any extra years needed for the plant to reach marketable size. Contact your local commercial grower to obtain a growing schedule for a container crop grown in your area.

WATERING

Watering becomes the most important consideration in growing container plants once they are placed in the growing beds. The plant has a smaller root growing area than field-grown plants and the container is subject to more heat and drying. A one-gallon container requires about one pint of water at each watering. In hot, dry summer weather an acre of container plants may require from 35,000 to 45,000 gallons of water a day. Water volume may exceed a flow rate of 300 gallons per minute. This volume of water requires a large, dependable water supply such as a pond, stream, or high-volume deep wells. Water must be analyzed for any pollutants that would be harmful to the plants, such as chemicals or salt.

Plants need watering when the surface of the medium in the container feels dry to the hand. At this stage they should be watered immediately. Some plants may be best watered before this stage, and a knowledge of the plant and its water requirements is needed.

How often to water (how many days between waterings) is dependent on a number of factors:

- Weather (wet or dry, hot or cold, windy). Plants require more water in hot, dry, windy weather.

- Plant growth. If the plants are in active growth, more water is needed. Plants also need more water when in bloom.

- Planting medium. A mix with a higher percentage of peat moss will hold water better than a sandy or high percent perlite or pine bark mix.

- Kind of plant. Plants such as broadleaf evergreens require more water than narrowleaf evergreens or succulents such as cactus.

- Size of container. The smaller the container, the faster it will dry out.

◆ Type of container. Plastic containers hold moisture longer than fiber or pressed paper containers.

◆ Container surface mulch. A mulch on the surface of containers or shade to container walls will keep the root area cooler and slow the loss of water.

A definite watering schedule should be set up. Only by examining containers by hand to determine dryness can the exact watering needs be found. It is important that the growing medium drains well (has large pore spaces) so that oxygen supply to the plant roots is good.

There are three methods for watering container plants in outdoor growing areas: by hand, by trickle tube, or by overhead irrigation (sprinklers). Bench growing allows use of a capillary mat under the containers. With all of these systems containers must be watered clear through from top to bottom.

HAND WATERING

This method is generally too costly as a regular watering system except in very small operations. It is used when plants are first planted in the container and to spot water areas that an automated system may have missed. Containers should always be watered until water runs through the bottom of the container.

TRICKLE TUBE SYSTEMS

These systems are the most efficient in saving water, although they are expensive. Water is delivered directly to each container by a small tube. Each container gets the same amount of water. Tubes are weighted on the end so they remain in place in the container top.

A typical layout for this system has a long plastic water main with secondary lines between every other row with trickle tubes coming out all along it to each pot, figure 24-11. Strainers must be placed in the main lines to prevent dirt from clogging the small tubes. Liquid fertilizer may be fed through these tubes to fertilize the plants as watering is being done.

The length of time the system is left on at each watering must be determined by watching the containers. Watering is stopped when water first starts to run through the drain holes in the bottom of the container.

One thing to watch for in trickle tube systems is that rodents, such as mice and rabbits, do not chew the small tubes. Treatment with a spray containing Arasan or Thiram and a sticker, which helps keep the material on the tubing longer, helps to prevent rodent damage. Tubes should be inspected each day for damage or clogging so that all plants are watered.

OVERHEAD IRRIGATION (ROTARY HEADS)

Overhead irrigation is perhaps the most often used system with container nursery crops. It is cheaper to install and operate. The biggest disadvantage is that it requires more water to do the same job. Some water is lost between containers, and also to evaporation as it is sprayed out through the air above the containers.

There are two different styles of sprinklers: overhead and oscillator. The overhead sprinkler distributes water in a circular pattern, figure 24-12. The oscillator spreads water in a rectangular pattern.

The procedure for establishing watering time is the same as for the trickle tube system. Sprinkle

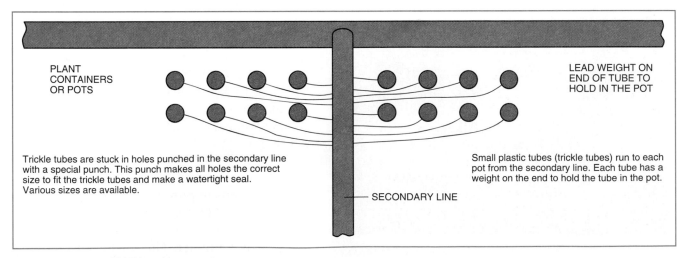

PLANT CONTAINERS OR POTS

LEAD WEIGHT ON END OF TUBE TO HOLD IN THE POT

Trickle tubes are stuck in holes punched in the secondary line with a special punch. This punch makes all holes the correct size to fit the trickle tubes and make a watertight seal. Various sizes are available.

Small plastic tubes (trickle tubes) run to each pot from the secondary line. Each tube has a weight on the end to hold the tube in the pot.

— SECONDARY LINE

FIGURE 24-11   Trickle tube system

FIGURE 24-12 Overhead irrigation used on container plants. Container growing requires a lot of water. (Ed Reiley, Photographer)

until the containers are wet through to the bottom. The time between waterings can be determined only by examining containers to see when water is needed. If this is not possible, good drainage of containers must be assured and watering be done often enough to prevent drying.

Each sprinkler head or oscillator should have a cutoff valve so sections of the system can be cut on and off as needed. Some areas of the system may be empty of plants and not need watering, or workers may need to enter the area.

FERTILIZERS

Fertilizers may be applied to container-grown plants through several methods. The planting medium probably had some fertilizer mixed in it prior to planting. This was explained under growing mixes.

Additional fertilizer will be needed as the plants grow.

LIQUID FERTILIZER

Fertilizer may be applied through siphoning devices into the irrigation water if plants are watered by hand or with trickle tubes. Fertilizers should not be applied with the irrigation water when overhead sprinklers or oscillator systems are used because much of it would fall between the containers and not get to the plants. This is expensive and wasteful and could pollute streams or ponds nearby.

If liquid fertilizer is used at each watering it should be applied at a rate of 100 to 150 ppm (parts per million) of nitrogen. Fertilizer ratio of nitrogen, phosphorus, and potassium will vary with the plant being grown.

Periodic feeding every two or three weeks with liquid fertilizer at 300 to 500 ppm is not as effective but is often used. A definite fertilizer schedule should be set up for each plant grown.

SLOW-RELEASE FERTILIZERS

New slow-release fertilizers, such as Magamp or Osmocote, are used with container plants in two ways. With Osmocote (18-6-12), for example, the material is mixed in the planting medium at a rate of 9 to 10 pounds per cubic yard for completely soilless mixes. (Do not steam the mix after the fertilizer is added.) The mix should be used within two weeks after the Osmocote is added.

Magamp (7-40-6) is another slow-release fertilizer that may be mixed into the growing medium. It feeds for one full growing season. Supplemental nitrogen may be needed.

Osmocote can also be used as a top dressing fertilizer on top of the medium in the container. Since the release of plant food (available to the plant) from Osmocote is related to soil temperature, it can be added at any time. As the temperature rises more nutrients are released, and as soil or medium temperature cools nutrient release slows. One application of Osmocote on top of the container medium will feed the plant for one full year. The plant food is washed down each time the container is watered.

Plants that are kept in containers for more than one year need to have more fertilizer added. For liquid fertilizer, granular fertilizers, or Osmocote, one teaspoon of fertilizer (18 percent nitrogen) should be added for each one gallon of container capacity. Liquid or granular fertilizers should be applied in the spring and Osmocote in the fall. Plants should be watered thoroughly immediately after fertilizing, if a granular or solid form is added.

The liquid fertilizer program is continued through the second or third season the same as for the first year. It should be started in early spring before growth starts and continued throughout the growing season. At the end of the growing season at least one watering should be done with no fertilizer.

Granular fertilizers should be started at the beginning of the growing season and repeated every eight weeks using one teaspoon per gallon of growing medium in the container. A maximum of three applications should be made per season.

Agriform 14-4-6 tablets for container use should be applied at the beginning of the growing season with repeat applications every ten weeks. One 12-gram tablet per gallon of growing medium in the container is the proper rate.

Growth of plants should be noted and fertilizer rates changed as plant foliage and growth rate show a need for more or less fertilizer. As the growing season comes to an end (August and September in the Mid-Atlantic states) nitrogen application should be reduced to slow growth and start the hardening off process in preparation for winter. Additional phosphorus fertilizer also helps plants to set flower buds and harden off for winter. Watering less also often tends to harden plants off for winter.

One advantage of Osmocote is that as temperatures lower at the end of the summer the plant nutrients are released more slowly. This assists in slowing growth and hardening plants for winter.

Fertilizer programs vary for each type of plant grown and the exact fertilizer schedule should be set up before plants are potted. Also, some growing media hold fertilizer better than others. Bark and sand do not hold nutrients well.

DISEASE AND INSECT CONTROL

Disease control is best started with a sterile planting medium, a sanitary growing site, and sanitary containers. If the site is clean and drains well, if the planting mix is sterilized, and if new or sterilized containers are used, the next source of disease or insects to check for is from the plants that are going into the containers. These plants should be examined prior to planting. Diseased plants should be destroyed and plants with insects should be sprayed immediately after placement in the growing area. If these precautions are taken, insect and disease problems should be minimal. Close inspection during the growing season will uncover problems and spray schedules or other control methods may be set up. Since chemicals used vary and new materials are constantly coming on the market, no specific recommendations will be made here. Consult your local extension office or state agriculture college for details.

PRUNING

Container-grown plants are pruned mainly to shape the plant or make it more compact. Plants where this is not a problem, such as groundcover plants, are usually not pruned. Plants should be pruned to help them develop the desired shape for that particular plant. The pruner must know if a plant should have one central leader stem or if it grows naturally as a many-stemmed plant. The pruner then prunes to assist the plant in that development.

Pruning is best done immediately after planting and during the growing season to remove long, unattractive shoots. After this first season pruning, plants with single main trunks usually have side branches shortened each spring to promote compact growth. Pines usually have the new shoots (candles) cut to one-half their length in late spring before the needles have expanded. If this is done on all shoots a compact tree will result. Plants such as boxwood or hedge-type plants may be sheared to the desired shape.

Pruning may be done in winter instead of spring to spread labor demands if cold injury to the pruned branches is not a problem in the growing area. Chemical pruning of azaleas was mentioned earlier under plant spacing.

WINTER STORAGE

In growing areas where winter temperatures drop much below freezing some sort of protection must be provided to container-grown plants.

The most serious winter injury is due to lack of root hardiness. The roots of most plants are not as hardy as the top part of the plant. Since the containers are up on top of the soil, the temperature around the roots is nearly the same as air temperature. Roots of many woody plants die when the soil or growing medium temperature drops below 20°F (–7°C). Plants with extensive root damage from cold usually die the following spring.

If plants are to survive temperatures at or below those listed in figure 24-13, some sort of protection must be provided. Some growers place plants in plastic-covered greenhouses where temperatures can be controlled with heaters to just above killing temperatures, figure 24-14, page 252. Other growers lay plants on their sides or stand them up under a low structure on the soil and stack containers as close together as possible without damaging the foliage. The plants are then covered with a single layer of clear, 4 mil polyethylene. Straw or other insulating material 4 to 6 inches thick is placed on top of this plastic and another

SCIENTIFIC NAME	COMMON NAME[1]	°F	°C
Buxus sempervirens	Common Box	27	−2.7
Cotoneaster congesta	Cotoneaster	25	−3.8
Ilex crenata 'Dazzler'	Dazzler Holly	25	−3.8
Cotoneaster dammeri	Lowfast Cotoneaster	23	−5
Daphne cneorum	Rose Daphne	23	−5
Euonymus fortunei vegetus	WinterCreeper Euonymus	23	−5
Hypericum species	St. Johns Wort	23	−5
Ilex cornuta 'Nellie Stevens'	Holly	23	−5
Ilex crenata 'Convexa'	Convexleaf Japanese Holly	23	−5
Ilex crenata 'Helleri'	Dwarf Holly	23	−5
Ilex crenata 'Hetzi'	Hetz Holly	23	−5
Ilex crenata 'Stokes'	Stokes Holly	23	−5
Ilex meserveae	Holly—Blue Girl and Blue Boy	23	−5
Ilex opaca	American Holly	23	−5
Magnolia X soulangeana	Saucer Magnolia	23	−5
Mahonia beali	Leatherleaf Mahonia	23	−5
Pyracantha cocinea 'Lalandi'	Firethorn	23	−5
Cornus florida	Dogwood	22	−5.6
Euonymus patens	Patens Euonymus	22	−5.6
Ilex crenata 'San Jose'	San Jose Japanese Holly	22	−5.6
Magnolia stellata	Star Magnolia	22	−5.6
Cotoneaster dammeri 'Skogholmen'	Skogholmen Cotoneaster	20	−6.7
Leucothoe catesbaei	Leucothoe	20	−6.7
Rhododendron prunifolium	Deciduous Azalea Prunifolium	20	−6.7
Viburnum plicatum tomentosum	Double-File Viburnum	20	−6.7
Euonymus alatus	Winged Euonymus	19	−7.2
Rhododendron 'Hino Crimson'	Azalea Hino-Crimson	19	−7.2
Stephanandra incisa	Cutleaf Stephanandra	18	−7.8
Cotoneaster horizontalis	Rock Spray Cotoneaster	17	−8.3
Cryptomeria japonica	Japanese Cryptomeria	17	−8.3
Rhodoendron 'Exbury Hybrid'	Deciduous Azalea Group	17	−8.3
Taxus media 'Hicksi'	Hicks Yew	17	−8.3
Cytisus X praecox	Warminster Broom	16	−8.9
Ilex glabra	Inkberry Holly	16	−8.9
Koelreuteria paniculata	Golden Rain Tree	16	−8.9
Acer palmatum 'Atropurpureum'	Japanese Maple	15	−9.4
Euonymus fortunei 'Argenteo-marginatus'	WinterCreeper Euonymus	15	−9.4
Euonymus fortunei 'Carrieri'	WinterCreeper Euonymus	15	−9.4
Hedera helix 'Baltica'	Baltic English Ivy	15	−9.4
Kalmia latifolia	Mt. Laurel	15	−9.4
Pachysandra terminalis	Japanese Spruce	15	−9.4
Pieris japonica	Japanese Andromeda	15	−9.4
Rhododendron 'Purple Gem'	Rhododendron Purple Gem	15	−9.4
Rhododendron schlippenbachi	Deciduous Azalea Species	15	−9.4
Viburnum carlesii	Korean Spicebush	15	−9.4
Vinca minor	Periwinkle	15	−9.4
Cotoneaster adpressa praecox	Cotoneaster	12	−11.1
Juniperus conferta	Blue Pacific Juniper	12	−11.1
Juniperus horizontalis 'Plumosa'	Andorra Juniper	12	−11.1
Juniperus squamata	Meyeri Juniper	12	−11.1
Taxus X media 'Nigra'	Media Yew	12	−11.1
Mahonia aquifolium	Oregon Grape, Holly	10	−12.2
Rhododendron X 'Gibraltar'	Deciduous Azalea Gibraltar	10	−12.2
Rhododendron X 'Hinodegiri'	Azalea Hinodegiri	10	−12.2
Thuja occidentalis	American Arborvitae	10	−12.2
Euonymus fortunei 'Colorata'	Purpleleaf Winter Creeper	5	−15
Leucothoe fontanesiana	Leucothoe	5	−15
Pieris floribunda	Floribunda Andromeda	5	−15
Juniperus horizontalis 'Douglasii'	Waukegan Juniper	0	−17.8
Rhododendron carolinianum	Rhododendron carolinianum	0	−17.8
Rhododendron catawbiense	Rhododendron Catawbiense	0	−17.8
Picea glauca	Colorado Spruce	−9	−23.3
Picea omorika	Serbian Spruce	−9	−23.3
Potentilla fruticosa	Cinquefoil	−9	−23.3
Rhododendron P.J.M. Hybrids	Rhododendron P.J.M.	−9	−23.3

Temperatures that will injure primary and possibly secondary roots, but will not result in 100% kill of the root systems in moist soils. Temperature data pooled from: John R. Havis. Root hardiness of woody ornamentals. *HortScience* 11(4), 1976, pp. 385–386, and Peter L. Steponkus, George L. Good, and Steven C. Wiest. Root hardiness of woody plants. *American Nurseryman* CXLIV(6), 1976, p. 16.

[1]Common names added by author.

FIGURE 24-13 Root killing temperatures of container-grown ornamentals in winter storage (Courtesy Dr. Francis R. Gouin, University of Maryland Horticulture Department)

FIGURE 24-14 A plastic-covered poly house used to overwinter container plants. Temperature is kept at or above 28°F (–2°C) at night and cooled during the daytime by opening vents and doors. (Ed Reiley, Photographer)

FIGURE 24-15 Container-grown plants are placed on the ground on their sides and covered with microfoam and white plastic to protect them from low temperatures damage to roots. (Courtesy Marshy Point Nursery)

layer of clear, 4 mil polyethylene is spread over the straw. Edges are sealed with soil and the plastic held down with old tires or other weight to keep the wind from blowing or tearing it.

Mice may get into the storage area and cause damage to plants over winter. It is important to use some sort of poison to control these rodents.

A thermo-blanket system may also be used to protect container plants from winter root damage, figure 24-15. The blanket is a microfoam made of white styrofoam. It has high insulating value and transmits about 50 percent light. Because it reflects 50 percent light, temperatures under it stay uniformly cool. Contact a local grower or your agricultural college for more information on this new material.

TAKING PLANTS FROM WINTER STORAGE In spring, when temperatures get above the danger level, uncover the plants. Set them upright and water well.

As mentioned earlier, some containers insulate better than others. This gives some root protection if the temperatures dropped below the root killing level for short periods of time such as a few hours or one night only. The main advantage of container insulating value, however, is to help keep the roots cooler in summer. This could help to control some root rot problems in plants such as rhododendron. Many root rotting organisms cause infection only at high soil temperatures.

If high summer temperatures are a problem with container crops, shade should be supplied to the growing area. Lath houses using wooden slats or a snow fence overhead give about 40 to 50 percent shade. Shade fabric, either plastic or cloth, may be used to provide varying degrees of shade. These materials require the support of posts and two-by-fours or other material to hold them in place above the plants.

Shade not only lowers temperature to the plant's top and to the roots in containers but also reduces the amount of water needed for irrigation as temperatures are lowered. Foliage burn is greatly reduced.

Shade in winter is also helpful in reducing foliage burn caused by the sun heating foliage and depleting moisture. This reduces leaf browning and results in a more attractive plant that is ready for sale the following spring.

MARKETING

Container-grown plants can be marketed and transplanted any time the ground is not frozen. Most plants are planted in spring and fall as these are the seasons for best survival. Rainfall and temperatures are usually better for transplanting at this time of year. Evergreens generally transplant best in early spring and deciduous plants in the fall. Figure 24-16 shows a rhododendron ready for market.

FIGURE 24-16 A container-grown rhododendron (Ed Reiley, Photographer)

When a container plant reaches the size specified in the standards chart (refer to figure 24-8), it must either be marketed or transplanted to the next larger container. If plants are allowed to stay in the same container for another year, the roots will grow into a tight spiral ball. This root-bound condition slows or even stops root extension out into the surrounding soil when the plant is transplanted. This results in stunted growth or even death of the plant.

When container plants are retailed through stores without trained personnel to tend them, much damage and loss of quality can result. The biggest problem is improper watering. Top quality plants can become almost worthless in a period of one to two weeks if roots are allowed to dry out and die. The plant must be properly cared for until planting in the permanent location is completed.

TRANSPLANTING FROM THE CONTAINER

If plants have been properly grown and are not seriously pot-bound, container-grown plants transplant easily. Prepare the soil so its texture is not too different from the medium in the container. This probably means digging in pine bark in all soils. Knock the plant from the container, keeping the root ball intact. Plant at the same depth the plant was growing in the container. Firm the soil around the root ball with your hands.

If the plant is pot-bound use a sharp knife to cut about 1 inch deep, down the full length of the root ball, from top to bottom. Do this at about four or five equally spaced spots around the root ball. This will encourage new root development at the cut ends and roots that will extend out into the soil. It is even better to pull the roots out of the root ball by straightening them out or digging away part of the potting medium. Place these roots out into the soil when transplanting and work soil around them. After the soil is firmed around the roots, water thoroughly to settle the soil around the roots. (Do not stomp with the feet or apply pressure that is strong enough to break roots.) Apply a mulch on the soil surface and stake tall growing plants to prevent the wind from blowing the root ball loose or out of the soil.

Student Activities

1) Transplant a small plant into a commercially used container and grow the plant to saleable size. Set up a growing schedule for such things as pruning, watering, fertilizing, pest control, and marketing date. Contact a local grower for help if needed.

2) Make an appointment to visit a local container grower and tour the operation. Discuss the tour in class and note the growing system used. If possible, obtain a few plants from the grower to be grown as a class project or individual projects.

   If the above trip is not possible, have a grower come in to speak to the class and demonstrate growing or sales techniques.

Self-Evaluation

A) Select the best answer from the choices offered to complete the statement or answer the question.

1) Container growing of plants may be defined as

    a) growing any plant in a pot.
    b) starting plants from seeds or cuttings in a container.
    c) growing a plant to marketable size in a container.
    d) bonsai culture of plants.

2) The initial cost of getting established in container growing of plants is

    a) lower than for field growing of plants.
    b) higher than for field growing of plants.
    c) about equal to field growing of plants.
    d) about triple the cost of field growing of plants.

3) To grow plants in containers (as compared to field growing), _____ irrigation water is required.

    a) more
    b) less
    c) the same amount of
    d) cleaner

4) The best site or growing area for container plants is one that

    a) has a flat, gentle slope.
    b) is close to a good water supply.
    c) has a good windbreak.
    d) all of these

5) A windbreak will slow the wind and is considered effective for a distance of about

    a) 4 times its height.
    b) 2 times its height.
    c) 7 times its height.
    d) none of these

6) When a container is too small for the plant growing in it,

    a) the plant should be sold or moved into a larger container.
    b) plant roots will circle around the container sides.
    c) the plant may turn brown and start to die.
    d) all of the above

7) A new fabric pot for growing trees in the field results in

    a) a more compact and fibrous root system.
    b) easier digging of trees.
    c) reduced labor costs.
    d) all of the above

8) A new copper chemical Spin Out is used to

    a) stop plant roots from circling the inside of the pot.
    b) kill plant roots.
    c) kill moss in the nursery.
    d) correct a mineral deficiency.

9) Growers who grow a variety of plants generally use at least two different kinds of growing media or mixes in order to

    a) provide proper drainage for different plants.
    b) provide proper pH for different plants.
    c) insure easy transplanting of different plants.
    d) have plants growing at different rates for different markets.

10) In order to sterilize a container mix it must be

    a) heated to 120°F for 30 minutes.
    b) heated to 180°F for 60 minutes.
    c) heated to 180°F for 30 minutes.
    d) dried completely before use.

11) Fertilizer or plant food added to plant growing mixes must be

    a) added when the mix is first mixed.
    b) added after soil sterilization by heat.
    c) all in a soluble form.
    d) applied in the container to the surface of the mix.

12) During and after planting the growing medium should be

    a) about ¾ inch down from the top of the container.
    b) watered in well to settle it around the plant roots.
    c) firmed around the plant.
    d) all of the above

13) Ideal spacing of container plants for continued growth is

    a) where the ends of branches almost touch the plant next to them.
    b) where the containers just touch one another.
    c) where branches of adjoining plants just overlap.
    d) all of the above

14) Once container plants are placed in the growing bed, _____ becomes the most important consideration.

    a) pruning   c) watering
    b) fertilizing   d) insect control

15) Plants need _____ water when in bloom.

    a) more   c) the same amount of
    b) less   d) very little

16) A small container will need to be watered _____ a large container.

    a) less often than
    b) more often than
    c) on the same schedule as
    d) none of the above

17) A watering schedule should be set up for container plants

    a) and followed exactly.
    b) and varied with temperature as needed.
    c) but only by examining containers by hand can the exact time to water be determined.
    d) and should be stopped when freezing temperatures come.

18) Select the method most often used for watering container plants.

    a) hand watering   c) capillary mat
    b) trickle tube   d) overhead irrigation

19) The watering system that uses the least amount of water is

    a) hand watering.   c) capillary mat.
    b) trickle tube.   d) overhead irrigation.

20) Fertilizers may be applied to container-grown plants using several methods. The newer, slow-release fertilizers are used two ways:

a) mixed in the planting medium and as a top dressing in the container.
b) through the irrigation system and as a top dressing in the container.
c) on top of the medium every eight weeks and mixed in the planting medium.
d) all of the above

21) Liquid fertilizer may be applied through the irrigation system if

a) plants are watered by hand or with trickle tube irrigation.
b) plants are watered by overhead sprinkler.
c) plants are watered by oscillators.
d) the fertilizer does not damage the irrigation system.

22) When granular fertilizers are used, the program is started at the beginning of the growing season and repeated

a) as the plants show a need.
b) every eight weeks.
c) every twelve weeks.
d) every fourteen weeks.

23) The best disease control program is started with

a) a good spray program.
b) a sterile planting medium and a sanitary site and container.
c) regular inspection for disease.
d) all of the above

24) Container-grown plants are pruned in order to

a) take cuttings for propagation of new plants.
b) keep them low and thick.
c) shape the plant or make it grow compactly.
d) thin the branches.

25) In growing areas with cold winter temperatures (below 20°F),

a) container plant roots are damaged most by cold.
b) container plant tops are damaged most by cold.
c) container plants survive as well as field-grown plants.
d) plants must be hardened off early for winter survival.

26) Container plants may be protected from winter cold by

a) placing them in plastic-covered greenhouses.
b) covering them with straw and plastic film.
c) using a thermo-blanket to cover them.
d) all of the above

27) Lath or shade fabric is used to shade plants. This shading protects the plant from

a) drying out and burning of foliage in the summer.
b) drying out and burning of foliage in the summer and winter.
c) sunburn.
d) none of the above

28) The American Association of Nurserymen sets up standards for nursery stock including container-grown plants. When a container plant reaches the size specified in the standards chart, the plant

a) must either be sold or thrown away.
b) can no longer be sold.
c) must either be sold or transplanted to a bigger container.
d) can be sold for a period of six weeks only.

29) If not seriously pot-bound, container-grown plants transplant easily. Pot-bound plants should

a) be thrown away.
b) have the roots cut so they will move out of the root ball and into the soil.
c) be transplanted to the next larger pot.
d) be transplanted without disturbing the roots.

B) Answer each of the following.

1) List four advantages of growing plants in containers.

2) List three disadvantages of growing plants in containers.

3) List five types of plants that are grown in containers.

4) List the ten things a good growing medium must provide.

5) List five container-grown plants whose roots are killed when the temperature reaches 15°F (−9.4°C) or below.

Using Plants in the Landscape

Annual Bedding Plants

OBJECTIVE

To plant and maintain annual flowers in the landscape.

Competencies to Be Developed

After studying this unit, you should be able to

◆ identify four uses of annual flowers.

◆ design a bed layout using annual flowers.

◆ explain the steps in preparing the soil for annual flowers.

◆ list the six steps in the aftercare of annual flowers.

◆ demonstrate proper transplanting techniques for annual flowers.

MATERIALS LIST

✔ sources of color pictures of annuals, such as nursery catalogs

✔ seed flats and growing medium in which to start plants

✔ seed and nursery catalogs (one for each two students)

✔ planting area in which to design a flower bed, using annuals

✔ all necessary ingredients for preparation of flower beds (peat moss, fertilizer, etc.)

Annuals are plants that complete their life cycle in one year. The plant starts from seed, grows, blooms, sets seed, and dies in one season. (The meaning of the word *annual* is "one year.") Both vegetable plants and flower plants can be classified as annuals. While both types bear flowers, the flowering plant is usually used for landscaping purposes.

The market for bedding plants (annuals) has grown in the last ten years. Today a wide variety of retail outlets such as roadside markets, department stores, garden centers, and grocery stores are marketing annuals.

When purchasing bedding plants it is important to make sure that the plants carry labels which give the price, required light exposure (shade or sun), and planting and care instructions.

FLOWERING ANNUALS

Flowering annuals are easy to grow and lend color to landscapes a short time after they are planted. They are usually started from seed indoors early in the spring and transplanted to the garden. The most popular annual flowering plants include marigolds, petunias, zinnias, ageratums, celosias, coleus, portulaca, pansies, and snapdragons. Figure 25-1 shows mass planting of petunias.

Flowering annuals are generally used to:

◆ provide a mass of color around a house foundation, in flower beds, or in front of evergreens, figure 25-2.

◆ fill spaces between shrub plantings or other perennials and give color when these plants are not blooming.

◆ provide color in bulb beds after the bulbs have bloomed.

◆ supply cut flowers.

◆ plant along fences or walks.

◆ cover bare spots between or in front of larger shrubs.

◆ create seasonal color, figure 25-3.

SELECTING FLOWERING ANNUALS

Great improvements have been made in the quality and vigor of annual flowering plants. One example is the Triploid hybrid marigold which blooms within six weeks from seed. It sets no seed and blooms

FIGURE 25-2 Impatiens used as a border planting to accent a garden walk in a park. (Courtesy George C. Ball Co.)

FIGURE 25-3 Annuals are valued and used for the seasonal color in the landscape. (Courtesy George C. Ball Co.)

longer than the traditional marigold, figure 25-4A. If seed is allowed to set on annual plants, the seed production causes strength to be sapped from the plants and reduces blooming. For this reason, faded blooms should be cut from all annual plants.

When deciding upon which annuals to grow, first consider the purpose of the plant and where it is to be planted. Be sure to consider height, keeping shorter plants in front of beds. Select plants

FIGURE 25-1 A planting of petunias adds a solid mass of color to the landscape. (Courtesy George C. Ball Co.)

FIGURE 25-4A The Triploid hybrid marigold. The marigold is one of the easiest annuals to grow.

FIGURE 25-4B Annuals such as gazania and dusty miller create an aesthetic, pleasing flower bed. (Courtesy George C. Ball Co.)

with colors that will blend in well with one another, figure 25-4B. Figure 25-5 details other information concerning various annuals.

Hanging baskets of annual flowers are very popular. These add new dimension to the display of plants and their use in the landscape. Baskets are often used to create an attractive floral display or focal point on the patio or at the main entrance of the home.

PLANTING HANGING BASKETS

Hanging baskets come in various sizes and shapes. The most commonly used hanging baskets range from 4 inches to a 14-inch size. All baskets used must provide proper drainage, therefore they should have drainage holes in order to promote proper plant growth.

The standard medium used to start annuals is also a very good source of medium for hanging baskets.

When planting hanging baskets, the standard rule of thumb is to plant one less plant than the size of the pot. For example, if using an 8-inch pot, select seven plants and place six of them around the outside perimeter of the pot and the other plant in the center of the basket. Proper care in watering, fertilizing, and deadheading (removal of dead and faded flowers) can help to produce an attractive hanging basket of annuals.

PROCEDURE

GROWING FLOWERING ANNUALS

For a summer of color with annuals:

1. Decide which annuals to grow. Outline the plan for the flower bed in writing. Probably the easiest plants for the beginner to grow are marigolds and zinnias.
2. Seed six to eight weeks prior to outside planting date (frost-free date).
3. Prepare the soil well ahead of the planting date. Add organic amendments, fertilizer, or lime as needed.
4. Plant according to the landscape plan after the frost-free date.
5. Water, weed, and mulch as needed.
6. Pinch off all faded blossoms.
7. Provide for insect control.
8. Enjoy cut flowers.

PLANNING AND DESIGNING THE FLOWER BED

Flower beds for annuals range in size and design according to individual needs and preferences. Figure 25-6, page 267, shows the design for a flower bed to be planted between a walkway and a home.

NAME	HEIGHT	SPACING	COLOR	USES	HOW TO START	REMARKS
Ageratum (ageratum)	6"–12"	6"–9"	blue, pink, white	edging	seed and transplant**	Compact, excellent bloom. Needs full sun.
Antirrhinum (snapdragon)	12"–36"	6"–8"	yellow, pink, white, blue, red, salmon	cut flowers, flower bed	seed and transplant	Must be staked for straight flower spikes. Needs full sun.
Aster	8"–36"	10"–12"	blue, red, salmon, pink, white, yellow	flower beds	seed and transplant	Excellent cut flowers.
Begonia (begonia)	6"–12"	6"–10"	pink, red, white	edging, potted plant, window box, hanging basket	cuttings or seed	May be used as a houseplant. Needs direct light.
Browallia (browallia)	12"–18"	8"–10"	blue, white	hanging basket, window box	cuttings or seed	Makes an attractive houseplant. Needs full sun.
Calendula (potted marigold)	12"–24"	10"–12"	yellow, orange	flower bed, hanging basket	seed and transplant	Flower petal used in cooking stews to add color. Needs full sun.
Callistephus (China aster)	12"–24"	10"–12"	blue, purple, white, yellow	flower bed, cut flowers	seed and transplant	Gives excellent summer color. Needs full sun.
Celosia (cockscomb)	12"–30"	10"–12"	red, orange, yellow, pink	flower bed, cut flowers	seed and transplant	Is excellent dried flower. Needs full sun.
Cleome speciosa (spider plant or spiderflower)	24"–36"	18"–36"	pink, white	flower bed	seed directly*	Makes an attractive houseplant.
Coleus (coleus)	12"–24"	10"–12"	red, bronze, yellow, pink foliage	flower bed, hanging basket	seed and transplant	Needs full to partial sun. Beautiful foliage plant.
Coreopsis (coreopsis)	12"–24"	6"–10"	yellow, red	flower bed	seed directly	Good for cut flowers.
Cosmos (cosmos)	48"–72"	12"–18"	red, pink, yellow	flower bed	seed and transplant	Keep near back of beds.
Dahlia variabilis (dahlia)	12"–24"	12"–15"	red, pink, yellow rose	flower bed, cut flowers	seed and transplant	Profuse bloomer. For maximum bloom, sow several weeks before other annuals.
Delphinium (larkspur)	18"–24"	8"–10"	blue, pink, white	flower bed, cut flowers	seed and transplant	Grows best in peat pots. Difficult to transplant.
Dianthus (pink)	6"–18"	6"–8"	white, pink	edging, cut flowers	cuttings or seed	Very fragrant flowers.

* seed directly—sow the seed directly into the soil where the plants are desired.
** seed and transplant—start the seed indoors (in greenhouse, hotbed, or portable seed germination case). As the seedlings develop, they are transplanted from the seed flat to other containers where they grow until ready to be set out.

FIGURE 25-5 Bedding plant chart for annual flowers

NAME	HEIGHT	SPACING	COLOR	USES	HOW TO START	REMARKS
Dimorphotheca (cape marigold)	12"–18"	8"–18"	yellow, orange, white	flower bed	seed and transplant	Grows well in dry areas.
Gaillardia (gaillardia)	12"–24"	8"–10"	red, orange	cut flowers, flower bed	cuttings	Loves seashore conditions. Does well in dry areas.
Gazania	6"–10"	8"–10"	orange, yellow, white, tangerine, bronze	flower bed	seed and transplant	Bright blooms.
Gomphrena (globe amaranth)	9"–24"	8"–10"	blue, pink, white	cut flowers, flower bed	seed and transplant	Makes good dried flower. Collect and hang in dry, dark place.
Gypsophila (baby's breath)	12"–18"	10"–12"	pink, white	flower bed, cut flowers	seed directly	This is the annual form; there is also a perennial plant.
Helianthus (sunflower)	18"–60"	12"–36"	yellow, brown	flower bed	seed directly	Seeds are edible.
Helichrysum (strawflower)	24"–36"	8"–10"	yellow, red, white	flower bed	seed and transplant	Makes good dried flower. Cut; hang in dry, dark place.
Hypoestes	8"–24"	10"–12"	burgundy, red, rose, white, pink	hanging baskets, window boxes, flower bed	seed and transplant	Foliage accented by splashed colored leaves.
Iberis (candytuft)	10"–18"	6"–8"	white, pink, red	edging	seed and transplant	Makes attractive ground cover.
Impatiens (impatience)	6"–18"	12"–18"	pink, red, white, multicolor	edging, hanging basket	seed and transplant	Does best in shaded area.
Ipomoea (morning glory)	60"–120"	10"–12"	blue	trellis	seed directly	Vine.
Lantana (lantana)	12"–36"	2"–15"	yellow, blue, red	hanging basket, flower bed, potted plant	cuttings	Makes excellent topiary plant for patio.
Lathyrus (sweet pea)	36"–60"	6"–10"	pink, blue, white	cut flowers	seed directly	Grows best in cool conditions.
Limonium (statice or sea lavender)	18"–30"	10"–12"	blue, pink, white	flower bed, cut flowers	seed and transplant	Makes good dried flower.
Lisanthus	6"–24"	12"–14"	blue, rose, orchid, white, pink, yellow	flower bed	seed and transplant	Large blooms, drought-tolerance.
Lobelia (lobelia)	3"–6"	4"–6"	blue, pink, white	edging, hanging basket, potted plant	seed directly	Very attractive in hanging basket.
Lobularia (sweet alyssum)	4"–6"	4"–6"	white, pink, blue	window box, edging, hanging basket	seed and transplant	Blooms all summer.

FIGURE 25-5 Bedding plant chart for annual flowers (continued)

NAME	HEIGHT	SPACING	COLOR	USES	HOW TO START	REMARKS
Matthiola (stock)	12"–36"	8"–10"	white, pink, yellow, red	cut flowers, flower bed	seed and transplant	Fragrant flowers.
Mirabilis (four-o'clock)	24"–36"	10"–12"	yellow, red, white	flower bed	seed and transplant	Withstands city conditions.
Myosotis (forget-me-not)	6"–12"	6"–8"	blue, pink	edging, cut flowers	seed directly	Makes attractive cut flowers.
Nicotiana (flowering tobacco)	18"–24"	10"–12"	rose, white	cut flowers, flower bed	seed and transplant	Fragrant flowers after dark.
Papaver (poppy)	18"–36"	10"–12"	red, yellow, white	cut flowers, flower bed	seed directly	Grows in masses.
Pelargonium (geranium)	12"–18"	10"–12"	red, white, pink, lavender	flower bed, window box, hanging basket	cuttings and seeds	Excellent flowering plant.
Petunia (petunia)	6"–18"	10"–12"	pink, red, white, blue	edging, flower bed, hanging basket	seed and transplant	Most widely used annual.
Phlox (phlox)	6"–18"	6"–8"	pink, blue, white, red	cut flowers, flower bed, window box, hanging basket	seed directly or transplant	Very intense colors.
Portulaca (rose moss)	4"–6"	3"–6"	red, orange, yellow	edging, hanging basket	seed directly	Reseeds by itself.
Salpiglossis (giant velvet flower)	18"–36"	10"–12"	gold, scarlet, rose, blue	flower bed, cut flowers	seed and transplant	Very small seed.
Salvia (scarlet sage)	10"–36"	6"–12"	blue, red, white	cut flowers, flower bed	seed and transplant	Very showy color.
Scabiosa (pincushion flower)	30"–36"	12"–14"	multicolor	cut flowers	seed and transplant	Variety of plants from which to select.
Snapdragon	12"–36"	8"–10"	pink, white, red, rose, bronze, orange, salmon	flower bed	seed and transplant	Excellent background plants.
Tagetes (marigold)	6"–48"	6"–12"	orange, yellow	flower bed, edging, potted plant, hanging basket	seed and transplant or seed directly	Wide range of varieties. No insect problems.
Thunbergia (black-eyed Susan vine)	24"–60"	10"–12"	orange	climbing vine, hanging basket	seed and transplant	
Tropaeolum (nasturtium)	12"–18"	10"–12"	orange, yellow	flower bed, window box, hanging basket	seed and transplant or seed directly	Flowers and leaves are edible.
Verbena (verbena)	6"–12"	10"–12"	purple, red, white	flower bed, edging	seed and transplant	Good rock garden plant.
Viola (pansy)	6"–8"	6"–8"	multicolor	edging, flower bed, cut flowers, hanging basket	seed and transplant	Gives excellent color in summer.
Zinnia (zinnia)	8"–36"	6"–12"	multicolor	edging, flower bed, cut flowers	seed and transplant or seed directly	Gives excellent color in summer.

FIGURE 25-5 Bedding plant chart for annual flowers (continued)

20 feet

(7)	(6)	(8)
(5)	(5)	(5)
(3)	(4)	(3)
(2)
(1)

Numbers refer to the plants shown below

(3) Dianthus—rosemarie

(1) Alyssum—snow crystal

(4) Salvia—red hot sally

(2) Ageratum—blue puffs

(6) Lisanthus

(5) Celosia—century

(7) Aster—master

(8) Snapdragon—Maryland

FIGURE 25-6   One way to beautify a landscape is to plant a flower bed between a walkway and a home. Such a planting could be designed according to this plan. (Courtesy George C. Ball Co.)

PLANT SCHEDULE FOR STARTING BEDDING PLANTS IN THE GREENHOUSE		
Name	Weeks to a Salable Plant	Sowing Date
FLOWERS		
Ageratum	12–14	Jan. 22
Begonia	13–14	Dec. 19
Celosia	5–6	Feb. 14
Coleus	6–8	Feb. 14
Dusty Miller	12–14	Jan. 22
Geranium	12	Jan. 22
Impatien	14	Dec. 19
Marigold	6–8	March 1
Petunia	12–14	Jan. 22
Salvia	6–8	March 1
Zinnia	6–8	March 10
VEGETABLES		
Broccoli	4–6	Feb. 1
Cabbage	4–6	Feb. 1
Cauliflower	4–6	Feb. 1
Canteloupe	3–5	March 20
Cucumber	4	March 20
Tomato	6–8	March 1

PREPARING THE SOIL

It is best to prepare the soil in the fall before planting the next spring. The soil should be spaded or tilled to loosen it. The addition of peat moss in dry well drainage soils or pine bark in heavier soil may be necessary. Peat moss adds organic matter to the soil and aids in moisture retention. Fine pine bark improves soil drainage and aeration.

When planting in beds that contain bulbs, do not dig deeply enough to damage the bulbs. Be sure that the bed does not dry out completely when dug. This drying could also damage the bulbs.

Before preparing new beds, test the soil for drainage. To do this, dig a 10-inch hole and fill it with water. Fill it with water again the following day. If all the water has not seeped into the soil within ten hours on the second day, drainage is a problem. To improve drainage, raise the bed level above the normal soil level and dig a shallow trench around it and add pine bark to the soil.

FERTILIZING

To fertilize the bed, add about 2 pounds of a 5-10-10 fertilizer per 100 square feet in the spring as the bed is being prepared. If the soil pH is below 6.0, add ground limestone at the rate of 5 pounds per 100 square feet and rake it into the soil surface. If the plants do not have a healthy, dark green color, sprinkle a 10-10-10 fertilizer lightly around each plant on the soil surface. Do not place fertilizer on or against the plant.

Many people are now growing beautiful annuals with soluble fertilizers. To make a soluble fertilizer, mix a complete fertilizer (20-20-20) with water at the rate of 1 tablespoon per gallon of water or according to directions. Fertilizer applied to plants in a solution is more readily available to the plants.

MULCHING AN ANNUAL FLOWER BED

The use of mulch for an annual bed is multifaceted. It creates the aesthetics that enhance the annuals planted. It is a source of organic matter to the plants, helps to control weeds, conserves moisture, and promotes even ground temperatures for improved plant growth.

When planting an annual flower bed, apply a two to three inch layer of mulch before this bed is planted. This procedure will allow for more even distribution of mulch as well as reducing mulching time for the planted bed.

Some of the more common mulches used in the industry today include hardwood bark, pine bark, cocoa bean hulls, buckwheat hulls, pine needles, well-rotted sawdust, and leafmold. Precaution when using some of these mulches is that they will influence the pH level of the soil in the annual flower bed. Therefore, it is important that the soil pH level is followed closely to ensure that the optimum growth medium is provided for the plants.

SEEDING

As mentioned, most annuals are seeded indoors and set outside as plants. Figures 25-7, 25-8, and 25-9 detail this process. The plants are set outside in the flower bed as soon as the danger of frost has passed. Some annuals may be seeded directly outdoors. Some examples of these are baby's breath, cornflower, globe amaranth, poppy, cleome, strawflower, sweet alyssum, and sweet pea. Seeds of these plants should be sown as early in spring as possible.

Wait until the soil warms to about 60°F before setting the started plants outdoors. Seedlings in peat pots are set directly into the soil. Very small seeds which are started in flats will probably require transplanting to a larger container before setting outside. These may be replanted to peat

FIGURE 25-7 To start seed indoors, fill the flat with properly prepared seeding medium. Small, square peat pots have been set in this flat, with one or two seeds placed in each pot. The peat pots are separated and transplanted individually to the flower bed after the danger of frost has passed. Very small seeds are planted in flats without peat pots and are transplanted to the pots at a later date.

FIGURE 25-8B Most annual seeds should be covered for proper germination.

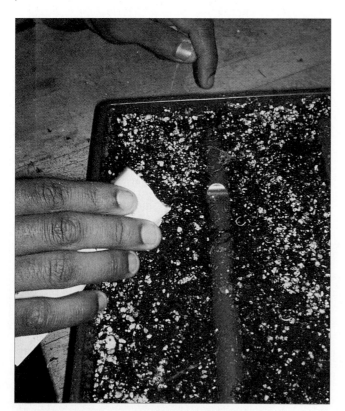

FIGURE 25-8A Annual seeds should be sown carefully to obtain an even distribution of the seeds in rows.

FIGURE 25-9 There are two methods for watering: (1) sprinkle the flat as shown or (2) set the entire flat in water and allow it to soak up from the bottom. Both methods allow even distribution of water on the seeds. Very small seeds, such as petunia, must not be watered from the top with any force, or they will be washed down into the planting medium too deeply.

pots so that they can establish better growth prior to being planted outside, figure 25-10.

Plant seedlings at about the same depth as the depth at which they grew in the starting container. Firm the soil around the roots and water well. It is helpful to add a starter solution of 1 tablespoon of a water soluble, high phosphate fertilizer in 1 gallon of water after transplanting. Space the plants according to the bedding plant chart (figure 25-5).

FIGURE 25-10 To transplant the small seeded annual to a larger pot, first grasp the seedling by its true leaves. Gently dislodge it from its original pot and place it in the new pot. Fill the pot with soil mixture to the soil level of the original pot. Firm the soil by pressing down lightly with the thumb and index finger.

PROCEDURE

CARING FOR THE PLANTS

Care for the rest of the season is relatively simple, figure 25-11.

1. Water if the soil becomes too dry.

2. Pull weeds or mulch for weed control.

3. Remove all faded blossoms.

4. If necessary, spray with malathion or Sevin, depending upon the insect to be controlled. If a chemical is used, read the label carefully before use.

5. Fertilize as needed to keep plants growing actively.

6. Cut flowers as desired for use in the home.

7. Prune or pinch when necessary for better growth. When the plants are herbaceous (soft stemmed), the stem may be shortened by pinching with the fingers. The pinching will result in a thicker, more attractive plant.

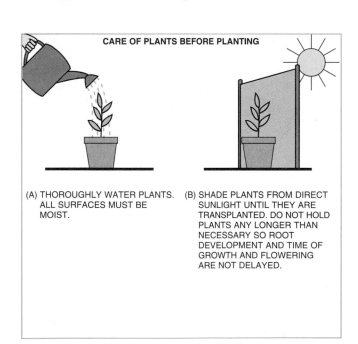

CARE OF PLANTS BEFORE PLANTING

(A) THOROUGHLY WATER PLANTS. ALL SURFACES MUST BE MOIST.

(B) SHADE PLANTS FROM DIRECT SUNLIGHT UNTIL THEY ARE TRANSPLANTED. DO NOT HOLD PLANTS ANY LONGER THAN NECESSARY SO ROOT DEVELOPMENT AND TIME OF GROWTH AND FLOWERING ARE NOT DELAYED.

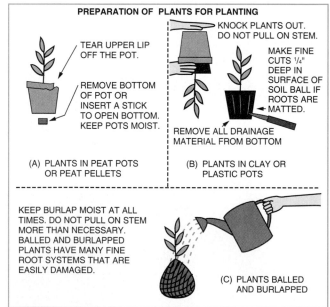

PREPARATION OF PLANTS FOR PLANTING

TEAR UPPER LIP OFF THE POT.

REMOVE BOTTOM OF POT OR INSERT A STICK TO OPEN BOTTOM. KEEP POTS MOIST.

(A) PLANTS IN PEAT POTS OR PEAT PELLETS

KNOCK PLANTS OUT. DO NOT PULL ON STEM.

MAKE FINE CUTS 1/4" DEEP IN SURFACE OF SOIL BALL IF ROOTS ARE MATTED.

REMOVE ALL DRAINAGE MATERIAL FROM BOTTOM

(B) PLANTS IN CLAY OR PLASTIC POTS

KEEP BURLAP MOIST AT ALL TIMES. DO NOT PULL ON STEM MORE THAN NECESSARY. BALLED AND BURLAPPED PLANTS HAVE MANY FINE ROOT SYSTEMS THAT ARE EASILY DAMAGED.

(C) PLANTS BALLED AND BURLAPPED

FIGURE 25-11 Care of plants before and after planting (Adapted from USDA)

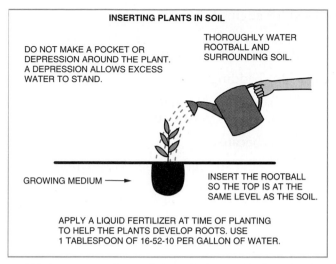

INSERTING PLANTS IN SOIL

DO NOT MAKE A POCKET OR DEPRESSION AROUND THE PLANT. A DEPRESSION ALLOWS EXCESS WATER TO STAND.

THOROUGHLY WATER ROOTBALL AND SURROUNDING SOIL.

GROWING MEDIUM ⟶

INSERT THE ROOTBALL SO THE TOP IS AT THE SAME LEVEL AS THE SOIL.

APPLY A LIQUID FERTILIZER AT TIME OF PLANTING TO HELP THE PLANTS DEVELOP ROOTS. USE 1 TABLESPOON OF 16-52-10 PER GALLON OF WATER.

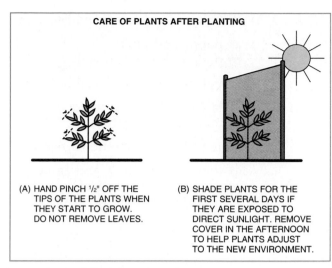

CARE OF PLANTS AFTER PLANTING

(A) HAND PINCH ½" OFF THE TIPS OF THE PLANTS WHEN THEY START TO GROW. DO NOT REMOVE LEAVES.

(B) SHADE PLANTS FOR THE FIRST SEVERAL DAYS IF THEY ARE EXPOSED TO DIRECT SUNLIGHT. REMOVE COVER IN THE AFTERNOON TO HELP PLANTS ADJUST TO THE NEW ENVIRONMENT.

FIGURE 25-11   Care of plants before and after planting (Adapted from USDA) (continued)

Student Activities

1) With the assistance of the instructor or a local landscaper, outline on paper a layout for a flower bed using annuals.

2) Start flowering annuals or vegetable annuals from seed according to directions in the text.

3) Plant the flowering annuals started on the school grounds or at home. Plant the annuals according to a plan approved by the instructor.

4) Sell annual bedding plants as a fund-raising class activity.

Self-Evaluation

A) Select the best answer from the choices offered to complete the statement or answer the question.

1) Which of the following are not annual plants?

a) asparagus and strawberry
b) petunia and geranium

c) marigold and ageratum
d) morning glory and scarlet

2) Seed for most annuals should be

a) started inside and transplanted outside later.
b) directly seeded in the flower bed.
c) soaked in water before planting.
d) planted only in peat pots.

3) Which two annuals listed below are best used for edging?

a) poppy and pink
b) pink and sweet alyssum
c) stock and strawflower
d) sunflower and pink

4) Which two annuals listed below are best used for bedding?

a) larkspur and morning glory
b) marigold and petunia
c) balsam and impatiens
d) summer-cypress and sweet pea

5) Faded blossoms should be removed from annuals because

a) they prevent the plants from setting seed and using strength for seed formation.
b) it improves the appearance of the plants.
c) new flowers develop faster when old ones are removed.
d) all of the above

6) Annuals which are easiest to grow are

a) marigold and zinnia.
b) ageratum and sweet pea.
c) pansy and coleus.
d) primrose and petunia.

7) Most annuals should be transplanted outdoors

a) as soon as the soil can be tilled in the spring.
b) in early June.
c) as soon as danger of frost has passed.
d) after spring shrubs have bloomed.

8) When transplanting annual plants from the flat to the bed

a) dip the roots in water to moisten them.
b) shake any excess soil from the roots before planting.
c) allow the roots to dry out before transplanting.
d) keep as much soil on the roots as possible.

9) When planting annuals, organic matter is added to soil to

a) give the soil better drainage.
b) hold moisture in the soil.
c) lighten the soil.
d) mulch the plants.

10) When planting annuals, pine bark is added to the soil to

a) give the soil better drainage.
b) hold moisture in the soil.
c) make the soil heavier.
d) help warm the soil in the spring.

11) On annual beds the mulch should be applied

a) after watering.
b) before the plants are planted.
c) after the annuals have been planted.
d) before the lime is applied.

B) Provide a brief answer for each of the following.

1) List the seven steps in maintaining a healthy annual planting.

2) List four uses of annuals in the landscape.

3) List five annuals that make attractive hanging baskets.

4) List four reasons for using mulch on annual beds.

5) List five of the most common mulches.

Perennials, Ornamental Grass, Vines, and Bamboo

OBJECTIVE

To design, plant, and maintain a perennial garden.

Competencies to Be Developed

After studying this unit, you should be able to

- ◆ list five perennial flowers.
- ◆ design a layout for a perennial border.
- ◆ list the environmental factors that affect perennial gardens.
- ◆ list three ornamental grasses, vines, and bamboos.

MATERIALS LIST

- ✔ nursery catalogs listing perennials
- ✔ planting area for which to design a perennial garden

PERENNIALS

Perennials are plants that live from year to year and that therefore do not require replanting. The tops may or may not die back in the winter or dry season. Some perennials bloom their first year, but most produce larger, more attractive flowers and develop stronger root systems in successive seasons. Perennials may be small flowering plants, shrubs, or flowering trees.

Perennials have made a resurgence in the United States in the past several years, and there are many hardy perennials that are native to this country. The demand for perennials has expanded with the growing interest of the general public in gardening and landscaping, and the desire to use plants that will come back each year. Perennials adorn gardens with color and foliage, and they result in a long-term garden meant to be enjoyed. This interest in perennials has even resulted in the formation of the Perennial Plant Association.

Many of the early perennial gardens contained the old favorites such as chrysanthemums, daylilies, irises, peonies, and delphiniums. Many of these plants have been in borders for over 100 years. With the introduction of new cultivars, perennials have been hybridized to give a variety of colors, forms, textures, fragrances, and hardiness. Europeans have always been strong in the use of perennials in their gardens, and they have introduced many of the new cultivars. As we learn more about these perennials and their uses, demand for them will grow.

As densely populated areas continue to grow and townhouses become more popular, modern perennial gardens are becoming smaller. With this scaled-down approach to landscaping, perennial plants have become an outdoor extension of indoor living featuring unique color design that creates a special relationship with nature and the environment.

PERENNIAL BEDS AND LOCATION

The perennial bed should be first designed on paper to fit the garden area for a particular bed. Some basic guidelines to follow are:

1. Use an island bed with an informal-shaped border, surrounded with a lawn area. The taller plants are planted in the center, with the design using a sequence of tall-to-short plants, with the shortest plants on the outer edge. The bed should be 5 feet wide. This will allow the individual to work the garden from all sides without walking over the bed. With the resulting reduction in soil compaction, damage to the plant roots will be minimal, thus lowering maintenance requirements. If a walkway through the bed is part of the design, then the size of the garden may be from 9 to 15 feet wide.

2. Use a perennial border planted along a fence or property line as a divider. It should be up to 5 feet wide and as long as desired by the designer and the gardener.

SOIL, LIGHT, AND VIEW When planting a perennial bed, one must consider other factors such as environment, soil type, drainage, depth of the top soil, organic matter content, and light exposure (full sun, partial shade, or full shade). The view from the house, patio, or lawn will also have a bearing on the location of the perennial bed or border.

SELECTION

Perennials should be selected by personal preference, color scheme, texture, shape, growth habit, and the microclimate of the local area. Examine plants grown in the area or ask for advice from a nursery or garden center, extension service, or horticultural instructor. The color pictures in seed and nursery catalogs are another source that can be helpful in plant selections. Check figure 26-1 to determine if the plants selected will grow under the light and soil conditions in the area in which they are to be planted, and to be sure they will bloom during the season when added color is needed in the landscape (mid-spring, late summer, and fall). Plant flowers that bloom at the same time together for the best display of color.

When choosing plants it is important to mass varieties in groups of three, five, nine, or more. A single variety of perennial may be used as long as the growth is full. For example, the use of daylilies as a single variety is good because they tend to grow in a clump and have several flowers blooming at one time. Plants with small growth habits should not be used individually as they give a sparse appearance in the garden.

To concentrate on early color, one must consider the use of spring flowering bulbs. No perennial garden should be without some small, early flowering bulbs that multiply freely over the years. Some bulbs that can be used are the grape hyacinth, snowdrop, and crocus. Another variety to use is the lovely miniature iris (*Iris reticulata*) with fragrant purple flowers that appear in March. Daffodils are also important to color an early spring perennial garden.

DESIGNING A PERENNIAL GARDEN

To design the perennial garden, you must consider the soil, time of blooming, and space and have a plant list, including the colors and varieties that are best suited for your geographic area. Using such design equipment as a T-square, a triangle, architect's scales, a #2 pencil, and a drafting board, lay out the desired perennial garden for the area. Keep in mind the height, color, and spread of the plants. A recommended rule to follow for planting distance is to have perennials planted at one-half of the mature height of the plant. For example, if a plant grows to 3 feet, then the distance between plants should be 18 inches O.C. (on center). See figure 26-2, page 279.

NAME	WHEN TO PLANT SEED	EXPOSURE	GERMINATION TIME (DAYS)	SPACING	HEIGHT	BEST USE	COLOR	REMARKS
Achillea mille folium (yarrow)	early spring or late fall	sun	7–14	36'	24"	borders, cut flowers	yellow, white, red, pink	Seed is small. Water with a mist. Easy to grow.
Alyssum saxatile (golddust)	early spring	sun	21–28	24"	9"–12"	rock garden, edging, cut flowers	yellow	Blooms early spring. Good in dry and sandy soils.
Anchusa italica (Alkanet)	spring to September	partial shade	21–28	24"	48"–60"	borders, background, cut flowers	blue	Blooms June or July. Refrigerate seed 72 hrs. before sowing.
Anemone pulsatilla (windflower)	early spring or late fall for tuberous	sun	4	35"–42"	12"	borders, rock garden, potted plant, cut flowers	blue, rose, scarlet	Blooms May and June. Is not hardy north of Washington, D.C.
Anthemis tinctoria (golden daisy)	late spring outdoors	sun	21–28	24"	24"	borders, cut flowers	yellow	Blooms midsummer to frost. Prefers dry or sandy soil.
Arabis alpina (rock cress)	spring to September	light shade	5	12"	8"–12"	edging, rock garden	white	Blooms early spring.
Armeria maritima alpina (sea pink)	spring to September	sun	10	12"	18"–24"	rock garden, edging, borders, cut flowers	pink	Blooms May and June. Plant in dry sandy soil. Shade until plants are well established.
Aster alpinus (hardy aster)	early spring	sun	14–21	36"	12"–60"	rock garden, borders, cut flowers	white	Blooms June.
Astilbe arendsii (false spirea) 'Europa' 'Fanal' 'Deutschland' 'Superba'	early spring	sun	14–21	24"	12"–36"	borders	pink, red, white	Blooms July and August. Gives masses of color.
Begonia evansiana (hardy begonia)	summer in shady, moist spot	shade	12	9"–12"	12"	flower bed	yellow, pink, white	Blooms late in summer. Can be propagated from bulblets in leaf axils.
Bergenia purpurascens (bergomot)	late winter	light shade	10	18"	2'–3'	medicinal	pink, red	Hummingbirds love it.

FIGURE 26-1 Various flowering perennials (From USDA Bulletin 114)

NAME	WHEN TO PLANT SEED	EXPOSURE	GERMINATION TIME (DAYS)	SPACING	HEIGHT	BEST USE	COLOR	REMARKS
Candytuft (*Iberis*)	early spring or late fall	sun	20	12"	10"	rock garden, edging, ground cover	white	Blooms in late spring. Prefers dry places. Cut faded flowers to promote branching.
Canterbury bells (*Campanula medium*)	spring to September (Do not cover.)	partial shade	20	15"	24"–30"	borders, cut flowers	white, pink, blue	Divide mature plants every other year. Best grown as a biennial.
Carnation (*Dianthus caryophyllus*)	late spring	sun	20	12"	18"–24"	flower bed, borders, edging, rock garden	pink, red, white, yellow	Blooms in late summer. Cut plants back in late fall and hold in cold frame.
Cerastium tomentosum (snow-in-summer)	early spring	sun	14–28	18"	6"	rock garden, ground cover	white	Blooms in May and June. Forms a creeping mat and is a fast grower. Prefers a dry spot.
Chinese lantern (*Physalis alkekengi*)	late fall	sun	15	36"	24"	borders, specimen plant	orange	Lanterns are borne the second year in the fall.
Columbine (*Aquilegia*)	spring to September	sun or partial shade	30	12"–18"	30"–36"	borders, cut flowers	wide color range	Blooms in late spring. Best grown as a biennial.
Coreopsis lanceolata	early spring	sun	5	30"	24"	borders	yellow	Blooms from June to fall if old flowers are removed.
Daisy, Shasta (*Chrysanthemum maximum*)	early spring to September	sun	10	30"	24"–30"	borders, cut flowers	white	Blooms June and July. Best grown as a biennial in well-drained location.
Delphinium elatum (delphinium)	spring to September	sun	20	24"	48"–60"	borders, background, cut flowers	blue, lavender, white, pink	Blooms in June. Best grown as a biennial. Needs dry location.
Dianthus deltoides (pink)	spring to September	sun	5	12"	12"	borders, rock garden, edging, cut flowers	pink	Blooms in May and June. Best grown as a biennial. Needs dry location.
Foxglove (*Digitalis purpurea*)	spring to September	sun or partial shade	20	12"	48"–60"	borders, cut flowers	pink, white, purple, rose	Blooms in June and July. Grown as a biennial. Shade summer plantings.

FIGURE 26-1 Various flowering perennials (From USDA Bulletin 114) (continued)

NAME	WHEN TO PLANT SEED	EXPOSURE	GERMINATION TIME (DAYS)	SPACING	HEIGHT	BEST USE	COLOR	REMARKS
Gaillardia grandiflora (gaillardia)	early spring or late summer	sun	20	24"	12"–30"	borders, cut flowers	scarlet, yellow	Blooms from July until frost.
Gypsophila paniculata (baby's breath)	early spring to September	sun	10	48"	24"–36"	borders, cut flowers, drying	white, pink	Blooms early summer until early fall. Needs lots of lime.
Hemerocallis (daylily)	late fall	sun or partial shade	15	24"–30"	12"–48"	borders, among shrubbery	pink, red	Plant several varieties for longer blooming season.
Hibiscus moscheutos (Mallow Rose)	spring or summer	sun or partial shade	15 or longer	24"	36"–96"	background, flower bed	white, pink, red, rose	Blooms July to September.
Iris	bulbs or rhizomes in fall	sun or partial shade	next spring	18"–24"	3"–30"	borders, cut flowers	blue, red, yellow, pink, bronze, wine	Blooms spring and summer if different varieties are used.
Liatris pycnostachya (gayfeather)	early spring or late fall	sun	20	18"	24"–60"	borders, cut flowers	rose-purple	Blooms summer to early fall. Easily started from seed.
Lupinus polyphyllus (lupine)	early spring or late fall (Soak before planting.)	sun	20	36"	36"	borders, cut flowers	white, yellow, pink, rose, red, blue, purple	Blooms most of summer. Needs excellent drainage. Does not transplant easily.
Menthia ssp. (mint)	late winter	sun	8	15"	6"–3'	culinary	white	Mint flavors.
Peony (Paeonia)	Plant tubers in late fall 2"–3" deep.	sun	variable	36"	24"–48"	borders, cut flowers, flower bed	pink, red, white, rose	Blooms late spring. Difficult to grow from seed.
Phlox paniculata (summer phlox)	late fall or early winter	sun	25 irregular	24"	36"	borders, cut flowers	red, pink, blue, white	Blooms early summer. Color of flower varies.
Phlox sublata (moss phlox)	grown from stolons	sun		8"	4"–5"	borders	blue, red, white, pink	Blooms in spring. Drought resistant.
Poppy, Iceland (Papaver nudicaule)	early spring	sun	10	24"	15"–18"	borders, cut flowers	white, pink, red	Blooms early summer. Does not transplant easily.
Poppy, Oriental (Papaver orientale)	early spring	sun	10	24"	36"	borders, cut flowers	pink, red, rose, orange, white, salmon	Blooms early summer. Does not transplant easily.

FIGURE 26-1   Various flowering perennials (From USDA Bulletin 114) (continued)

NAME	WHEN TO PLANT SEED	GERMINATION TIME (DAYS)	SPACING	HEIGHT	BEST USE	COLOR	REMARKS
Primrose (*Primula polyantha* and *P. veris*)	January, in a flat on surface. Allow to freeze; then bring in to germinate.	25 irregular	12"	6"–9"	rock garden	white, yellow, pink, red, blue	Blooms April and May. May be seeded in fall.
Pyrethrum roseum (painted daisy)	spring to September	20	18"	24"	borders, cut flowers	various, including gold, pink, and lavender	Blooms May and June. Prefers well-drained soil.
Rudbeckia (*Echinacea purpurea*) (cone flower)	spring to September	20	30"	30"–36"	borders, flower bed, cut flowers	white, pink, red, rose	Blooms midsummer to fall. Shade summer plantings.
Salvia (*Salvia azurea grandiflora* and *S. farinacea*)	spring	15	18"–24"	36"–48"	borders	red	Blooms August until frost.
Sea lavender (*Limonium latifolia*)	early spring	15	30"	24"–36"	flower bed, cut flowers, drying	pink, yellow, mauve	Blooms in July and August.
Sedum spectabile (sedum)	late winter	10	10"	4"–15"	ground cover	pink, white	Fall foliage.
Stokesia cyanea (Stokes' aster)	early spring to September	20	18"	15"	borders, cut flowers	white, blue	Blooms in September if started early.
Sweet pea (*Lathyrus latifolius*)	early spring	20	24"	60"–72"	background	pink, white, purple, red	Blooms June to September. Easily grown as a vine on fence or trellis.
Sweet William (*Dianthus barbatus*)	spring to September	5	12"	12"–18" (Dwarf form also available.)	borders, edging, cut flowers	red, pink, white	Blooms May and June. Very hardy. Needs well-drained soil.
Veronica spicata (speedwell)	spring to September	15	18"	18"	borders, rock garden, cut flowers	purple	Blooms June and July. Easy to grow.

FIGURE 26-1 Various flowering perennials (From USDA Bulletin 114) (continued)

(D) Coreopsis—early sunrise

(G) Pennisetum ornamental grass

(A) Sedum stonecrop

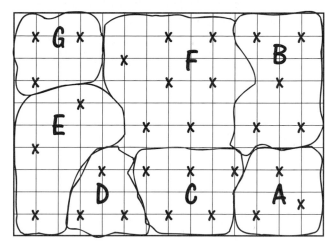

SCALE: 2 BLOCKS = 1'

(F) Rudbeckia Compter's gold

(B) Hollyhock

(C) Veronica

(E) Liatris floristan

FIGURE 26-2   A perennial garden plan (Courtesy George C. Ball Co.)

In designing the perennial garden, the use of some deciduous shrubs as well as evergreens creates interest. The shrubs create a background for, and contrast with, the perennials.

You could also implement the border design that Gertrude Jeckyll (an artist and gardener of the 1800s) used. She was interested in perennials and in using them in borders. She had the unique ability to blend and contrast perennials having different sizes, shapes, and foliage. She did not limit her gardens to the use of perennials only, but worked with biannuals and annuals as well. In designing the garden, she would create a sequence of blooming flowers from spring to fall by arranging the garden on a north-south axis, using blue flowers (cool colors) on the north and, moving south, yellow and then red flowers (warm colors). About halfway down the garden the colors reversed to range back to the blues (cool colors) at the opposite (south) end of the garden.

SOIL PREPARATION AND FERTILIZING

Remember that perennials live in the same soil for more than one year. It is necessary, therefore, to prepare the soil well. It will then provide medium conducive to root development. Organic matter should be dug in and drainage improved if necessary. Adequate fertilizer should be dug into the soil along with the organic matter to a depth of at least 1 foot. When the bed is being prepared use 2 pounds of 5-10-10 (commercial fertilizer) or 3 pounds of bone meal (organic fertilizer) per 100 square feet of bed. Careful preparation prior to planting results in a much more attractive floral display.

MULCHING A PERENNIAL FLOWER BED

The use of mulch for a perennial bed is multifaceted. It creates the aesthetics that enhance the perennial planted. It is a source of organic matter to the plants, helps to control weeds, conserves moisture, and promotes even ground temperatures for improved plant growth.

When planting a perennial flower bed it is best to apply a two to three inch layer of mulch before the bed is planted. This will allow for more even distribution of mulch as well as reducing mulching time for the planted bed.

Some of the more common mulches used in the industry today include hardwood bark, pine bark, cocoa bean hulls, buckwheat hulls, pine needles, well-rotted sawdust, and leafmold. A precaution when using some of these mulches is that they will influence the pH level of the soil. Therefore, it is important that the soil pH level is followed closely to ensure that the optimum growth medium is provided for the plants.

TRANSPLANTING

Perennials are transplanted in the spring or the fall. When they are transplanted, the crown of the plant must be at the correct level. (The *crown* is the point where the top is connected to the roots, and where the new growth comes from.) Be sure the crown is set at the original depth; as the plant is set, firm in around the roots with your hands. Water to settle the soil in place, and to prevent drying out of the roots. If the plants flower in the spring, they should be divided and replanted in the fall. If they flower in the fall, they should be divided and transplanted in the spring. See figure 25-11.

It is usually best to start perennial seeds indoors six to eight weeks prior to the transplanting date. Plants that do not transplant well must be started in peat pots or direct seeded in the permanent location. The procedures for starting seeds and transplanting perennials are the same as those used for annuals.

ORNAMENTAL GRASSES

Ornamental grasses are excellent for use as accents and, therefore, make attractive edging. They are also good ground covers. They have unique characteristics with their various colors, shapes, and textures of plumage. The leaves of ornamental grasses are excellent in dried arrangements as the foliage will vary from dark to variegated with seasonal changes in the plants. Most of these grasses prefer a hot, sunny, and well-drained location. Figure 26-3 lists some ornamental grasses and their characteristics.

VINES

Vines are important in today's landscape. Many types of vines are used to mask various landscape features and create a flowing effect throughout the garden. They are used to soften the appearance of walls, fences, porches, trellises, and pergolas. They are also very effective in small gardens to provide flowers and foliage where space is limited. Vines on a wall have a microclimatic effect. For example,

NAME	EXPOSURE	SPACING	HEIGHT	BEST USE	REMARKS
Carex morrowii variegata (sedge)	semishade	9"–12"	12"–15"	border, bank	Leaves narrow, white margin.
Calamagrostis	sun	4'	3'–5'	accent	Looks great with black-eyed Susan.
Festuca glauca (blue fescue)	sun	4"–6"	10"–12"	border, bank	Blue-gray foliage.
Miscanthus floridulus (giant miscanthus)	sun	9'–12'	12'–15'	accent, screen, windbreak	Effective fall and winter.
Panicum virgatum	sun	2'–3'	3'–4'	accent	Fall interest. Open panicles.
Pennisetum	sun	4'	3'–8'	accent	Excellent fall color.
Phalaris arundinaccea picta (ribbon grass)	sun	12"–15"	18"	accent	White, variegate leaves. Spread by rhizomes.
Chasmanthium latifolium (northern sea oats)	sun	18"–24"	3'	accent	Winter interest.

FIGURE 26-3 Ornamental grasses

the use of a deciduous vine on a wall creates a cooling effect in the summer, while in the winter it allows the sun to be exposed to the wall for a warming effect.

Vines climb in three different ways. English ivy forms small, rootlike holdfasts to a wall as a means of support. Sometimes these are modified tendrils with small circular discs at the tips. Other vines, like clematis and grape, climb by winding tendrils or leaflike appendages which act as tendrils around the object on which they are growing. The third group, like bittersweet and wisteria, climb by twining. It is obvious that one must know in advance how each vine climbs so that a proper means of support can be provided for those vines selected for use.

Figure 26-4 lists some typical vines and their characteristics.

NAME	REMARKS
Campsis radicans (trumpet vine)	This vine has a rich-looking orange flower. It is a very fast grower and will cling to most surfaces. It will grow in most soils in full sun.
Celastrus scandens (American bittersweet)	Bittersweet is fast growing, with orange and red berries on the female plants. It is very effective in the fall and grows well in most soils in full sun. With such bright-colored berries it is excellent for holiday decorations.
Clematis hybrida (clematis)	Clematis is best known for its excellent flowers, but must be trained on some support trellis, fence, or even a porch post. It needs well-drained soil that is slightly alkaline. The flowers like a sunny location, while the roots prefer a cool, shaded location. There are many cultivars on the market featuring striking colors of pink, purple, white, just to name a few. It is best to prune in late winter or very early spring.
Parthenocissus tricuspidata (Boston ivy)	This vine is a very rapid grower. The foliage opens red in the spring, then turns a rich green in the summer. It does very well on walls.
Polygonum auberti (silverlace vine)	A very rapid grower, with white flowers in the summer. It needs a well-drained soil in full sun or in shade.
Wisteria sinensis (wisteria)	This vine is the oldtimer, with fragrant blue-violet pendulous racemes (looks like a bunch of grapes) in late May. It was introduced into this country in 1816 and it requires deep, moist, well-drained soil.

FIGURE 26-4 Vines used in today's landscape

BAMBOO

Bamboo is being used more extensively now in landscaping, especially to create an oriental touch. Bamboo can range in height from 6 inches to 120 feet. Although it is not really winter hardy, when it is killed back by cold winter temperatures it will often sprout back from the root when spring comes.

Bamboo has a rhizome root system. In some cases, when the landscape area calls for contained bamboo, it may be necessary to confine the roots to a given area. This can be done by planting the bamboo in a container, thus keeping the plant from spreading into other parts of the landscape. Bamboo is very difficult to contain in the open soil.

Metal inserted several feet into the soil will slow spread until the metal rusts out.

The following are types of bamboo used in the landscape industry.

- *Arundinaria variegata*—The leaves of this variety are grasslike with variegated foliage. It grows to 18 inches high in full sun to semishade.

- *Phyllostachys*—This variety grows tall, in a range of 30 to 70 feet with feathering foliage. It needs full sun exposure and must be forced to grow in clumps.

- *Arundinaria pygmaea* (pygmy bamboo)—This bamboo grows low with grassy green foliage. It needs full sun and spreads very rapidly by rhizomes. It should be contained to the designated area.

Student Activities

1) Draw a layout for a perennial flower bed to be planted on the school grounds or at home. If possible, plant the flower bed. Observe and record changes in the perennials as they enter the winter season.

2) Visit a local garden center that specializes in perennials.

3) Visit Longwood Gardens in Pennsylvania or White Flower Farm in Connecticut.

4) Have a guest speaker who produces perennials for resale.

5) Visit a nursery/landscape trade show.

6) Visit flower shows held in the early spring.

Self-Evaluation

A) Select the best answer from the choices offered to complete the statement or answer the question.

1) A perennial is a plant that lives

a) from year to year without replanting.
b) for only one year.
c) for only two years.
d) all of these

2) Soil that is to be planted with perennials should be well prepared because

    a) perennials are particular about soil conditions.
    b) perennials grow in one location for many years.
    c) all perennials require rich soil.
    d) none of the above

3) Important considerations when selecting perennials for planting are

    a) height and color.   c) soil fertility and drainage.
    b) time of bloom and color of flower.   d) all of these

4) The best way to select flowering perennials for a particular area is to

    a) observe plants that are growing there.
    b) judge according to what looks nicest in the seed catalog.
    c) purchase plants with colors that blend well together.
    d) all of the above

5) Which of the following is *not* a part of soil preparation for flowering perennials?

    a) adding organic matter
    b) providing proper drainage
    c) adding mulch
    d) adding fertilizer and digging it into the soil

6) When starting flowering perennials from seed, how many weeks before the transplanting date should they be started?

    a) two to three months   c) one to two weeks
    b) six to eight weeks   d) three to four weeks

7) Plants are often seeded in peat pots. This is especially recommended for plants

    a) that are difficult to transplant.   c) that require good drainage.
    b) that are to be sold.   d) that have tap root systems.

B) Provide a brief answer for each of the following.

    1) List three perennials that have blue flowers.

    2) List three perennials that grow well in full sun.

    3) What color flowers does the astilbe produce?

    4) List three varieties each of ornamental grasses, vines, and bamboo used in the landscaping industry.

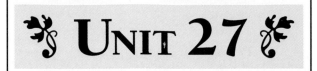

UNIT 27

Narrowleaf Evergreens

OBJECTIVE

To identify and plant narrowleaf evergreens in the landscape.

Competencies to Be Developed

After studying this unit, you should be able to

- ◆ differentiate between the two types of narrowleaf evergreen leaves.
- ◆ list four reasons for using narrowleaf evergreens in the landscape.
- ◆ describe the primary purpose of a lath house.
- ◆ describe the proper fertilizer practices for narrowleaf evergreens.
- ◆ demonstrate the proper procedure for planting narrowleaf evergreens.
- ◆ calculate fertilizer's three active ingredients of nitrogen, phosphorus, and potassium.

MATERIALS LIST

- ✔ two different narrowleaf evergreen trees or tree branches
- ✔ several deciduous narrowleaf trees or tree branches
- ✔ 5-10-10 fertilizer
- ✔ round-point shovel
- ✔ nursery spade
- ✔ garden hose

Narrowleaf evergreens have long been popular landscape plants. This may be because they generally remain green year-round, thereby adding color to the landscape when other plants are dormant. Another advantage is that they are adaptable to various soil types and weather conditions.

The leaves of narrowleaf evergreens may be needlelike or scalelike. Pines have needlelike leaves attached to the branch in bundles (known as sheaths or *fascicles*). For example, the white pine has five needles per bundle or sheath. These evergreens carry their needles throughout the year, with old needles dropped in the fall and new needles produced in the spring. Spruces have single needles attached to the branch which are square, stiff, and sharp.

Narrowleaf evergreens with scalelike leaves have leaves arranged so that each scale overlaps another to form a flat spray. An example of a narrowleaf evergreen with scalelike leaves is the juniper.

USES

The narrowleaf evergreen has long been a standard choice for foundation plantings because of its year-round green foliage. Figures 27-1 through 27-3 show several narrowleaf evergreens used as

FIGURE 27-1 Mugho pine—Pinus mugo mughus

FIGURE 27-2 Plume false cypress—Chamaecyparis pisifera "Plumosa"

FIGURE 27-3 Hemlock—Tsuga canadensis

part of foundation plantings. Narrowleaf evergreens are also used extensively as boundary plants to accent borderlines of property, figure 27-4A and B. The narrowleaf evergreen may also be used to screen a view of a less attractive area or to help control air currents. For example, in the winter months, a windbreak of narrowleaf evergreens can cause the temperature on the protected side of the windbreak to be 5 to 10 degrees warmer than on the unprotected side. See figure 27-5.

CARE

Narrowleaf evergreens are grown in nurseries. They are shipped three different ways: bare root (BR), balled and burlapped (B&B), or as container stock (C).

Bare root narrowleaf evergreens are shipped while the plants are seedlings. When the seedlings arrive at their destination, the quality of the nurs-

ery stock must be maintained, and therefore the seedlings must be handled with care. Until the ground permits planting, the seedlings should be stored in a building that will maintain high humidity and a cool temperature range of 33° and 40°F. The roots must be covered and kept damp with material such as straw, shingle tow, burlap, or sphagnum moss. Another good storage method is to heel-in the seedlings into well-rotted sawdust or soil outside. The roots must be prevented from drying out until they are planted in the ground.

Balled and burlapped narrowleaf evergreens may be moved at almost any time of the year except when the plant has shoots of new growth or when the ground is frozen. The new shoots must be allowed to harden before the plant is moved. If these plants are to be out of the ground any length of time, it is best to mulch them heavily with well-rotted sawdust, pine bark, peat moss, straw, shingle tow, salt marsh hay, or hardwood bark to retain

FIGURE 27-4A Blue Colorado spruce—Picea pungens "Glauca" (winter).

FIGURE 27-4B Blue Colorado spruce—Picea pungens "Glauca" (spring).

FIGURE 27-5 Black Japanese pine—Pinus thunbergii

moisture around the roots. This process is called *heeling-in*.

To maintain healthy plants, narrowleaf evergreens should be watered regularly and stored in a lath house until they are planted on the site. A *lath house* gives protection from the sun by reducing the light available to the plant. This reduces the transpiration rate (water loss) of the plant.

PLANTING BALLED AND BURLAPPED PLANTS

The prepared hole should be one and one-half to two times as large as the ball of soil around the tree being planted. This allows room to set the tree in the hole and mix peat moss with the soil surrounding the hole. The ball of the tree is set 1 inch above the original soil line, since the tree will settle somewhat in the hole. Remove the burlap, twine, and wires from the top third of the root ball. It is not necessary to pull the burlap out from under the ball, unless burlap made of plastic is used. Plastic will not rot and will restrict root development. Add prepared soil in and around the root ball until it is covered. When the hole is two-

thirds filled with soil, fill it to the top with water and let it soak in. Then finish filling the hole with soil. With the remaining soil, form a saucerlike shape (known as a *berm*) around the tree to hold water that will be added later.

PLANTING CONTAINERIZED TREES

Containerized trees are planted in the same way as balled and burlapped trees, except for the following differences.

1. The container is removed from the tree roots. Care must be taken not to break the root ball apart.

2. After the container is removed, check to see if the roots are pot-bound (the root ball will be excessively tight). If this has happened then it is necessary to cut or loosen the roots to encourage growth from the root ball. This procedure stops the circular root growth and allows the roots to make their growth into the soil. If the growth pattern remains circular, the plant growth will be restricted. In severe cases, trees can be killed if this growth pattern is allowed to continue.

STAKING

Staking evergreens is important because it gives the tree needed support to allow the root system to become established and to prevent the tree from swaying in the wind. For trees 10 feet or less in height, the stakes should be 5 to 6 feet long and 1¾ inches square. Stakes should be made of wood that is capable of standing in the ground for at least two years.

To stake a tree, drive two stakes into the ground, with one of the stakes into the prevailing wind side of the tree. Drive the other stake directly opposite the first stake. Using a pliable 12- or 14-gauge soft steel wire and piece of garden hose or commercially prepared material (about 1 foot long), slip the wire through the hole in the hose or commercial sleeve and place it around the trunk of the tree about halfway up the tree. Pull the two ends of the wire around the first stake, making sure the section inside the rubber hose is around the trunk of the tree. Twist the wire around the stake. Repeat the process on the other stake. Using a pair of fencing pliers, twist the wire until it is tight enough to hold the tree in place. Repeat this procedure on the opposite stake, making sure the tree is secure from high wind and will grow straight.

FERTILIZING

The general recommendation for fertilizing a narrowleaf evergreen is 3 to 6 pounds of 5-10-10 fertilizer per 100 square feet placed in a circle under the drip line at the outer end of the branches. Fertilize before new growth starts in the spring. Fertilizer with the analysis of 5-10-10 means 5 percent nitrogen, 10 percent phosphorus, and 10 percent potassium. If you apply 6 pounds per 100 square feet, you will apply 0.3 pounds of nitrogen (N), 0.6 pounds of phosphorus (P), and 0.6 pounds of potassium (K).

> 5 percent of 6 pounds = 0.3 pounds N
> 10 percent of 6 pounds = 0.6 pounds P
> 10 percent of 6 pounds = 0.6 pounds K

After applying the fertilizer, water it in with a garden hose sprinkler. Organic fertilizers such as cottonseed meal or soybean oil meal may also be used. Although there is an advantage to the use of organic fertilizers in that there is no danger of burning the plant, they are usually more expensive.

WATERING

Narrowleaf evergreens need to be watered every ten to fourteen days during the first year after planting. Thoroughly soak the soil to 6 inches or deeper when they are planted. Watering after planting will depend on the weather conditions of the geographic area.

PRUNING

There are two methods of pruning narrowleaf evergreens.

1. **Pruning** individual branches gives the plant a natural, informal appearance. (Large branches may be removed from specific locations on the plant.) Spreading junipers and *Taxus* should be pruned by removing the ends of long twigs. This gives an informal appearance to these plants.

2. **Shearing** is done as when shaping a hedge. The ends of all small branches are clipped to shape the edges of the plant in straight lines, giving a more formal appearance. Hicks yew (*Taxus x media* "Hicksii"), columnar juniper, and globe arborvitae may be sheared in this manner.

CHARACTERISTICS OF SPECIFIC NARROWLEAF EVERGREENS

Figures 27-6, 27-7, and 27-8 list narrowleaf evergreens used in landscaping. The plants are listed in sequence according to their height. The general breakdown includes one group of those plants 3 feet or less in height; a second group of those plants ranging from 3 to 9 feet in height; and a third group of those plants greater than 10 feet in height.

NAME	HARDINESS	FOLIAGE COLOR	PERIOD OF INTEREST	LANDSCAPE USE	OTHER REMARKS
Erica carnea (Spring heath)	5	bright green	small, rosy red spikes in April	ground cover	Attractive spring colors.
Juniperus chinensis sargentii (Sargent juniper)	4	steel blue	blue berries in fall	ground cover	Excellent for planting along seashore. Does well on steep banks to prevent erosion.
Juniperus horizontalis wiltonii (Blue rug juniper)	2	steel blue	blue berries in fall	ground cover	Excellent rock garden plant.
Taxus baccata "repandens" (Spreading English yew)	4	dark green	fall	foundation planting	Excellent plant; low maintenance; good color all year. Requires well-drained soil.
Taxus cuspidata aurencens (Dwarf Japanese yew)	4	light green	fall	foundation planting	Compact plant which produces red berries. Requires well-drained soil.

FIGURE 27-6 Narrowleaf evergreens (3 feet or less in height)

NAME	HARDINESS	FOLIAGE COLOR	PERIOD OF INTEREST	LANDSCAPE USE	OTHER REMARKS
Chamaecyparis obstusa "nana"	3	deep dark green	all year	accent plant, specimen plant	Compact form; pyramidal shape.
Juniperus squamata Meyer (Meyer's juniper)	4	blue	all year	foundation planting	Needs good management and care in dry areas.
Juniperus virginiana tripartia	2	dark green	all year	foundation planting	Good in dry areas.
Pinus mugo mughus "compacta" (Mugho pine)	2	dark green	all year	foundation planting	Global shape; slow growing.
Picea abies varietus (a) Dwarf Alberta spruce	2	light green	all year	foundation planting, specimen plant	Fine-textured plant.
(b) Bird's nest spruce	2	light green	all year	rock garden	Requires well-drained soil.
Juniperus chinensis pfitzeriana (Pfitzer juniper)	4	blue green	all year	foundation planting, screen plant	Plant in full sun. Requires well-drained soil. Control of bag worms necessary.
Juniperus chinensis Hetzii	4	blue green	all year	screen plant, foundation planting	Plant in full sun; does well in dry areas.
Taxus x media "Hicksii" (Hicks yew)	4	dark green	all year; fall fruit	foundation planting, screen plant	Plant in full sun. Excellent in formal gardens; columnar shape. Requires well-drained soil.

FIGURE 27-7   Narrowleaf evergreens (3 to 9 feet in height)

NAME	HARDINESS	FOLIAGE COLOR	PERIOD OF INTEREST	LANDSCAPE USE	OTHER REMARKS
Cypressocyparis leylandi (Leyland cypress)	4	light green	all year	screen or hedge plant	Fast grower, excellent windbreak; columnar shape.
Taxus baccata (English yew)	6	dark green	all year	screen or hedge plant, foundation planting on large buildings	Female plant has berries. Will stand shade. Good background shrub.
Taxus cuspidata (Japanese yew)	4	dark green	all year	screen or hedge plant, foundation planting on large buildings	Pyramidal form. Tolerates shade. Rapid grower. Produces red fruit on female.

FIGURE 27-8   Narrowleaf evergreens (over 10 feet in height)

The following information is given for each plant.

◆ The **hardiness** of each plant. This number refers to one of the areas shown on the hardiness zone map at the end of the text. It indicates where the plant can be grown in the United States.

◆ **Foliage color** is the color of the narrowleaf evergreen. Some types vary from a very light green to dark green.

◆ **Period of interest** is that time of year during which the plant is most attractive. This may be when it flowers, bears fruit, or changes foliage color. All of these create interest during a particular season of the year.

FIGURE 27-10 Sargent juniper—Juniperus chinensis sargenti

FIGURE 27-9 Dwarf Hinoki false cypress— Chamaecyparis obtusa "Nana"

FIGURE 27-11 Blue Rug juniper—Juniperus horizontalis "wiltonii"

♦ **Landscape use** indicates how a particular plant is used in the landscape. The term *ground cover* refers to plants, other than grass, that are used to cover the ground. Ground covers also have special uses, such as preventing erosion by holding steep banks in place.

   Foundation planting refers to plants which are used around buildings to help accent and tie the building into the landscape. *Rock garden* refers to plants used in gardens which are planted in a rocky environment, whether natural or artifi-

cially reproduced. See figures 27-9 through 27-11. *Specimen plants* refer to plants that are used alone for their own beauty or as an accent. *Screen plants* and *hedges* refer to those plants that confine certain areas, used to reduce noise or an unappealing sight.

♦ **Other remarks** are given to help the individual make a more educated selection of plants.

   Figures 27-12 through 27-18 identify narrow-leaf evergreens by their leaf characteristics.

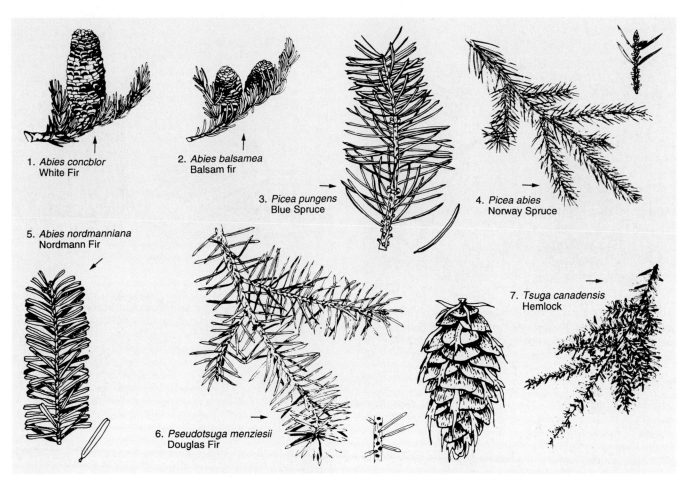

FIGURE 27-12   Narrowleaf evergreens with single needles. (Adapted from Extension Bulletin 238, University of Maryland)

1. *Abies concblor* White Fir
2. *Abies balsamea* Balsam fir
3. *Picea pungens* Blue Spruce
4. *Picea abies* Norway Spruce
5. *Abies nordmanniana* Nordmann Fir
6. *Pseudotsuga menziesii* Douglas Fir
7. *Tsuga canadensis* Hemlock

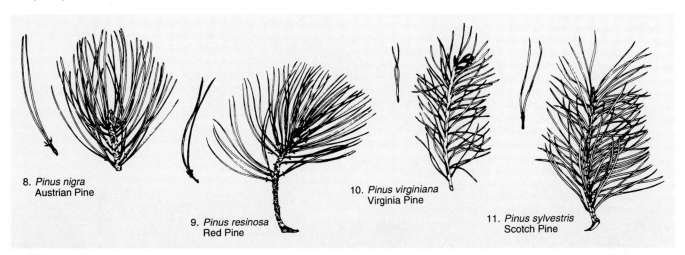

FIGURE 27-13   Narrowleaf evergreens with two needles per cluster. (Adapted from Extension Bulletin 238, University of Maryland)

8. *Pinus nigra* Austrian Pine
9. *Pinus resinosa* Red Pine
10. *Pinus virginiana* Virginia Pine
11. *Pinus sylvestris* Scotch Pine

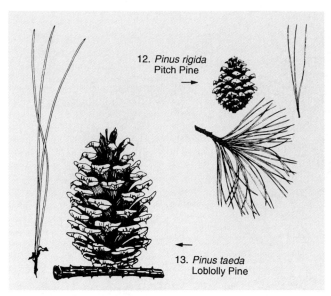

FIGURE 27-14   Narrowleaf evergreens with three needles per cluster. (Adapted from Extension Bulletin 238, University of Maryland)

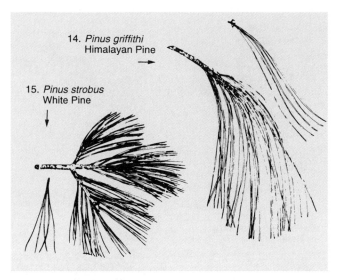

FIGURE 27-15   Narrowleaf evergreens with five needles per cluster. (Adapted from Extension Bulletin 238, University of Maryland)

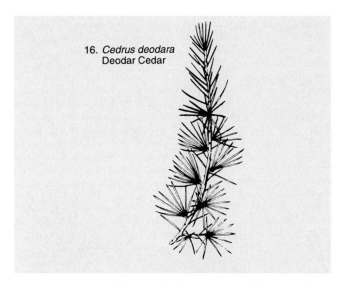

FIGURE 27-16   Narrowleaf evergreen with more than five needles per cluster. (Adapted from Extension Bulletin 238, University of Maryland)

FIGURE 27-17   Narrowleaf evergreens that are deciduous. This group includes those evergreens that drop their needles in the winter. (Adapted from Extension Bulletin 238, University of Maryland)

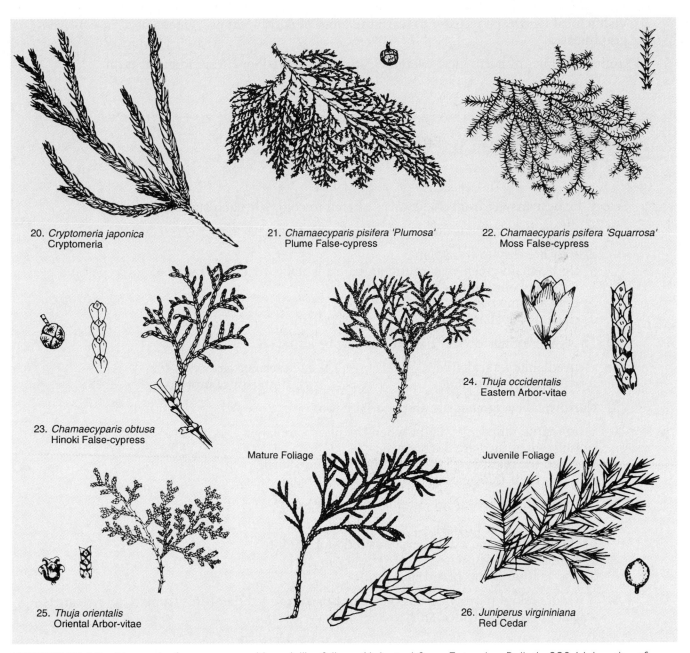

20. *Cryptomeria japonica*
Cryptomeria

21. *Chamaecyparis pisifera 'Plumosa'*
Plume False-cypress

22. *Chamaecyparis psifera 'Squarrosa'*
Moss False-cypress

23. *Chamaecyparis obtusa*
Hinoki False-cypress

24. *Thuja occidentalis*
Eastern Arbor-vitae

25. *Thuja orientalis*
Oriental Arbor-vitae

Mature Foliage

Juvenile Foliage

26. *Juniperus virgininiana*
Red Cedar

FIGURE 27-18   Narrowleaf evergreens with scalelike foliage. (Adapted from Extension Bulletin 238, University of Maryland)

Student Activities

1) Produce ten cuttings of selected narrowleaf evergreens and propagate them. When propagating, follow the procedure outlined in Section 3.

2) Transplant a bare root, balled and burlapped, and containerized narrowleaf evergreen or explain the correct procedure for transplanting.

3) Demonstrate or explain the difference between pruning for an informal appearance and shearing.

4) Identify scalelike and needlelike narrowleaf evergreens from specimen twigs and branches.

5) Visit a local landscape nursery to learn the types of narrowleaf evergreens used in your geographic area.

6) Collect samples of narrowleaf evergreens to be preserved for future identification.

Self-Evaluation

Select the best answer from the choices offered to complete each statement.

1) Narrowleaf evergreens are popular landscape plants because

    a) they are easy to transplant.
    b) they remain green year-round and are adaptable to a wide range of soil types and weather conditions.
    c) they are inexpensive.
    d) none of the above

2) The leaves of narrowleaf evergreens may be either

    a) needlelike or scalelike.
    b) whorled or clustered.
    c) alternate or opposite.
    d) simple or compound.

3) Narrowleaf evergreens are shipped bare root

    a) to reduce shipping weight when the plants are large.
    b) only when they are dormant.
    c) when the plants are still seedlings and quite small.
    d) when their foliage drops and they are dormant.

4) Narrowleaf evergreens are often chosen for

    a) their attractive flowers.
    b) their edible fruit which serves as food for wildlife.
    c) bare stems in winter.
    d) use as foundation plantings.

5) A windbreak of narrowleaf evergreens can result in temperatures from _____ degrees warmer on the protected side.

    a) 10 to 15
    b) 5 to 10
    c) 25 to 30
    d) 0 to 5

6) Balled and burlapped evergreens may be moved at any time of year except when

    a) new shoots are growing.
    b) they are dormant.
    c) they are setting seed.
    d) the soil is too wet.

7) Plastic burlap around the root ball should be removed

    a) entirely from the root ball.
    b) before the plant is set in the hole.
    c) by rolling it down the sides and leaving it under the ball.
    d) only from the top of the ball.

8) Balled and burlapped plants are set in the planting hole at a depth

    a) of one-half the depth of the root ball.
    b) equal to the depth at which it grew in the nursery.
    c) 3 inches above the surrounding soil line.
    d) 1 inch above the original soil line.

9) Three to six pounds of 5-10-10 fertilizer per 100 square feet should be applied annually

    a) in a circle at the drip line of the plant.
    b) in holes punched into the soil around the plant.
    c) under the mulch around the plant.
    d) on top of the mulch around the plant.

10) The best time of year to apply a 5-10-10 fertilizer is

    a) in early fall.
    b) in early spring before new growth starts.
    c) in midsummer to encourage a longer growth period.
    d) in late summer.

11) Informal pruning to remove individual branches is the technique usually used to prune

    a) Hicks yew.   c) columnar juniper.
    b) spreading juniper.   d) globe arborvitae.

12) Two narrowleaf evergreens that grow to a height of 3 feet or less are

    a) Sargent juniper and blue rug.
    b) dwarf Alberta spruce and Hicks yew.
    c) English yew and Japanese yew.
    d) all of these

13) 100 pounds of 5-10-10 has _____ pounds N, _____ pounds P, and _____pounds K.

14) 30 pounds of 0-20-20 has _____ pounds N, _____ pounds P, and _____ pounds K.

15) 50 pounds of 10-20-10 has _____ pounds N, _____ pounds P, and _____pounds K.

Broadleaf Evergreens

OBJECTIVE

To identify the cultural requirements, planting techniques, care, and uses of broadleaf evergreens in the landscape.

Competencies to Be Developed

After studying this unit, you should be able to

- ◆ list four ways in which broadleaf evergreens are used in the landscape.
- ◆ list four cultural requirements of broadleaf evergreens.
- ◆ describe the soil and fertilizer requirements of broadleaf evergreens.
- ◆ explain the procedure for transplanting broadleaf evergreens.
- ◆ list three pests which attack broadleaf evergreens and one control for each.

MATERIALS LIST

- ✔ branches and leaves of broadleaf evergreens grown in the local area
- ✔ one broadleaf evergreen properly balled and burlapped

The description of broadleaf evergreens is just as the name implies— *evergreen* plants (plants which hold their leaves all year) with broad leaves rather than the needlelike leaves of narrowleaf evergreens such as pines, yews, and junipers. Figure 28-1 illustrates some typical broadleaf evergreens.

USES OF BROADLEAF EVERGREENS

Broadleaf evergreens are used in the landscape in several ways.

FIGURE 28-1 Leaves and twigs of three broadleaf evergreens: (left) American holly; (middle) English holly; (right) Japanese holly.

◆ **Around foundations.** The smaller-leaved broadleaf evergreens are commonly used as foundation plantings. Some of these are the Japanese holly, azalea, andromeda, dwarf English holly, and dwarf rhododendron.

◆ **As specimen plants.** A large rhododendron is beautiful as a singular specimen plant, figure 28-2. American holly, some of the larger varieties of English holly, and magnolia also make beautiful specimen plants, figure 28-3.

◆ **As hedges.** Privet, a broadleaf evergreen, is often used as a hedge. Japanese holly and boxwood are also used in hedges.

◆ **Along sidewalks.** Broadleaf evergreens are frequently used to lead the way to the home entrance. A privet hedge, if cut short, and azalea also work well as plantings along sidewalks.

◆ **In woodland plantings.** Azaleas make beautiful understory plants in open deciduous woods, figure 28-4.

FIGURE 28-3 The larger southern magnolia (Magnolia grandiflora), an impressive planting when covered with fragrant 8- to 10-inch blossoms (inset). (Photo courtesy of Ferrell M. Bridwell)

FIGURE 28-4 Azaleas used to add color to the spring landscape. (Ed Reiley, Photographer)

Figure 28-5 summarizes growing information and uses for some of the more common broadleaf evergreens.

FIGURE 28-2 Rhododendron in full bloom, a beautiful specimen plant.

NAME	SIZE AT MATURITY	SOIL AND SOIL pH PREFERENCE	PRUNING DATE	BLOOMING DATE	USES	OTHER REMARKS
Andromeda (*Pieris japonica*)	8'–10'	loam pH 5.0–6.5	after blooming in spring	early spring	foundation, specimen, center of group	Gives very early white blooms. Excellent for background. Hardy in Zones 6 to 9.
Azalea (*Rhododendron* varieties)	from creeping to 15' according to variety	loam pH 5.0–6.0	after blooming in spring	early spring through late spring depending on variety	foundation, beds, edging, woodland	Many variations in size, color, bloom date. Has a variety of uses. Should be sprayed for best results. Hardy in Zones 6 to 8.
Barberry (*Berberis darvinii*)	8'	broad range	anytime	spring flowering; berries in fall	hedge, foundation, specimen	Few thorns, bright orange flowers, purple berries. Needs full sun. Hardy in Zones 4 to 9.
(*Berberis wilsoniae*)	5'	broad range	anytime	spring flowering; berries in fall	hedge, foundation, specimen	Bright scarlet berries. Hardy in Zones 4 to 9.
(*Berberis thunbergii*)	4'	broad range	anytime	spring flowering; berries in fall	hedge, foundation, specimen	Rich purple foliage turns scarlet in autumn.
(*Berberis verruculosa*)	4'	broad range	anytime	spring flowering; berries in fall	specimen, hedge	Yellow flowers in spring. Foliage white underneath. Hardy in Zones 4 to 9.
Boxwood, Common (*Buxus sempervirens*) (*B.s. arborescens*)	fast growing up to 15'	heavy loam pH 6.0–7.0	early spring	no flowers	hedge	Grows very fast. Full sun or shade.
Boxwood, English (*Buxus sempervirens*) (*B.s. suffruticosa*)	slow growing up to 3'	heavy loam pH 5.0–7.0	early spring	incon- spicuous	edging walkways, topiary designs, inside corners at foundation	Suffers windburn in strong winter winds. Grows in full sun or shade. Grows slowly (1–2 inches each year). Hardy in Zones 6 to 8.
Camellia (*C. sasaqua*)	8' to 15'	loam, high in organic matter pH 5.0–6.5	after flowering or early spring	autumn	specimen, border, hedge	White to pink single and double flowers. Will grow in full sun or partial shade. Hardy in Zones 7 to 9.
(*C. japonica*)	15' to 30'	loam, high in organic matter pH 5.0–6.5	after flowering	early spring	specimen	Dark green foliage; white to deep red single or double flowers. Needs sun to flower. Hardy in Zones 7 to 9.
Firethorn (*Pyracantha coccinea* "Lalandi")	20'	loam wide range of pH (6.0–8.0)	early spring	June	specimen	Many varieties. Difficult to transplant (use container plants). Orange fruit.

FIGURE 28-5   Common broadleaf evergreens

NAME	SIZE AT MATURITY	SOIL AND SOIL pH PREFERENCE	PRUNING DATE	BLOOMING DATE	USES	OTHER REMARKS
Holly, American (*Ilex opaca*)	20' to 30'	sandy loam pH 6.0–7.0	winter holiday season	spring	specimen	Red berries in fall and winter; spiny leaves.
Holly, Chinese (*Ilex cornuta*)	from 4' to 15' depending on variety	sandy loam pH 6.0–7.0	throughout the year except late summer	spring	foundation, hedge, specimen	Many new excellent varieties. Red or yellow berries. Full sun or shade. Needs heavy fertilization.
Holly, English (*Ilex aquifolium*)	2' to 10'	sandy loam pH 6.0–7.0	winter holiday season	spring	specimen	Glossy foliage. Yellow and red berries.
Holly, Japanese (*Ilex crenata*)	from 3' to 15' depending on variety	loam pH 5.0–6.0	throughout the year except late summer	early summer	hedge, topiary design, specimen, foundation	Small spineless leaves and black berries. Needs fertilizer each year.
Magnolia (*Magnolia grandiflora*) Southern magnolia	30' to 50'	loam pH 6.0–6.5	after flowering	mid-summer	specimen	Native to the United States. 6" to 10" fragrant white flowers. Hardy in Zone 7 and in southern areas.
Sweet Bay Magnolia (*Magnolia virginiana*)	30' to 50'	loam	after flowering	early summer	back-ground, specimen	Evergreen in south, Zones 7 to 10. Deciduous in Zones 5 and 6. Hardy in Zones 5 to 10.
Nandina (*Nandina domestica*)	6' to 8'	loam pH 5.0–6.5	after flowering	mid-summer	shrub borders, front of taller evergreens	White flowers; red berries in autumn and winter. Hardy in Zones 7 and 8.
Privet (*Ligustrum japonicum*)	9' to 18'	wide range	after flowering	June–July	hedge	Black berries. Hardy in Zones 6 to 7.
Glossy Privet (*Ligustrum lucidum*)	8' to 15'	wide range	after flowering	August–September	hedge, borders	Easy to grow. Hardy in Zones 7 to 10. Grows in shade.
Rhododendron varieties	3' to 10' for most commercial varieties	varies with variety pH 5.0–6.0	early spring	spring	specimen, foundation	Requires mulch at all times and good drainage. High organic matter needed in soil. Hardy in Zones 4 to 8.
Viburnum (Many varieties; select for your area)	5' to 12' depending on variety	varies with variety	after flowering	spring	foundation, specimen	Fertilize each year. Attractive fruit, especially to birds.

FIGURE 28-5 Common broadleaf evergreens (continued)

CULTURAL REQUIREMENTS

◆ **Ample moisture** is required by most broadleaf evergreens for best growth. This may be supplied by the addition of organic matter to the soil. Organic matter holds moisture well. Sandy soils need more organic matter such as pine bark or sphagnum peat moss that could make up as much as 50 percent of the planting mix.

◆ **Good soil drainage** is also necessary. If drainage is a problem, the planting may be raised above soil level. The addition of fine pine bark improves soil drainage by improving soil structure.

◆ **Mulching** is needed for the proper supply of moisture. Since these plants have green leaves year-round, a ready supply of moisture to the roots must be available at all times to replace moisture lost through the leaves due to transpiration. If the soil freezes in winter, the roots cannot pick up moisture, causing the plants to dry out. A mulch prevents the soil from freezing as deeply and the roots are better able to supply lost moisture. In summer, a mulch keeps the soil cooler and saves moisture loss due to heat and drying winds. Mulching should *not* be done on boxwood and used at only 2 or 3 inches of coarse material on rhododendrons. The roots of these plants need a lot of oxygen.

◆ **A location which is sheltered from strong winds** is best for broadleaf evergreens. Cold winter winds and winter sun do the most damage by removing moisture from the leaves faster than the roots can replenish it in some cases. Drying out in winter is one of the most serious problems for broadleaf evergreens.

◆ **Enough sunlight** is needed to develop a full, compact plant. Plants tend to grow tall and leggy with too much shade. Full sunlight or partial shade is best for most of the broadleaf evergreens. For example, rhododendrons grow best in 40 percent shade. If plants do not flower well, they are probably not getting enough light.

SOIL AND FERTILIZER

Any good loam soil is satisfactory for most broadleaf evergreens. An acidic soil is usually the ideal choice. The rhododendron, azalea, and Japanese andromeda require a pH of from about 5.0 to 6.0 for best growth. A soil pH higher than this results in an iron deficiency.

Fertilizer should be used in small amounts. Too much fertilizer, especially nitrogen, causes soft, fast growth and results in open, straggly plants which are unattractive. A slower growing, compact plant makes a much better addition to the landscape picture. Also, if a fertilizer is applied in smaller amounts, the plant will not outgrow its space in the landscape as quickly. If all other cultural conditions are right, short twig growth and small, pale green leaves may indicate a need for fertilizer.

Fertilizer is added in the spring. It is applied in a circular fashion around the plant at the drip line. (The *drip line* is an imaginary line directly below the outer edge of the branches.) A few handfuls carefully sprinkled around the plant are sufficient for small plants. For larger trees, use 1 to 1½ pounds per inch of tree trunk. Any type of fertilizer that is not alkaline may be used for "acid-loving" broadleaf evergreens; cottonseed meal is a safe fertilizer for these plants.

The varieties of Chinese and Japanese holly require more fertilizer than most broadleaf evergreens. Two to three pounds of 10-5-5 fertilizer per 100 square feet should be used. On individual plants, use 1½ pounds per 3 feet of height or spread of plant. Chinese holly grows best at a soil pH of 6.0 to 7.0.

TRANSPLANTING

Since these plants are evergreen, the demand for moisture is always high. The roots must always be in close contact with moist soil. For this reason, broadleaf evergreens are always moved with the *root ball*. This means that the soil is left intact around the roots and moved with the plant, figure 28-6.

The hole for broadleaf evergreens should be wide enough so that 10 to 12 inches of space exists all the way around the root ball. The plants are always planted at the same depth or slightly higher than they were growing before; they are never planted deeper, as is done with many deciduous trees. No soil should be placed on top of the root ball, figure 28-7. Be sure that the soil in the bottom of the hole is firm so that the root ball does not sink after it is planted.

Plants must be planted in raised beds above the soil line in poorly drained or alkaline soils. Bring in enough good, well-drained topsoil of the proper pH, or plant entirely in a soilless mix of two parts pine bark, one part peat moss, and one part sand. Raise the bed so the plant root ball sets on top of the existing, poorly drained soil.

FIGURE 28-6 This azalea was dug from the ground with the root ball intact in preparation for transplanting. Careful handling is important so that the soil is not broken loose from the roots. (Ed Reiley, Photographer)

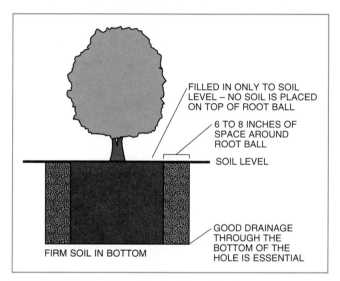

FILLED IN ONLY TO SOIL LEVEL – NO SOIL IS PLACED ON TOP OF ROOT BALL

6 TO 8 INCHES OF SPACE AROUND ROOT BALL

SOIL LEVEL

GOOD DRAINAGE THROUGH THE BOTTOM OF THE HOLE IS ESSENTIAL

FIRM SOIL IN BOTTOM

FIGURE 28-7 Transplanting broadleaf evergreens

Firm the soil around and under the root ball with the hands, and water with a garden hose at low water pressure. Do not tramp with feet or pack too tightly. Mulch around the plant with wood chips or pine bark to a depth of 2 to 3 inches and out beyond the branch ends.

PRUNING

Broadleaf evergreens may be pruned for a special effect or cut back to thicken the plant. Pruning is most often done in early spring or immediately after blooming. Holly is often pruned during the winter holiday season and the prunings used for holiday decorations.

The rhododendron should be pruned at the point where a flush of growth stops and buds are more plentiful. This usually occurs at the point where branches extend from the plant.

INSECTS AND DISEASES

If varieties of broadleaf evergreens that are adapted to the area are selected, chances are good that few insect and disease problems will develop. Chewing insects may feed on leaves, causing some disfigurement. Except on small plants, this is seldom serious enough to require spraying. If a spray is needed, use malathion and Sevin, or orthene. (Read the label.)

Small sucking insects and mites, such as the lacebug, are the most serious pests that attack broadleaf evergreens. These insects attack the lower surface of the leaf and suck the chlorophyll from the leaf. An attack causes leaves to appear white or yellow. Afterwards, the plants do not grow as well. Small mites or insects can be seen on the undersurface of the plant's leaf with a 10-power microscope (one which magnifies the insects ten times their normal size.) For control of these pests, spray with malathion if the pest can be hit with the spray. (Malathion will kill only if the insect comes in direct contact with the spray.) Orthene, a systemic, will kill without coming into contact with the pests and do a better job. However, systemics may be more toxic to warm-blooded animals, including humans, and proper precautions must be taken.

Fungus diseases such as leaf spots, rusts, or dieback (caused by tips of branches being killed by a fungus) may be a problem. If these diseases are present, spray just after bloom and again in ten days with zineb, ferbam, captan, or alliette. Some new systemic fungicides such as Subdue used as soil drenches give excellent plant protection. (Check the label for application information.)

Use a spreader sticker to apply all sprays to broadleaf evergreens. A *spreader sticker* is a chemical added to a pesticide which causes the spray to break into small droplets and spread evenly and thinly over the leaf surface. Shiny leaves are hard to get completely wet, and sprays tend to run off the plant without giving the proper protection. A spreader sticker helps to overcome this problem by causing the spray to stick to the plant more easily. Ordinary soap or detergent is acceptable as a spreader sticker if nothing else is available. Use only enough soap to produce light suds in the water. There are many insect and disease resistant varieties which require no spraying.

Most broadleaf evergreens are propagated by semihardwood cuttings. See Unit Seven for details.

Student Activities

1) Draw up a chart such as the one shown in figure 28-5. Insert local broadleaf evergreens that are different from the ones listed in the text. Supply the information in each column for plants listed.

2) Tour the school grounds or part of the local community and identify various broadleaf evergreens sighted. Also note their use in the landscape.

3) Decide what broadleaf evergreens would be best in an area around your school. Draw a plan using these evergreens.

Self-Evaluation

A) Select the best answer from the choices offered to complete the statement or answer the question.

1) Broadleaf evergreens are plants that

   a) hold their leaves all year round.
   b) have broad leaves rather than needles.
   c) require a constant supply of moisture.
   d) all of the above

2) Which broadleaf evergreen is often used as a hedge?

   a) privet
   b) rhododendron
   c) magnolia
   d) American holly

3) Which of the following are good choices as specimen plants in a landscape?

   a) Japanese holly, privet, magnolia
   b) magnolia, rhododendron, American holly
   c) boxwood, magnolia, English holly
   d) magnolia, privet, rhododendron

4) Broadleaf evergreens that work well in lining sidewalks and other entranceways are

   a) English boxwood, azaleas, Japanese holly.
   b) English boxwood, azaleas, magnolia.
   c) rhododendron, magnolia, Chinese holly.
   d) American holly, magnolia, Chinese holly.

5) The four most important cultural requirements of broadleaf evergreens are

　　a) ample moisture, good soil drainage, mulch, and a windbreak.
　　b) dry soil, strong sunlight, heavily fertilized soil, and air circulation.
　　c) ample moisture, dry soil, strong sunlight, and a windbreak.
　　d) dry soil, good drainage, strong sunlight, and a windbreak.

6) Many of the broadleaf evergreens should be fertilized with only small amounts of nitrogen fertilizer because

　　a) nitrogen makes the leaves turn yellow.
　　b) nitrogen causes soft growth which is easily killed by frost.
　　c) nitrogen causes long shoot growth and open, unattractive plants.
　　d) nitrogen tends to dwarf the plants.

7) When dug from the ground, broadleaf evergreens should always be moved with soil around the roots because

　　a) this makes it easier to plant correctly.
　　b) the heavy root ball helps stabilize the plant.
　　c) the roots will otherwise fall off the plant.
　　d) the roots are less disturbed and are able to continue supplying moisture to the leaves.

8) Most broadleaf evergreens require mulching for best growth. The most important function performed by the mulch is to

　　a) keep the soil cool in summer, thus conserving moisture.
　　b) control weeds.
　　c) prevent deep soil freezing in winter allowing the roots to obtain moisture easier.
　　d) all of the above

9) Which of the following is *not* a reason for drying out of broadleaf evergreens due to lack of available moisture?

　　a) freezing of soil around roots in winter
　　b) transplanting with a root ball
　　c) lack of mulch in summer
　　d) planting in areas having strong winds

10) Two broadleaf evergreens that require relatively high amounts of fertilizer are

　　a) Japanese and Chinese holly.　　　　c) English and Chinese holly.
　　b) the rhododendron and azalea.　　　d) the boxwood and andromeda.

11) A safe chemical spray used to control chewing insects on broadleaf evergreens is

　　a) Sevin.　　　　　　　　　　　　　c) zineb.
　　b) sulfur.　　　　　　　　　　　　　d) captan.

12) A spray chemical used to control fungus diseases on broadleaf evergreens is

　　a) Sevin.　　　　　　　　　　　　　c) captan.
　　b) malathion.　　　　　　　　　　　d) Cygon.

13) A spray chemical used to control sucking insects on broadleaf evergreens is

　　a) Sevin.　　　　　　　　　　　　　c) captan.
　　b) zineb.　　　　　　　　　　　　　d) malathion or orthene.

B) Provide a brief answer for each of the following.

1) Describe in writing how you would plant a broadleaf evergreen in a wet, poorly drained soil.

2) Which broadleaf evergreen should have no mulch over its roots?

Deciduous Trees

OBJECTIVE

To properly position and plant deciduous trees in the landscape.

Competencies to Be Developed

After studying this unit, you should be able to

- ◆ list the six functions of trees in the landscape.
- ◆ select at least two specific trees to fulfill each of these six functions.
- ◆ demonstrate the proper planting technique for bare root, balled and burlapped, and containerized trees.

MATERIALS LIST

- ✔ deciduous or flowering trees, at least one of each in a container, bare root, and balled and burlapped
- ✔ round-point shovels for preparing the holes
- ✔ fertilizer
- ✔ peat moss
- ✔ mulching material (hardwood or pine bark)
- ✔ tree stakes, tree wrap, 12- to 14-gauge soft steel wire, small sections of garden hose
- ✔ a bar for punching holes to fertilize an established tree

THE USES OF TREES IN THE LANDSCAPE

Trees are often positioned in the landscape to serve a specific purpose. The following are some of the functions of trees in the landscape. (Figures 29-1 through 29-10 are from the University of

FIGURE 29-1

FIGURE 29-2

FIGURE 29-3

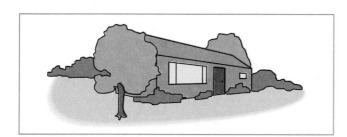

FIGURE 29-4

FIGURE 29-5

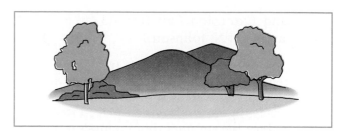

FIGURE 29-6

FIGURE 29-7

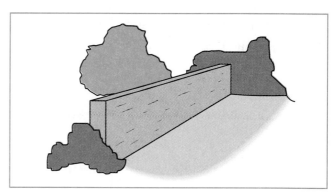

FIGURE 29-8

FIGURE 29-9

FIGURE 29-10

Maryland Extension Bulletin 183 by Robert L. Baker and Carl N. Johnson.)

◆ **To provide shade.** Shade trees keep temperatures inside houses cooler in summer as well as providing outdoor shade, figure 29-1. A well-placed tree can reduce the summer room temperature of a frame house as much as 20 degrees in certain areas. To serve this purpose, trees may be planted as close as 15 feet to the house. Larger spreading trees, such as oaks, must be given more generous spacing.

◆ **To frame the house.** Trees are sometimes used to make houses the center of interest in the view from a street, figure 29-2. Generally, the farther the house is set back from the street, the more effective the tree is as a framing device.

◆ **To soften lines.** A tree placed at the back of a home is effective in softening the lines of the house, figure 29-3.

◆ **To relieve bare spots.** The end walls of many houses often have a bare look. This bareness can be relieved by planting a small or medium-sized tree near the side of the house, figure 29-4. This creates a softening effect that a planting of shrubs alone may not accomplish.

◆ **To screen an object.** A single tree or a group of trees may be used to partially screen an object that would otherwise stand out too strongly in the landscape, figure 29-5.

◆ **To frame a view.** Tall trees with an underplanting of shrubs or small trees may be used to frame a view from a house or terrace, figure 29-6.

◆ **To ensure privacy.** A carefully positioned tree can create privacy by screening a porch or terrace, figure 29-7.

◆ **To accent the landscape.** Flowering trees or trees with graceful and interesting branching habits are often used as accents in the landscape, figure 29-8. They can be used to direct attention to certain areas or to act as a terminal planting for a hedge or wall.

◆ **To break monotony.** Trees and shrubs can be used to break the monotony of an obtrusive architectural feature or enclosure, figure 29-9. The planting, however, should be placed off center so that the feature does not appear to have been cut in half. White brick walls, fencing, and other architectural means of enclosing property may be attractive in themselves, but can become monotonous unless softened by planting materials.

◆ **As windbreaks.** Tall evergreens are used as windbreaks, protecting the house from north and northwest winds in the winter, figure 29-10. These should be placed about 50 feet from the house to be effective.

MULTIPLE USE TREES

Some trees have multiple uses in the landscape. One of the best examples is the pecan tree *Carya illinoinensis*. The pecan is very popular in the southern United States both as a nut tree and as a shade and large accent tree in the landscape. Its natural range in the United States is from Texas east to Alabama and from the Gulf Coast north to Iowa and southern Wisconsin. New cold hardy selections have extended the area it is now planted in to include states as far north and east as New York. It is found throughout all the southern states. The pecan is the largest member of the hickory family. Occasionally, trees are found that are 170 feet tall with trunk diameters of 6 feet, but a more general size is 90 to 100 feet tall with a diameter of 2 to 4 feet. The pecan tree grows best in the bottomlands with rich, moist soils where they are long-lived trees.

When used in the landscape, pecans need lots of space to spread out and up, and when given this space add a touch of symmetrical grace to the landscape. The fact that they bear nuts, which drop to the ground, restricts their use in lawn areas. They are best planted at some distance from the house where their stately form can be viewed.

TREE TYPES

Figure 29-11 lists deciduous trees that may be used in landscaping for shade and ornamental flowering. The trees in this chart are separated according to their mature height, with small deciduous trees classified as those up to 35 feet in height; medium deciduous trees as those 36 to 75 feet; and large deciduous trees as those 75 feet and over in height.

The trees in the chart are classified according to various characteristics, under the headings *Form, Flowering, Color, Texture, How to Propagate, Height, Hardiness*, and *Ornamental Use*.

SMALL DECIDUOUS TREES (UP TO 35 FEET IN HEIGHT)

Name	Form	Flowering	Color		Texture	How to Propagate	Height	Hardiness	Ornamental Use	Other Remarks
			Summer	Fall						
Acer palmatum (Japanese maple)	rounded moundlike		red to green	scarlet	fine	budding	20'	5	specimen	Many varieties are available. Needs well-drained soil.
Acer ginnala 'Flame' (Amur maple)	rounded upright	purplish	green	scarlet	fine	seed	20'	2	specimen, screening	Fragrant flowers.
Cercis canadensis (Redbud or Judas tree)	rounded	purplish pink mid-May	green	yellow	coarse	seed	35'	4	border	Blooms same time as dogwood.
Chionanthus virginicus (Old Man's Beard)	rounded	white feathery late May	green	yellow	coarse	seed, softwood cuttings	30'	4	specimen	Does well in moist, shady site.
Cornus florida (Flowering dogwood)	rounded	white mid-May	green	red	medium	seed	35'	4	specimen	Does well in shade or full sun. Horizontal branching.
Cornus kousa (Chinese dogwood)	rounded	late May	green	red	medium	seed	30'	5	accent, specimen	Bright red fruit. Excellent flower quality.
Crataegus crus-galli (Cockspur hawthorne)	rounded	small white	green	scarlet	medium	seed	35'	4	border, specimen	Bright red fruit in fall. Needs sandy soil.
Crataegus oxycantha pauli (Paul's scarlet hawthorne)	dense rounded	double scarlet late May	green	—	medium	seed	35'	4	border	Thorny.
Crataegus phaenopyrum	dense rounded	white mid-June	green	scarlet	medium	seed	30'	4	specimen, accent	Bright red fruit in fall—good for wildlife.
Magnolia soulangeana (Saucer magnolia)	rounded open	white purple May	green	—	coarse open	seed, cuttings, grafting, layering	25'	5	specimen, border	—
Magnolia stellata (Star magnolia)	rounded open	double white to mid-April	green	yellow	coarse open	seeds, cuttings, grafting, layering	20'	5	specimen	Best color appears in direct sun.

FIGURE 29-11 Common deciduous trees

SMALL DECIDUOUS TREES (UP TO 35 FEET IN HEIGHT) (continued)

Name	Form	Color			Texture	How to Propagate	Height	Hardiness	Ornamental Use	Other Remarks
		Flowering	Summer	Fall						
Malus floribunda (Japanese flowering crabapple)	dense	white early May	green	—	fine	T-budded, grafted	30'	4	specimen	Flowers yearly.
Prunus americana 'thundercloud' (Thundercloud plum)	rounded	white May	red	dark purple	medium	T-budded	20'	3	specimen	Has dark purplish fruit.
Prunus serrulata (Oriental cherry)	rounded	white to pink mid-May	green	—	coarse	T-budded	25'	5	specimen	Many varieties are available.
Prunus subhirtella pendula (Weeping cherry)	weeping	double pink early April	green	—	medium	grafted	25'	5	specimen	Excellent as accent plant.

MEDIUM DECIDUOUS TREES (36 TO 75 FEET IN HEIGHT)

Name	Form	Color			Texture	How to Propagate	Height	Hardiness	Ornamental Use	Other Remarks
		Flowering	Summer	Fall						
Aesculus carnea (Red horsechestnut)	pyramidal	pink to red spikes mid-May	green	brown	coarse	seed	75'	3	specimen	—
Acer platanoides 'Crimson King' (Crimson King maple)	rounded dense	—	red-purple	red	medium	budding	50'	4	accent	Brilliant foliage.
Acer platanoides 'Emerald Queen' (Emerald Queen maple)	upright oval	—	dark green	yellow	medium	budding	40'	4	accent	Needs full sun.
Acer rubrum 'October Glory' (October Glory maple)	oval	—	dark green	red	medium	budding	50'	3	accent	Excellent fall color.
Acer rubrum 'Red Sunset' (Red Sunset maple)	rounded	—	dark green	red	medium	budding	45'	3	accent	Excellent fall color.
Betula pendula "Laciniata" (Cutleaf weeping birch)	oval	—	bright green	yellow	medium fine	budding	50'	2	accent	Weeping branches.

FIGURE 29-11 Common deciduous trees (continued)

MEDIUM DECIDUOUS TREES (36 TO 75 FEET IN HEIGHT) (continued)

Name	Form	Flowering	Color Summer	Color Fall	Texture	How to Propagate	Height	Hardiness	Ornamental Use	Other Remarks
Betula pendula (Weeping European birch)	pyramidal	—	green	yellow	fine	seed	60'	2	specimen	White bark in winter. Bronze birch borer can be a problem.
Carya illinoinensis (Pecan)	inverse pyramidal	—	green	yellow	medium	seed and grafting	160'	6	specimen	Excellent nut tree.
Cercidphyllum japonicum (Katsura tree)	broad	—	green	yellow/ orange	medium/ fine	seed	60'	4	accent	Excellent leaf and bark habit.
Fagus sylvatica "Atropurpurea" (Copper beech)	conical	—	dark purple	copper	medium/ fine	budding	70'	5	accent	Excellent foliage.
Franklinia alatamala (Franklinia tree)	rounded	white September	green	red	medium	—	36'	6	border, specimen	Late fall flower.
Fraxinus pennsylvanica lanceolata "Marshall" (Green ash)	rounded	—	green	yellow	coarse	seed	60'	2	streets, border	Attractive form.
Gleditisa triacanthos "shademaster" (Shademaster honeylocust)	rounded	—	light green	yellow	fine	budding	50'	4	massing	Open branching.
Magnolia virginiana (Sweet bay)	rounded	white fragrant late May	gren	green	coarse	seeds, cuttings, grafting, layering	60'	5	specimen	Excellent flower quality.
Quercus virginiana (Live oak)	rounded wide-spreading	—	green	green	medium	seed	60'	7	specimen, street	Evergreen in southern United States.
Oxydendrum arboreum (Sourwood or lily-of-the-valley tree)	rounded	white mid-July	green	red	medium	seed	75'	4	specimen	Late summer flower; good fall color.
Sorbus aucuparia (European mountain ash)	pyramidal	clusters of white	green	reddish	medium	graft	60'	5	specimen	Bright red berries appear in a cluster in fall.

FIGURE 29-11 Common deciduous trees (continued)

LARGE DECIDUOUS TREES (OVER 75 FEET IN HEIGHT)

Name	Form	Flowering	Color		Texture	How to Propagate	Height	Hardiness	Ornamental Use	Other Remarks
			Summer	Fall						
Acer platanoides (Norway maple)	rounded	small yellow before leaves	green	yellow	coarse	seed	100'	3	shade	Provides dense shade. Fast grower.
Acer rubrum (Red maple)	rounded	small red early April	green	brilliant red	medium	seed	120'	3	border, specimen	Several new varieties are available.
Acer saccharinum (Silver maple)	rounded	small green	green	green to yellow	medium	seed	120'	3	wildlife area	Fast growing; weak wooded.
Acer saccharum (Sugar maple)	rounded	green	green	yellow	coarse	seed	125'	3	specimen, shade	Used to produce maple syrup.
Fagus grandifolia (American beech)	pyramidal	—	green	bronze	medium	seed	90'	3	specimen	Wildlife food.
Fraxinus americana (White ash)	rounded	—	green	yellow	coarse	seed	120'	3	border	Attractive form.
Ginkgo biloba (Ginkgo)	pyramidal	inconspicuous	green	yellow	coarse	grafted	120'	4	street, specimen, border	Male form has fragrant fruit. Fan-shaped leaf.
Gleditisia triacanthos "inermis" (Thornless honeylocust)	rounded open	greenish pea-like June	green	yellow	fine	grafted	120'	4	lawn, specimen, shade	No thorns.
Juglans nigra (Black walnut)	rounded open	catkins	green	yellow	coarse	seed	150'	4	nut producing	Wood is used to make furniture.
Liquidambar styraciflua (Sweet gum)	pyramidal	greenish clusters	green	red to scarlet	medium	seed, leafy soft softwood cutting	125'	4	lawn, specimen, street, shade	Difficult to transplant. Good form. Corky ridges on stem.
Liriodendron tulipifera (Tulip tree)	rounded	greenish yellow tulip-shaped mid-June	green	yellow	coarse	seed	180'	4	specimen	Timber tree and fall color.

FIGURE 29-11 Common deciduous trees (continued)

LARGE DECIDUOUS TREES (OVER 75 FEET IN HEIGHT) (continued)

Name	Form	Flowering	Color		Texture	How to Propagate	Height	Hardiness	Ornamental Use	Other Remarks
			Summer	Fall						
Magnolia acuminata (Cucumber tree)	pyramidal	greenish yellow early June	green	—	coarse	layering, seed, cuttings, grafting	90'	4	specimen, border	Red to pink cucumber-shaped fruits.
Nyssa sylvatica (Black gum)	rounded	inconspicuous May-June	green	red	medium	seed	90'	4	specimen	Small blue berries appear in midsummer.
Platanus occidentalis (Sycamore)	pyramidal	yellow-green	green	yellow-green	coarse	seed	120'	4	specimen	Natives tend to defoliate.
Populus nigra italica (Lombardy poplar)	narrow tall	—	green	—	medium	seed, hardwood cuttings	90'	2	quick temporary screen	Short-lived tree.
Quercus alba (White oak)	rounded	—	green	purple red	medium	seed	150'	4	specimen	Excellent lumber tree. Slow grower.
Quercus borealis (Red oak)	rounded	—	green	red	coarse	seed	76'	4	specimen, street	Transplants easily.
Quercus coccinea (Scarlet oak)	rounded	—	green	bright scarlet	coarse	seed	100'	4	specimen, street	Difficult to transplant.
Quercus palustris (Pin oak)	pyramidal	—	green	scarlet	coarse	seed	100'	4	specimen	Very graceful. Excellent street tree.
Tilia americana (American linden)	rounded open	white June	green	—	medium	layering, seed, mound	100'	2	border	Wood is valuable for lumber.
Tilia cordata (Little leaf linden)	rounded open	white to yellow mid-July	green	yellow	medium	layering, seed, mound	90'	3	shade	Excellent street tree.
Ulmus americana (American elm)	vase	yellow	green	yellow brown	medium	seed, softwood cuttings	120'	2	street	Possible problem with Dutch elm disease.

FIGURE 29-11 Common deciduous trees (continued)

◆ **Form** is the shape of the tree. This is important to consider when choosing a tree for a particular location. For example, a tall, columnar tree is used to fill a particular landscaping need and fits in a smaller space. A broad, spreading tree requires more room and has a different appearance and use in the landscape.

◆ **Flowering and color.** If and when a tree flowers and the color of the blossoms are important factors in tree choice. Some trees have very large, attractive flowers while others have small, unnoticeable ones. Some trees vary in their leaf color from one season to the other, while others do not change at all. In some cases, a tree may be selected for its fall color.

When landscaping with deciduous trees it is important to consider foliage color display. Fall colors will change with the macroclimate of the area. Soil type, rainfall, and climate factors (e.g., temperature, sunlight value) are important, too. The quality of fall color will vary from year to year, depending on the season. In order for the trees to have attractive fall color, the autumn season must have warm, sunny days and cool nights. Another factor in choosing deciduous trees is the large selection of new cultivars on the market that will create an array of colors. The landscape architect must keep abreast of these new cultivars in order to recommend them for the client.

◆ **Texture** deals with the size of the leaves. Trees with large leaves appear to have a coarse texture while small leaves give an appearance of fine texture.

◆ **How to propagate** describes the way in which the tree is reproduced.

◆ **Height** is the average height to which that particular tree grows.

◆ **Hardiness** is the ability of the tree to live in a particular climate zone. This is influenced by the temperature, rainfall, and the soil of that particular area. Some trees can tolerate lower temperatures than others. (In the chart, hardiness refers to U.S. Hardiness Zones.) Hardiness zones give the average low temperature range for a particular area of the country.

◆ **Ornamental use** describes common uses of the tree in the landscape. Specimen trees are grown solely for special characteristics they possess which create interest in a landscape. Street trees are those trees which are able to withstand the special conditions occurring along roadways and streets.

Figures 29-12 through 29-15 illustrate four deciduous trees commonly used in landscapes. Examine the trees and identify their physical characteristics, such as shape and density.

PLANTING DECIDUOUS TREES

WAYS TREES CAN BE PURCHASED

Deciduous trees may be purchased in three different forms: bare root (BR), balled and burlapped (B&B), or containerized (C), figure 29-16.

BARE ROOT　Trees in this form should be purchased from a reliable nursery. The deciduous trees are dug in the nursery when the tree is dormant. No soil is left with the roots, hence the name bare root. Many nurseries purchase trees in this form

FIGURE 29-12　Flowering dogwood (Cornus florida)

FIGURE 29-13 Red maple "October Glory" (Acer rubrum) (Photo courtesy of Ferrell M. Bridwell)

FIGURE 29-14 Weeping Japanese maple (Acer palmatum)

for transplanting in the late fall or early spring (lining-out stock). It is best to work with only small bare root trees, since they are easier to handle. Bare root stock is usually less expensive. The cost of shipping large quantities of these trees is smaller as compared with balled and burlapped and container-grown stock.

FIGURE 29-15 Saucer magnolia (Magnolia soulangeana) (Photo courtesy of Ferrell M. Bridwell)

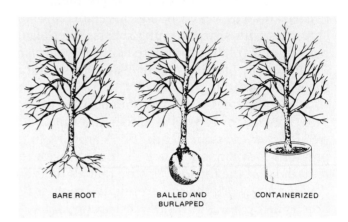

FIGURE 29-16 Ways in which deciduous trees are purchased

BALLED AND BURLAPPED Trees in this form have a ball of soil around the roots. Almost all ornamentals can be transplanted in this form. The soil is held around the roots by burlap which is secured by twine and nails. Deciduous trees moved with a root ball larger than 20 inches should be drum laced to provide additional support for the root ball.

CONTAINER STOCK Trees in this form are planted in a container such as a basket, or plastic or metal can. Many trees are grown today in this fashion because they can be transplanted at any time of the year without disturbing the root system and the grower has an increase in production per acre. Container stock becomes root-bound in the container if allowed to grow there too long.

PLANTING BARE ROOT TREES

Bare root trees are planted only when they are dormant. Since the soil is removed from the roots, the tree cannot replenish moisture rapidly enough to support leaves. Only after the roots have again made good soil contact can they supply the amount of water needed for a tree. The dormant tree requires adequate moisture to survive transplanting.

The bare roots should never be allowed to dry out before planting. Keep them covered with damp peat moss, sawdust, burlap, or some material so that the roots remain moist and away from the wind and sun. Some nurseries use a moisture-retention agent (e.g., Liquigel) on the root system to prevent dry-out of the roots.

The hole should be large enough to receive the roots without crowding them. The roots should be spread in a natural manner without twisting them in the hole. Long roots should be cut off and not twisted around the hole. Be sure to remove any broken roots with sharp pruning shears. After planting, one-third of the top growth of the original plant is removed by pruning, figure 29-17. This reduces the water loss from the plant.

The hole should be large enough so that at least 4 to 5 inches of topsoil is under the roots. The tree is set at the same depth it was growing in its previous location, and the hole filled in with the remainder of the topsoil. Be sure that all roots are covered and work the soil around the roots as the hole is being filled. Firm the soil around the root system to hold the tree in correct growth position.

WATERING Use a garden hose (with the water running slowly) to settle the soil around the roots and to eliminate air pockets, which can cause the roots to dry unnecessarily. Water until the soil is saturated with water.

To ensure a good supply of water, form a ring (berm) around the base of the tree about 3 inches high. Fill the depression (saucer) with water, figure 29-18. The water gradually soaks in and settles the soil around the roots. The berm also catches rainwater and keeps it from running off. Drive a stake on two sides of the tree and anchor the tree to the stake with a piece of rubber hose and wire. Mulch with 2 to 3 inches of bark or some other coarse material that will not blow away. Be sure to put the mulch about 1 inch away from the trunk of the tree. Never allow it to touch the tree; this discourages pests.

PLANTING BALLED AND BURLAPPED TREES

Select the area in which the tree is to be planted. Dig a hole one and one-half to two times larger across than the ball of the tree to be planted. The hole should be dug so when the plant is set in the hole, the top of the ball sets slightly above the ground level. This allows for sinking of the tree rootball caused by settling of the loose soil in the bottom of the prepared hole. Set the tree in the hole and loosen the burlap around the root ball, figure 29-19. Fill the area around the root ball with soil. Follow the same procedure for watering as for bare root stock.

PLANTING CONTAINERIZED TREES

Containerized trees are planted in the same way as balled and burlapped trees except for the following differences.

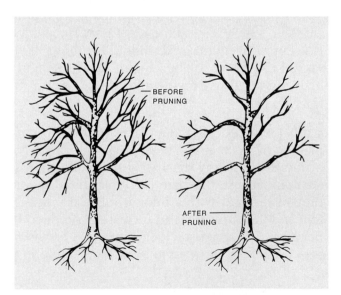

FIGURE 29-17 Pruning the bare root tree after planting

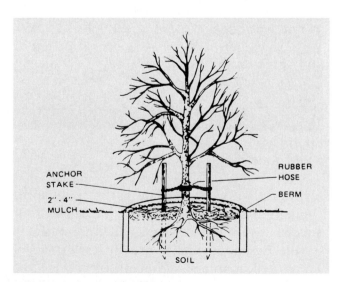

FIGURE 29-18 Formation of berm around base of tree

FIGURE 29-19 Transplanting a balled and burlapped tree. The tree after it has been placed in its new location, a prepared soil mixture of peat and soil is distributed around the base of the tree.

1. The container is removed from the tree roots. It should be cut or otherwise removed carefully so that the soil is not broken from the tree.

2. After the container is removed, roots which may be growing around the outside of the root ball in a circular fashion are straightened out before planting. If the roots cannot be straightened, they are cut off at the edge of the root ball (called butterflying). This will stop the circular root growth which tends to girdle root systems and restrict growth. In severe cases, it can kill trees.

WRAPPING

It is necessary to wrap the trunk of a transplanted tree to prevent sun scald, reduce water loss, and reduce chances of pest infestation. The wrapping material is burlap or special tree wrap paper. On large trees, it is best to wrap from the ground line up to and including several lower branches. Each turn of wrap spiral should overlap about one-half the width of the previous wrap.

STAKING/GUYING

Staking or guying the tree prevents the wind from swaying it and loosening the roots. It also helps to keep the tree standing straight.

When staking, use a piece of rubber garden hose and wrap it around the tree trunk above the first set of branches. Then slip a piece of 12- to 14-gauge soft steel wire through the rubber garden hose. Twist the wire around and fasten it to each of the stakes, figure 29-20. On large trees,

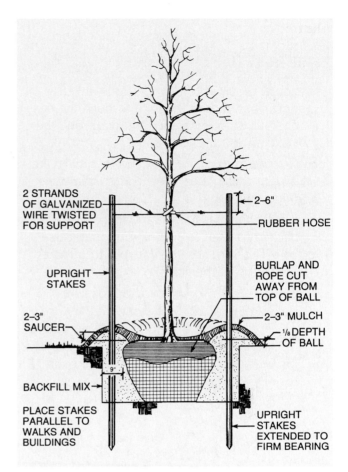

FIGURE 29-20 Bracing the tree by staking. (Courtesy LCA, Rockville, Maryland)

the stakes should remain about two years. It is best to remove the plant ties after this time since the hose may girdle the tree. As the tree grows older, it grows laterally and if the ties are not removed, the tree will grow over the hose, thus weakening the tree. In a bad storm, the top of the tree could break completely off at the weakened point.

PROCEDURE

TREES BRACED BY STAKING

1. Staking shall be completed within forty-eight hours of planting the tree. Place stakes parallel to walks and buildings.

2. Space stakes evenly and vertically on the outside of the tree ball and drive firmly into the ground. Stakes shall be driven at an angle and drawn to vertical. **Note:** Never drive a stake through the tree ball, as it will damage the tree's root system.

3. Cut pieces of hose long enough to loop around the trunk of the tree.

4. Place the hose around the trunk at the height required to provide optimum support. Then thread the wire through the hose and pull both ends horizontally beyond the stake by about two feet.

5. Cut the ends of the wire and then twist the wire at the rubber hose to keep it in place.

6. Wind both ends of the wire together around the stake twice and then twist wire back onto itself to secure. Cut off excess wire. The wire shall be within 2 to 6 inches of the top of the stake.

7. Steps 1–6 are to be followed for each stake, keeping the tree straight at all times. There should be a 1- to 3-inch sway to the tree (the wires should not be pulled tight) for best establishment. (Procedure courtesy LCA, Rockville, Maryland)

Procedure

TREES BRACED BY GUYING

1. Choose the correct size and number of stakes and size of hose and wire according to Tree Support Schedule. Guying shall be completed within forty-eight hours of planting the tree. See figure 29-21.

2. Cut lengths of staking hose to extend 2 inches past tree trunk when wrapped around.

3. Space stakes evenly on outside of the tree saucer and drive each firmly into the ground. Stakes shall be driven at a 30-degree angle with the point of the stake toward the tree until 4 or 5 inches are left showing.

4. Place the hose around the trunk just above the lowest stout branch.

5. Thread the wire through the hose and past the stake, allowing approximately two feet of each of the two ends beyond the stake before cutting the wire.

6. Twist wire at rubber hose to keep it in place.

7. Pull the wire down and wind both ends around the stake twice. Twist wire back onto

itself to secure it before cutting off the excess.

8. Steps 1–7 are to be followed for each stake, keeping the tree straight at all times. There should be a 1- to 3-inch sway in the tree (the wires should not be pulled tight) for best establishment.

9. Flag the guy wires with surveyor's flagging tape or approved equal. (Procedure courtesy LCA, Rockville, Maryland)

FERTILIZING

When planting trees, it is recommended that a well-prepared soil be used for backfilling. A slow-release fertilizer such as osmocote or Mag-Amp may be mixed in the soil at planting time according to manufacturer's recommendations. Recent research shows that this may not be necessary on some tree species.

Another way to fertilize is to apply 2 pounds of a 5-10-5 fertilizer for each 1 inch of diameter of the tree trunk at a height 3 feet above ground level. Use a soil auger, wrenching bar, digging iron, or post-hole digger to form holes that measure 15 to

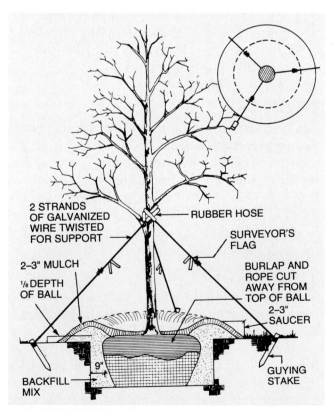

2 STRANDS OF GALVANIZED WIRE TWISTED FOR SUPPORT

RUBBER HOSE

SURVEYOR'S FLAG

2–3" MULCH

⅛ DEPTH OF BALL

BURLAP AND ROPE CUT AWAY FROM TOP OF BALL

2–3" SAUCER

9"

BACKFILL MIX

GUYING STAKE

FIGURE 29-21 Bracing the tree by guying. (Courtesy LCA, Rockville, Maryland)

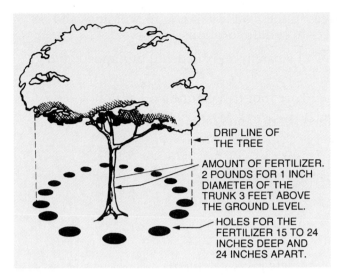

FIGURE 29-22 Fertilizing the tree. (Courtesy USDA)

24 inches deep and about 2 feet apart around the drip line of the tree, figure 29-22. The fertilizer is distributed in equal amounts in the holes around the drip line of the tree. After the fertilizer has been applied, finish filling with soil the holes in which the fertilizer was placed.

MULCHING

Mulching is needed for moisture retention, weed control, supplying organic matter, and moderating the temperature for the roots. Mulching also improves the aesthetics of the area. Trees in the landscape may be mulched with very attractive materials such as hardwood bark, pine bark, coconut husks, and tanbark.

Mulch applied around deciduous trees should be spread evenly and neatly, 3 to 4 inches deep. When mulching trees, it is important to keep the mulch away from the tree trunk. This can be done by running your hand down beside the tree trunk and pulling the mulch away from the trunk. This will help to prevent pests from entering the tree. The size of the mulched area under deciduous trees will vary according to personal preference. A rule of thumb to follow is apply mulch two-thirds of

FIGURE 29-23 It is important to water after transplanting, especially during the first growing season.

the distance from the tree trunk to the drip line of the tree.

AFTERCARE

The tree may require watering during dry conditions, especially during the first year after planting, figure 29-23. Never add less than 1 inch of water at any one time. This is a sufficient amount for one week with no rain. Repeat every week from April through October if there is not sufficient natural rainfall to provide 1 inch of rain per week.

Student Activities

1) Visit a local park or other landscaped area and practice tree identification. Identify at least ten local trees.

2) Draw a simple landscape plan. Place specific trees in the plan in relation to buildings and other structures. List reasons for placing the trees in certain locations.

3) Demonstrate the techniques used in the planting of a bare root, balled and burlapped, and container-grown tree.

4) Visit a local nursery in the spring to observe their bare root tree planting operation.

5) Plant a landscape plan that uses bagged and balled (B&B) and container nursing stock.

Self-Evaluation

Select the best answer from the choices offered to complete each statement.

1) Deciduous trees may be purchased with their roots prepared in three different ways. These are

    a) bare root, root pruned, and containerized.
    b) balled, burlapped, and bare root.
    c) bare root, balled and burlapped, and containerized.
    d) root pruned, container grown, and root ball.

2) The form of a tree is a characteristic which describes

    a) the shape of the tree.
    b) the way the roots are treated for transplanting.
    c) whether it is evergreen or deciduous.
    d) how it is pruned.

3) Bare root deciduous trees are planted only when they

    a) can be dug into the ground.   c) are beginning to leaf.
    b) are dormant.   d) are in full leaf.

4) Bare root trees are dug when the trees are dormant and are shipped

    a) early in the spring.   c) for fall planting.
    b) with no soil on the roots.   d) with a layer of soil on the roots.

5) Balled and burlapped trees are shipped

    a) with a ball of soil around the roots wrapped in burlap.
    b) with bare roots.
    c) in early spring for immediate planting.
    d) in early fall for immediate planting.

6) Container-grown trees become root-bound if

    a) the roots are not pruned each year.
    b) they are left in the same container too long.
    c) planted in the fall of the year.
    d) all of these

7) In planting bare root trees, all broken and badly damaged roots are

    a) patched with tree wound dressing.
    b) pulled aboveground to heal.
    c) pruned with pruning shears.
    d) given extra fertilizer.

8) About one-third of the top is pruned off a newly planted tree to

    a) shape the new tree.
    b) thin the tree and let in more light.
    c) select and keep only the best branches.
    d) reduce water loss from the plant.

9) A bare root tree should be planted as deeply as

    a) necessary to keep it from blowing over.
    b) it was growing before being transplanted.
    c) possible for good anchorage.
    d) the level of the topsoil.

10) Trees must be watered well when transplanting to settle the soil and

    a) keep the roots moist so that they can grow.
    b) prevent the roots from drying out.
    c) close any air pockets around the roots.
    d) all of the above

11) A berm is built around the newly planted tree to

    a) help make watering the tree easier and to hold rainwater.
    b) keep rainwater from reaching the tree.
    c) prevent erosion.
    d) hold mulch in place.

12) When planting balled and burlapped trees, a hole is dug

    a) to fit the root ball.
    b) exactly as deep as the root ball.
    c) one and one-half to two times larger than the width of the root ball.
    d) with slanted sides.

13) Trees are staked and tied when planted to

    a) keep animals from running over them.
    b) mark the location so that they can be recognized.
    c) keep them from blowing over.
    d) hold them still and prevent movement which can cause root damage.

14) When applying a fertilizer such as a 5-10-5, a general rule of thumb is to apply about _____ pounds for each inch of trunk diameter.

    a) 4   c) 6
    b) 5   d) 2

15) Newly planted trees are generally fertilized by punching holes in the ground and placing the fertilizer in the hole. These holes are placed

    a) in a circle 3 feet from the tree with the holes 24 inches apart.
    b) at the drip line circling the tree with the holes 24 inches apart.
    c) at the drip line circling the tree with the holes 10 inches apart.
    d) close to the tree trunk where the feeding roots are located.

16) Watering newly planted trees is not necessary if natural rain supplies at least _____ of rain per week.

    a) ½ inch   c) 1 inch
    b) 2 inches   d) ¼ inch

17) If watering the tree is necessary, no less than _____ of water should be applied at any one time.

    a) 1 inch   c) 2 inches
    b) ½ inch   d) ¼ inch

Deciduous Shrubs

OBJECTIVE

To select, use, and care for deciduous shrubs.

Competencies to Be Developed

After studying this unit, you should be able to

- ◆ identify at least five deciduous shrubs.
- ◆ list at least two uses of deciduous shrubs in the landscape.
- ◆ explain orally or in writing the three ways deciduous shrubs are purchased.
- ◆ outline the planting procedure for deciduous shrubs.

MATERIALS LIST

- ✔ one bare root deciduous shrub
- ✔ one balled and burlapped deciduous shrub
- ✔ one container-grown deciduous shrub
- ✔ hardiness map
- ✔ round-point shovel
- ✔ nursery spade
- ✔ mulch (hardwood pine bark)
- ✔ 10-6-4 and 5-10-10 fertilizers
- ✔ hand pruners

USES IN THE LANDSCAPE

Deciduous shrubs are used widely in the landscape to form borders, screens, background plantings, and foundation plantings. They offer an especially attractive feature in that they change in color as seasons change. Deciduous shrubs are selected for the various effects that are created by their flower color, form, fruit they bear, color and shape of the stem, and color and height of the foliage. Through careful selection, an interesting and attractive landscape design can de developed with deciduous shrubs, either alone or in combination with other plant materials.

The following are some important factors to consider when selecting deciduous shrubs for the landscape.

PERIOD OF INTEREST

A *period of interest* is that time of year in which plant materials create the most interest in the landscape. For example, flower color or fragrance of the flower might be a notable feature determining the period of interest. One example is the mock orange (*Philadelphus*). When in bloom, the fragrance of the mock orange is often apparent before the plant is seen. The period of interest for many shrubs is during the fall months, when they bear colorful fruit. An example is the cranberry bush (*Viburnum opulus*). It produces attractive red fruit in the fall that also provides food for wildlife in the winter.

FOLIAGE

The foliage of the plant has a definitive influence on its selection for a particular role in the landscape. For example, some shrubs, such as the oakleaf hydrangea (*Hydrangea quercifolia*), have large, coarse foliage. Others, such as white baby's breath spirea (*Spiraea thunbergi*), are very fine in texture. The two textures create a contrast in a landscape planting.

HEIGHT

When selecting deciduous shrubs, it is important to consider the height of the mature plant. For example, tall growing shrubs should not be planted in foundation plantings where they may block windows.

HARDINESS

The hardiness rating of a plant is based on the lowest minimum temperature that the plant can endure without noticeable damage to flower buds or wood. The adaptability of a plant to an area is further influenced by the amount of rainfall, wind velocity, exposure to sun, and soil type. Figure 30-1 gives essential information for choosing shrubs to fill a particular role in the landscape.

PURCHASING PLANTS

After particular shrubs are identified for use in the landscape, a source must be located. It is always best to purchase plants from a reliable nursery. Deciduous shrubs may be purchased as bare root, balled and burlapped, or container grown.

BARE ROOT

A bare root shrub is transported and transplanted with no soil around the roots. Only small shrubs should be purchased bare root. After plants reach three of four years of age or older, the survival rate is better if the plant is moved and planted with a root ball. Bare root plants are moved only when they are dormant. This allows enough time for the roots to establish themselves prior to leafing out (followed by a high demand for moisture).

BALLED AND BURLAPPED

Balled and burlapped plants are purchased with a ball of soil around the roots. The root ball is wrapped tightly in burlap to hold the soil firmly around the roots and to prevent the soil from breaking apart and separating from the roots. The ball should be large enough to include most of the plant's root system, or as large a ball as can be lifted and moved. Balled and burlapped plants can be transplanted any time during the year.

CONTAINER GROWN

Container-grown shrubs are cultivated in various containers, such as pots, baskets, and tubs. The root system is damaged less during transplanting, so these plants can be transplanted at any time during the year. The roots of root-bound plants should be straightened after removal from the container. If twisted or curled roots cannot be straightened, they should be cut off at the edge of the ball to encourage the development of new roots as the plant continues to grow.

PLANTING TIME

The best time to plant bare root deciduous shrubs is when they are in their dormant stage. Balled and burlapped and containerized plants, on the other hand, may be transplanted any time during the growing season. This usually occurs in fall or early spring. In areas with mild winters, however, deciduous shrubs may be planted year-round. This includes areas south of Chesapeake Bay through the coastal plains of Texas and the coasts

DECIDUOUS SHRUBS LESS THAN 3 FEET TALL

Name	Color and Date of Flowering	Period of Interest	Hardiness	Exposure	Soil pH	Propagation	Remarks
Chaenomeles japonica (Japanese quince)	orange to red	April	4	sun	6.0 to 7.0	cuttings	Border shrub. Grows in partial shade.
Cotoneaster horizontalis (Cotoneaster)	small, pink mid-June	fall	5	southern slopes	6.0 to 7.0	cuttings	Good for rock gardens. Low spreading branches.
Daphne genkwa (Daphne, blue daphne)	blue April	April	5	sun	6.5 to 7.5	cuttings	Difficult to transplant. Slow growing.
Deutzia gracilis (Slender deutzia)	white May	May	4	sun	6.0 to 7.0	cuttings	Showy, white.
Forsythia viridissima (Dwarf forsythia)	yellow April	April	5	sun	6.0 to 7.0	layering, cuttings	Rock garden plant.
Hypericum sp (St. Johnswort)	yellow summer	summer	5	sun	5.5 to 6.5	cuttings	Many forms; yellow flowers throughout summer; dense.
Potentilla fruticosa (Cinquefoil)	yellow single rose mid-May	summer	2	sun	6.0 to 7.0	stem cuttings	Fine textured. Good border plant.
Spiraea bumalda "Anthony waterei" (Anthony Waterer spirea)	pink	June	6	sun	well drained 6.0 to 7.0	cuttings	Blooms throughout summer. Compact and bushy.

DECIDUOUS SHRUBS 3 TO 6 FEET TALL

Name	Color and Date of Flowering	Period of Interest	Hardiness	Exposure	Soil pH	Propagation	Remarks
Abelia grandiflora (Abelia)	pink summer	summer	6	sun	6.0 to 8.0	cuttings	Flowers all summer. Good border or foundation plant.
Berberis thunbergi (Green leaf Japanese barberry)	yellow	fall, winter	6	dry sun	6.0 to 7.5	cuttings	Small hedge. Grows well under all conditions. Green foliage in summer; red foliage in fall.
Berberis thunbergi Atropurpurea (Red leaf Japanese barberry)	yellow	fall, winter	6	dry sun	6.0 to 7.5	cuttings	Red all summer. Good specimen or hedge.
Buddleia davidii (Butterfly bush)	white, pink, and blue spikes	summer	6	sun	6.0 to 7.0	cuttings	Cut back to ground in spring. Fast growing, arching plant.

FIGURE 30-1 Common deciduous shrubs

DECIDUOUS SHRUBS 3 TO 6 FEET TALL (continued)

Name	Color and Date of Flowering	Period of Interest	Hardiness	Exposure	Soil pH	Propagation	Remarks
Euonymus alatus compactus (Dwarfwinged euonymus)	greenish May	fall	4	sun	5.5 to 7.0	cuttings	Brilliant red leaf in fall. Dense shrub; hedge plant.
Fothergilla monticola (Fothergilla)	white	fall, May	5	sun	5.0 to 7.0	cuttings	
Hydrangea quercifolia (Oakleaf hydrangea)	white mid-July	summer, fall	6	sun	moist soil 6.0 to 7.0	division, cuttings	Naturalistic effect when used to edge wooded lots.
Kerria japonica (Japonica)	yellow	winter, May		sun	6.0 to 7.0	division, cuttings	Green twigs in winter. Large perennial borders. Sometimes called *Mt. Vernon shrub.*
Prunus glandulosa (Flowering almond)	double pink	early May	4	sun	6.0 to 7.0	cuttings	Double blossoms.
Rhododendron calendulaceum (Flame azalea)	yellow to red	late May	5	sun	4.5 to 5.5	layering, cuttings	Acid soil. Open growth habit.
Spiraea thunbergi (Baby's breath spirea)	white	April	6	sun	6.5 to 7.0	cuttings	Informal hedge. Fine textured.
Spiraea prunifolia (Bridalwreath spirea)	white	April	5	sun	6.0 to 7.0	cuttings	Arching branches; background plant.
Spiraea vanhouttei (Vanhoutte spirea)	cascade of white, flat flowers	May	5	sun, shade	6.0 to 7.0	cuttings	Good hedge. Should not be sheared; best to prune to base.
Weigela sp (Weigela)	pink to red	May	5	sun	6.0 to 7.0	cuttings	Abundance of flowers. Used in boundary plantings.

DECIDUOUS SHRUBS 6 TO 10 FEET TALL

Name	Color and Date of Flowering	Period of Interest	Hardiness	Exposure	Soil pH	Propagation	Remarks
Aesculus parviflora (Bottlebrush buckeye)	white spikes	July	5	semishade	6.0 to 7.0	cuttings	Green mass of foliage in summer.
Elaeagnus umbellata (Autumn olive)		fall	3	sun	5.0 to 6.0	cuttings	Provides wildlife food. Pink-red fruit; silvery foliage. Used in border plantings.

FIGURE 30-1 Common deciduous shrubs (continued)

DECIDUOUS SHRUBS 6 TO 10 FEET TALL (continued)

Name	Color and Date of Flowering	Period of Interest	Hardiness	Exposure	Soil pH	Propagation	Remarks
Forsythia intermedia spectabilis (Border forsythia)	yellow	spring	5	sun	6.0 to 7.5	cuttings	Good in borders. Easy to transplant. Fast growing.
Ilex verticillata (Winterberry)	inconspicuous	winter	4	sun	5.5 to 6.5		Showy red fruit in winter. Grows well in swamps.
Kolkwitzia amabilis (Beauty bush)	pink	June	5	sun	6.0 to 7.5	cuttings	Boundary or specimen plant. Requires sun.
Ligustrum ovalifolium (California privet)	white	May	4	shade, sun	6.0 to 7.0	cuttings	Good hedge material.
Lonicera tatarica (Tatarian honeysuckle)	white to yellow	May	4	sun	6.5 to 8.0	cuttings	Red berries in June or July.
Philadelphus sp (Mock orange)	white	June	6	sun	6.0 to 8.0 well drained	cuttings	Many varieties are fragrant.
Syringa chinensis (Chinese lilac)	lavender	April	5	sun	6.0 to 7.5 moderately drained	cuttings	Varieties available with white or red flowers.
Syringa vulgaris (Common lilac)	purple	April May	3	sun	6.0 to 7.5	cuttings	Powdery mildew is sometimes a problem. Suckers are pruned back to stimulate blooming.
Viburnum dentatum (Arrowwood)	flat white clusters	late May	3	wet to damp woods	6.0 to 7.5	cuttings	Excellent wildlife food source and border plant.
Viburnum macrocephalum (Chinese snowball)	large white flower heads	May	6	sun	6.0 to 7.5 rich soil	cuttings	Largest flower heads of all—snowballs.
Viburnum opulus (Cranberry bush)	white	May, fall	3	sun	6.0 to 7.5	cuttings	Park plantings. Red fruit. Purplish brown fall color.
Viburnum plicatum tomentosum (Double file viburnum)	white	May, fall	5	sun	6.5 to 7.5	cuttings	Flowers and fruit develop on upper side of branch.

FIGURE 30-1 Common deciduous shrubs (continued)

of Oregon and central California. Where winters are severe, spring is the best time to plant.

PLANTING SITE

Most shrubs prefer a well-drained soil. A good garden loam is the best soil type for most shrubs. Soil may be improved upon by the addition of sand for drainage or organic matter to help hold moisture in dry soils. Soil pH preference and the effects of exposure to sun and wind vary for each shrub type. The selection of the proper site is important since deciduous shrubs are long-lived plants. Deciduous shrubs should be fertilized in the spring with a commercial 10-6-4 fertilizer at 2 to 5 pounds per 100 square feet for foliage-type shrubs, and a commercial 5-10-10 fertilizer at 2 to 5 pounds per 100 square feet for flowering-type shrubs.

PROCEDURE

PLANTING

Note: Plant roots should never be allowed to dry prior to planting. Bare root shrubs should be packed in moist peat moss or sawdust. Balled and burlapped and container-grown plants should be kept properly watered until planting time.

1. Dig a hole large enough to allow roots to spread in a normal manner without cramping or twisting. Dig deeply so that 2 to 4 inches of topsoil can be placed under the roots. The sides of the hole should be straight up and down and the bottom flat. This flat surface encourages straight growth of the root system.

2. Prune diseased and broken roots.

3. Plant the shrub at the same depth as that at which it grew before.

4. Work the soil in and around the roots and apply 2 to 5 pounds per 100 square feet of 10-6-4 (foliage-type) or 5-10-10 (flowering-type) fertilizer.

5. Build up a slight berm around the outer edge of the hole to hold water.

6. Water well to settle the soil around the roots, eliminate any air pockets, and moisten the soil. Water regularly for the first year any time that the plant receives less than 1 inch of natural rainfall per week.

7. Mulch with chips or bark to a depth of 2 to 4 inches. Apply coarse material more heavily than fine material.

8. Prune the top to remove any broken branches and to help shape the plant.

Holes for balled and burlapped and container-grown plants should be at least 25 percent larger in diameter than the root balls to be planted. After the plant is placed in the hole, the container is cut away and removed or the burlap rolled down to expose the root ball as much as possible without loosening or breaking apart the ball. Soil is worked in around the root ball and the plant is watered and mulched.

CARE OF THE PLANT

DISEASE AND INSECT CONTROL

Some deciduous shrubs require spraying for control of insects or diseases. Consult your local agriculture experiment station or extension service for details.

Most of the diseases that affect deciduous shrubs are caused by microscopic organisms such as bacteria and fungi. These small organisms infect parts of the plant structure such as roots, leaves, or stem.

PRUNING

Pruning is used to thin out old or dead wood, to shape plants, and to control plant size. The following are general rules that apply to pruning shrubs.

1. Shrubs that bloom on wood grown the previous season should be pruned immediately after flowering. (Examples are forsythia and spirea.)

2. Shrubs that bloom on the current year's growth should be pruned in fall or early spring. (One example is the rose.)

3. Some shrubs require annual pruning to thin out old wood or wood that is shaded out and killed. (Examples are hydrangea, privet, spirea.)

4. If shrubs send up shoots or suckers from the roots or at the base of the plant, at least some of them should be removed so that the growth on the plant does not become too thick.

Deciduous shrubs are a long-lived investment in the landscape. Their uses are as many as are the rewards in year-round beauty, figure 30-2.

(A) Oakleaf Hydrangea

(B) Exbury Azalea

(C) Little Princess Spirae

(D) Korean Spice Viburnum

FIGURE 30-2 Four deciduous shrubs used in landscapes

Student Activities

1) Make ten cuttings of selected deciduous shrubs in the area and propagate them. Observe the growth habit of the resulting plants. Practice pruning, fertilizing, and insect and disease control. If possible, transplant the shrubs to school grounds.

2) Design and develop a small rock garden using shrubs discussed in this unit that are common to the local area.

3) Ask a representative from a local nursery to speak to the class on the uses of deciduous shrubs in the landscape.

Self-Evaluation

Select the best answer from the choices offered to complete each statement.

1) When used to describe a plant, the word *deciduous* means that

   a) the plant is perennial.
   b) the plant sheds its leaves each year.
   c) the plant is a shrub.
   d) all of the above

2) The period of interest for a deciduous shrub is

   a) the time of year during which it creates the most interest in the landscape.
   b) the time of year during which it sheds its leaves.
   c) pruning time.
   d) blooming time.

3) Pruning is used to

   a) shape plants.
   b) reduce the size of the top of the plant when transplanting.
   c) thin out dead wood.
   d) all of the above

4) A bare root plant is one that must be planted

   a) in the fall of the year.   c) when it is dormant.
   b) in the spring.   d) after severe pruning of the top.

5) One advantage in purchasing container-grown plants is that

   a) they may be planted during any season of the year.
   b) they are less expensive than other forms.
   c) they are more readily available.
   d) they have better root information.

6) If the roots of a container-grown plant are twisted around in the pot,

   a) they are left undisturbed before planting.
   b) they are straightened out or pruned before planting.
   c) the plant is placed back in the container for growing.
   d) the root ball is broken apart.

7) Deciduous shrubs prefer a soil that is

    a) a well-drained loam.
    b) sandy and moist.
    c) poorly drained.
    d) underlaid with a clay subsoil.

8) The planting hole for deciduous shrubs is formed with straight sides and a flat bottom

    a) to make it more attractive.
    b) to simplify packing soil around the plant roots.
    c) to encourage root growth through the sides of the hole.
    d) because this kind of hole is easier to dig.

9) The planting depth for shrubs is

    a) 2 inches deeper than the depth at which they previously grew.
    b) 4 inches deeper than the depth at which they previously grew.
    c) 2 inches above ground level.
    d) the same depth at which they were growing before.

10) Deciduous shrubs should be fertilized with _____ to _____ pounds per 100 square feet of 10-6-4 foliage-type fertilizer or 5-10-10 flowering-type fertilizer.

    a) 10 to 20
    b) 8 to 12
    c) 1 to 2
    d) 2 to 5

11) Shrubs that bloom on previous season's wood are pruned

    a) immediately after blooming.
    b) in the fall.
    c) in the spring.
    d) in midsummer.

12) Shrubs that bloom on the current season's growth are pruned

    a) immediately after blooming.
    b) in the fall or early spring.
    c) in the spring.
    d) in the fall.

Ground Covers

OBJECTIVE

To select, establish, and maintain ground covers in the landscape.

Competencies to Be Developed

After studying this unit, you should be able to

- ◆ identify the three major types of ground covers.
- ◆ list three uses of ground covers.
- ◆ describe the cultural requirements of ground covers.
- ◆ identify five factors that must be considered when selecting ground covers.

MATERIALS LIST

- ✔ samples of different types of ground covers
- ✔ 5-10-5 fertilizer
- ✔ peat moss
- ✔ pine bark or hardwood mulch

Ground covers are low-growing plants that cover the ground in place of turf. When used functionally, ground covers fill in bare spots in the landscape, help to prevent erosion of soil on steep banks, and fill in shady areas under trees where other plants have difficulty growing.

TYPES OF GROUND COVERS

There are three principal types of ground covers: broadleaf evergreens, deciduous plants, and narrowleaf evergreen plants. Most ground covers are perennial, but a few are annual plants. Figure 31-1 lists common ground covers and planting information for each.

NAME	COLOR & DATE OF FLOWERING	HEIGHT	LIGHT	FOLIAGE	PROPAGATION	HARDINESS	REMARKS
Ajuga reptans (Bugleweed)	blue to purple April or May	4" to 8"	shade or sun	narrow leaves, 4" long	seed and division	4	Gives dense coverage.
Arabia alpina (Rock-cress)	white April to May	4" to 10"	full sun	evergreen	seed and division	3	Spreads by creeping rootstock; useful in small areas.
Asarum caudatum (Wild ginger)	brownish purple June	6" to 8"	shade	heart-shaped leaves with pungent taste	division	4	Grows best in moist soil which is high in organic matter.
Asperula odorata (Sweet woodruff)	white May to June	6" to 8"	shade	fine	division	4	Scented leaves in whorls of six to eight.
Chrysogonum	yellow May to June	4" to 6"	sun	dense	division	4	Bright flowers.
Convallaria majalis (Lily-of-the-valley)	white spikes mid-May to mid-June	8"	shade	large oval leaves	division	2	Spreads by underground rootstock.
Cornus canadensis (Bunchberry)	small, yellow flowers May to June	9"	shade	evergreen	layering	2	Spreads by underground rootstock.
Coronilla varia (Crown vetch)	pinkish white June to September	1' to 2'	full sun	compound leaves	division or seed	3	Excellent cover for steep slopes.
Dianthus deltoides (Maiden pink)	red to pink May to June	2" to 6"	full sun	dense, glossy carpet	division or seed	2	Needs well-drained, acid soil.
Epimedium grandiflorum (Barrenwort)	white, yellow or lavender May to June	12" to 15"	semi-shade	dense	division	3	Good in any soil.
Hosta undulata (Plantain lily)	lavender August	2' to 3'	full sun to shade	dense, variegated	division	3	Dies to ground in fall.
Hypericum calucinum (Aaronsbeard, St. Johnswort)	bright yellow July to August	1'	semi-shade	dense after established	seed, division, or cuttings	6	Spreads freely by stolons.
Lamiastrum	yellow May	10"	sun, part shade	dense	division	4	Silver-edged foliage.
Iberis sempervirens (Evergreen candytuft)	white late May	12"	full sun	narrowleaf evergreen	division, cuttings, or seed (Roots easily.)	4	Needs moist soil. Prune only after flowering in the spring.

FIGURE 31-1 Common ground covers

NAME	COLOR & DATE OF FLOWERING	HEIGHT	LIGHT	FOLIAGE	PROPAGATION	HARDINESS	REMARKS
Liriope muscari (Lilyturf)	lavender to purple August	10" to 15"	sun or shade	glossy	division	4	Spreads underground stem forming a mat or sod. Prune in the spring.
Lysimachia nummularia (Moneywort)	yellow summer	2" to 3"	sun or shade	leafy, soft stem	division	3	Spreads quickly. May become pest in lawns if not controlled.
Ophiopogon japonicus (Dwarf lilyturf)	violet July to August	10"	sun or shade	glossy	divison	6	Forms a mat or sod.
Pachysandra terminalis (Japanese spurge)	small, white spikes early May	6"	shade	dark green, leafy	cuttings or division	4	Excellent with evergreens.
Sedum (Stonecrop)	yellow, white or pink summer	2" to 8"	full sun	evergreen	division or cuttings	3 or 4	Good for dry areas. Excellent new varieties.
Vinca minor (Periwinkle or myrtle)	blue violet or white late April	6"	partial to full shade	evergreen	division, root or stem cuttings	4	Needs fertile soil.
Arctostaphylos uva-ursi (Bearberry)	white to pink May	6" to 12"	sun or partial shade	dark green	cuttings	2	Hard to transplant unless pot grown.
Calluna vulgaris (Heather)	white to red summer	4" to 24"	sun	small needles	cuttings	4	Needs acid soil high in organic matter.
Cotoneaster adpressus (Cotoneaster)	white June	6" to 12"	sun	small, dark green leaves	cuttings	5	Useful on banks.
Erica carnea (Heath)	rosy red early April	6" to 12"	shade	small needles	cuttings	5	Needs loose, well-drained soil.
Euonymus fortunei (Wintercreeper)	none	4"	shade or sun	dark green	cuttings or division	5	Good on banks. Holds leaves most of winter.
Gaultheria procumbens (Wintergreen)	white mid-May	3"	shade	small, shiny green leaves	cuttings or division	3	Needs moist soil high in organic matter.
Hedera helix (English ivy)	small, greenish September	6" to 8"	full or partial shade or full sun	dark green clinging vine	cuttings	5	Many cultivars available. Excellent plant.

FIGURE 31-1   Common ground covers (continued)

NAME	COLOR & DATE OF FLOWERING	HEIGHT	LIGHT	FOLIAGE	PROPAGATION	HARDINESS	REMARKS
Helianthemum nummularium (Sunrose)	yellow, pink or white June to July	6" to 12"	full sun	evergreen	seed, cuttings, division	5	Does best in dry soil.
Mahonia repens (Creeping mahonia)	yellow early May	10"	sun	dull bluish green, leathery and spiny-like holly	root cuttings or division	5	Vigorous when established.
Mitchella repens (Partridgeberry)	white	4" to 6"	shade	dark green	division or seed	5	Tolerant of dense shade and slow growing.
Paxistima canbyi (Paxistima)	small reddish early May	12"	sun or shade	small, dark green, and spreading	cuttings, layering, division	5	Needs fertile, acid soil.
Sarcococca hookeriana humilis (Sarcococca)	fragrant white mid-spring	1' to 2'	light to heavy shade	glossy green	seed, division, cuttings	5	Prune in early spring to keep plant compact. Slow grower.
Juniperus horizontalis		12" to 18"	full sun	summer— greenish blue winter— purple	cuttings	2	Excellent for banks and steep slopes.
Juniperus horizontalis 'Bar Harbor'		10" to 12"	full sun	greenish blue	cuttings	2	Excellent for banks and steep slopes.
Juniperus horizontalis 'Douglasi' (Waukegan)		12" to 18"	full sun	summer— steel blue winter—tinge of purple	cuttings	2	Excellent for banks and steep slopes.
Juniperus horizontalis 'Wiltoni' (Blue rug)		4" to 6"	full sun	steel blue	cuttings	2	Excellent for banks and steep slopes.
Taxus baccata 'Repandens' (Spreading English yew)		36"	full sun to partial shade	deep dark green	cuttings	2	Excellent foliage and foundation plant. Expensive.

FIGURE 31-1 Common ground covers (continued)

(A) Periwinkle (myrtle)

(B) Japanese Spurge

(C) Sweet Woodruff

(D) Chrysogonum

(E) Lamiastrum

FIGURE 31-2 Some commonly used ground covers

- ◆ **Broadleaf evergreen ground covers** retain their leaves year-round.

- ◆ **Deciduous ground covers** lose their leaves during the fall and winter season.

- ◆ **Narrowleaf evergreen ground covers** have needlelike or scalelike leaves. These plants retain their color throughout the year and thereby make excellent ground covers.

USES

Ground covers play an important role in landscaping because they can be used in many areas where soil is not suitable for growing grass, figure 31-2. These areas may be too steep, rocky, shaded, shallow soiled, or eroded for the proper growth of grass. However, ground covers are also used simply for the beauty they contribute to the landscape.

Low-growing, dense ground covers that grow relatively slowly make attractive foreground plantings for shrubbery borders. They are also placed between plantings of broadleaf evergreens where they serve to keep the soil cool and shade the roots of certain plants, such as rhododendrons.

SELECTION

Before selecting ground covers for use in the landscape, the following questions must be answered.

- In what type of soil does the plant grow best?

- Is it suitable for the locality?

- How is it propagated?

- How long will it take for the plants to cover the area in which they are planted?

- What is the mature height of the plant?

- Are diseases and insects a problem?

- How expensive are the plants?

There may be other questions which require answering, depending upon the particular landscape in question. Is the ground cover to serve a specific role in the landscape? If so, even more care must be taken when choosing the plant. If steep banks are to be covered for control of soil erosion, a plant that keeps its stem and plant top year-round should be selected. An evergreen would probably be best for this use.

Remember, also, that ground covers, as other plants, differ somewhat according to the soil type, moisture conditions, and amount of light they require. In a lightly shaded area, English ivy, pachysandra, and vinca minor would be good choices. For a sunny bank, a prostrate juniper might be chosen. For planting in an area with continuous shade on the north side of the building, English ivy, *ajuga reptans*, or lily-of-the-valley are wise choices. If the effect in the landscape is desired only for certain seasons of the year, select a plant that is most attractive during that season.

PLANTING GROUND COVERS

When preparing the soil for ground covers, consider any special problems in the area. On a steep slope, very little can be done except to add topsoil if needed. No tilling or digging in of organic matter is possible on a very steep slope. If tilling can be done, add organic matter in the form of peat moss or rotted manure and 2 to 5 pounds of 5-10-5 fertilizer spread evenly over each 100 square feet. Dig the materials into the top 6 inches of soil before planting.

Groundcover plants are spaced according to their size, growth rate per year, and time allowed for plants to spread and cover the area. On steep slopes, the area may need to be covered more quickly. In this case, place the plants closer together. On more level areas, the plants may be spaced farther apart. This reduces original plant costs. It will, of course, take longer for the ground cover to grow together and completely cover the soil surface. Local nursery employees may be able to give spacing directions for the area to be planted. Water well after planting to settle soil around the roots.

CARING FOR GROUND COVERS

FERTILIZING

After the initial application of fertilizer at planting time, fertilizer should be applied only as needed to keep plants healthy. If fertilizing is necessary, it should be done in early spring. There are two ways in which it can be applied: by means of a dry, granular fertilizer which is scattered over the area and watered in, or in the form of a soluble fertilizer that is mixed with water and siphoned through a garden hose and sprinkler system.

Once ground covers are established, further soil cultivation is not necessary. However, it is good practice to mulch between newly positioned plants to aid in weed control and moisture retention until the ground cover has traveled over the entire soil surface. When mulching, cover the entire surface with the material before the plants are placed in the ground. This method is less time-consuming than if mulch were applied after the plants were installed. Push the mulch from the area where the hole is to be made and pull it back around the plant as soon as the plant is set in the soil. Pine bark, hardwood bark, pine needles, well-rotted sawdust, peat moss, and decomposed leaves are suitable mulching materials for ground covers.

PRUNING

Pruning is necessary only to confine the planting to the area in which it is desired. This is usually accomplished by mowing around the edges or otherwise cutting back the outer perimeter.

WATERING

Newly planted ground covers should be watered as needed during the first year. Any time the soil becomes dry or plants start to wilt, apply at least 1 inch of water to the entire planting area. If the proper plants are selected, watering should not be necessary after the first year.

CONTROLLING INSECTS AND DISEASES

Insects and diseases are rarely a serious problem with ground covers. If control becomes necessary, identify the disease or insect and spray with a recommended pesticide. (Be sure to use protective clothing as required by the pesticide container label.)

 # Student Activities

1) Draw a landscape plan depicting an area around the school or at home. Position ground covers where they could be used to best advantage. A class project might include actually planting and cultivating one type of ground cover.

2) Propagate groundcover cuttings in sand, in 2¼-inch peat pots.

3) Collect samples of five different ground covers.

4) Determine the amount of ground cover needed to cover 150 square feet.

 # Self-Evaluation

Select the best answer from the choices offered to complete each statement.

1) Ground covers selected for use should be

   a) adapted to the soil in which they are to be planted.
   b) adapted to the local hardiness zone.
   c) relatively free of disease and insect problems.
   d) all of the above

2) For steep banks with soil erosion problems, a ground cover should be selected that

   a) is an annual.
   b) is deciduous.
   c) keeps its stem and top cover aboveground year-round.

3) Before planting groundcover plants, _____ pounds of a 5-10-5 fertilizer should be applied to every 100 square feet of area.

   a) 5 to 10
   b) 10 to 20
   c) 1 to 2
   d) 2 to 5

4) The best time to apply fertilizer to an established ground cover is

   a) late fall.
   b) midsummer.
   c) early spring.
   d) midwinter.

5) The best time to apply mulch to an area planted to ground covers is

    a) before any plants are planted in the area.
    b) after any plants are planted in the area.
    c) as soon as weeds appear.
    d) every fall.

6) Pruning of ground covers is necessary to

    a) keep them from becoming too tall.
    b) confine the plants to the area set aside for them.
    c) renew growth.
    d) thin the plants.

7) A ground cover that is properly selected for the area planted should be watered

    a) every ten days.
    b) each year, in the heat of summer.
    c) every spring and fall.
    d) only during the first year after planting.

8) In the cultivation of ground covers, insects and diseases

    a) are almost always a problem and must be carefully controlled.
    b) never attack and therefore never require control.
    c) are generally not a serious problem, but sometimes require control.
    d) are a very limiting factor.

Bulbs

OBJECTIVE

To force bulbs indoors and to use bulbs in the landscape.

Competencies to Be Developed

After studying this unit, you should be able to

- ◆ list four uses of bulbs in the landscape.
- ◆ describe the soil and fertilizer used in the flowering of bulbs.
- ◆ explain how planting depth and spacing of bulbs are determined.
- ◆ describe how to care for bulbs after they have flowered.
- ◆ list the steps in the forcing of bulbs.

MATERIALS LIST

- ✔ assorted bulbs and corms, including tulip, daffodil, crocus, hyacinth, amaryllis, begonia, calla, ismene, and gladiolus
- ✔ bulb planter and nursery spade
- ✔ bag of 5-10-5 fertilizer
- ✔ cold frame or cold treatment storage
- ✔ bulb bed for demonstration
- ✔ greenhouse space, if available

Many people use the term *bulb* in a general way to refer to bulbs, corms, tubers, and rhizomes—all of which are structures containing an embryonic plant and the necessary stored food for plant growth. However, these structures are different in appearance and in their methods of propagation. Review Unit 9 for information on types of bulbs and descriptions of each.

USES OF BULBS IN THE LANDSCAPE

Many people purchase flowers which grow from bulbs, such as tulips, hyacinths, and daffodils, but are unsure about how to use them most effectively in the landscape. As a general rule, bulbs are most striking when they form a *massing* (grouping of color). This is usually more attractive than mixing various colors in the same bed. Bulbs are not attractive when planted in thin narrow rows. Circular groups or masses of a dozen or up to hundreds in one spot are better. See figure 32-1A and B.

FIGURE 32-1A Flowering bulbs massed in the landscape add a burst of color early in the spring. (Courtesy Netherlands Flower Bulb Institute)

FIGURE 32-1B Even a small bed of tulips accents a lawn area in the garden. (Ed Reiley, Photographer)

IN WOODED AREAS

Bulbs grow well and produce beautiful color in natural wooded areas. Some bulbs grow well with evergreen ground covers in these areas. Growing taller than the ground covers, they give an excellent show of color. After they have flowered, the tops turn brown and die back. However, the ground cover remains to fill this area in the landscape until the next season's blooming of bulbs.

IN ROCK GARDENS

Rock gardens are excellent places to use flowering bulbs. Bulbs that produce flowers that can grow low to the ground, such as crocus, dwarf daffodils, and species tulips, can be worked in very easily. To be most appealing, the flowering bulbs should be massed together. Using color masses throughout the garden accents the areas in which they are planted; this is an effective way to emphasize certain areas.

WITH EVERGREEN SHRUBS

Many people use bulbs to add color around evergreen foundation shrubs. Bulbs planted in fall give early spring color to the landscape. Summer annuals can be planted to replace the color when the bulbs die back. When planting annuals, be careful not to damage the bulbs by digging too deeply.

AS CUT FLOWERS

Tulips and narcissus are used extensively by florists in January, February, and March to create an early touch of spring in arrangements. When bulbs are to be used in these months, they must be forced to bloom out of season, a process which will be discussed later in this unit.

If bulb flowers are to be used as a cutting garden, the beds for cutting should be located in a special area so that when flowers are cut, the appearance of the landscaped area is not spoiled.

SELECTING AND PREPARING SOILS

Bulbs grow well in a good garden loam which is well drained. The organic matter content in the soil can be improved by adding sphagnum moss, leaf mold, garden compost, pine bark, or well-rotted manure. To apply these materials, spread them over the ground in a layer 3 to 4 inches thick and work in with a Rototiller the spring before planting in the fall.

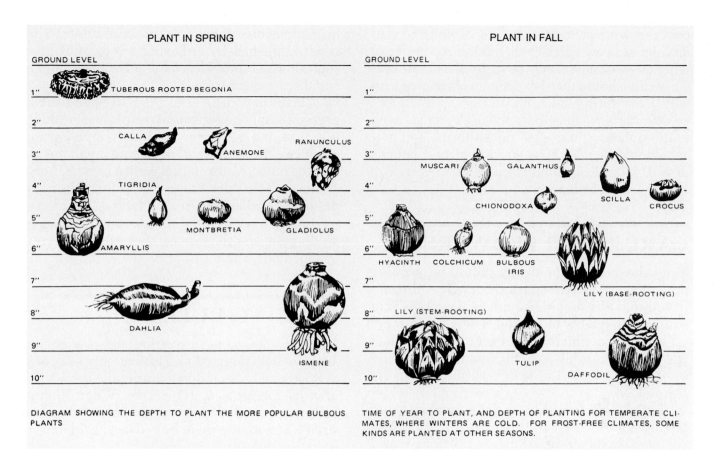

PLANT IN SPRING

GROUND LEVEL

TUBEROUS ROOTED BEGONIA

CALLA
ANEMONE
RANUNCULUS

TIGRIDIA

MONTBRETIA
GLADIOLUS

AMARYLLIS

DAHLIA

ISMENE

DIAGRAM SHOWING THE DEPTH TO PLANT THE MORE POPULAR BULBOUS PLANTS

PLANT IN FALL

GROUND LEVEL

MUSCARI
GALANTHUS
SCILLA
CHIONODOXA
CROCUS

HYACINTH COLCHICUM
BULBOUS IRIS
LILY (BASE-ROOTING)

LILY (STEM-ROOTING)
TULIP
DAFFODIL

TIME OF YEAR TO PLANT, AND DEPTH OF PLANTING FOR TEMPERATE CLIMATES, WHERE WINTERS ARE COLD. FOR FROST-FREE CLIMATES, SOME KINDS ARE PLANTED AT OTHER SEASONS.

FIGURE 32-2 Planting depth of bulbs. (Courtesy Brooklyn Botanic Garden)

Bulbs grow better if the soil is neutral, that is, neither acidic nor alkaline. If the pH value is below 6.0, ground limestone should be added at the rate of 10 pounds per 100 square feet to raise the pH to 7.0. A small amount of 5-10-5 fertilizer should be dug into the bottom of the bed.

PLANTING BULBS

Bulbs such as the crocus, narcissus, hyacinth, and grape hyacinth are planted in the fall. Others, such as the dahlia, amaryllis, and gladiolus, are planted in the spring. It is important to consider the height of the various flowers since they are most attractive when they grow in sequence of height. When planting bulbs, use a bulb planter, nursery spade, or hand trowel.

Each type of bulb has a recommended planting depth and spacing, figure 32-2. Spacing of the plants may be varied to achieve different effects. If a bed is to be showy, the bulbs are spaced closer together, figure 32-3. As a general rule, bulbs should be spaced the same distance apart as their planting depth. This gives a more natural appearance to the landscape. Water the bed thoroughly after planting.

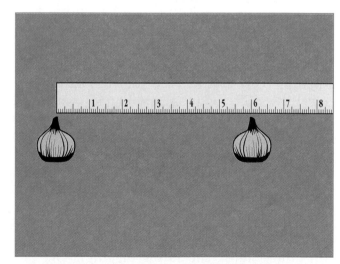

FIGURE 32-3 Tulip bulbs are planted 6 inches to center for maximum production of floral display.

FERTILIZING BULBS

Bulbs should be fertilized when planted by digging a small amount of 5-10-5 fertilizer into the bottom of the bed. This fertilizer should be covered with about 1 inch of soil before the bulbs are planted. Do not place the bulbs in direct contact with the fertilizer. After the bulbs are planted and covered with at least 2 inches of soil, a small amount of a new, special bulb food (not bone meal) with a 9-9-6 analysis may be sprinkled on the bed. The bulbs should then be completely covered with soil. Each spring after the first year, just before growth starts, a light application of a complete bulb food containing nitrogen, phosphorus, and potash should be used. The equivalent of 5 pounds of 5-10-5 fertilizer should be applied to every 100 square feet of bed area.

CARE AFTER FLOWERING

Bulbs flowering in the spring should be dug *after the foliage turns yellow and dies back.* Tulip and hyacinth bulbs should be replanted every year to insure high quality blooming. There are now perennial tulips which do not require replanting. Daffodil, crocus, lily, and colchicum bulbs should be dug up every 3–5 years, thinned, and replanted.

Bulbs flowering during the summer should be dug after the first fall frost has killed the foliage. Some in this group are dahlia, tuberous-rooted begonias, calla, and gladiolus. It is important that the foliage of the bulbs has no green color in it. This condition indicates that the bulb has completed its growth cycle for the season and all the food manufactured is stored in the bulb.

When digging bulbs use a nursery spade or spading fork to lift them from the earth. Very carefully spade next to the brown foliage, lifting the bulbs and soil. Gently remove the soil from the bulbs and store them at 50°F (10°C) in a dim light.

Dusting with a pesticide prevents insects and rodents from attacking the bulbs while they are in storage. To prevent the loss of moisture store them in sawdust or peat moss. Remove any bulbs that show signs of disease, insect damage, or mechanical injury (those cut during digging). This prevents the spread of disease, insect damage, and the rotting of healthy bulbs. Do not cut the tops off any bulb plants until they have died naturally. Flowering the next year will be greatly reduced if tops are removed too soon.

BULB PESTS AND DISEASES

Bulbs require special protection from certain diseases and insects. Many pests can be controlled by maintenance of good cultural conditions. This is accomplished by removing weeds and other rubbish which provide a natural home for insects and disease organisms.

Chemical control is also effective and usually necessary. Applying chemicals before placing bulbs in storage is a good practice. Before applying any chemical, check the label to be sure it will not injure the bulbs. Consult local garden centers, nurseries, or hardware stores for information on pesticides available for use on bulbs.

Figure 32-4 lists the most common types of bulbs, pests and the damage they cause, and control methods which can be used.

FORCING BULBS

Forcing bulbs is an interesting and challenging aspect of bulb production. The fragrance and color of flowering bulbs indoors during the late winter months can be very refreshing. When buying bulbs, be sure that they are the best quality and size by purchasing them from a local garden center florist, figure 32-5, page 342.

After bulbs are potted, they are placed outside in a cold frame and covered with straw, or stored in a refrigerator at 40°–50°F (5°–10°C) for ten to twelve weeks. (A *cold frame* is a frame, usually glass covered, which is used to protect plants without artificial heat.) Roots will start to develop and the pots should be left in the cold frame for eight weeks, then brought into a cool, partially lit room. This is done so that the bulbs will develop a root system, break dormancy, and establish some growth before they are brought into a warmer room for flowering.

PROCEDURE

FORCING BULBS

1. Identify bulbs to use for forcing. Select Number 1 (large) bulbs.
2. Select a potting medium that is well drained and high in organic matter: one-third soil, one-third sand, and one-third peat moss.
3. Place drainage material (a stone or piece of clay pot) in the bottom of the pot.
4. Place a layer of the medium in the pot. Set the bulb in the pot and fill in around the

HOST, PEST, DISEASE	DAMAGE	CONTROL
	Amaryllis	
Spotted Cutworm	Feeds on flowers at night.	Scatter cutworm bait or spray with Sevin.
Bulb Mites	Rotting bulbs. See *Hyacinth*.	Discard soft bulbs.
Narcissus Bulb Fly	Decaying bulbs. See *Narcissus*.	Discard soft bulbs.
Leaf Scorch, Red Blotch	Reddish spots on flowers, leaves, bulb scales; stalks deformed.	Discard bulbs or remove diseased leaves. Avoid heavy watering.
	Gladiolus	
Thrips	Leaves silvered, flowers streaked, deformed.	Spray with lindane in spring. Dust corms before storing.
Botrytis and other flower blights	Flowers, leaves, stalks spotted, then blighted.	Spray with zineb (Dithane Z 78 or Parzate).
Corm Rots, Scab	Lesions on corms, spots on leaves.	Dust with Arasan before planting.
Yellows (due to a soil fungus)	Plants infected through roots, turn yellow and wilt.	Choose resistant varieties.
	Hyacinth	
Bulb Mites	Minute; less than 1/25 inch, white mites in rotting bulbs.	Discard infested bulbs.
Aphids, several species	Leaves are curled; virus diseases may be transmitted.	Spray with malathion, rotenone, or nicotine.
Bulb Nematode	Dark rings in bulbs.	Discard.
Soft Rot	Vile-smelling bacterial disease; often after mites.	Discard.
	Iris (Bulbous)	
Tulip Bulb Aphid	See *Tulip*.	See *Tulip*.
Gladiolus, Iris Thrips	Leaves russeted or flecked, flowers speckled or distorted.	Spray or dust with malathion or lindane.
Leaf Spot	Light brown foliage spots with reddish borders.	Spray with zineb or bordeaux mixture; clean up old leaves.
	Lily	
Aphids (Lily, Bean, Melon, Peach, other species)	Curl leaves, transmit mosaic and other virus diseases.	Spray with malathion, being sure to cover underside of leaves.
Botrytis Blight	Oval tan spots on leaves, which turn black, droop.	Spray with bordeaux mixture.
Mosaic and other virus diseases	Plants mottled, stunted.	Rogue infected plants. Start lilies from seed in isolated portion of garden.
	Narcissus	
Narcissus Bulb Fly	Fly resembling bumblebee lays eggs on leaves near ground in early summer. Larva, fat, yellow maggot 1/2 to 3/4 inch long, tunnels in rotting bulb.	Sprinkle naphthalene flakes around plants to prevent egg-laying. Before planting dust trench with 5% chlordane and dust over bulbs after setting.
Bulb Nematode	Dark rings in bulb.	Discard bulbs. Commercial growers treat with hot water, adding formalin to prevent rot.
Basal Rot	Chocolate-colored dry rot at base of bulbs.	Inspect bulbs before planting.
Smoulder (Botrytis Rot)	Plants stunted or missing; masses of black sclerotia on rotting leaves or bulbs. Yellow, red, or	Remove diseased plants. Put new bulbs in new location.
Scorch	brown spots blight tips of leaves	Spray or dust with zineb, maneb, or copper.
	Tulip	
Tulip Bulb Aphid	Powdery white or grayish aphids common on stored bulbs.	Dust with 1% lindane before storing.
Green Peach, Tulip Leaf, and other Aphids	Transmit viruses to growing plants.	Spray or dust with malathion or lindane.
Botrytis Blight, Fire	Plants stunted, buds blasted, white patches on leaves, dark spots on white petals, white spots on colored petals, gray mold, general blighting. Small, shiny black sclerotia formed on petals, foliage rotting into soil and on bulbs.	Discard all infected bulbs. Plant new tulips in new location. Spray with ferbam or zineb, starting early spring. Remove flowers as they fade, remove all tops as they turn yellow.
Cucumber Mosaic	Yellow streaking or flecking of foliage.	Do not grow near cucurbits or gladiolus.
Lily Mottle Viruses	Cause broken flower colors, mottled foliage, in tulips.	Do not plant near lilies. Control aphids.

FIGURE 32-4   Common bulb pests and diseases and their control. (Courtesy Brooklyn Botanic Garden)

FIGURE 32-5 Bulbs being graded different sizes by machine. Note the bins at the end of the chutes to catch different sized bulbs. Bulbs fall through smaller holes to larger holes as they move toward the end of the machine. (Courtesy Netherlands Flower Bulb Institute)

FIGURE 32-6 The bulbs are set in a 6-inch azalea pot or bulb pan. Medium is added so that it almost covers the entire bulb. (Courtesy USDA)

bulb with the growing medium so that the top of the bulb is exposed, figure 32-6.

5. Water by setting the pot in a pan of water.

6. Be sure to label the pot correctly according to the variety of bulb.

7. Set the planted pot outside in a cold frame with a mulch of straw at a temperature of 50°F (10°C) from November 25 until January 15. Bulbs may be protected against disease by treatment with a fungicide. A refrigerator may be used for this cold treatment.

8. Remove the pot from the cold frame or refrigerator about January 15, figure 32-7, and place in the greenhouse. Check for root development. Water well. Maintain a temperature of 60°F (16°C).

9. Tulip bulbs should bloom five weeks from the date of removal from the cold frame. Other plants may require more or less time to bloom.

FIGURE 32-7 After eight weeks of cold treatment, the bulbs are brought inside to wait for blooming. (Ed Reiley, Photographer)

outstanding beauty are now available which are disease resistant enough to persist in the environment. "Enchantment" is one of the best to start with. Purchase others based on your own preferences, but look for varieties that are vigorous, disease resistant, and multiplying from year to year. The lily bulb is soft and never dormant, and must be handled carefully in order not to bruise it, nor should it be allowed to dry out.

LILIES

A bulb crop now becoming much more important in the landscape is the lily. Many new hybrids of

NOTES ABOUT BULBS

1. Always plant bulbs at the depth prescribed. Do not plant in shallow soil.

2. Always plant bulbs in a well-drained soil.

3. Provide protection from mice where they are a problem. Mice love lilies and tulips especially. They will not eat daffodils or crown imperial (*Fritillaria imperialis*). Crown imperial will repel mice if interplanted with other bulbs. Cages made with ¼-inch mesh wire and sunk into the bed with bulbs planted in them are also effective. The cage should be tight at corners and extend above the soil level for 1 or 2 inches.

4. All bulbs have planting instructions supplied with them when purchased. Be sure to follow these directions.

Student Activities

1) Visit a local garden center and note what types of bulbs are available. If possible, select and purchase bulbs for home or school use.

2) Force several types of bulbs, including tulips, daffodils, hyacinths, and crocuses. Develop a bulb-forcing schedule and produce the flowers for sale on a certain date.

3) Under the supervision of the instructor, build a cold frame or use a refrigerator for use in producing and forcing bulbs.

4) Design a flower bed using bulbs for a home or school garden.

5) Attend a local garden show and observe the use of bulbs in the landscape. If possible, discuss the design of the plantings with the landscaper.

6) Cut and care for bulbs used in arrangements.

7) Create a bulletin board showing different types of bulbs. (Bulbs may be placed in small plastic bags and fastened to the board.)

8) Choose one type of bulb and research further information on its culture and uses.

9) Compile a list of different bulbs and record their natural blooming dates. In this way, the sequence of their flowering can be determined.

Self-Evaluation

Select the answer from the choices offered to complete each statement.

1) Flowers which are grown from bulbs include

   a) zinnias, hyacinths, and petunias.
   b) hyacinths, tulips, and daffodils.
   c) tulips, marigolds, and begonias.
   d) all of these

2) When landscaping with bulbs, the most striking effect is achieved by placing them in a

   a) single row.
   b) massing.
   c) grouping of mixed colors.
   d) all of these

3) Flowers grown from bulbs can be used

   a) as cut flowers.
   b) in rock gardens.
   c) in wooded areas.
   d) all of these

4) Bulbs planted in fall give color to the landscape in

a) late summer.
b) early spring and summer.
c) mid-fall.
d) early summer.

5) The organic matter in soil used for bulb production can be increased by the addition of

a) sand.
b) 5-10-10 fertilizer.
c) well-rotted manure.
d) all of these

6) The best soil for bulbs has a pH of

a) 6.0.
b) 5.0.
c) 7.0.
d) 7.5.

7) Fertilizer used for bulbs should have an analysis of

a) 10-10-5.
b) 10-10-10.
c) 5-10-5.
d) 10-6-4.

8) Tulip bulbs should be planted

a) 4 inches apart.
b) 10 inches apart.
c) 12 inches apart.
d) 6 inches apart.

9) The best time to fertilize bulbs is

a) one year after planting.
b) after they have flowered.
c) after it has rained.
d) none of these

10) Bulbs that are dug and stored should be kept at a temperature of

a) 50°F.
b) 60°F.
c) 70°F.
d) 80°F.

11) To protect bulbs from insects, diseases, and rodents,

a) place them in the soil with additional lime.
b) dust them with pesticide.
c) wrap them in paper.
d) none of these

12) Begonia, calla, gladiolus, and dahlia bulbs should be planted in the

a) summer.
b) spring.
c) fall.
d) winter.

13) Muscari, hyacinth, and crocus bulbs should be planted in the

a) fall.
b) summer.
c) spring.
d) winter.

14) The cold treatment for forcing tulip bulbs to bloom takes about

a) ten to twelve weeks.
b) twenty weeks.
c) twenty-two weeks.
d) four weeks.

15) Bulbs for forcing should be

a) large.
b) medium.
c) small.
d) any of these

16) Cold treatment for root formation can be accomplished by placing potted bulbs in a

a) hot bed.
b) greenhouse.
c) nearing frame.
d) cold frame.

Techniques of Pruning

OBJECTIVE

To select proper pruning techniques for specific plants and demonstrate their use in the landscape.

Competencies to Be Developed

After studying this unit, you should be able to

- ◆ list five reasons for pruning.
- ◆ describe four types of pruning.
- ◆ demonstrate pruning a stem at the proper angle.
- ◆ explain how the correct time to prune is determined.
- ◆ give three examples of each of the following types of plants and describe the pruning technique used for each type.
  —deciduous spring flowering shrubs
  —summer flowering shrubs
  —broadleaf evergreens
  —conifers

MATERIALS LIST

✔ lopping shears	✔ spring flowering shrubs
✔ chain saw	✔ summer flowering shrubs
✔ anvil pruners	✔ conifers
✔ hand pruners	✔ broadleaf evergreens
✔ pruning saw	✔ gas-powered hedge shears
✔ stepladder	✔ pole pruner
✔ protective equipment	

Plants are pruned to:

- ◆ remove dead, diseased, insect-infested, or broken branches. This keeps the plant healthy by removing any parts which might hinder further growth and stimulates new growth.

(A) Shows an unpruned landscape rose. Note the length of rose canes.

(B) Select the larger rose canes for removal to 10 inches or 12 inches above the ground level.

(C) Using lopper shears to remove the large rose canes.

(D) Demonstrates the proper height to prune landscape roses.

FIGURE 33-1   Pruning a landscape rose

♦ change the size or proportion of the plant. This is necessary when plants become overgrown in their landscape sites. This sometimes occurs when the mature height and growth rate of the plant are not considered before placing it in the landscape.

♦ develop a special form or shape. Hedge pruning and topiary and espalier work are examples of this type of pruning. *Topiary* is a practice used in formal gardens in which plant material is trimmed into different forms and shapes. *Espalier* is a

(A) Pruning tools. From left to right: Hand pruners, lopping shears, folding saw.

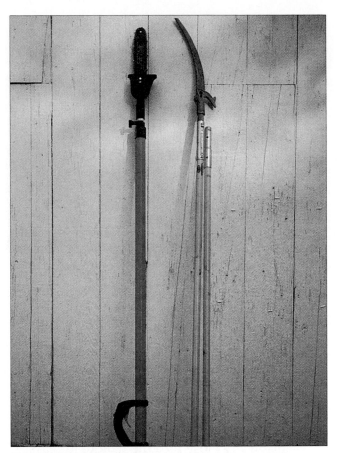

(B) Left to right: Power pruning saw and pole pruning saw/hook.

(C) Chain saw.

FIGURE 33-2 Tools used for pruning

pruning method of training plant material to grow flat against a trellis or wall.

◆ remove wilted or dead flowers and seed pods. Wilted or dead flowers give plants an unattractive appearance and should be removed. Seed pods use the plant's energy to develop seeds. Therefore, seed pods are removed to encourage further growth of the plant itself. For example, on rhodo-dendrons and lilacs, seed pods develop rather than new shoots; the seed pods must be pruned to encourage new growth.

◆ rejuvenate a declining shrub. New growth is stimulated by pruning older wood. After pruning, the plant usually produces better quality flowers and develops a better form, figure 33-1.

EQUIPMENT FOR PRUNING

There are a few basic pruning tools needed for all pruning, figure 33-2. These tools must be kept

(A) Using a pole pruner to remove a lower branch.

(B) Pruning to the collar at the base of the branch.

FIGURE 33-3   Thinning involves removing certain branches from a plant to open up the plant.

clean and sharp so that when used they will give a good, clean cut.

♦ **Pruning saw.** This saw is a folding pruning saw. It has a coarse teeth setting. The saw has a special setting to cut through the (live) green wood or dead wood without pinching the saw. The folding pruning saw blade will close into the handle for easier, safer, and better storage.

♦ **Lopping shears.** These shears have long handles to give more leverage for cutting larger branches. The type shown in figure 33-2 is a scissor-action lopper.

♦ **Hand shears.** Hand shears are clippers consisting of a single blade that cuts against another piece of metal (as shown in figure 33-2), or two blades that work like scissors.

♦ **Pole pruner.** This tool has a saw and hook (which is used to pull, cut, or break dead branches). It is used to remove branches 10 to 12 feet overhead.

♦ **Chain saw.** A power chain saw is used extensively to remove large branches that are 3 inches or more in diameter. Chain saws can be dangerous—use caution while operating them.

♦ **Power pole pruner.** The power pole pruner is a small chain saw on an extended handle that is used to remove overhead branches. It may be used to elevate lower branches to create improved vision.

TYPES OF PRUNING

There are four basic methods of pruning. Before deciding which pruning method to use, consider the type of plant to be pruned and the desired finished effect.

THINNING

Thinning involves removing certain branches from a plant to open up the plant. This method is usually used when the horticulturist wishes to keep the natural shape of the plant. Thinning allows more light to reach inner branches so that they can develop new growth more easily. The effect of thinning does not have to be noticeable, figure 33-3.

Fruit trees are a good example of plants that are commonly thinned. These trees are pruned while the plants are dormant (in winter or dry season). All cuts are made to the collar of the limb so that the tree wound will heal properly. See figure 33-3. In pruning young fruit trees, it is important to thin limbs to leave about four to six lateral branches that are about 8 inches apart on the main trunk. The lowest branch should be about 2 feet from the ground. It is best to proceed in a spiral around the main trunk when pruning, keeping the lateral branches evenly spaced.

When thinning deciduous shade trees it is important to remove one third of the top of the tree during transplanting. Figure 33-4 demonstrates the correct procedure for thinning.

HEADING BACK

Heading back is the removal of the end section of the plant branches at the same height, figure 33-5.

(A) An unpruned shade tree.

FIGURE 33-4 Thinning deciduous shade trees

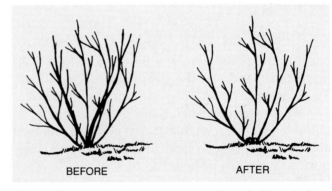

(B) The same shade tree properly pruned by thinning.

FIGURE 33-5 The pruning technique of heading back

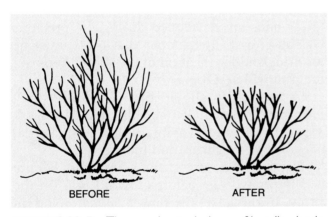

FIGURE 33-6 Renewal pruning before (left) and after (right)

This causes new shoots to develop from dormant buds, therefore making the plant thicker. The plant will grow in the direction the bud is facing, so it is important to prune back to a live bud. The new growth will begin from that point. Heading back is usually considered the easiest of the four techniques discussed.

RENEWAL PRUNING

In *renewal pruning*, old branches that are large and unproductive are removed. The oldest branches are removed by cutting them back to ground level, figure 33-6. This results in the development of new branches and better flowering on the new branches. Renewal pruning is usually effective on flowering shrubs.

ROOT PRUNING

Root pruning, an important step in transplantation of plants, involves cutting off all the lateral roots of the plant with a sharp spade in a circle around the stem. In commercial nurseries, root pruners are mounted on tractors to allow for a more efficient operation. A general rule for root pruning is to cut in a circular motion around the plant so that 1 inch of stem diameter equals 10 inches of

FIGURE 33-7   The proper distance for root pruning deciduous trees

the circle diameter, figure 33-7. It is best to root prune one growing season prior to transplanting the plant. This results in the formation of many small, fibrous roots. Early fall is the best time to accomplish root pruning.

HOW AND WHEN TO PRUNE

CUTTING AT THE PROPER ANGLE

Cutting the plant stem at a 90-degree angle is an important aspect in the pruning process. Figure 33-8 shows how pruning cuts should be made.

A cut that is made at too great an angle, as shown in figure 33-8, leaves too great an exposed area for the plant to heal properly. It is also important that the cut be made the correct distance from the bud, figure 33-9. Cutting too far from

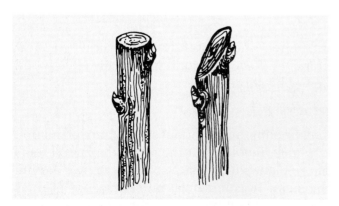

FIGURE 33-8   Pruning at the proper angle of the stem. The stem on the left has been cut at the correct angle; the stem on the right has been cut at too great a slant.

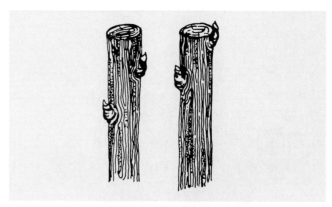

FIGURE 33-9   Pruning at the correct distance from the bud. On the left, the cut has been made too far from the bud; on the right, too close to the bud.

the bud leaves too much stub above the cut, causing the branch tip to dry out and die. It may also provide a home for insects, and results in an ugly appearance. Cutting too close to the bud may injure it.

WHEN TO PRUNE

The time schedule for pruning is usually dependent upon when the plant flowers or bears fruit. For example, if a plant flowers in the spring, it should be pruned after it flowers. This means that flower buds are developed during the previous season's growth. Remember that there must be flowering and fertilization of the flower buds before fruit will develop.

Shrubs which should be pruned after they flower are forsythia, deutzia, lilac, weigela, mock orange, bridal wreath, spirea, beauty bush, kerria, magnolia, sweet shrub, and Juneberry.

Plants that flower in the summer or fall are usually pruned just before growth begins in the spring. In this case, the flower buds develop on the new growth (current season's growth). Some examples of plants which may be pruned in summer or fall are abelia, crape myrtle, Russian olive, bush honeysuckle, roses, five-leaf aralia, Norfolk Island pine, bamboo, aucuba, heather, boxwood, bittersweet, croton, dogwood, winter hazel, viburnum, blueberry, hemlock, and bald cypress.

PLANTS WITH SPECIAL PRUNING NEEDS

Some plants require special attention at pruning time. Listed below are some examples and their pruning requirements.

NANDINA Prune the oldest canes to the ground to start new growth. Prune to thin out the plant.

OLEANDER Prune the faded flower clusters. Remove the top branches to induce new growth and to control the size of the plant.

PRIVET Prune to control shape and size.

RHODODENDRON Pinch the flower truss after blooming to prevent the formation of seed pods. If they are permitted to set seed, additional energy is required from the plant to produce new growth. After pinching, the plant will produce new growth faster and set more flower buds for blooming the following year.

HEMLOCK Prune in the early spring before new growth develops. This stimulates thicker growth and helps to control plant size and shape.

SPRUCE Prune to one-half of their new growth candles in August to result in a fuller tree. *Candles* are new growth on the ends of branches of pine species.

FIR Prune the new growth in August to one-half the length of the candle. Growth may be retarded by removing the tip bud in the spring before the candle develops. Usually, the only pruning done is to shape the tree. Firs have multiple buds on the stem for regrowth, whereas pine produces growth from the terminal end.

YEW Prune after the new growth has hardened in the spring. Yews may be pruned in the summer if necessary to control plant growth.

ANDROMEDA Prune to remove dead faded flower clusters in late spring. The new shoots should be pinched back to shape the plant. This will cause the plant to develop new growth.

AUCUBA Pinching back results in a thicker plant with more shape. If some of the leaves have turned black from winter burn, remove them.

ARBORVITAE Prune before new growth starts in the spring. Do not prune beyond the green leaves, since this will result in a permanent brown area.

PINE Prune about one-half of each candle in June, figure 33-10. This promotes thicker plant growth and the production of new side shoots.

AZALEA Prune after flowering to promote the production of new growth. Pinching back new top growth stimulates heavier blooming.

BOXWOOD Pruning boxwood is important to keep the plant compact and full. Prune in the fall/early winter by removing short branches with hand pruners. Remove diseased branches, but be sure to dip the hand pruners in a 70-percent alcohol solution between each cut to prevent the spread of the disease. Do not use hedge shears, since this causes the leaves to turn brown on the end where the leaf is cut.

(A)

(B)

(C)

FIGURE 33-10 Pruning a pine tree

HOLLY Prune various types of holly in December so that the branches can be used for holiday decorating. Select and cut the branches to develop the plant's shape and form. Holly may also be pruned in early spring before new growth begins.

MOUNTAIN LAUREL Prune to encourage growth of new shoots. Oddly shaped, tall, or leggy plants may be pruned to the ground to start new growth. When pruning, be sure that the natural shape and form of the plant is maintained.

Student Activities

1) Collect various branches of trees and shrubs and bring them to class. Take turns demonstrating the proper cutting technique to the rest of the class.

2) Under the supervision of your instructor, prune shrubs on the school grounds in accordance with the procedures specified in the text. If possible, visit a local orchard before pruning to observe professional techniques.

3) Invite an arborist to speak to the class.

4) Have a demonstration on chain saw safety by a factory representative.

5) Develop a safety program for other students to follow.

6) Enter the national FFA safety contest.

Self-Evaluation

Select the best answer from the choices offered to complete each statement.

1) Pruning is done to
   a) remove dead wood.
   b) remove diseased wood or create special tree forms.
   c) remove insect-infested or broken wood.
   d) all of the above

2) The pruning method in which all the terminal ends of the plant branches at the same height are removed is called
   a) thinning.
   b) heading back.
   c) root pruning.
   d) renewal pruning.

3) In the root pruning process, lateral roots are cut, resulting in the formation of
   a) small fibrous roots.
   b) large branches.
   c) better quality flowers.
   d) none of these

4) The pruning procedure that removes old and unproductive branches is called
   a) heading back.
   b) root pruning.
   c) renewal pruning.
   d) none of these

5) Pruning plants to result in a special shape or form is called
   a) heading back.
   b) thinning.
   c) root pruning.
   d) topiary pruning.

6) Pruning plant branches on an angle promotes better healing. Branches are pruned at an angle of

    a) 42 degrees.
    b) 36 degrees.

    c) 50 degrees.
    d) 90 degrees.

7) Deciduous plants that flower in spring are pruned

    a) before flowering.
    b) after flowering.

    c) in late spring.
    d) none of these

8) New growth on pines is called

    a) needles.
    b) branches.

    c) candles.
    d) none of these

9) The spruce and fir are pruned after the new growth has hardened off in

    a) late summer.
    b) early spring

    c) fall.
    d) early winter.

10) Yews are pruned in the spring after the new growth has

    a) fully developed.
    b) hardened off.

    c) become green.
    d) none of these

Principles of Landscaping, Maintenance, and Xeriscaping

OBJECTIVE

To apply the principles of landscaping, maintenance, and xeriscaping to an actual setting to understand the goals of the landscape profession.

Competencies to Be Developed

After studying this unit, you should be able to

◆ describe the three major career fields within the landscape profession.

◆ list the main objectives of good residential landscaping.

◆ list the five principles of landscape design and examples of an application of each principle.

◆ list three hard paving and three soft paving materials.

◆ list the technical procedures for landscape maintenance.

◆ calculate the volume of mulch needed to cover a landscape bed.

◆ list the basic concepts of xeriscaping.

◆ list ways the soil can be improved to conserve water.

◆ identify plants that can be used in a xeriscape setting.

◆ explain why mulches are important.

MATERIALS LIST

✔ photographs or drawings of well landscaped property
✔ felt, construction paper, scissors

THE LANDSCAPE INDUSTRY

The *landscape industry*, also known as the *green industry*, functions to improve our natural environment to meet the needs and desires of people. *Landscape architects* are professionals who integrate the disciplines of art and science, and know how plants and landscape factors will react to the environment around them. They prepare designs which show what, where, and how objects are installed in the landscape, and which provide for ease and comfort for use as well as appearance. In addition, today's landscape architects are using the new technologies of mechanized equipment and computers in conjunction with the design process. This is known as computer aided design (CAD).

It is in the mind's eye of the landscape architect that the landscape features first take shape. The successful architect must be able to look at undeveloped, scarred, and ugly land, and recognize its potential for becoming more attractive and/or functional for environmental/human uses.

Many states set high standards and legal requirements for those individuals calling themselves landscape designers or landscape architects. Often, four to five years of college training are required. State licensing may also be necessary. Only through training can the designer's full imagination be developed. Only through on-the-job work experience can the designer channel creativity and imagination in the development of practical design solutions.

The language of the landscape designer is *graphic art*. Through this art form (graphics), designers are able to reduce the actual dimensions of the area to be landscaped to a size that can be illustrated to the client in the form of a reduced drawing. The landscape plan is a collection of graphic symbols, which represent trees, shrubs, flowers, buildings, and other constructed materials necessary to the proposed landscape, figure 34-1. When the land dimensions and the symbols are reduced in the same proportion, the drawing is said to be done to *scale*. Two examples of commonly used landscape scales are 1 inch on the drawing equal to 10 feet of real land (expressed as 1" = 10'), and 1 inch on the drawing equal to 20 feet of real land (expressed as 1" = 20').

By means of landscape plans, designers are able to show their ideas to clients in such a form that they can be understood and discussed.

Once the design has been approved by the client, it must be installed and brought into reality. *Landscape contracting* is the career field which deals primarily with the installation of landscapes. Landscape contractors are the main link between the

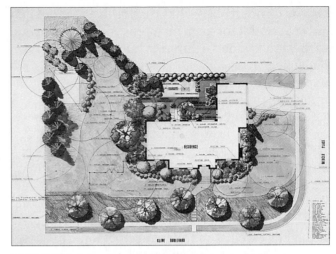

FIGURE 34-1 A completed landscape design. (Courtesy Greene Expectations Landscape and Design Consulting, Andy Daymunde, Designer)

design and implementation of a landscape and, to be successful, must have knowledge of plant materials and proper planting techniques, and engineering and building skills. It is as important that they know how to lay a patio or build a brick wall as how to plant a tree or shrub. The landscape contractor must be able to read and follow the scaled plans prepared by the landscape architect. If the landscape architect is responsible to the client for the actual development of the landscape, he or she may be the one who hires the landscape contractor to build the landscape. This practice is known as *subcontracting*. A sizeable landscape may involve numerous subcontractors, all responsible to the landscape architect. The landscape architect is, in turn, accountable to the client.

The care of the landscape after it has been installed is the job of the *landscape maintenance contractor*. Professionals in this career field are sometimes called gardeners or professional horticulturists. Some landscape maintenance contractors spend their time totally on a single landscape, such as a large estate, shopping center, cemetery, or park, while others serve numerous landscapes, accomplishing all or part of the maintenance tasks required by the landscape. Typical landscape maintenance tasks include lawn care (fertilizing, weeding, mowing); tree and shrub care (fertilizing, pruning, mulching); flower care; repair of walls, fences, walks, and drives; painting; and snow plowing. The fastest-growing part of the business is the management of landscapes once they have been installed.

The simplicity or complexity of the landscape design will make the maintenance of landscape easy or difficult. It is vital that both the architect and the contractor keep maintenance requirements of the landscape in mind while developing the landscape. It is a weak design that contains so many high-maintenance features that the cost of upkeep is prohibitive.

Likewise, the landscape maintenance contractor must know how to read a landscape plan. As the landscape grows over a period of years, certain plants may need to be replaced. Others may require special pruning techniques. Flower plantings often require special care to obtain the desired color patterns year after year. A poorly trained or careless maintenance person can cause the best landscapes to appear shabby.

While there are different levels of involvement by various professionals in the landscaping business, all are interrelated. Neither the architect, the contractor, nor the landscape maintenance contractor can work without an appreciation for the role of others if landscapes that both serve and satisfy the client are to be created.

THE OBJECTIVES OF RESIDENTIAL LANDSCAPING

The objectives of residential landscaping are evident in the definition of the term *landscaping*: that is, to serve the needs and desires of people in development of the outdoor environment. Specifically, the goals of residential landscaping are:

- to determine the exact landscape needs and desires of the homeowners.

- to determine the capabilities of the land (site) to fulfill those needs and desires.

- to develop the outdoor living areas of the landscape in a manner similar to the way indoor living areas are developed.

- to design the landscape in such a way that maintenance practices do not exceed that which the homeowner is willing to do.

- to keep costs within the budget of the homeowner.

The needs and desires of the homeowner are best determined by an interview. A direct conversation between the designer and the client promotes trust and confidence between the two parties. This helps the designer to recognize the client's actual desires much more easily and allows

for that personalized approach to landscaping. During the interview, questions such as the following should be answered:

a. How many family members are there? What are their ages?

b. How much does the family use the outdoor areas around the home?

c. Does the family entertain frequently? Large groups or small?

d. How much privacy from the neighbors and passing cars do they desire?

e. How much maintenance are they willing to do in the upkeep of the landscape?

f. Are there certain plants they are fond of, dislike, or are allergic to?

g. What service needs will the landscape be expected to accommodate? (Examples: clothesline, trash cans, pets, vegetable garden, compost pile, garden tool storage)

h. Will the family be using the garden after dark?

i. How much does the family want to spend on the total development of the landscape? Do they want to spread the cost over several years?

j. Is the family willing to wait several years for the plants to reach maturity or do they want large plants installed for an immediate effect?

To determine the capabilities of the site requires a thorough *site analysis* by the landscape architect. Becoming familiar with the site helps the designer to determine how easily and how many of the client's needs and desires can be met. The site analysis usually involves several visits to the property by the designer. A good site analysis often suggests additional possibilities for development that had not occurred to the homeowner, but which might be of interest once identified.

The following are some of the things to look for and take note of during a site analysis.

- dimensions of the property
- topography of the site (how flat or rolling it is)
- quality of the topsoil and subsoil
- condition of the lawn areas
- types and condition of existing plants
- location of utility lines, meters, and utility easements
- good and bad views from the site

- locations of glass areas in the house and where they open onto in the landscape
- architectural style of the neighborhood
- environmental setting of the site and the neighborhood
- existing natural features such as streams, rock outcroppings, specimen plants, and wildlife habitat areas

THE CONCEPT OF THE OUTDOOR ROOM

When developing the design for a landscape, it is helpful to visualize the outdoors in the same manner as the indoors. The home is a series of connecting living units called rooms. When the landscape designer sees the outdoor areas as rooms, the landscape becomes more familiar and easier to work with. Outdoor rooms have walls, ceilings, and floors, just as indoor rooms do, figure 34-2. The main difference between indoor and outdoor rooms is in the materials used to construct them. Outdoor walls may be developed using shrubs, fences, brick or stone, exterior walls of buildings, or trellises. Outdoor floors may be the natural earth, sand, crushed stone, poured concrete, brick,

FIGURE 34-2 An outdoor room. Notice the materials that act as the walls, floor, and ceiling.

FIGURE 34-3   An outdoor wall constructed of natural stone materials

decking, turf grass, or many other similar surfacing materials. Outdoor ceilings can be developed with trees, awnings, canopies, or other overhead structures.

Trees are excellent materials for use in developing the outdoor ceiling. If an intimate patio setting is desired, a small tree branching overhead can create the low-ceiling effect. Also important to consider is the shade that the tree provides on a hot summer day. The southwest corner of a house will benefit greatly from the cooling effect of a medium to large tree planted nearby. Deciduous trees will allow the sun to pass through and warm the house with solar energy during the cold winter months.

The material selected for outdoor walls is usually determined by how much privacy and security is needed in the landscape. When total privacy or security is desired, the outdoor wall should be at least 6 feet tall and solid. When less privacy is needed or security is not a factor, the outdoor wall can be lower and more open.

The main function of the outdoor wall is to define the shape and limits of the outdoor room. Plants or constructed materials used alone may

accomplish this; however, the combination of natural and constructed materials often creates the most attractive effect of all, figure 34-3.

There are several groups of materials from which to select for the outdoor floor. *Hard paving* includes concrete, flagstone, tile, decking, and brick, among other materials. Hard paving is expensive to install but inexpensive to maintain. *Soft paving* includes crushed stone, wood chips, marble chips, washed river gravel, pine needles, and other loose materials. Soft paving has a moderate cost for both installation and maintenance. It requires periodic replacement of worn areas. Turf grass is the most popular of all surfacings. Turf has a low installation cost but a high maintenance cost, since it must be cared for regularly. *Ground covers* are good for surfacing areas where no one will be walking, such as on slopes and directly under trees. Figure 34-4 describes the various outdoor room surfacings.

The type of material chosen for the outdoor floor is certainly affected by what the client can afford. However, there are other needs of the individual client to consider:

Does the material absorb noise and/or dust?

How well does it blend in with the rest of the landscape?

Does it produce a glare from the sun?

Does it track easily?

Does it become slippery when wet?

THE PRINCIPLES OF LANDSCAPE DESIGN

There are five basic principles which guide the landscape architect or designer in planning the landscape's outdoor rooms. These principles are:

TYPE OF SURFACING	COST OF INSTALLATION	COST OF MAINTENANCE	DURABILITY
Hard paving	high	low	high
Soft paving	moderate	moderate	moderate
Turf grass	low	high	moderate
Ground covers	moderate	high at first	low

FIGURE 34-4   Comparison chart of various outdoor room surfacings

1. simplicity
2. balance
3. focalization of interest
4. rhythm and line
5. scale and proportion

The principle of simplicity is very important to the overall unity of the design. When this principle is correctly applied, the landscape is understood and appreciated by the viewer. The principle of simplicity is accomplished by repeating specific plants throughout the design, and by massing plant types or colors into groups rather than spacing them so that each plant or color is seen separately. The fewer different objects there are for the eye to focus upon, the simpler the design will seem. Finally, straightlined or gently curving bedlines around shrub plantings, rather than fussy, scalloped bedlines, add to the design's simplicity, figure 34-5.

The principle of balance is applied by imagining the area of landscape placed on a seesaw. If properly balanced, the left side of the landscape should have no more visual weight than the right side. Balance may be either symmetrical or asymmetrical. *Symmetrical* balance is attained when one side of the landscape is an exact duplicate of the other side, figure 34-6. This is a form of balance common in formal designs. It is sometimes ap-

plied to modern residential design, but more frequently asymmetrical balance is used. With *asymmetrical* balance, one side of the landscape has the same visual weight as the other side, but they are not duplicates, figure 34-7.

The principle of focalization of interest recognizes that the viewer's eye wants to see only one feature as being most important within any given view. All other elements complement that important feature (the focal point) but do not compete with it for attention. When looking at a house from the street, the viewer's eye should go quickly to the front door. When sitting on the patio looking out across the backyard, the viewer may have no focal point at which to look unless one is created by the designer. Focal points may be created using especially attractive plants (*specimen* plants), statues, fountains, pools, and flower masses. Once created, all bedlines and plant arrangements should be designed to lead the eye of the viewer to the focal point, figure 34-8.

The principle of rhythm and line also contributes to the overall unity of the landscape design. This principle is responsible for the sense of continuity among different areas of the landscape. One way in which this continuity can be developed is by extending planting beds from one area to another. For example, shrub beds developed

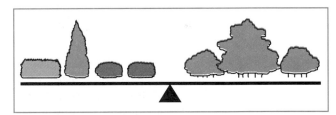

FIGURE 34-7   Asymmetrical balance. One side of the landscape provides as much interest as the other, but does not duplicate it exactly.

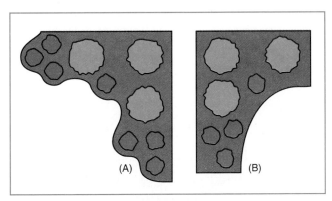

FIGURE 34-5   A fussy bedline (A) and a simple bedline (B). Bedline (B) was more thoughtfully planned and would be much easier to maintain.

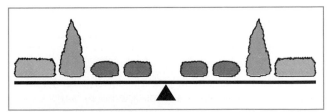

FIGURE 34-6   Symmetrical balance. Each side of the landscape duplicates the other.

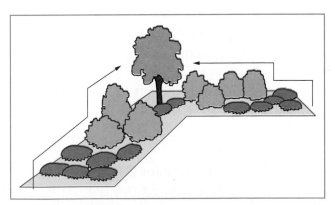

FIGURE 34-8   A corner planting that easily leads the eye to the focal point.

around the entrance to the house can be continued around the sides and into the backyard. Such an arrangement helps to tie the front and rear areas of the property together. Another means by which rhythm is given to a design is to repeat shapes, angles, or lines between various areas and elements of the design.

The principle of scale and proportion helps to keep all elements of the landscape in the correct size relationship without towering over the building when fully grown, figure 34-9A and B. Plants selected for the landscape should add to human comfort in the setting. For example, plants and other materials used around a children's play area should be small so that the children can relate to them. In a world of giant adults, it is nice to feel as "tall as a tree," even when child and tree are only 3½ feet tall.

The principle of unity is the master principle of landscape design. Unity creates the flow among scale and proportion, balance, accent, rhythm, and simplicity, figure 34-9C. The landscape design will complement the surroundings and create an aesthetic appeal that is pleasing and beautiful.

LANDSCAPE MAINTENANCE

This is one of the fastest growing divisions of the green industry. It is involved in caring for the grounds after the installation of the plant material. Landscape maintenance is important because it maintains the aesthetics of the landscape area and the proper care of the plant materials and other landscape accessories (patios, pools, mulches, fertilizers, and organic/inorganic chemicals) throughout the year.

Landscape maintenance involves the technical knowledge of the following skills:

- ◆ Replacement of plant material
- ◆ Application of mulch to proper depth around plants
- ◆ Application of soil analysis (for proper pH levels and fertility needs)
- ◆ Pruning
- ◆ Weed control
- ◆ Planting and caring for flower beds
- ◆ Proper mowing procedures
- ◆ Maintenance of landscape accessories (i.e., pools, fountains, and lighting)

(A)

(B)

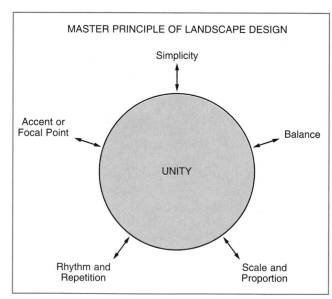

(C)

FIGURE 34-9 The trees in (A) overpower the structure to such an extent that it is dehumanizing and uncomfortable to view. The landscape in (B) is more carefully planned, with plantings in proportion to the structure. The master principles of landscape design are illustrated in (C).

POWER TOOLS	USES
Walk Behind Mower	Professional lawn mowing equipment.
Backpack Blower	Removes leaves and trash from walks and lawn areas.
Line and Blade Trimmer	To trim grass and plant material from areas unable to be mown with the walk behind mower.
Lawn Edger	To trim the grass along the edge of the sidewalk.

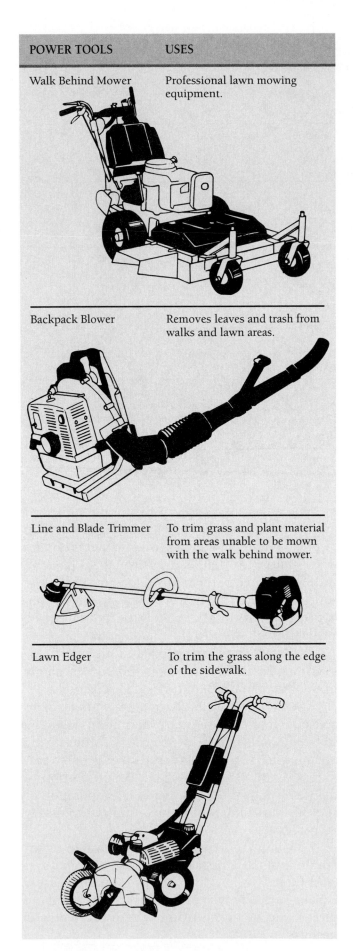

In today's world of landscape maintenance, it is vitally important that the landscape maintenance personnel be able to determine the volume of the mulch to cover a given area. In order to determine the volume, multiply the length in feet by the width in feet by the depth in feet in the area you are mulching. This will give the total cubic feet. Mulch is sold by the cubic yard. There are 27 cubic feet per cubic yard. To determine the cubic yards of mulch to be used divide cubic feet by 27. This will give the total cubic yards needed.

$$\frac{\text{length in feet} \times \text{width in feet} \times \text{depth in feet}}{27 \text{ cubic feet/cubic yard}} = \text{cubic yards}$$

On the average most mulch is applied to the depth of three inches. Therefore 3 inches is what part of 12 inches? $^3/_{12}$ reduced to $^1/_4$ or .25 of a foot.

For example:

10 ft × 100 ft × .25 =

$$\frac{10' \times 100' = 1000 \text{ sq ft} \times .25 = 250 \text{ cubic feet}}{27 \text{ cubic feet/cubic yard}}$$
$$= 9.25 \text{ cubic yards}$$

The most common type of mulch used is in the form of wood chips. The two major types are hardwood bark and pine bark. The bark will vary in size from fine to coarse. Each of these mulches has its special aesthetic appearance, water-holding capacity, organic matter content, and weed control quality, which enhances the quality of the plant and the landscape. Mulch can be purchased in bags or in bulk (which means it is loose and usually sold by loader scoop. Loader scoops vary in size, so it is important to check the volume of the bucket to insure you are receiving the correct amount of mulch).

Landscape maintenance requires the use of hand and power tools. In the chart below is a list of common tools, drawings of them, and their uses.

XERISCAPING

This is a technique used to practice water conservation in creative landscapes. This application ensures water efficiency in all future landscape practices, especially for areas where water is scarce and expensive. This basic landscape practice has been used extensively in the southwestern United States. The demand for water has increased rapidly.

HAND TOOLS		USES
Hand Trowel		Used to transplant seedling plants.
Nursery Spade		Used for edging planting beds and digging nursery stock.
Hand Pruners		Used to prune branches 3/4" or less in diameter.
Lopping Shears		Used to prune branches 3/4" to 1 1/2" in diameter.
Pruning Saw		Used to prune branches 1 1/2" or greater in diameter.
Pole Pruner		Used to prune branches 1 1/2" or greater in diameter 10' to 12' above the reach of the professional grounds maintenance worker.

Therefore, landscape practices must change to meet these new water requirements.

The basic concepts of xeriscaping involve the following applications of landscape principles:

- ◆ good design and planning
- ◆ improving the soil
- ◆ turf areas
- ◆ use of ground covers
- ◆ low water use
- ◆ plant selection
- ◆ use of mulches
- ◆ use of low-volume irrigation
- ◆ proper landscape maintenance

Good design and planning is important to proper xeriscaping because it is necessary to group plants in relation to their water needs. This grouping practice allows one to grow high-water-use plants and still maintain the water conservation needed. When designing a garden, use mass plantings with borders of lower plants in the front and taller plants in back, figure 34-10.

Soil can be improved to provide optimum plant growth and conservation of water. The addition of organic matter will improve the aeration and water supply for the plants. Organic matter is an excellent source of nutrients to plants. The use of soil amendments, such as gypsum, lime, and hydro-gels, aids in water retention and improved availability of water to the planted material.

Turf areas should be seeded in high-drought-tolerant grasses. The turf areas should be separate from other plantings and irrigated separately.

Mulches in the landscape are important to conserve moisture, reduce evaporation, keep the soil cool, and reduce weed growth. Mulches are organic and inorganic materials that provide aesthetic value to the landscape. A high portion of the water that falls is lost through evaporation or run-off. Mulching will reduce this water loss by 90 percent.

Low-volume irrigation is important to incorporate into the landscape. The use of water systems that deliver spray patterns for defined areas is necessary. The use of drip irrigation is a more direct way of pinpointing where the water is needed.

URBAN CONDITIONS	POOR/DRY SOILS	HEAVY CLAY SOILS
Shrubs		
Acanthopanax sieboldanus	Acanthopanax	Acanthopanax sieboldianus
Aronia arbutifolia	Acer ginnala	Aralia spinosa
Berberis thunbergii and cvs.	Berberis x mentorensis and cvs.	Aronia spp.
Celastrus scandens	Berberis thunbergii and cvs.	Berberis, most
Chaenomeles spp. and cvs.	Chaenoleles spp. and cvs.	Chaenomeles, most
Chionanthus virginicus	Cornus racemosa	Cornus, shrub types
Cornus alba	Cotinus coggygria and cvs.	Deutzia, most
Cornus mas	Cytisus spp. and cvs.	Elaeagnus, most
Cornus stolonifera	Elaeagnus angustifolia	Euonymus, most
Elaeagnus umbellalus	Erica carnea	Forsythia, most
Euonymus spp. and cvs.	Genista x "Lydia"	Hamamelis spp.
Forsythia spp. and cvs.	Hamamelis spp. and cvs.	Juniperus, most
Hamamelis spp. and cvs.	Juniperus, most	Kolkwitzia amabilis
Hibiscus syriacus and cvs.	Kolkwitzia amabilis	Ligustrum, most
Hydrangea spp.	Ligustrum spp. and cvs.	Lonicera, shrubby types
Ilex crenata and cvs.	Myrica pennsylvanica	Myrica pennsylvanica
Ilex glabra and cvs.	Potentilla spp. and cvs.	Physocarpus opulifolius and cvs.
Ilex x meserveae and cvs.	Rhamnus spp. and cvs.	Pyracantha, most
Leucothoe spp. and cvs.	Rhus, all	Rosa rugosa
Ligustrum spp. and cvs.	Ribes alpinum and cvs.	Rhodotypos scandens
Lindera benzoin	Rosa rugosa and cvs.	Rhus, most
Lonicera spp. and cvs.	Viburnum lentago and cvs.	Ribes alpinus and cvs.
Magnolia stellata and cvs.	Yucca filamentosa and cvs.	Salix, most
Mahonia aquilohorn and cvs.		Spiraea, most
Myrica pennsylvanica		Stephanandra incisa "Crispa"
Philadelphus virginalis		Symphoricarpos x chenaultii "Hancock"
Physocarpus opulifolius		Taxus, most
Potentilla		Thuja occidentalis and cvs.
Pyracantha		Viburnum dentatum
Rhamnus spp. and cvs.		Viburnum lentago
Rhodotypos scandens		Viburnum opulus and cvs.
Rhus spp. and cvs.		Viburnum prunifolium and cvs.
Ribes alpinum and cvs.		Viburnum sargentii and cvs.
Rosa rugosa and cvs.		
Rosa species type		
Spiraea vanhouttei		
Syringa x prestoniae		
Syringa reticulata		
Syringa vulgaris		
Taxus spp. and cvs.		
Viburnum, most		
Wisteria sinensis and cvs.		
Yucca filamentosa and cvs.		
Trees		
Acer x "Celebration" P.A.F.	Acer saccharinum	Acer campestre
Acer ginnala	Betula spp. and cvs.	Acer rubrum and cvs.
Acer platanoides and cvs.	Celtis occidentalis	Acer saccharinum
Acer saccharinum	Cercis canadensis	Betula nigra
Chamaecyparis nootkatensis and cvs.	Corylus corlurna	Crataegus spp. and cvs.
Crataegus spp. and cvs.	Elaeagnus spp.	Elaeagnus angustifolia
Elaeagnus angustifolia	Eucommia ulmoides	Eucommia ulmoides
Fagus sylvatica and cvs.	Gleditsia triacanthos inermis and cvs.	Fraxinus pennsylvanica and cvs.
Fraxinus pennsylvanica and cvs.	Gymnocladus dioica	Gleditsia triacanthos inermis and cvs.
Ginkgo biloba and cvs.	Kerria japonica and cvs.	Koelreuteria paniculata
Gleditsia triacanthos inermis and cvs.	Koelreuteria paniculata	Liquidambar styraciflua and cvs.
Koelreuteria paniculata	Ostrya virginiana	Malus, most
Liquidambar styraciflua and cvs.	Ulmus parvifolia	Ostrya virginiana

FIGURE 34-10   Plant selection

URBAN CONDITIONS	POOR/DRY SOILS	HEAVY CLAY SOILS
Trees (continued)		
Liriodendron tulipifera		Picea spp. and cvs.
Malus spp. and cvs.		Platanus x acerifolia "Bloodgood"
Picea spp. and cvs.		Phellodendron amurense
Pinus spp.		Pyrus, most
Platanus spp. and cvs.		Salix, most
Populus spp., hybrids and cvs.		Taxodium distichum
Pyrus calleryana and cvs.		Ulmus spp. and cvs.
Taxodium distichum and cvs.		
Ulmus x hollandica and cvs.		
Ground Covers		
	Aegopodium podagraria "Variegata"	Aegopodium podagraria "Variegatum"
	Hosta spp. and cvs.	Ajuga spp.
	Parthenocissus spp.	Euonymus fortunei "Colorata"
	Sedum	Grasses, ornamental
		Hemerocallis, all
		Houttunynia cordata "Chameleon"
		Juniperus, low types
		Lonicera japonica "Halliana"
		Pachysandra terminalis and cvs.
		Parthenocissus spp.
		Rhus aromatica and cvs.

FIGURE 34-10   Plant selection (continued)

XERISCAPE GROUND COVERS

Euonymus fortunei "Colorata"
Grasses, ornamental
Hemerocallis, all
Juniperus, low types
Parthenocissus spp.
Rhus aromatica and cvs.

FIGURE 34-11

Proper landscape maintenance is needed to preserve the beauty of the landscape and conserve water. Weeding, pruning, pest control, and adjustments in the irrigation system help to conserve moisture in the landscape.

Ground covers give the landscape a wide variety of texture and color and retain soil moisture. Large turf areas transpire large quantities of moisture. Using ground covers in beds which are properly mulched will help to conserve moisture. The ground covers in figure 34-11 may be used in a xeriscape setting.

FIGURE 34-12   Properly edged and mulched plant materials create a unique aesthetics in the landscape.

Select plant material with low water requirements. There are a wide range of plants that survive well on minimal watering. The varieties include natives, exotics, cultivated ground covers, trees, and shrubs, figure 34-12.

Student Activities

1) Use cutouts of symbols on a felt board to practice arranging different types of plantings. Prepare the symbols (to represent various plantings and structures) beforehand. Try some arrangements which will (a) draw attention to a corner, (b) give partial screening, (c) give total screening, and (d) provide an interesting silhouette for the viewer's eye.

2) Invite someone who is in a landscaping profession (landscape architect, landscape contractor, or professional horticulturist) to the class to discuss his or her occupation.

3) As a class, install a planting around the school or in a nearby park. Discuss the function of the planting before the actual work is begun.

Self-Evaluation

Provide a brief answer for each of the following.

1) What are the three major career fields in the landscape profession?

2) What is the fastest-growing profession in the landscape industry today?

3) List the objectives of residential landscaping.

4) List the principles of design. Explain each.

5) What is the main function of an outdoor wall?

6) Give three examples of hard paving and soft paving materials.

7) The major types of wood chip mulches are _____ and _____ .

8) Mulches are used for the following purposes:

   a) _____   c) _____
   b) _____   d) _____

9) Determine the cubic yards of mulch for the following areas:

   a) 20' x 45' x 3"
   b) 125' x 6' x 4"
   c) 36' x 34' x 3"
   d) 250' x 250' x 4"
   e) 98' x 25' x 2"

Complete the following statements.

10) Xeriscaping is a landscaping process which incorporates

   a) no water in the landscape.
   b) draining water from the root zone of the plants.
   c) water conservation for the future of our environment.
   d) arid landscaping.

11) Xeriscaping basic concepts involve the following landscape principles:

    a) good design and planning/low-volume irrigation.
    b) improved soil conditions/using mulches.
    c) plant selection/proper landscape maintenance.
    d) all of the above

12) Good design in xeriscaping is important because

    a) it allows plants to be grouped according to their water needs.
    b) plants that require high water usage can be planted together.
    c) it helps to improve water conservation.
    d) all of the above

13) Organic matter will

    a) aerate the soil.
    b) reduce the availability of water.
    c) restrict available nutrients.
    d) none of the above

14) Turf areas in xeriscaping designs should be seeded in

    a) bentgrass.
    b) drought tolerant grass.
    c) perennial ryegrass.
    d) dicondia.

15) Organic mulches are important

    a) to conserve moisture.
    b) as sources of organic matter.
    c) to reduce weeds/improve aesthetic value.
    d) all of the above

16) Mulching can reduce water loss by

    a) 50 percent.   c) 90 percent.
    b) 35 percent.   d) 25 percent.

17) Ground covers are excellent xeriscape plants because

    a) they come in a wide variety of colors.
    b) they conserve moisture.
    c) they come in a variety of textures.
    d) all of the above

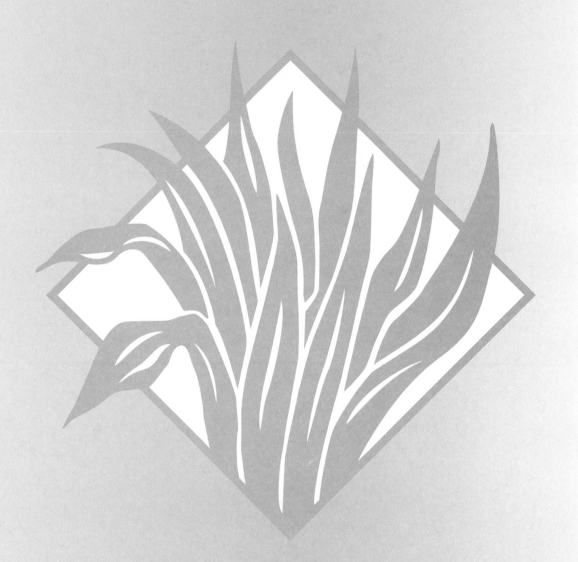

Lawn and Turfgrass Establishment and Maintenance

◆

Establishing the Lawn

OBJECTIVE

To start a lawn by seeding, sodding, and sprig or strip planting.

Competencies to Be Developed

After studying this unit, you should be able to

- ◆ list three reasons for establishing and maintaining a lawn.
- ◆ describe two methods of establishing proper drainage in a lawn.
- ◆ list three materials that are used to increase the organic matter in a new lawn.
- ◆ name eight items that must be included on a seed label.
- ◆ describe three ways turf grasses are started in the United States.
- ◆ demonstrate the five steps in seeding a lawn.

MATERIALS LIST

- ✔ bag of lawn seed
- ✔ bag of lawn fertilizer, lime, and spreader
- ✔ sample of drainage tile
- ✔ samples of peat moss, well-rotted sawdust, and weed-free manure
- ✔ samples of commercial fertilizer and lime
- ✔ samples of cool-season and warm-season turf grasses
- ✔ several steel rakes
- ✔ Rototiller
- ✔ straw
- ✔ lawn roller

Lawns are a major part of most home landscapes. Basically, lawns are established for three reasons.

- ◆ They add beauty to the landscape. A well-kept lawn is very appealing and inviting.

- ◆ They are used as play areas for sports such as baseball, football, basketball, golf, or lacrosse, or for relaxation. Since sports are tough on lawns, it is important to select a lawn grass that can take wear.

- ◆ They provide an excellent cover to help control soil erosion while allowing the movement of air and water to roots of trees and shrubs in the soil below.

Turf grasses are grasses that act as vegetative ground cover. The turf areas they produce are usually mowed regularly. Turf grasses serve a functional purpose by preventing soil erosion but have an aesthetic purpose as well. A properly managed turf area is very attractive. Turf grasses are able to withstand hard use, and some provide an ideal surface for sports fields and other recreational facilities.

SOIL AND GRADING

The first thing to consider in establishing a new lawn is the present condition of the soil. Is this an area in which the builder has graded off all the topsoil? Is the slope too steep to establish a lawn and mow it safely? Is the drainage adequate? These are all questions that must be answered before establishing a lawn.

The builder usually establishes the rough grade. The lot is graded so that the land slopes away from the foundation of the house, to help prevent any water from entering the basement. After the rough grade is established, topsoil which the builder has set aside may be spread over the subsoil which was kept in the rough grading. Six inches of topsoil should be spread evenly over the surface and tilled to loosen and break up clods, figure 35-1.

The general slope for the lawn after the topsoil is spread should not exceed 15 percent, that is, no more than a 15-foot drop for every 100 feet of distance. Slopes greater than 15 percent are unsafe to mow. If a slope steeper than 15 percent cannot be avoided, it should be covered with plants that do not require mowing.

FIGURE 35-1 Topsoil being spread on an area that will be seeded to turf grass (Ed Reiley, Photographer)

DRAINAGE

Good drainage ensures a balance between the air and water in the soil. This, in turn, encourages proper root growth. Grasses can endure relatively wet soil, but not bog or swamp conditions. There are two ways of establishing proper drainage in lawns. One method is to install drainage tile about 3 feet below the surface of the soil to drain the subsoil. Another method is to make use of the slope of the land to drain surface water away.

PREPARATION OF THE SOIL

As mentioned, 6 inches of topsoil should be spread over the rough graded subsoil. A good garden loam is the best medium for most grasses. If quality topsoil is not available, the organic matter content of the soil should be increased by adding well-rotted sawdust, weed-free, well-rotted manure, or commercial peat moss at the rate of 6 cubic feet per 1,000 square feet of land. Work the material well into the soil with a Rototiller. It may be necessary to remove stones or dirt clods. Use a hand rake or a stone rake mounted on a tractor to remove stones. A good seedbed should have a firm and smooth, but not powder fine, surface texture.

FERTILIZER

A complete fertilizer with a high phosphorus content is recommended for use in establishing new lawns. Some companies manufacture a special fertilizer known as a *starter fertilizer* which is especially high in phosphorus.

Before applying fertilizer, the soil should be tested to determine the correct amount of fertilizer and lime to add. This can be done in one of two ways: with a special portable soil test kit, or by sending a sample of the soil away to a soil testing laboratory. When test results are returned from these laboratories, they should contain a recommendation regarding the amount and analysis of fertilizer to apply.

The results of soil tests sometimes indicate a need for the addition of lime to the soil. Lime reduces the acidity of soil and encourages root development. There are different forms of lime available for use on lawns. They include ground limestone (calcium carbonate), burned lime (calcium oxide), and hydrated lime (calcium hydroxide). When choosing a lime for use in a lawn, pick the one which gives the most calcium for the money. If the soil pH is too high or alkaline, sulfur or iron sulfate may be used to lower the pH.

Most lawn grasses grow best in a well-limed soil with a pH level from 6.0 to 6.5. If the soil test calls for lime, add the recommended amount evenly over the entire soil surface. Work it into the ground 4 to 6 inches before seeding.

SPREADING THE FERTILIZER

Fertilizer should be spread in two different directions on new lawns. Spread one-half the recommended amount in a north-south direction and the rest in an east-west direction. This gives a better distribution of the fertilizer.

Fertilizers are usually applied most efficiently by use of a spreader. It is important that the fertilizer be applied uniformly. There are two basic types of spreaders available. Figure 35-2 shows a rotary spreader, which spins the fertilizer out over a relatively wide path. Figure 35-3 shows a conventional spreader which spreads fertilizer in a strip as wide as the spreader. Both spreaders can be adjusted so that the proper amount of fertilizer is spread. The fertilizer is then worked into the soil surface with a hand rake, figure 35-4.

STARTING THE LAWN

Lawns are started in one of two ways: (1) seeding, (2) vegetatively by sodding, plug, strip, or sprig planting.

FIGURE 35-2 Rotary or cyclone spreader used for applying fertilizer or seeding grass (Ed Reiley, Photographer)

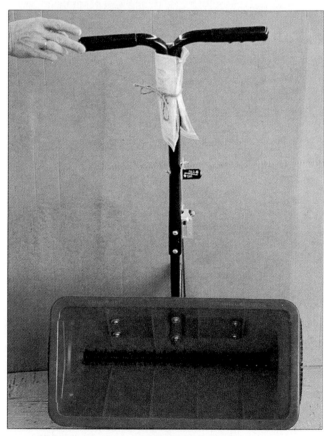

FIGURE 35-3 Conventional spreader (Ed Reiley, Photographer)

SELECTING THE SEED

It is very important to purchase the best lawn seed. All lawn grass seed is required by law to have a label with the following information.

NAME The seed package must give the correct name of all the seeds included in the package. Information on where this type of seed grows best—

FIGURE 35-4 The fertilizer is worked into the topsoil with a hand rake. (Photo courtesy of Michael Dzaman)

in a shaded area, partial shade, or full sun—is also provided.

PURITY These figures give the makeup of the seed by percent of the different grasses included in the package.

PERCENT OF GERMINATION This figure indicates the percentage of seeds in the package that will sprout when planted. The higher this percentage, the more seed there is that will germinate. (A germination percentage of 85 means that 85 seeds out of 100 will germinate.)

OTHER CROP The percentage of other crop seed is important to consider. Crop seeds such as wheat, barley, orchard grass, and timothy are undesirable seeds for lawns.

INERT MATTER This is material that will not grow. It includes pieces of seed, sand particles, and chaff from the seed. All of these things add weight to the package, but they do not germinate. This type of material is undesirable.

WEED SEED A weed is any plant that is growing where it is not wanted. Weed seeds are not desirable in a lawn grass mixture. Some weed seeds are very small and difficult to remove from the desirable seed. Most packages of seed contain some weed seed, but the manufacturer is required by law to list the percentage present in each container.

NOXIOUS WEEDS Certain weeds have been declared by each state as noxious weeds because they are particularly difficult to control. In some cases, these weeds may be harmful. This material is listed on the package by a specific number, for example, one listing might read, *four garlic per ounce.*

YEAR TESTED Commercial seed must be tested every year for the correct germination percentage. The month and year that the seed was tested are sometimes given also.

COMPANY NAME The name of the company selling the seed and its correct address must be given on the label. This is done so that anyone having problems with the seed can contact the grower.

There are some much improved varieties of cool-season grass seeds especially for use where heavy traffic problems exist. Some of the best are tall fescues which are drought and disease resistant, and take a lot of wear. Check with your local lawn specialist for varieties best suited to your area.

Some grass varieties are resistant to insect attack. Recent research discovered that many of these grasses have a fungus growing in their leaves. This fungus, called an endophyte, does not harm the grass but protects it against insect attack. Insects will not eat an endophyte-infected plant. Endophytes are found in tall fescues and red fescue and are naturally transmitted through the seed. Researchers are attempting to place endophytes in other grass varieties to improve insect resistance. There is also genetic resistance to insects found in grasses. This is different from the resistance caused by endophytes.

Insect resistant varieties should be used whenever they are available. The tall fescues Titan, Shenandoah, and Mesa are resistant because of endophyte infection.

SEEDING

SOWING SEED Seed may be planted by hand or with a mechanical seeder such as the seeder shown in figure 35-2. To obtain uniform distribution, the seed is mixed with small amounts of a *carrier*, such as sand. The mixed material is divided into two equal parts; one part is sown in one direction, and the other part crosswise to the first sowing. Figure 35-5 illustrates hand sowing. Figure 35-6 lists seeding rates for cool-season grasses.

FIGURE 35-5 Lawn seed is sown by hand or by means of a mechanical seeder. A rotary spreader is shown here. (Ed Reiley, Photographer)

GRASS VARIETY	POUNDS PER 1,000 SQUARE FEET
Tall fescues	6 to 8
Red or fine fescues	3 to 5
Bluegrass mixes	1½ to 2
Bluegrass fescue mix	2 to 3
Perennial ryegrass	5 to 6

FIGURE 35-6 Lawn seeding rates for cool-season grasses

COVERING THE SEED The seed is lightly covered by hand raking. Large seeds are covered with ¼ to ⅜ inch of soil and small seeds with ⅛ to ¼ inch of soil. It is important that all seed is covered by and in close contact with the soil, figure 35-7.

MULCHING Mulching with a light covering of weed-free straw or hay helps to hold moisture and prevent the seed from washing away during watering or rainfall. Straw also helps to hide the seed from birds. One 60- to 80-pound bale of straw or hay mulch will cover 1,000 square feet of area. Mulches applied evenly and lightly may be left in place and the grass allowed to grow through. Peat moss or other fine material does not make satisfactory mulch. These materials become packed too tightly, resulting in the seed being planted too deep, figure 35-8.

FIGURE 35-7 A hand rake is used to cover the seed. (Ed Reiley, Photographer)

FIGURE 35-8 Straw being spread over a newly seeded lawn (Ed Reiley, Photographer)

On terraced areas or sloping banks, cheesecloth, burlap, or commercial mulching materials helps to hold in moisture and keep the seeds in place. Grass is able to grow through these mulching materials, which may be left to rot.

WATERING New seedlings should be kept moist until they are well established, figure 35-9. Once seeds have begun to germinate, they must not be allowed to dry out, or they will die. Avoid saturating the soil, however; excessive moisture

FIGURE 35-9 The newly seeded lawn should be given at least 1 inch of water. Notice the newly seeded lawn is mulched with straw to shade and keep the soil moist. Seed comes up faster in moist soil. (Ed Reiley, Photographer)

is favorable for the development of *damping off*, a fungus disease.

VEGETATIVE PLANTING

There are some grasses for which seed is not available, or the seed that is available does not produce plants that are true to type. These grasses must be planted by one of several vegetative methods, such as spot or plug sodding, strip sodding, sprigging, or stolonizing. Grasses planted by vegetative methods include zoysia, improved strains of Bermuda grass, St. Augustine grass, centipede grass, creeping bentgrass, and velvet bentgrass.

Whether plugged, strip sodded, sprigged, or stolonized, the planted material must be kept moist until well established. During the first year, light applications of a nitrogenous fertilizer every two to four weeks during the growing season helps to speed the spread of the grass.

SODDING

Sod consists of grass and grass roots in a thin layer of soil which is removed from the areas in strips. It is rolled and transported to the area to be sodded, figure 35-10. Unless good quality sod is available and complete coverage is needed immediately,

FIGURE 35-10 Sod cut at the sod farm. It is rolled ready for shipment to the planting area. (Courtesy Dr. Kevin Mathias)

the expense of sodding is justified only on steep slopes or terraces where erosion may be a serious problem.

Sod should not be cut more than 1 inch thick. Sod which is cut 3/4 inch thick will knit to the underlying soil faster than thicker sod. The sod pieces are laid in the same way as bricks are and fitted together as tightly as possible. After the first strip is laid, a broad board is placed on the sodded strip. The board is knelt on and moved forward as the job progresses. This eliminates tramping on the prepared seedbed.

The planting area is prepared and fertilized in the same manner for sodding as for seeding. A roller is used to firm the seedbed after final hand raking and leveling. After the sod is laid it is tamped or rolled lightly and topdressed with a small amount of topsoil to fill in any cracks between the pieces of sod. The back of a rake or a straw broom may be used for this job. It is important that the sod be kept moist until the roots have grown well into the soil.

SPOT SODDING OR PLUGGING *Spot sodding* or *plugging* is the planting of small plugs or blocks of sod at measured intervals. The plugs are spaced from 8 inches to 1 foot apart depending on how rapidly the area is to be completely covered. The closer the plugs are set, the faster the lawn will be completely covered with grass. The plugs are fit tightly into prepared holes, figure 35-11. They are then tamped firmly into place. Zoysia grass is started in this way, figure 35-12.

STRIP SODDING *Strip sodding* is the planting of strips of sod end to end in rows that are 1 foot

FIGURE 35-11   A turf plugger is sometimes used to make the holes into which plugs of grass are planted. (Photo courtesy of The Scotts Company, Mavysville, OH)

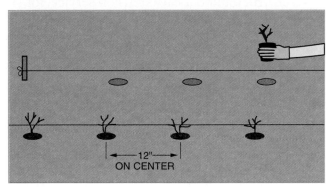

FIGURE 35-12   Zoysia grass plugs spaced 12 inches on center. A line is being used to keep the rows straight.

apart. The sod strips should be 2 to 4 inches wide. Firm contact with surrounding soil is necessary.

SPRIGGING   *Sprigging* is the planting of individual plants, runners, cuttings, or stolons at certain spaced intervals. The *runner* (sprigs of grass) are obtained by shredding solid pieces of sod. The spacing of the runners is determined by how fast that particular grass grows, how fast the area is to be covered, and the amount of plant material available. Runners may also be planted end to end in rows. See figure 35-13.

STOLONIZING   In *stolonizing*, shredded stolons are spread over the area with mechanized equipment. The spreading is followed by disking or rolling the planted area and topdressing with fertilizer. (*Topdressing* is the spreading of material on top of the soil.) Stolonizing is usually done only when large areas are to be planted or when the area to be planted is highly specialized, such as golf course putting greens.

FIGURE 35-13   Turf area planted with sprigs

Large areas of Bermuda grass may be established by spreading shredded stolons with a manure spreader and disking lightly to firm into the soil. (*Disking* is the mixing of soil with a disc harrow.) This method requires 90 to 120 bushels of stolons per acre. Creeping bentgrass and velvet bentgrass are stolonized by spreading shredded stolons at the rate of 10 bushels per 1,000 square feet, topdressing with topsoil to a depth of ¼ inch, and rolling to firm the stolons into the topdressing.

CHOOSING THE GRASS TYPE

When deciding upon which grass to plant in a certain area, the most important factors to consider are the climate conditions, such as the temperature of the area and the available moisture. Figure 35-14 shows a map of the continental United States which is divided into areas suitable for the planting of certain grass types. Following is a list of the various grasses suitable for the climatic regions illustrated in figure 35-14.

Region 1. Common Kentucky bluegrass, Merion Kentucky bluegrass, red fescue, and Colonial bentgrass. Tall fescue,

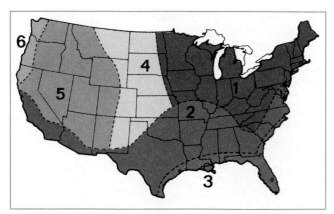

FIGURE 35-14 Choosing a grass type for a certain area

Bermuda grass, and zoysia grass in southern portion of the region.

Region 2. Bermuda grass and zoysia grass. Centipedegrass, carpetgrass, and St.

Augustine grass in southern portion of the region with tall fescue and Kentucky bluegrass in some northern areas.

Region 3. St. Augustine grass, Bermuda grass, zoysia grass, carpetgrass, and bahiagrass.

Region 4. Nonirrigated areas: Crested wheatgrass, buffalograss, and blue gramagrass. Irrigated areas: Kentucky bluegrass and red fescue.

Region 5. Nonirrigated areas: Crested wheatgrass. Irrigated areas: Kentucky bluegrass and red fescue.

Region 6. Colonial bentgrass and Kentucky bluegrass.

Before planting, also check other sources, such as your instructor and local extension service, for varieties that are recommended for your locality.

Student Activities

1) Create a display showing labeled samples of lawn seed.

2) Establish small yard-square plots of various lawn grasses. Practice leveling the area with a steel rake to establish a seedbed.

3) Practice using a lawn spreader to obtain uniform coverage of areas to be seeded. If the seed and the necessary ground are not available, sand can be used in place of seed, and applied to marked areas of a parking lot.

Self-Evaluation

A) Select the best answer from the choices offered to complete each statement.

1) The finished grading around a home is done so that rainwater will

   a) stay in place and soak into the soil.
   b) gradually drain away from the house.
   c) drain toward the house.
   d) quickly run off the lawn area.

2) For easiest and safest mowing, the grade or slope of the lawn should not exceed

   a) 5 percent.
   b) 10 percent.
   c) 25 percent.
   d) 15 percent.

3) After the rough grade of a lawn is established, topsoil is added to a depth of

    a) 6 inches.   c) 3 inches.

    b) 2 inches.   d) 12 inches.

4) A complete fertilizer which is high in _____ is best for seeding grass to start new lawns.

    a) phosphorus   c) potash

    b) nitrogen   d) lime

5) A good seedbed for sowing grass consists of topsoil prepared so that it is

    a) loose and porous.

    b) coarse and open.

    c) powdery fine.

    d) firm, smooth, and relatively free of rocks.

6) When establishing new lawns, the fertilizer is spread evenly over the soil surface and

    a) watered in.

    b) seeded immediately.

    c) worked into the soil surface with a rake.

    d) dug deeply into the soil.

7) The percentage of germination listed on seed container labels indicates

    a) the purity of the seed.

    b) the percentage of seeds that will sprout or germinate.

    c) the percentage of weed seeds in the mixture.

    d) the amount of inert matter in the seed.

8) Grass seed should be spread evenly over the soil surface to reduce the chance of missing areas. To accomplish this,

    a) sow half the recommended amount of seed in one direction, and the other half at a 90-degree angle to the first seeding.

    b) always sow by hand.

    c) always sow using a rotary seeder.

    d) always rake the seed into the soil.

9) Lawn seed should be carefully covered with soil to a depth of

    a) ¼ inch.   c) ½ inch.

    b) ⅛ to ⅜ inch, depending on seed size.   d) 1 inch.

10) A newly seeded area is often mulched with clean, weed-free straw or other material to

    a) help shade out weeds.

    b) make it possible to walk on the area immediately after seeding.

    c) hold moisture in the soil and prevent seed from washing away.

    d) eliminate the appearance of bare soil.

11) Once grass seed germinates and new seedlings begin to grow, it should

    a) never be allowed to dry out.

    b) be mowed regularly.

    c) be fertilized regularly.

    d) be rolled to firm the roots against the soil.

12) Planting grass by use of a solid covering of sod is an expensive method of establishing a lawn and is generally used

    a) only by wealthy people.

    b) on steep slopes and terraces where erosion is a serious problem.

    c) by contractors.

    d) in the off-season, when seed would not grow otherwise.

13) In establishment of lawns, spot sodding is a process in which

    a) small plugs or blocks of sod are placed at measured intervals.
    b) sod is used to patch bare spots in an old lawn.
    c) sod is used on steep slopes.
    d) sod is used on terraces.

14) Sprigging is

    a) cutting sprigs of grass to keep it short.
    b) sowing a new lawn.
    c) chopping up a lawn to provide better drainage.
    d) planting individual plants, runners, cuttings, or stolons at spaced intervals to establish a new lawn.

15) Stolonizing is

    a) spreading shredded stolons of grass to establish a lawn.
    b) rolling a seedbed to make it firm.
    c) removing stolons from grass to thicken the lawn.
    d) all of the above

B) List the three reasons lawns are established.

Maintaining the Lawn

OBJECTIVE

To use proper lawn maintenance techniques.

Competencies to Be Developed

After studying this unit, you should be able to

- ◆ list the seven factors of good lawn maintenance and explain each orally.
- ◆ describe the analysis of a good turf fertilizer.
- ◆ determine the best time to apply fertilizer to a lawn.
- ◆ demonstrate how to set the mowing height of a rotary mower.
- ◆ list the three causes of fungus disease in lawns.

MATERIALS LIST

- ✔ lime and fertilizer
- ✔ tape measure and enough string to lay out 1,000 square feet of lawn area
- ✔ one dull and one sharp mower blade
- ✔ knapsack sprayer for applying weed killer
- ✔ crabgrass weed killer (in combination with fertilizer)
- ✔ 2,4-D for broadleaf weed control
- ✔ rotary mower

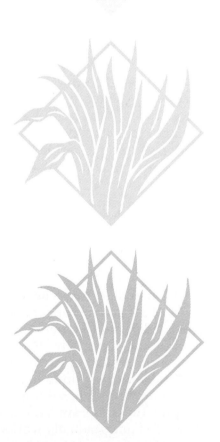

After the lawn has been seeded or planted vegetatively, care must be taken to keep it healthy. Proper lawn maintenance is dependent upon seven factors:

- ◆ planting the proper variety or species of grass.
- ◆ applying fertilizer and lime or sulfur at the proper time and in the proper form and amount.

◆ mowing to the proper height at the correct time.

◆ watering properly.

◆ using chemicals for weed, insect, and disease control if necessary.

◆ using the lawn in such a way that the traffic is not too heavy.

◆ providing the proper amount of light.

PLANTING THE PROPER VARIETY OR SPECIES

Different varieties or species of grasses grow best under different temperature, light, and moisture conditions. If planting is to be done in a cool climate, for example, cool-season grasses should be selected. Some grasses, such as the creeping red fescues, grow better in shade than in full sun. They also grow better in dry areas. And some grasses are disease resistant and require less fertilizer. No grass does well in dense shade, however. Bluegrass grows well in cool climate areas receiving full sun and adequate moisture. Bermuda grass grows well in hot climates.

Another factor to consider when choosing a variety of grass is the way in which the lawn is to be used and the amount of traffic to which the lawn will be exposed. The more a lawn is used as a recreation or work area, the more stress there is on the grass. If heavy use is anticipated, select a hardy grass that can withstand more traffic. (The proper placement of walks and rest or play areas reduces traffic on grass plants.) Check with a local agricultural experiment station for the varieties of grass that grow best in your area.

APPLYING LIME, SULFUR, AND FERTILIZER

pH ADJUSTMENT

Lime should be used whenever necessary to keep the soil pH in the proper range. A pH of 6.0 to 6.5 is a good pH range for most grasses. Finely ground limestone is generally the best and least expensive form of lime to use. Late fall and winter are generally the best times to apply lime.

Lime moves down through the soil very slowly. Applications are spread on top of the soil as needed. Lime gradually works its way down into the soil at a rate of ½ to 1 inch a year.

Sulfur should be used on soils that are alkaline to adjust pH downward.

FERTILIZER

If a healthy lawn is to be maintained, annual applications of a turf fertilizer are needed. Nitrogen is leached from soil and must be replaced regularly.

Lawns require a fertilizer high in nitrogen (indicated by the first number on the fertilizer label). Of the three analysis numbers on fertilizer labels, the first number is always highest on high quality maintenance turf fertilizers.

Many turf fertilizers supply nitrogen in an organic form. This form of nitrogen is released slowly and thus does not burn the grass. It also supplies nutrients over a longer period of time. The urea formaldehyde form of nitrogen is often used as a slow-release fertilizer for turf grasses. If an inorganic form of nitrogen which is released rapidly (such as nitrate of soda) is used, it must not be applied to wet, actively growing grass. Such an application would burn the grass badly.

APPLYING FERTILIZER

Fertilizer may be applied by hand if uniform coverage can be obtained. However, it is better to use a spreader.

Fertilizer should be applied just prior to the active growing season. For cool-season grasses, this is in the very early spring or better in the fall. (Two fall applications, six weeks apart, are best.) These grasses grow in spring and fall and are either dormant or very slow growers in the hot, dry summer months. To fertilize them in summer would encourage weeds that grow in hot weather and could use the fertilizer for growth. Where grass remains active all summer, fertilizer should be applied at least once during this time.

Care must be taken not to skip spaces or overlap with the fertilizer spreader. Difference in green shades and growth rate of the grass will result, ruining the appearance of the lawn. Fertilizer rates depend upon such things as soil type and the amount of rainfall available in the area. Check local recommendations by contacting your local extension service for a lawn care calendar which gives the proper dates for fertilizing as well as spraying and seeding.

Tips: Proper fertilizer use is one of the most important factors in maintaining a good lawn.

MOWING THE LAWN

There are two types of mowers used for mowing lawns, the reel mower and the rotary mower. The type of mower that is used generally depends upon individual preference; either does a good job. Close mowing of ½ to 1 inch is done most efficiently with a reel mower.

Most lawns are cut too short because the homeowner believes that the grass looks better when short. A very short cutting reduces the leaf area of the grass to such an extent that it cannot manufacture enough food. Also, grasses that are cut too short encourage weed growth since the grass plants cannot kill weed seeds by "shading them out." Cool-season grasses should not be cut shorter than 2 to 3 inches, figure 36-1. Warm-season grasses are cut shorter, from ½ to 1¼ inches, depending upon the grass variety. Since warm-season grasses grow rapidly during warm weather, they are better able to compete with weeds growing during the same season. Bermuda grass should be clipped to ½ to 1 inch and zoysia ¾ to 1¼ inches.

Lawns should be mowed often enough so that no more than one-third of the top is cut off in any one mowing. For example, if grass is kept at a height of 2 or 3 inches, only 1 inch should be cut off. This means that when the grass reaches 3 or 4 inches in height, it is time to mow again. This may require mowing two or three times a week when the grass is in periods of most active growth. This is a very important point to remember, since if grass is allowed to grow to a height of 5 to 6 inches and is cut back to 3 inches in one mowing, most of the food-producing leaf blade is removed. This causes the lawn to appear yellow, an indication that it cannot manufacture enough food until more new leaves are formed. This constantly drains food reserves from the roots, weakening the root system. Tall weeds also have a better chance to establish a root system if the lawn is not mowed often enough.

Cutting the lawn to the proper height requires adjustment of the mowing height of the mower. This is done by setting the mower on a flat, smooth surface and measuring from the surface up to the blade with a yardstick, figure 36-2. Adjust the height of the mower according to the height of the grass being mowed.

> *Caution: Remove the spark plug wire before turning the blade to measure it.*

The mower should be kept sharp at all times so that grass blades are cut and not torn off. Jagged, split ends cause the lawn to appear uneven and provide open wounds for the attack of fungus diseases.

The new improved mulching mowers cut grass blades into very fine pieces which do not have to be removed from the lawn. This saves mowing time and also reduces fertilizer needs by about 50

FIGURE 36-1   This lawn of creeping red fescue grass has been mowed to a height of about 2½ inches, ideal for this cool-season grass. When it reaches a height of 3½ inches, it will be ready for another mowing. Thus, no more than one-third of the leaf area is cut at any one mowing. (Ed Reiley, Photographer)

FIGURE 36-2   Adjusting the mower height to the height required by the grass. This blade will cut the grass to a height of about 2 inches. (Ed Reiley, Photographer)

percent. As the grass blades rot, plant food elements are released into the soil.

MOWING TECHNIQUE

The lawn should be mowed in such a pattern that it is cut in one direction at one mowing and at right angles to that direction the next mowing. This helps to eliminate soil compaction and gives the lawn a more even appearance. It may also help to reduce thatch buildup in the lawn. A slight overlap is necessary on each pass over the lawn. This overlap prevents missed areas and picks up any grass that the mower wheel may have only pushed down on the previous trip.

If the lawn becomes too tall and it is necessary to cut off a large part of the grass blades, the grass should be collected in a catch bag attached to the mower or raked up and removed. Heavy accumulation of grass clippings shades out light and kills grass. It also mats down the grass and causes a buildup of thatch on the soil surface. (*Thatch* is a layer of dead grass.) Thatch reduces soil aeration, thus damaging roots and also providing a breeding place for insects and diseases. It is better to mow two or three times in close succession, taking off only 1 inch each time, until the desired height is reached.

New growth regulators sprayed onto the leaf blades of grass slow down the rate of growth of turf grasses. This cuts down on the number of times the lawn must be mowed. One of these growth regulators, named PRIMO, works by temporarily inhibiting or stopping the use of gibberellic acid by the turf grass. As explained in Unit 5, gibberellic acid stimulates cell division and lengthening, causing plants to grow taller. PRIMO reduces mowing frequency by one half. Grass becomes greener in color and more compact. These plant growth regulators work best on lawns having a single grass type. Mixed grasses tend to grow back unevenly, one grass type may regrow faster than others. PRIMO is relatively safe to use and breaks down in the soil into carbon dioxide and water within 3 months. There is not space in this text to give more detail on the new growth regulators, but you may want to research them further.

WATERING THE LAWN

A great deal of water is required to give lawns the moisture they need. An inch of water equals about ½ gallon of water on each square foot of lawn. Unless an adequate supply is available, the lawn should not be watered at all. Shallow watering does more harm than good. This is because shallow watering causes the grass roots to move to the surface to absorb water. Here, they are more easily dried out and require still more frequent watering. Surface roots are more easily torn loose from the soil during winter freezing and thawing. Frequent watering also encourages the growth of fungus disease on the grass blades.

At least 1 inch of water should be applied at each application. An inch of water penetrates a loam soil about 16 inches. To determine when enough has been applied by a sprinkler, set a rain gauge or any container with straight sides on the lawn on the area to be watered. When the container has 1 inch of water in it, 1 inch has been applied to the lawn.

Do not apply water faster than it can soak into the soil surface; runoff is wasted water. Do not water until the grass wilts and needs water, and then water only if you have enough water to do the job well.

A healthy lawn can become dormant and withstand a great deal of dry weather without being permanently damaged. The grass will become green and active again with the first good rain. The appearance may not be as pleasing when the grass turns yellow or brown, but little actual damage is being done.

SOLVING PROBLEMS IN THE LAWN

CONTROLLING WEEDS

If a lawn is heavily infested with weeds, chemicals should be used to eliminate the problem. However, development of a thick, healthy turf is the best way to guard against a serious weed problem. Weeds in a lawn are usually an indication of an unhealthy lawn caused by poor maintenance practices. When weeds are spotted, first check maintenance practices and correct them if necessary. If weeds are still a problem, consider chemical control.

Two types of chemical weed killers are used on lawns. One is a preemergence weed killer, which kills germinating seeds and very tiny weed seedlings before the weeds become established. This is the type of weed killer which is applied in the spring to cool-season grasses for the control of crabgrass. It is often mixed with a fertilizer before application. Since crabgrass is a grass, the mature plant cannot be controlled without killing the lawn grass. However, the tender seedlings can be killed with small amounts of weed killer which are not strong enough to kill the lawn grass. Do not use on a newly seeded lawn.

Postemergence weed killers are applied after the weeds sprout and begin to grow. A sprayer is generally used to apply these types of weed killers, figure 36-3. Notice that the sprayer is labeled to avoid using the container for any other purpose. Since many weed killers are selective and kill only certain types of plants, it is possible to select a chemical that does not kill grass, but does kill broadleaf weeds. Once chemical of this type is 2,4-D. It is used to kill many broadleaf weeds in lawns, such as dandelions and plantain. For specific weed identification and control, contact your local extension service to obtain bulletins written for the local area.

FIGURE 36-3 Small pressurized sprayer used to apply 2,4-D weed killer to broadleaf weeds in a lawn. Notice that the sprayer is marked "weed killer" to guard against its use on ornamental plants. (Ed Reiley, Photographer)

LAWN DISEASES AND THEIR CONTROL

Most turf diseases are caused by parasitic plants called fungi (plural of *fungus*). These parasites live in and on dead grass and in the soil where they attack the green grass and rob the soil of nutrients. This weakens and can kill the lawn grass plant. Fungus diseases are spread easily by mowing or simply walking through the lawn, especially if the grass is wet. The tiny seedlike spores spread rapidly.

For fungus diseases to cause a serious problem, there must be, in the area:

- grass plants on which the fungus can live.
- fungus spores and a means of spreading them to the grass.
- temperature and moisture conditions favorable for the growth of fungi.

DISEASE PREVENTION

The best control measure for fungus is prevention. To reduce the chance of attack by fungus spores on lawns,

1. do not overuse nitrogen fertilizer.
2. add lime or sulfur as needed to maintain a pH of 6.0 to 6.5. (This helps to control thatch buildup.)
3. avoid thatch buildup by collecting clippings or by raking and removing thatch.
4. water only when necessary, and then water deeply. Do not repeat watering for one week. Water at night. Most fungus diseases require fourteen to sixteen hours of wet grass to become infectious or spread. The grass may be wet at night from dew, and if watering is done in the morning the wet period is extended long enough for disease to spread.
5. mow frequently, and remove only one-third of the top growth at any one mowing. Cut grass to the proper height and keep it within 1 inch of that height throughout the entire growing season.
6. keep trees pruned to allow sufficient light for good growth.

INSECT CONTROL

Insects can cause serious damage to lawns. There are insects that attack the top of the plant, such as the chinch bug, sod webworm, and the flea beetle. Serious damage is also done by insects that attack grasses below the soil level by eating roots and stems. Some below-the-surface feeders are the grubs (larvae) of the Japanese beetle and many other beetles and cutworms. See figure 36-4.

When insect attacks are serious, spraying or application of granular insecticides may be necessary. Consult local extension service bulletins for pictures to assist in identification of the pests and control recommendations.

THATCH CONTROL

Thatch is a layer of dead stems, leaves, and roots of grass which builds up on the soil surface and under the green leaf area of grass. A thick layer of thatch may prevent water from penetrating the soil, prevent proper soil aeration, and provide a breeding area for insects and disease. Heavy thatch buildup and reduced aeration of the soil result in shallow root systems.

(A)

(B)

FIGURE 36-4 (A) Typical turf damage caused by underground grubs eating roots off the grass. (B) Grubs in the soil in the grass root area (Courtesy Dr. Kevin Mathias)

The amount of thatch buildup varies somewhat according to the variety of grass that is planted. For example, Merion Kentucky bluegrass builds up thatch faster than common bluegrass. Bentgrass and some red fescues also tend to build up thatch.

Other factors that may cause thatch to build up are:

♦ not adjusting the mower blade properly, resulting in too much of the grass blades being cut, or not using one of the newer thatching mowers.

♦ returning clippings to the lawn.

♦ heavy fertilization.

♦ heavy clayey soil.

♦ acidic soil (apply lime).

On small areas, thatch may be removed with a hand rake. A better method is to use a power-driven thatch machine; it is well worth the rental fee. Early spring is usually the best time to remove thatch from the lawn. Another solution to the problem of thatch is a light application of topsoil ($1/8$ inch). This buries the thatch, causing it to rot. However, the best thatch control measures are preventive. Proper mowing, fertilizing, and liming, and removal of clippings if a mulching mower is not used usually provide adequate control.

COMPACTED SOIL

One of the most needed lawn maintenance activities, especially for heavy soils, is aeration. As lawns get older or experience heavy pedestrian traffic, the soils get packed down. This closes soil pores and keeps out both air and water. The grass roots are robbed of two very necessary elements for good growth, oxygen and water.

The use of a machine known as a plugger is the best treatment for compacted soils, figure 36-5. This roller-type machine pulls out plugs of soil about $1/2$ inch to $3/4$ inch in diameter and up to 2 inches long, figure 36-6. This leaves holes in the lawn soil for air and water to enter.

Pluggers come in small, walk-behind sizes and large, tractor-mounted sizes, figure 36-7.

LOOSENED GRASS

Rollers are used to press down grass that may have been loosened, pushed, or heaved out of the ground by freezing and thawing during the win-

FIGURE 36-5   A plugger used for soil aeration. Notice the hollow tubes which pick up and push out plugs of soil to produce holes in the lawn. (Ed Reiley, Photographer)

FIGURE 36-7   A well-cared for lawn being aerated to stimulate deep root growth and moisture intake. (Courtesy Dr. Kevin Mathias)

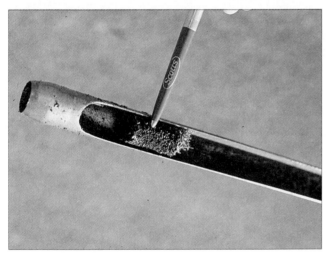

FIGURE 36-6   A plug of soil removed from the lawn. It is about ³/₄ inch in diameter and about 1 ¹/₂ to 2 inches long. (Photo courtesy of The Scotts Company, Mavysville, OH)

PROBLEM	SOLUTION
too much shade	Reduce shade or plant a variety of grass that is shade tolerant.
poor growth of grass	Fertilize properly. (A thick turf prevents the growth of moss.)
acidic soil conditions	Test soil and add lime.
a wet area (Moss will grow in soil too wet for grass.)	Fill low, wet areas with topsoil or drain the area.

FIGURE 36-8   Conditions which can result in the growth of moss in the lawn

ter. Rolling should be done in early spring after the last freezing and thawing. The roller should be heavy enough to push the grass firmly against the soil. Too much weight may make the soil too compact and reduce aeration in the soil. This is especially true for heavy clayey soils.

MOSS

Moss sometimes grows in shady, wet areas of lawns. It is not a weed. Moss grows only where grass has died for some reason. There are four problem areas which can result in the growth of moss in the lawn, figure 36-8.

Student Activities

The following activities are to be accomplished on an area of lawn assigned by the instructor.

1) Examine the area and make a list of problems found in the lawn. Write down possible reasons for the problem areas and report to the class. Topics included in the report might be weed control, fertilizer needs, and insect and disease problems.

2) Develop a lawn maintenance and/or renovation schedule for the lawn area. Have the schedule approved by the horticulture teacher.

3) Locate a good source of lime or sulfur and turf fertilizer. Apply each to the lawn area at the proper time and rate using both a rotary and conventional fertilizer spreader. Calculate how much fertilizer to spread on 1,000 square feet and set the spreader for that application. Check your accuracy by seeing how much fertilizer remains after the job or if you run out of fertilizer before the job is finished.

4) Adjust a rotary mower blade to the best mowing height for the grass to be mowed. Repeat with a reel mower.

5) Examine a sharp and a dull mower blade. Sharpen or observe the proper sharpening technique of the dull blade as demonstrated by the instructor. Balance the blade after sharpening.

> *Caution: Be careful of the sharp cutting edge.*

6) Identify weeds in the lawn area. Determine if a weed killer is needed for control and, if so, which type. Apply the chemical. Observe and record the results.

7) Visit a local golf course. Ask the greens keeper to discuss the type of grass used on the greens, height of the grass and adjustment of the blade, and maintenance schedule used to keep the greens in order. The student can expect this schedule to be more time consuming than that used on the home lawn.

Self-Evaluation

A) List the six management practices that reduce or help to control turf grass diseases.

B) Proper lawn maintenance depends on seven factors. List them.

C) Select the best answer from the choices offered to complete the statement or answer the question.

    1) A cool-season grass that thrives in partial shade is

        a) creeping red fescue.   c) Bermuda grass.
        b) Kentucky bluegrass.   d) centipedegrass.

    2) Lime moves slowly through the soil to correct soil acidity. When applied to the soil surface, lime moves down through the average loam soil at a rate of about

        a) 2 inches a year.   c) ½ to 1 inch a year.
        b) 3 inches a year.   d) 6 inches a year.

    3) Grasses feed heavily on the fertilizer element

        a) phosphorus.   c) potash.
        b) nitrogen.   d) calcium.

    4) What form of nitrogen in fertilizer is released slowly?

        a) organic   c) chemical
        b) inorganic   d) soluble

5) Fertilizer should be applied to a lawn grass

a) when it is available at the lowest price.
b) in midsummer for cool-season grasses.
c) when the grass begins the dormant rest cycle.
d) just before the beginning of the active growth cycle of the grass.

6) Cool-season grasses should be cut to a height of

a) 3 to 4 inches.
b) ½ to 1 inch.
c) 2 to 3 inches.
d) ½ to ¾ inch.

7) Warm-season grasses such as Bermuda grass should be cut to a height of

a) 3 to 4 inches.
b) ½ to 1 inch.
c) 2 to 3 inches.
d) 4 to 5 inches.

8) Frequent shallow watering of the lawn

a) helps to keep the lawn looking fresh and green.
b) helps to control insects in the lawn.
c) causes shallow root development of the grass.
d) helps to control fungus diseases in the lawn.

9) Preemergence weed killers are applied before weeds become established. This type of weed killer is used to control

a) dandelions.
b) broadleaf weeds.
c) all lawn weeds.
d) crabgrass.

10) Postemergence weed killers are used after weeds are present in the lawn. The weed killer 2,4-D is a selective weed killer used to kill

a) annual grasses in the lawn.
b) crabgrass in the lawn.
c) broadleaf weeds in the lawn.
d) all lawn weeds.

11) Most lawn diseases are caused by parasitic plants called

a) toadstools or mushrooms.
b) molds.
c) fungi.
d) slime.

12) Insects that attack grass below the soil level and eat the grass roots are

a) chinch bugs.
b) beetle larvae and cutworms.
c) flea beetles.
d) sod webworms.

13) Thatch buildup in a lawn causes all of the following except

a) faster growing areas that require more frequent mowing.
b) slow penetration of water.
c) poor soil aeration.
d) development of a shallow root system in the lawn.

14) Rolling the lawn each spring may be needed to

a) level the lawn.
b) push grass roots into close contact with the soil which have heaved loose because of frost.
c) provide a firm surface for mowing.
d) roll down any big clumps of grass that might be present.

15) A light application of topsoil (about ⅛ inch) on a lawn is often used to

   a) help control thatch.
   b) help to level the lawn.
   c) fill in the small cracks in the lawn.
   d) add depth to the topsoil.

16) When mowing the lawn, no more than _____ of the top of the grass plants is cut at any one time.

   a) one-half   c) seven-eighths
   b) three-quarters   d) one-third

17) The best weed control for any lawn is

   a) a good, healthy, thick turf.
   b) a combination of preemergence and postemergence weed killers.
   c) a weed killer/fertilizer combination.
   d) none of the above

18) Which of the following does not encourage moss growth?

   a) too much shade
   b) inadequate fertilizer, resulting in poor growth of grass
   c) acid, wet soil conditions
   d) too much lime is used

19) If there are 43,560 square feet in an acre, and recommendations call for 2 pounds of fertilizer per 100 square feet of surface, how much fertilizer would you put on an acre?

   a) 520 pounds   c) 871.2 pounds
   b) 615 pounds   d) 851.3 pounds

Renovating the Lawn

OBJECTIVE

To select and use the proper renovation technique for a specific problem in the lawn.

Competencies to Be Developed

After studying this unit, you should be able to

- ◆ inspect a lawn area and determine if it requires renovation.
- ◆ determine which of the four methods of renovation should be used on a lawn.
- ◆ use the step-by-step approach to renovate a lawn.

MATERIALS LIST

- ✔ grass seed
- ✔ fertilizer and lime, if needed
- ✔ weed killer and sprayer or other applicator
- ✔ garden rake
- ✔ thatch rake or thatching machine
- ✔ lawn mower
- ✔ shovel or sod-lifting tool
- ✔ straw or burlap for mulching seeded areas

When good maintenance practices are not carried out or weather conditions have prevented the growth of healthy grass plants, it is time to consider lawn renovation.

WHY DO LAWNS FAIL?

THE WRONG SPECIES OR VARIETY OF GRASS

Be sure to check the type of grass best suited for the area. It is especially important to consider soil and light conditions, and the amount of traffic on the lawn. Before purchasing seed, check the label for

percentages of the different varieties of seed it contains. Good seed is expensive, but it saves money over a period of time since the lawn will require less reseeding. Seed cost is a small part of the total cost of establishing a lawn.

IMPROPER MOWING

The lawn may have been cut too short. Some people prefer a lawn that is cut very short, but cool-season grasses cannot survive if mowed to heights of ½ to 1½ inches. Grass should not be allowed to grow to be 3 inches high and be cut back to 1 inch. Only one-third of the top should be removed at any one mowing. Remember that close mowing does not reduce the number of times the lawn needs to be mowed.

IMPROPER FERTILIZING

The most common mistakes in fertilizing are applying the fertilizer too late in spring for cool-season grasses or too late in fall for warm-season grasses. Fertilizer should be applied a few weeks before the grass begins its active growing stage. Use a slow-release fertilizer that is high in nitrogen. Apply fertilizer at the proper rate.

IMPROPER WATERING

The most common mistake in watering is failure to apply enough water to soak into the soil to a depth of from 4 to 6 inches. This requires at least 1 inch of water at each watering. One inch of water penetrates 24 inches in sandy soil, 16 inches in loamy soil, and 11 inches in clayey soil. As a general rule, lawns should be watered no more than once a week, and then only when the soil is dry.

HEAVY TRAFFIC

Too much traffic on lawns, especially on those planted with a variety not known for its durability, can result in bare spots. To solve this problem, reduce traffic or change the grass variety. The new, tall fescue grasses take heavy traffic.

EXCESSIVE SHADE

Dense shade and competing shallow roots of trees such as maples kill lawn grasses. In these cases, the tree, and thus the source of the shade, must be removed, or a shade-tolerant grass or other ground cover must be planted under the tree.

SOIL NOT PREPARED PROPERLY FOR PLANTING

A good topsoil, proper fertilizer and pH adjustment, and proper planting techniques are essential for a healthy lawn. (Test the soil to determine fertilizer and pH adjustment needs.)

INFESTATION OF WEEDS, DISEASES, AND INSECTS

Weed, disease, and insect problems are usually the result of one or more of the above problems. A thick, healthy lawn turf is seldom overcome by insects, diseases, or weeds.

THATCH

When old grass leaves, roots, and stems accumulate on the soil surface faster than they rot, the buildup is called thatch. Thatch buildup on the soil surface prevents air from circulating in the soil—a very necessary condition for root growth. Thatch also causes water to run off rather than soak into the soil, and harbors insects and disease organisms. To help speed up thatch rotting, maintain a soil pH of 6.2 to 6.5 and apply more fertilizer. Another solution is to remove the thatch with a thatch removing machine or a special thatching rake.

RENOVATION TECHNIQUES

If the lawn still fails to prosper, renovation, or rebuilding of the lawn, must be considered. Lawns should be renovated at the time of year best suited for starting a new lawn, that is, just prior to the most active growing season for the type of grass being planted in the renovation. The entire lawn may be reseeded, resodded, or planted by other vegetative means such as strip planting or with plugs, or a technique known as spot patching may be used.

WEED-INFESTED LAWNS

When more than one-third of the lawn grass plants are dead and large bare spots exist, it may be better to kill the existing grass with a weed killer such as Roundup (glyphosate) and reseed the entire area, figure 37-1. More than one application may be needed to kill some grasses. After the grass is killed, the entire lawn area is dug up, the old sod is removed, and the lawn is reseeded. The procedure after the soil is cleared of all old grass is the same as for seeding a new lawn. (Refer to Unit 35 for details.)

THINLY COVERED LAWNS

If the lawn grasses cover the entire lawn area but do not produce a thick turf, the lawn may be sal-

FIGURE 37-1 A section of lawn with less than ⅓ of the area producing healthy lawn grass. If the entire lawn suffers from such severe infestation, it is best to kill all vegetation and reseed. (Ed Reiley, Photographer)

FIGURE 37-2 Although the lawn in this area is thin, the bare spots are very small and the plants that are present are healthy ones. There are few weeds present and drainage is good. The best treatment for such an area is an application of fertilizer. Lime may be added if a soil test indicates a need. (Ed Reiley, Photographer)

vaged as follows, providing good topsoil is present and drainage is not a problem.

Procedure

RENOVATING A THINLY COVERED LAWN

1. If a heavy layer of thatch exists, remove it. A thatch-removing machine should be used to dig up the thatch and vertically slice the root area for improved aeration.
2. Add seed on the condition that it will come in close contact with the soil surface. If seed cannot rest firmly on the soil surface, it will not grow. A roller helps to bring seed into contact with the soil.
3. Apply weed killer to control broadleaf weeds if they are present. Do not apply weed killer if new seed was sown as directed in step 2. Weed killer should not be used for six months after a new seeding is made.
   A decision must be made as to whether step 2 or step 3 should be followed; both cannot be accomplished on the same lawn area.
4. Fertilize the lawn with a good turf fertilizer (a slow-release, high-nitrogen fertilizer), figure 37-2.
5. Mow the lawn properly. Remove no more than one-third of the top with any one mowing. Mow at the proper height for the grass species planted. See "Improper Mowing."
6. Follow all good maintenance practices.

SPOT SEEDING

When there are isolated spots of dead or weak grass, the lawn may be renovated by a "patchup" of these spots. Any spot that measures 1 foot and over should be treated.

The same procedure is used to renovate these spots as when seeding a new lawn. The only difference is that the process is used on a much smaller area.

Procedure

RENOVATING ISOLATED SPOTS IN THE LAWN

1. Remove any dead grass or weeds from the area to be reseeded. Dig up the spot with a steel rake and remove any old sod or weeds. Loosen the soil well to prepare a shallow seedbed 1 or 2 inches deep.
2. Add fertilizer as needed. Use a fertilizer with extra phosphorus, such as a 5-10-5. Dig the fertilizer into the top 1 to 2 inches of the soil with the rake.
3. Level the soil and firm the seedbed. Remove any rocks or large clods of soil.
4. Spread the seed uniformly over the surface by hand.
5. Lightly rake the seed into the soil surface. Cover with ⅛ to ¼ inch of soil.

FIGURE 37-3 A small, weedy area in which most of the grass is dead. The weedy area is dug out and pieces of sod (one of which is shown) are placed in the hole. (Ed Reiley, Photographer)

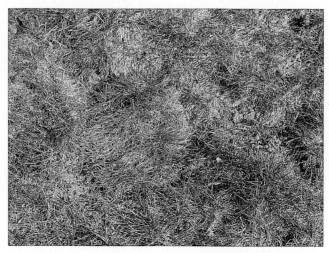

FIGURE 37-4 Sod piece in place. The sod is tamped in place so that is makes good contact with the soil. (Ed Reiley, Photographer)

6. Lightly cover with straw or burlap mulch.
7. Tamp with the feet or roller to bring the seed in close contact with the soil.
8. Practice proper maintenance techniques.

SPOT PATCHING

When good sod can be taken from inconspicuous areas, small spots in the lawn may be patched up by digging out the weak area about 1 inch deep and replacing it with a piece of sod cut to fit the spot, figure 37-3. For large spots, more than one piece of sod may be fitted into the hole. Tamp the sod so that it rests tightly against the loosened soil in the bottom of the hole. Fertilize and water well to stimulate new root growth. It is important that the sod knit well and quickly to the new soil area, and that it not be allowed to dry out until its roots have grown into the soil, figure 37-4.

Student Activities

1) Take a field trip to observe lawns in the local area. Note lawns that are in need of renovation and ones that appear to be healthy. Inquire about the present condition of the lawns and their maintenance schedules. Determine why the lawns are in varying conditions and report to the class.

2) Establish a 1,000-square foot plot for use in practicing lawn renovation. Keep notes of the problems found, treatment, and results of treatment.

3) Invite a nurseryman, golf course superintendent, or other lawn specialist to the class to discuss the most common lawn problems in the local area and the methods used to correct them.

4) Note any soil drainage, shade, or nutrient deficiencies that exist in the local area.

Self-Evaluation

Select the best answer from the choices offered to complete each statement.

1) The most common mistake in mowing cool-season grasses is to

a) mow them too short.

b) mow too often.

c) use a dull mower blade.

d) mow them too high.

2) The renovation technique for weak, bare lawn areas under dense shade is to

a) fertilize more heavily.

b) let more sun in and/or plant a shade-tolerant grass or other ground cover plant.

c) lime the area heavily.

d) mow more often.

3) To renovate a lawn means to

a) dig up the lawn completely.

b) fertilize the lawn.

c) rebuild or restore a strong stand of grass to the lawn.

d) mow the lawn properly.

4) When over one-third of the lawn grass is dead and large bare spots exist, the best renovation practice is to

a) kill all existing plants with weed killer and reseed the lawn.

b) spray with 2,4-D and fertilize to stimulate plant growth.

c) patch up the dead areas and fertilize the entire lawn.

d) none of the above

5) Where lawn grasses are thin but cover the entire lawn area and few weeds are present, the best renovation practice is to

a) remove the thatch layer if it is too thick.

b) either add seed or use weed killer—whichever is needed most.

c) fertilize the lawn and establish a good mowing schedule.

d) all of the above

6) When mowing the renovated lawn, no more than _____ of the top of the grass plant should be removed at any one mowing.

a) ½ inch

b) one-third

c) 2 inches

d) 3 inches

7) When patching spots in the lawn, it is best to treat spots 1 foot across and larger by

a) reseeding, sodding, or sprigging, as if starting a new lawn.

b) applying fertilizer so that the grass will spread to these spots.

c) sodding.

d) mowing the lawn properly until the spots are again covered with grass.

8) Sod is an expensive method of starting a lawn and is usually used in renovation work to

    a) resod all weak and weedy area, no matter how extensive the work is.
    b) patch up all small spots in the turf.
    c) lay out areas for gardens.
    d) all of the above

9) Lawns are often mowed too short because

    a) it results in mowing the lawn fewer times.
    b) it is difficult to adjust lawn mowers to higher cutting heights.
    c) insect problems are better controlled.
    d) some people prefer a close-cut lawn.

The Vegetable Garden

Planning and Preparing the Garden Site

OBJECTIVE

To select and prepare a vegetable garden site.

Competencies to Be Developed

After studying this unit, you should be able to

- ◆ list four items to consider when choosing the location of a vegetable garden.

- ◆ draw to scale a garden plan that includes at least four vegetables. Also include plans for successive plantings of two vegetables which are planted early and harvested early, and two vegetables which are planted after earlier crops and harvested in the fall.

- ◆ explain the difference between preparing a heavy, clayey garden soil and a sandy garden soil.

- ◆ take a soil sample and have it tested. Write a fertilizer program for your garden plan with this type of soil.

MATERIALS LIST

✔ graph paper, ruler, and pencil
✔ seed catalogs

No matter where you live, there is probably space for at least a small vegetable garden. Gardening is a good hobby which provides exercise, the satisfaction of growing some of your own food, and a savings on grocery bills. For the beginner, it is probably best to start with a small garden and work up in size as skill increases. The best techniques of planning and planting a garden are useless unless proper weed control, watering, and insect and disease control are practiced. Many of the skills developed in growing a small garden can be transferred to large-scale commercial vegetable production.

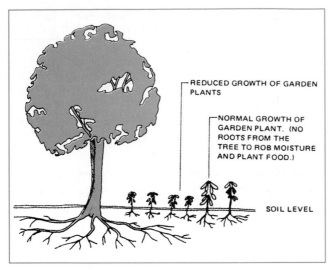

REDUCED GROWTH OF GARDEN PLANTS

NORMAL GROWTH OF GARDEN PLANT. (NO ROOTS FROM THE TREE TO ROB MOISTURE AND PLANT FOOD.)

SOIL LEVEL

FIGURE 38-1   The effect of large trees on growth of garden crops. As shown in the illustration, tree roots usually extend farther than the limbs.

LOCATION

If irrigation of the garden is necessary, the vegetable garden must be located near a supply of water. More importantly, the location should have a healthy, loamy, well-drained soil and plenty of sunshine. Some vegetables will grow in partial shade, but most need full sun. A location in the shade of large trees is especially bad, since the tree roots may extend into the garden and rob garden crops of needed plant food and moisture, figure 38-1. If some of the garden plants must be grown in shade, leafy green vegetables and pumpkins suffer less from shade than other vegetables.

The garden should not be located on a steep hill where rainwater will run off rapidly, since the runoff water is lost for use by the plants and also causes soil erosion. If the garden is located on a slope, form the rows along or around the hill instead of up and down the hill. This helps to prevent soil erosion and slows down water runoff, allowing it to soak into the soil.

THE GARDEN PLAN

Deciding what to plant and how much to plant are important steps in vegetable gardening. First, list the vegetables you and your family like to eat. Decide if you want to plant enough to freeze or can for winter use, or only grow enough to eat during the summer. A garden plan saves time and work. Draw a sketch of the plot to scale, prefer-

ably on graph paper. Keep in mind that a small, well-kept garden gives better returns than a large, weedy one.

The following are some points to consider when planning the garden.

◆ Plant perennials such as asparagus, rhubarb, horseradish, and sorrel together on one side of the garden where they will not interfere with the plowing and working of the rest of the garden.

◆ Group quickly maturing crops together, or plant them between rows of crops that mature later. When they are harvested, the area can be either replanted or left to provide more room for the later maturing crops.

◆ Plan the distance between rows according to the type of cultivation methods that will be used. Hoeing and mulching by hand allows the rows to be closer together. Double rows and wide row planting may also be used.

◆ Crops such as snap beans and sweet corn can be planted at intervals of every two weeks so that they can be harvested at different times during the season.

◆ Replant areas where early crops, such as peas and lettuce, are harvested with fall crops, such as kale or turnips.

Beets, carrots, leaf lettuce, onions, early peas, radishes, and spinach can be planted in wide rows or in rows a foot apart (although 18 inches is preferable), but other vegetables must have more room. Plants set 18 inches apart with rows 2 feet apart is a good rule of thumb for most other vegetables except staked tomatoes, which must be planted about 2 feet apart in rows 3 feet apart. Sweet corn should be either in hills 3 feet apart each way or in rows 2½ feet apart with plants spaced 1 foot apart. Vine crops (cucumber, melons, and squash) require 2½ to 5 feet of space in all directions. For space savers, remember that cucumbers can be grown on a fence and bush varieties of squash require less room than vine types. There are now new bush types of most melons, cucumbers, and squash.

WIDE ROW

A relatively new growing system for gardeners is the raised bed, wide row system. Planting areas are raised above the existing soil level and the entire top of the wide row is planted in vegetable

FIGURE 38-2 A garden laid out using beds for wide row planting. The entire top of each of these beds will be planted with several rows of vegetable plants. (Courtesy Gardenway, Inc.)

crops. See figure 38-2. The result is up to three or four times as much production per foot of row. Other advantages are that the raised beds dry out and warm up quicker in the spring allowing earlier planting, there is deeper topsoil for plant roots, especially important for root crops such as carrots, figure 38-3, and wide rows are easier to weed.

All of these spacing rules are for gardens that are hand cultivated. If a tractor or cultivator is to be used, more room must be allowed between rows. In a location with a considerable slope, make the rows across the grade to help prevent erosion during heavy rains. Use of mulch for weed control helps to control soil erosion on steep slopes and no cultivation is necessary. Plant tall-

FIGURE 38-3 Carrots grown in a raised bed. Notice the carrots are much longer in the raised bed, thus producing more pounds of vegetables per foot of row. (Courtesy Gardenway, Inc.)

growing vegetables on the north or west side of the plot to prevent shading of lower-growing varieties.

If your garden is tiny, a good garden plan would be a half row each of radishes and leaf lettuce (to be followed by a succession planting of snap beans), a row of snap beans planted at the regular time, and a few tomato plants. With more space, plan to include snap beans, lima beans, beets, broccoli, carrots, cabbage, a hill or two of cucumbers, greens such as spinach, Swiss chard, and tampala, leaf and head lettuce, a few early peas, and tomatoes. If there is plenty of room, include Brussels sprouts, cauliflower, sweet corn, eggplant, melons, onions, peas, peppers, squash, and any other varieties of vegetables you especially like.

CROP ROTATION

If possible, plan a crop rotation program for the garden. This will greatly reduce insect and disease problems. A four-year rotation is best. The following areas are suggested: (1) a green manure crop, (2) high nitrogen feeders such as corn, (3) root crops, (4) tomatoes, potatoes, eggplant, and other related plants. Plant corn or leafy vegetables such as lettuce after the green manure crop and follow in sequence as numbered.

Mixed plantings of vegetables also help with insect and disease control. Bugs often zero in on plants by detecting odors associated with a specific plant. Mixed crops confuse bugs by the great number of odors or chemical stimuli.

This also works with underground pests. Researchers believe that the roots of pest-resistant varieties release compounds which repel or even kill insects' larvae, protecting their roots as well as the roots of susceptible plants interplanted with them.

A garden guide from the W. Atlee Burpee Company is given in figure 38-4 to help in the planning of a vegetable garden. The guide is designed to provide vegetables for four people.

DEFINITION OF TERMS IN GUIDE

◆ **Hardiness** is a measure of the crop's resistance to frost. See notes under growing suggestions.

◆ **Days to germinate** is the number of days from planting of the seed to the date on which the plant surfaces.

◆ **Quantity to grow** is how much to plant for four. Increase or reduce to fit personal needs.

VEGETABLES AND TYPES	HARDINESS	DAYS TO GERMINATE	GROWING SUGGESTIONS	QUANTITY TO GROW	DAYS TO HARVEST	SATISFACTORY pH RANGE	USES
BEAN, snap bush and pole green, yellow	T	7–14	Sow bush types every 2 weeks until midsummer. Support pole types.	50 ft. bush; 8 hills pole	50–70	5.5–6.7	Fresh, frozen, canned. Vitamins A, B, C.
BEAN, bush shell red, white, green	T	7–14	Fava or English Broad Bean hardier than other types. Sow as early in spring as soil can be worked.	50 ft.	65–103	5.5–6.7	Fresh shell beans, or use dried for baking, soup, or Spanish or Mexican dishes. Vitamins A, B, C.
BEAN, lima bush and pole	T	7–14	Wait until ground is thoroughly warm before planting. Bush types mature earlier. Support pole varieties.	70 ft. bush; 8 hills pole	65–92	6.0–6.7	Fresh, frozen, canned, dried for baking. Vitamins A, B, C.
BEETS, red, golden, white	HH	10–21	For continuous harvest, make successive sowings until early summer. Do not transplant; this may cause forked or split roots.	25 ft.	55–80	6.0–7.5	Fresh, pickled, canned. Cook "thinnings" first, and tops later on for delicious greens. Vitamins A, B, C.
BROCCOLI	H	10–21	Plant again in midsummer for fall harvest. Grows best in cool weather.	25–40 plants	60–85*	5.5–6.7	Fresh, frozen. Vitamins A, B, C.
BRUSSELS SPROUTS	H	10–21	Pick lowest "sprouts" on stem each time; break off accompanying leaves but do not remove foliage.	25–40 plants	80–90*	5.5–6.7	Light frost improves flavor. Sprouts delicious fresh or frozen. Vitamins A, B, C.
CABBAGE, early, late, red, green	HH	10–21	Do not plant where any of the cabbage family grew the previous year.	25–40 plants	60–110*	5.5–6.7	Fresh, salads, coleslaw, sauerkraut. Winter storage. Vitamins A, B, C.
CARROT, long, short	HH	7–14	Short root types best for shallow or heavy soil. Plant again in midsummer for fall harvest.	25–30 ft.	65–75	5.2–6.7	Salads, relish, juice. Stews, soup. Vitamins A, B, C.
CAULIFLOWER, white, purple	HH	10–21	Tie leaves over heads to whiten.	16–24 plants	50–85*	6.0–6.7	Fresh, frozen, salad, relish. Vitamins A, B, C.

Note: See key on the last page of this figure.

FIGURE 38-4   Burpee vegetable garden guide (Courtesy W. Altee Burpee Company)

VEGETABLES AND TYPES	HARDINESS	DAYS TO GERMINATE	GROWING SUGGESTIONS	QUANTITY TO GROW	DAYS TO HARVEST	SATISFACTORY pH RANGE	USES
CELERY	HH	10–21	To whiten, mound soil up around mature stalks.	30–36	115–135*	5.5–6.7	Raw in salads and as relish. Cooked and creamed, soups. Vitamin A.
CHARD red, white, stalked	HH	7–14	Pick frequently to encourage fresh leaves. Stands summer heat.	20–30 ft.	60	6.0–6.7	Cook leaves for greens; midribs and stalks like asparagus. Vitamins A, B, C.
COLLARDS	H	7–14	Easily grown, nonheading, cabbagelike leaves.	20–30 ft.	80	5.5–6.7	Cook leaves for greens. Popular in southern states. Vitamins A, B, C.
CRESS, garden and water	H	3–14	Sow in garden every 2 weeks for continuous supply. Also grows well on sunny windowsill. Grow watercress in moist, shady spots or along a shallow stream.	20–30 ft.	10–50	6.0–7.0	Salads, sandwiches, garnish, seasoning. Vitamins A, B, C.
CUCUMBERS, slicing, pickling	T	7–14	Grow on fence to save space. Keep picking to encourage new fruit.	8–12 hills	53–65	5.5–6.7	Salad, relish, pickles. Vitamin A.
EGGPLANT	T	10–21	Needs warm temperature—70° to 75°F for good germination. Pick fruits when skin has high gloss.	8–12 plants	62–75*	5.5–6.7	Delicious fried, sauteed, or in casseroles. Vitamin A.
ENDIVE	H	7–14	Grows best in cool weather.	20–30 ft.	90	6.0–7.0	Salad, greens. Hearts can be cooked and served with cream cheese or grated cheese. Vitamins A, B, C.
KALE	H	14–21	Mature plants take cold fall and winter weather. Frost improves flavor.	25–30 ft.	55–65	5.5–7.0	Chop young leaves for salads and sandwiches. Cook for greens. Vitamins A, B, C.
KOHLRABI	HH	14–21	Grow for spring or fall crop; thrives in cool weather.	16–20 ft.	55–60	5.5–6.7	Fresh, frozen; cooked like turnips. Vitamins A, B, C.
LEEK	H	14–21	Whiten and improve flavor by mounding soil around mature plants.	25–40 ft.	130	5.5–6.7	Fresh in salads. Cooked in soups, stews, or creamed.

FIGURE 38-4 Burpee vegetable garden guide (Courtesy W. Altee Burpee Company) (continued)

VEGETABLES AND TYPES	HARDINESS	DAYS TO GERMINATE	GROWING SUGGESTIONS	QUANTITY TO GROW	DAYS TO HARVEST	SATISFACTORY pH RANGE	USES
LETTUCE, leaf	H	7–14	Make successive sowings in spring and another in late summer. Keep seedbed moist to get good germination for a fall crop.	25–40 ft.	40–47	6.0–7.0	Salad, sandwiches, garnish. Vitamins A, B, C.
LETTUCE, head	H	7–14	Needs cool weather in spring or fall to head well.	25–30 ft.	65–90	6.0–7.0	Salad, sandwiches, garnish. Vitamins A, B, C.
MUSTARD GREENS, fringed, smooth leaves	H	7–14	Grows as fall, winter, and spring crop in mild winter areas; spring and fall in north.	25–30 ft.	35–40	5.5–6.5	Greens. Vitamins A, B, C.
MELONS, cantaloupe, crenshaw, casaba, honeydew, watermelon	T	7–14	Very sensitive to frost. Black plastic mulch speeds maturity. Needs warm sunny weather when ripening for good flavor.	12–20 hills	75–120	6.0–6.7	Fresh, frozen. Ripe cantaloupes slip easily from stems. Ripe watermelons sound dull and hollow when tapped. Vitamins A, B, C.
OKRA	T	7–14	Needs hot weather to mature well. Pick pods young.	16–20 ft.	52–56	6.0–7.0	Soups, stews. Vitamins A, B, C.
ONIONS, yellow, white	H	10–21	Grows best in fine, well-drained sandy loam soil.	50–100 ft.	95–120	5.5–6.7	Fresh, salads, pickled. Vitamins B, C.
PARSLEY, curled or plain leaves	H	14–28	Attractive edging for flower garden; pot herb on sunny windowsill in winter.	10–20 ft.	72–90	6.0–7.5	Salad, garnish, seasoning. Dries or freezes well. Vitamins A, B, C.
PEA, dwarf, tall	H	7–14	Plant as early as ground can be worked.	40–100 ft.	55–79	5.5–6.7	Fresh, frozen, canned, dried. Vitamins A, B, C.
PEPPER, sweet, hot	T	10–21	Needs warm temperature—70° to 80°F for good germination.	8–10 plants	60–77*	5.5–6.5	Salad, stuffed, relish, seasoning. Vitamins A, B, C.
PUMPKIN, large, small bush, vine	T	7–14	For huge "contest" pumpkins, let only 1 or 2 grow per plant.	12–20 hills	95–120	5.5–6.5	Fresh, canned, frozen. Vitamins A, B, C.

FIGURE 38-4 Burpee vegetable garden guide (Courtesy W. Altee Burpee Company) (continued)

VEGETABLES AND TYPES	HARDINESS	DAYS TO GERMINATE	GROWING SUGGESTIONS	QUANTITY TO GROW	DAYS TO HARVEST	SATISFACTORY pH RANGE	USES
RADISH, red, white, black	H	7–14	Make successive sowings until early summer; again a month before fall frost.	15–30 ft.	22–60	5.2–6.7	Relish, salad. Vitamins B, C.
RUTABAGA	H	14–21	Grows best in cool weather.	20–30 ft.	90	5.2–6.7	Fresh. Winter storage. Vitamins B, C.
SPINACH, crinkled, smooth	H	7–14	New Zealand and Malabar take hot weather; other varieties cool.	20–40 ft.	42–70	6.0–6.7	Greens, frozen, canned. Vitamins A, B, C.
SQUASH, summer, bush, vine	T	7–14	Keep fruits picked so plants produce more.	8–12 hills	48–60	5.5–6.5	Fresh, frozen. Vitamin A.
SQUASH, winter, bush, vine	T	7–14	Black plastic mulch speeds maturity.	8–12 hills	80–120	5.5–6.5	Fresh, frozen, canned. Winter storage. Vitamin A.
SUNFLOWER	T	7–14	Use for screen plant. Protect maturing heads with bags to prevent bird damage.	25–50 ft.	80	6.0–7.5	Bird, poultry seed.
SWEET CORN, white, yellow	T	7–14	Plant in blocks of short rows for good pollination and well-filled ears.	50–100 ft.	63–90	5.2–6.7	Fresh, frozen, canned. Vitamins A, B, C.
TOMATO, red, pink, yellow	T	7–14	Hybrids especially need warm temperature—70° to 80°F for good germination.	16–20 plants	52–68*	5.2–6.7	Fresh, salad, canned, juice, pickled. Vitamins A, B, C.
TURNIPS, white, yellow	H	14–21	Grow best in cool weather.	20–30 ft.	35–60	5.2–6.7	Fresh, raw, or cooked. Leaves of some types for greens.

KEY

H — HARDY VARIETIES OF VEGETABLES AND FLOWERS—Tolerate cool weather and frost. Plant fall to early spring in zones G, H, I, J. Two to four weeks before last killing spring frost in all other zones.

HH — HALF HARDY VARIETIES OF VEGETABLES AND FLOWERS—Tolerate cool weather and very light frost. For earlier maturity, varieties that transplant well can be started inside or in a cold frame 6–10 weeks before last expected light frost.

T — TENDER VARIETIES OF VEGETABLES AND FLOWERS—Cannot stand frost; plant in spring after last frost date.

* — Time from when plants are set into garden.

FIGURE 38-4 Burpee vegetable garden guide (Courtesy W. Altee Burpee Company) (continued)

- ◆ **Days to harvest** is the number of days from planting of the seeds or plants in the garden to the date on which the crops are gathered.
- ◆ **pH range** is the soil acidity at which the plants grow best. Check soil with a soil test. Add lime or sulfur as recommended.
- ◆ **Uses** gives the best use for each vegetable.

THE GARDEN LAYOUT

Figure 38-5 shows a garden plan for spring and summer planting. Figure 38-6 shows a garden plan for the same garden later in the summer. Notice

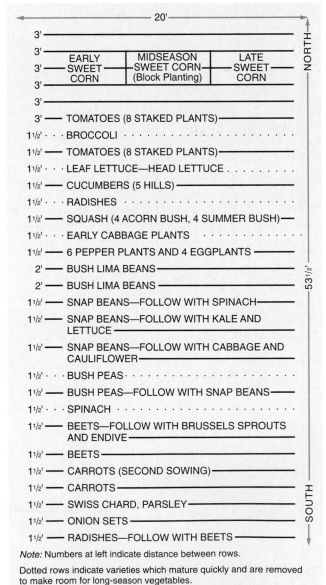

FIGURE 38-5 Suggested plan for a vegetable garden, 20 feet × 53½ feet, spring and summer. Corn is planted in blocks by variety for better pollination.

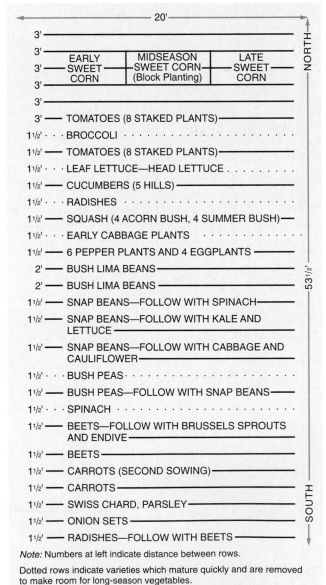

FIGURE 38-6 Suggested plan for a vegetable garden, 20 feet × 53½ feet, late summer and fall. This is how the garden in figure 38-5 appears after quickly maturing varieties have been harvested and successive plantings have been made.

that the early maturing vegetables have been harvested and the same areas replanted with late summer and fall crops. Crops that were planted between tomatoes, such as broccoli, are harvested; however, no other crop is replanted in these places because the tomatoes need all the space.

GARDEN SOIL

The best garden soil is a healthy loamy one. It must have good drainage so that oxygen is available to roots, enabling the roots to penetrate the soil. A good supply of organic matter should be available to hold moisture and provide plant food. The soil should also have a good supply of plant food. The pH range (soil acidity) should be from 6.3 to 7.0 for most vegetables.

PREPARING THE SOIL

CLAYEY SOILS   It is best to plow heavier garden soils (those containing large amounts of clay)

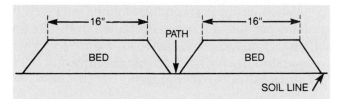

FIGURE 38-7 Raised beds allow earlier planting, especially in heavy clay soils. A 16-inch bed allows for wide planting (e.g., two rows of corn, two or three rows of cabbage, multiple rows of lettuce).

in the fall of the year. Fall plowing allows the heavier soils to be planted earlier in the spring because plowed soil dries more quickly than soil left unplowed. Preparing raised beds will also encourage drying out in spring, figure 38-7. Barnyard manure, crop residue, or leaves should be plowed under in these soils to add organic matter. Where a steep slope could result in erosion, it is best to wait until spring to plow the garden.

SANDY SOILS Sandy soils are best planted with a fall cover crop such as rye or vetch to grow green manure and then plowed in early spring. This increases the organic matter in these soils. Sandy soils require a high level of organic matter to assist in holding moisture and plant food. These soils dry more quickly in spring and can be plowed earlier in the spring than heavy soils. Gardens in dry land areas should be plowed in the fall and left rough to absorb and retain moisture that falls during the winter.

Never plow a soil when it is wet. A good way to test the wetness of soil is to squeeze a handful of the soil. If it sticks together in a tight ball and does not easily crumble under slight pressure by the thumb and finger, it is too wet for plowing or other working. When checking for wetness, check at the lower plow depth, since soil may be dry on top and too wet on the bottom.

A soil test of the garden plot should be taken each year. Results of this test will tell if lime is needed to raise soil pH and how much of the three major plant food elements (nitrogen, phosphorus, and potash) are needed. Follow the recommendations given by the soil laboratory or the extension agent. Fall is a good time to take soil tests.

LIMING (pH ADJUSTMENT)

If a soil test indicates a need for lime, it is best to *broadcast* (distribute) the lime on top of the plowed soil and harrow or otherwise mix it into the soil. When the pH is close to the level needed for best growth, an application of 50 pounds of ground

limestone per 100 square feet every two or three years should keep the pH near the proper level. This application will vary according to the soil type and the crop being grown. Plants that require high pH levels, such as beans, peas, and onions, need extra lime for good crop production. In a small garden where crops are rotated, the ideal pH cannot be obtained for each type of plant.

If soil pH is too high (alkaline), sulfur may be used to lower the pH in the same way in which lime is used to raise it.

FERTILIZING

Vegetable gardens need fertilizers added each year for best production. As stated, a soil test will indicate the level of nitrogen, phosphorus, and potash available in the soil. From the soil test, the gardener can determine how much of each plant food element must be added to the garden soil.

To supply the major plant food elements, apply a good commercial fertilizer made for vegetable crops. The percentage of each of the three elements is given in the analysis on the bag. A 5-10-5 analysis indicates 5 percent nitrogen, 10 percent phosphorus, and 5 percent potash. The plant food elements are always listed in this order. Good vegetable fertilizers are those with an analysis of 5-10-5, 5-10-10, 5-10-15, 10-10-10, and 10-6-4.

In general, leafy vegetables and corn require larger amounts of nitrogen. A 10-10-10 or 10-6-4 analysis is good, depending upon how much phosphorus and potash the soil contains. The pod or fruit crops need more phosphorus—a fertilizer with a 5-10-5 or 1-2-1 ratio is good. Root crops need extra potash; 5-10-15 or 5-10-10 is a good analysis for these crops. Unless marked plant food deficiencies exist in the soil, it is not necessary to have special fertilizer for different crops in a small garden. In large commercial production, the analysis should match the crop needs as noted.

APPLYING FERTILIZER

BROADCASTING AND SIDEDRESSING

Vegetable crops may require fertilizer application as high as 5 pounds per 100 square feet. If this great an amount is added, it might be best to broadcast or spread evenly over the entire soil surface and mix half the amount into the soil before planting. Apply the rest later in the season as a *sidedress* on top of the soil on each side of the plants in the

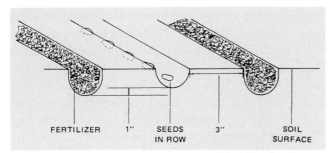

FIGURE 38-8 Banding fertilizer. The fertilizer is placed in bands 3 inches from each side of the seed row and 1 inch deeper than the depth of the seeds. Both the fertilizer and the seeds are covered with soil.

row, about 3 to 4 inches from the stem of the plant. (*Sidedressing* means placing fertilizer on the top of the soil around each plant.)

BANDING Another method for applying fertilizer is to place the fertilizer in rows dug 3 inches from each side of the row of seeds or plants and slightly deeper than the depth at which the seeds are planted. This is called *banding* fertilizer, figure 38-8.

PLOWING UNDER If a soil tests very low in phosphorus and/or potash, it is best to plow a fertilizer high in these elements under the soil. If applied on top of the soil, phosphorus and potash do not leach or wash down through the soil fast enough to be effective the same year. (If plowed under one year, these elements remain in the soil for the following year.)

When fertilizer recommendations are given in pounds per acre, it is often difficult to translate the amounts into those for small rows. Figure 38-9 gives conversion figures for determining how much fertilizer to use on smaller plots.

SUMMARY

1. Decide what is to be planted.

2. Draw a garden plan showing how much of each crop is to be planted and the exact location in the garden of each crop.

3. Determine the amount of space between rows and whether or not interplanting between rows with early maturing crops is to be done.

4. Locate the garden close to water and in the sun on a loamy soil with good drainage.

5. Prepare the soil with organic matter by plowing under manure, compost, or cover crops.

6. Plow in fall or spring when the soil is dry enough to crumble.

7. Use lime or sulfur to correct pH. Dig into plowed or spaded soil.

8. Apply a good garden fertilizer. Apply the amount recommended in the soil test report.

WEIGHT OF FERTILIZER TO APPLY WHEN THE WEIGHT TO BE APPLIED PER ACRE IS:				
MEASUREMENT	**100 POUNDS**	**400 POUNDS**	**800 POUNDS**	**1,200 POUNDS**
Space between rows, and row length (in feet)	Pounds	Pounds	Pounds	Pounds
2 wide, 50 long	0.25	1.0	2.0	3.0
2 wide, 100 long	.50	2.0	4.0	6.0
2½ wide, 50 long	.30	1.2	2.4	3.6
2½ wide, 100 long	.60	2.4	4.8	7.2
3 wide, 50 long	.35	1.4	2.8	4.2
3 wide, 100 long	.70	2.8	5.6	8.4
Area (in square feet)				
100	.25	1.0	2.0	3.0
500	1.25	5.0	10.0	15.0
1,000	2.50	10.0	20.0	30.0
1,500	3.75	15.0	30.0	45.0
2,000	5.00	20.0	40.0	60.0

FIGURE 38-9 Approximate rates of fertilizer application per 50 or 100 feet of garden row, and per 100 to 2,000 square feet of garden area, corresponding to given rates per acre.

Student Activities

1) Draw a garden plan for a family garden. Include successive crops of early and late plantings of short-season crops.

2) Study a vegetable variety recommendation fact sheet from a local university or extension agent. Write to an extension service in some other area of the country and request a fact sheet. Compare the two.

3) Study a garden seed catalog. Select and list the best vegetable varieties for your area.

4) Select a garden site to transform your plan into reality. Actually plant and care for the garden if possible. This may be a group or class project.

Self-Evaluation

A) Select the best answer from the choices offered to complete the statement or answer the question.

1) Perennial plants such as asparagus should be

a) left out of the family garden.
b) planted together on one side of the garden.
c) rotated each year.
d) none of the above

2) Two short-season crops that may be used in successive plantings are

a) radishes and lettuce.
b) horseradish and beets.
c) tomatoes and snap beans.
d) peppers and corn.

3) What is added to acidic soil (with a low pH) to raise the pH?

a) organic matter
b) fertilizer
c) lime
d) sand

4) Select the three long-season crops that should remain in the garden all season with no replanting.

a) broccoli, snap beans, spinach
b) spinach, kale, turnips
c) beets, radishes, head lettuce
d) tomato, eggplant, squash

5) Which of the following makes the best garden soil?

a) sand
b) loam
c) clay
d) silt

6) The best pH (soil acidity) range for most garden plants is

a) 4.5 to 5.5.
b) 5.5 to 6.5.
c) 6.3 to 7.0.
d) 6.5 to 8.5.

7) Heavier (clayey) soils are best plowed in the

a) spring.
b) summer.
c) fall.
d) dry season.

8) Sandy soils are best plowed in the

a) spring.
b) summer.

c) fall.
d) dry season.

9) Which vegetable crops use large amounts of nitrogen fertilizer?

a) leaf
b) root

c) fruit
d) all of these

10) Which vegetable crops use large amounts of potash fertilizer?

a) pod
b) leaf

c) fruit
d) root

11) Lime is applied

a) as a sidedress after vegetables are planted.
b) on top of plowed soil and dug in.
c) on top of soil and plowed under.
d) in the planting hole.

12) Three vegetables that require extra lime and a relatively high soil pH are

a) beans, peas, and onions.
b) corn, celery, and beets.
c) eggplant, tomatoes, and corn.
d) corn, spinach, and peas.

13) Which vegetable crops use large amounts of phosphorus fertilizer?

a) all
b) root

c) leaf
d) pod or fruit

14) Which of the fertilizer analyses below is higher than the others in phosphorus?

a) 5-10-5
b) 10-5-5

c) 5-5-10
d) 2-3-4

15) Which of the fertilizer analyses below is highest in nitrogen?

a) 5-10-5
b) 10-5-5

c) 5-5-10
d) 16-8-4

B) Answer each of the following as instructed.

1) Explain in writing how to tell if a soil is too wet to plow.

2) Draw a sketch showing how to band fertilizer at planting time.

3) List the four most important items to consider when selecting a garden site.

4) One hundred pounds of a 5-10-5 fertilizer provides

a) _____ pounds of potash.
b) _____ pounds of total plant food.

UNIT 39

Planting the Vegetable Garden

OBJECTIVE

To demonstrate techniques for the planting of a small vegetable garden.

Competencies to Be Developed

After studying this unit, you should be able to

- ◆ list the best varieties of ten vegetables for planting in the local area, including at least one disease-resistant variety for each.

- ◆ use the frost-free map and planting charts to determine planting dates for the ten vegetables.

- ◆ list five vegetables which are direct seeded in the garden and five which must be seeded indoors for transplanting.

- ◆ describe the steps in seeding vegetables for transplanting outdoors.

MATERIALS LIST

- ✔ listing of vegetable varieties recommended for the local area
- ✔ seed catalogs
- ✔ garden seeds
- ✔ containers for starting seeds to grow plants for transplanting (flats, Jiffy 7, etc.)
- ✔ garden rake, garden spade, garden hoe
- ✔ string for lining out rows
- ✔ garden hose
- ✔ aluminum foil or cardboard for cutworm collars

SELECTING VARIETIES

One of the most important steps in the preparation of any garden is selecting the proper variety of vegetables to plant. For example, seed catalogs list dozens of varieties of sweet corn or tomatoes. Which is best for the area in which you live? An individual who sells seeds locally would probably be willing to offer advice concerning varieties. A local extension agent or horticulture teacher can also help to determine which varieties are recommended for the area. Consider the use of hybrid varieties for more vigorous plants and a higher yield.

It pays to purchase seed from a reputable professional and not to depend on home supplies. Homegrown seed may carry diseases, or cross-pollination may result in seeds that do not come true to variety. The seed from hybrid vegetables does not produce plants that are true to seed. The cost of seed is very low compared with the total cost of producing vegetables; it pays to start with the best.

Disease-resistant varieties should be used whenever a disease is known to be a problem in the area. The selection of disease-resistant varieties can eliminate spraying and low yields at very little, if any, extra cost to the grower. Hybrid varieties are generally more vigorous and give higher yields for the same effort and cost. Hybrids should be used whenever possible.

Fresh seed should always be used. Read the date on the package to be sure that the seed was packed for the current year. Some seeds retain their vitality longer than others, and may grow after being held over for a year. Percent of germination will probably be lower than that listed on the package for first-year planting, however.

Vegetable seeds may be divided into three classes: (1) *short-lived* (do not germinate well after one or two years)—corn, leek, onion, parsley, rhubarb, and salsify; (2) *moderately long-lived* (often good for three to five years)—asparagus, beans, Brussels sprouts, cabbage, carrot, cauliflower, celery, kale, lettuce, okra, peas, pepper, radish, spinach, turnip, and watermelon; and (3) *long-lived* (may be good for more than five years)—beet, cucumber, eggplant, muskmelon, and tomato. Any seed over one year old should be given a germination test to see how many seeds actually sprout vigorously. This should be done before any of the seeds are planted in the garden. *Always buy seeds that are dated for planting that year.*

Seed stored in a dry, cool, dark place remains good longer than seed that is stored in an open container in a warm, moist place. How seeds are stored helps to determine the length of active life and causes the figures given for the three classes of seed (short-lived, moderately long-lived, and long-lived) to vary considerably. These figures pertain to seed stored under good storage conditions.

DETERMINING PLANTING TIME

Planting seeds or transplanting plants at the proper time for the locality is very important. Temperatures can differ greatly between areas that are not many miles apart, causing the planting date for a particular vegetable to differ by as much as one or two weeks.

Vegetable crops may be grouped according to their hardiness (tolerance to cold). Figure 39-1 is a timetable listing the planting dates for some vegetable crops. The frost-free date in the spring of any locality is about the time that the oak trees in that area leaf out about 1 inch long or the American dogwood blooms.

Dr. R. J. Happ of the University of Vermont had a guide based on the Persian Lilac. When the lilac was in its first leaf stages, cool-season crops such as lettuce, peas, and root crops could be safely planted; full bloom stage warm-season plants like beans and cucumbers also could be safely planted. (Taken from the Vermont Agricultural Experiment Station pamphlet number 36, by Dr. R. J. Happ et al.) By being observant you can find plants in your area to use as frost-free guides.

It is important that the first planting of vegetables be as early in the spring as possible without danger of cold damage. Some vegetables can be planted before the frost-free date; others cannot. Those vegetables indicated for planting earlier than the frost-free date grow best in cool weather, and cannot tolerate summer heat.

A gardener anywhere in the United States can determine the safe planting date for his or her particular area for different crops by using the information found in figures 39-2, 39-3, 39-4, and 39-5, pages 411 through 418. The maps show the average dates of the last killing frost in spring (figure 39-2) and the first killing frost in fall (figure 39-3). Specific planting dates are determined by plugging these dates into figures 39-4 and 39-5.

COLD-HARDY PLANTS FOR EARLY SPRING PLANTINGS		COLD-TENDER OR HEAT-HARDY PLANTS FOR LATER SPRING OR EARLY SUMMER PLANTING			
Very Hardy (Plant 4 to 6 weeks before frost-free date.)	Hardy (Plant 2 to 4 weeks before frost-free date.)	Not cold-hardy (Plant on frost-free date.)	Requiring hot weather (Plant 1 week or more after frost-free date.)	Medium heat-tolerant (Good for summer planting.)	Hardy plants for late summer or fall planting except in the northern region (Plant 6 to 8 weeks before first fall freeze.)
Broccoli	Beets	Beans, snap	Beans, lima	Beans, all	Beets
Cabbage	Carrots	Okra	Eggplant	Chard	Collards
Lettuce	Chard	New Zealand	Peppers	Soybeans	Kale
Onions	Mustard	spinach	Sweet potatoes	New Zealand	Lettuce
Peas	Parsnip	Soybeans	Cucumbers	spinach	Mustard
Potatoes	Radishes	Squash	Melons	Squash	Spinach
Spinach		Sweet corn		Sweet corn	Turnips
Turnips		Tomatoes			

FIGURE 39-1   Common vegetables grouped according to the approximate times they can be planted and their relative requirements for cool weather

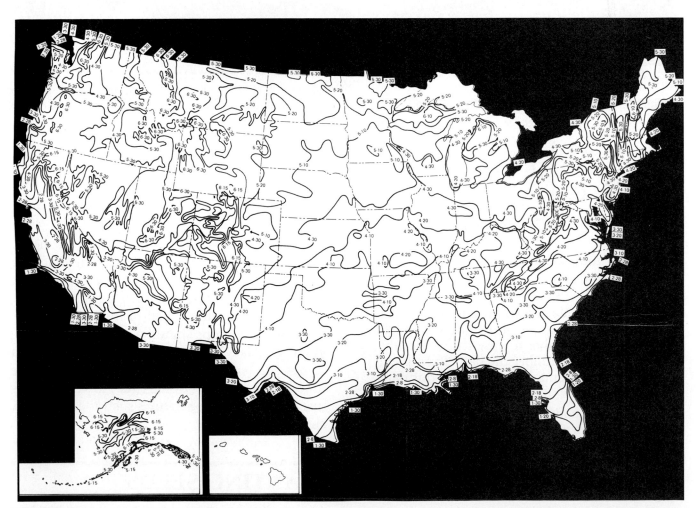

FIGURE 39-2   Average dates of the last killing frost in spring (USDA Home and Garden Bulletin 202) Note: Numbers such as 3-30 mean the third month (March) 30th day.

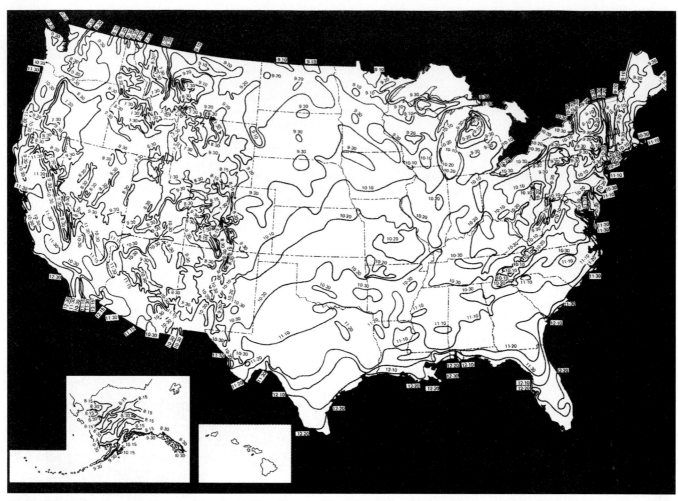

FIGURE 39-3 Average dates of the first killing frost in fall (USDA Home and Garden Bulletin 202)

HOW TO USE THE MAP AND TABLES

To determine the best time for spring planting of any vegetable in your locality:

1. Find your location on the map in figure 39-2 and then find the solid line on the map that comes closest to that location.

2. Find the date shown on the solid line. This is the average date of the last killing frost. The first number represents the month; the second number the day. (Example: 5-10 represents May 10.) Note this date and go to figure 39-4.

3. Find the column with this date at the top of it. This is the only column you will use. It gives dates for all vegetable crops listed for your area.

4. In this column, locate the vegetable that you wish to plant. The dates by the vegetable show the period of time during which the crop can be safely planted. The best time is on or soon after the first date. The second date is the latest date for planting. Planting time becomes less desirable as you move toward the second date.

Figure 39-5 is used with the map in figure 39-3 in the same way to find planting dates for fall planting. A date about halfway between the two dates given in figure 39-5 is usually best.

Along the northern half of the Pacific Coast, warm-weather crops should not be planted as late as the frost date and table indicate. Although frost comes late, very cool weather prevails for some time before frost, retarding late growth of crops that prefer warm temperatures such as corn, beans, and tomatoes.

PLANTING SEEDS

Assume that the garden area is plowed and soil pH and fertilizer requirements necessary prior to

SPRING PLANTING DATES FOR LOCALITIES IN WHICH AVERAGE DATE OF LAST FREEZE IS

CROP	Jan. 30	Feb. 8	Feb. 18	Feb. 28	Mar. 10	Mar. 20	Mar. 30
Asparagus[1]	Feb. 1–Apr. 15	Feb. 10–May 1	Mar. 1–May 1	Mar. 15–June 1	Jan. 1–Mar. 1	Feb. 1–Mar. 10	Feb. 15–Mar. 20
Beans, lima	Feb. 1–Apr. 1	Feb. 1–May 1	Mar. 1–May 1	Mar. 10–May 15	Mar. 20–June 1	Apr. 1–June 15	Apr. 15–June 20
Beans, snap	Jan. 1–Mar. 15	Jan. 10–Mar. 15	Jan. 20–Apr. 1	Feb. 1–Apr. 15	Mar. 15–May 15	Mar. 15–May 25	Apr. 1–June 1
Beet	Jan. 1–30	Jan. 1–30	Jan. 1–30	Feb. 1–June 1	Feb. 15–June 1	Feb. 15–May 15	Mar. 1–June 1
Broccoli, sprouting[1]	Jan. 1–30	Jan. 1–30	Jan. 15–Feb. 15	Feb. 1–Mar. 1	Feb. 15–Mar. 15	Feb. 15–Mar. 15	Mar. 1–20
Brussels sprouts[1]	Jan. 1–30	Jan. 1–30	Jan. 15–Feb. 15	Feb. 1–Mar. 1	Feb. 15–Mar. 15	Feb. 15–Mar. 15	Mar. 1–20
Cabbage[1]	Jan. 1–15	Jan. 1–Feb. 10	Jan. 1–Feb. 25	Jan. 15–Feb. 25	Jan. 25–Mar. 1	Feb. 1–Mar. 1	Feb. 15–Mar. 10
Cabbage, Chinese	(2)	(2)	(2)	(2)	(2)	(2)	(2)
Carrot	Jan. 1–Mar. 1	Jan. 1–Mar. 1	Jan. 15–Mar. 1	Feb. 1–Mar. 1	Feb. 10–Mar. 15	Feb. 15–Mar. 20	Mar. 1–Apr. 10
Cauliflower[1]	Jan. 1–Feb. 1	Jan. 1–Feb. 10	Jan. 10–Feb. 10	Jan. 20–Feb. 20	Feb. 1–Mar. 1	Feb. 10–Mar. 10	Feb. 10–Mar. 20
Celery and celeriac	Jan. 1–Feb. 1	Jan. 10–Feb. 10	Jan. 20–Feb. 20	Feb. 1–Mar. 1	Feb. 10–Mar. 20	Mar. 1–Apr. 1	Mar. 15–Apr. 15
Chard	Jan. 1–Apr. 1	Jan. 10–Apr. 1	Jan. 20–Apr. 15	Feb. 1–May 1	Feb. 15–May 15	Feb. 20–May 15	Mar. 1–May 25
Chervil and chives	Jan. 1–Feb. 1	Jan. 1–Feb. 1	Jan. 1–Feb. 1	Jan. 15–Feb. 15	Feb. 1–Mar. 1	Feb. 10–Mar. 10	Feb. 15–Mar. 10
Chicory, witloof					June 1–July 1	June 1–July 1	June 1–July 1
Collards[1]	Jan. 1–Feb. 15	Jan. 1–Feb. 15	Jan. 1–Mar. 15	Jan. 15–Mar. 15	Feb. 1–Apr. 1	Feb. 15–May 1	Mar. 1–June 1
Corn, sweet	Feb. 1–Mar. 15	Feb. 10–Apr. 1	Feb. 20–Apr. 15	Mar. 1–Apr. 15	Mar. 10–Apr. 15	Mar. 25–May 1	Mar. 25–May 15
Cornsalad	Jan. 1–Feb. 15	Jan. 1–Feb. 15	Jan. 15–Feb. 15	Jan. 15–Feb. 15	Jan. 1–Mar. 1	Jan. 1–Mar. 15	Jan. 15–Mar. 15
Cress, upland	Jan. 1–Feb. 1	Jan. 1–Feb. 15	Jan. 1–Feb. 15	Feb. 1–Mar. 1	Feb. 10–Mar. 15	Feb. 20–Mar. 15	Mar. 1–Apr. 1
Cucumber	Feb. 15–Mar. 15	Feb. 15–Apr. 1	Feb. 15–Apr. 15	Mar. 1–Apr. 15	Mar. 15–Apr. 15	Apr. 1–May 1	Apr. 10–May 15
Eggplant[1]	Feb. 1–Mar. 1	Feb. 10–Mar. 15	Feb. 20–Apr. 1	Mar. 10–Apr. 15	Mar. 15–Apr. 15	Apr. 1–May 1	Apr. 15–May 15
Endive	Jan. 1–Mar. 1	Jan. 1–Mar. 1	Jan. 15–Mar. 1	Feb. 1–Mar. 1	Feb. 15–Mar. 15	Mar. 1–Apr. 1	Mar. 10–Apr. 10
Fennel, Florence	Jan. 1–Mar. 1	Jan. 1–Mar. 1	Jan. 15–Mar. 1	Feb. 1–Mar. 1	Feb. 15–Mar. 15	Mar. 1–Apr. 1	Mar. 10–Apr. 10
Garlic	(2)	(2)	(2)	(2)	(2)	(2)	Feb. 10–Mar. 10
Horseradish[1]						Feb. 20–Mar. 15	Mar. 1–Apr. 1
Kale	Jan. 1–Feb. 1	Jan. 1–Feb. 1	Jan. 20–Feb. 10	Feb. 1–20	Feb. 10–Mar. 10	Feb. 20–Mar. 15	Mar. 1–20
Kohlrabi	Jan. 1–Feb. 1	Jan. 1–Feb. 1	Jan. 20–Feb. 10	Feb. 1–20	Feb. 10–Mar. 1	Feb. 20–Mar. 15	Mar. 1–Apr. 1
Leek	Jan. 1–Feb. 1	Jan. 1–Feb. 1	Jan. 1–Feb. 15	Jan. 15–Feb. 15	Jan. 25–Mar. 1	Feb. 1–Mar. 1	Feb. 15–Mar. 15
Lettuce, head[1]	Jan. 1–Feb. 1	Jan. 1–Feb. 1	Jan. 1–Feb. 1	Jan. 15–Feb. 15	Feb. 1–20	Feb. 15–Mar. 10	Mar. 1–20
Lettuce, leaf	Jan. 1–Feb. 1	Jan. 1–Feb. 1	Jan. 1–Mar. 15	Jan. 1–Mar. 15	Jan. 15–Apr. 1	Feb. 1–Apr. 1	Feb. 15–Apr. 15
Muskmelon[2]	Feb. 15–Mar. 15	Feb. 15–Apr. 15	Feb. 15–Apr. 15	Mar. 1–Apr. 15	Mar. 15–Apr. 15	Apr. 1–May 1	Apr. 10–May 15
Mustard	Jan. 1–Mar. 1	Jan. 1–Mar. 1	Jan. 1–Mar. 1	Feb. 1–Mar. 1	Feb. 10–Mar. 15	Feb. 20–Apr. 1	Mar. 1–Apr. 1
Okra	Feb. 15–Apr. 1	Feb. 15–Apr. 15	Mar. 1–June 1	Mar. 10–June 1	Mar. 20–June 1	Apr. 1–June 15	Apr. 10–June 15
Onion[1]	Jan. 1–15	Jan. 1–15	Jan. 1–15	Jan. 1–Feb. 1	Jan. 15–Feb. 15	Feb. 1–Mar. 1	Feb. 15–Mar. 15
Onion, seed	Jan. 1–15	Jan. 1–15	Jan. 1–15	Jan. 1–Feb. 15	Feb. 1–Mar. 1	Feb. 15–Mar. 15	Feb. 20–Mar. 15
Onion, sets	Jan. 1–15	Jan. 1–15	Jan. 1–15	Jan. 1–Feb. 1	Jan. 15–Mar. 1	Feb. 1–Mar. 1	Feb. 15–Mar. 20
Parsley	Jan. 1–30	Jan. 1–30	Jan. 1–30	Jan. 15–Mar. 1	Jan. 15–Mar. 10	Feb. 15–Mar. 10	Mar. 1–Apr. 1
Parsnip			Jan. 1–Feb. 1	Jan. 15–Feb. 15	Jan. 15–Mar. 1	Feb. 1–Mar. 1	Mar. 1–Apr. 1
Peas, black-eye	Feb. 15–May 1	Feb. 15–May 15	Mar. 1–June 15	Mar. 10–June 15	Mar. 15–July 1	Mar. 15–July 1	Apr. 15–July 1
Peas, garden	Jan. 1–Feb. 15	Jan. 1–Feb. 15	Jan. 1–Mar. 1	Jan. 15–Mar. 1	Jan. 15–Mar. 15	Feb. 1–Mar. 15	Feb. 10–Mar. 20
Pepper[1]	Feb. 1–Apr. 1	Feb. 15–Apr. 15	Mar. 1–May 1	Mar. 15–May 1	Apr. 1–June 1	Apr. 10–June 1	Apr. 15–June 1

[1] Plants.
[2] Included in this section because no appropriate coverage is available elsewhere.

FIGURE 39-4 Earliest dates, and ranges of dates, for safe spring planting of vegetables in the open (USDA Home and Garden Bulletin 202)

SPRING PLANTING DATES FOR LOCALITIES IN WHICH AVERAGE DATE OF LAST FREEZE IS (continued)

CROP	Jan. 30	Feb. 8	Feb. 18	Feb. 28	Mar. 10	Mar. 20	Mar. 30
Potato	Jan. 1-Feb. 15	Jan. 1-Feb. 15	Jan. 15-Mar. 1	Jan. 15-Mar. 1	Feb. 1-Mar. 1	Feb. 10-Mar. 15	Feb. 20-Mar. 20
Radish	Jan. 1-Apr. 1	Jan. 1-Apr. 1	Jan. 1-Apr. 1	Jan. 1-Apr. 1	Jan. 1-Apr. 15	Jan. 20-May 1	Feb. 15-May 1
Rhubarb[1]							
Rutabaga							Feb. 1-Mar. 1
Salsify	Jan. 1-Feb. 1	Jan. 1-Feb. 1	Jan. 15-Feb. 20	Jan. 1-Feb. 1	Jan. 15-Mar. 1	Jan. 15-Mar. 1	Mar. 1-15
Shallot	Jan. 1-Feb. 1	Jan. 10-Feb. 10	Jan. 1-Feb. 20	Jan. 1-Mar. 1	Jan. 15-Mar. 1	Feb. 1-Mar. 1	Feb. 15-Mar. 15
Sorrel	Jan. 1-Mar. 1	Jan. 1-Mar. 1	Jan. 15-Mar. 1	Jan. 15-Mar. 1	Feb. 1-Mar. 1	Feb. 10-Mar. 20	Feb. 20-Apr. 1
Soybean	Mar. 1-June 30	Mar. 1-June 30	Mar. 10-June 30	Mar. 20-June 30	Apr. 10-June 30	Apr. 10-June 30	Apr. 20-June 30
Spinach	Jan. 1-Feb. 15	Jan. 1-Feb. 15	Jan. 1-Mar. 1	Jan. 1-Mar. 1	Jan. 15-Mar. 10	Jan. 15-Mar. 15	Feb. 1-Mar. 20
Spinach, New Zealand	Feb. 1-Apr. 15	Feb. 15-Apr. 15	Feb. 15-Apr. 15	Mar. 15-May 15	Mar. 20-May 15	Apr. 1-May 15	Apr. 10-June 1
Squash, summer	Feb. 1-Apr. 15	Feb. 15-Apr. 15	Feb. 15-Apr. 15	Mar. 15-May 15	Mar. 15-May 1	Apr. 1-May 15	Apr. 10-June 1
Sweet potato	Feb. 15-May 15	Mar. 1-May 15	Mar. 1-May 15	Mar. 20-June 1	Apr. 1-June 1	Apr. 10-June 1	Apr. 20-June 1
Tomato	Feb. 1-Apr. 1	Feb. 20-Apr. 10	Feb. 20-Apr. 10	Mar. 10-May 1	Mar. 20-May 10	Apr. 1-May 20	Apr. 10-June 1
Turnip	Jan. 1-Mar. 1	Jan. 1-Mar. 1	Jan. 10-Mar. 1	Jan. 20-Mar. 1	Feb. 1-Mar. 1	Feb. 10-Mar. 10	Feb. 20-Mar. 20
Watermelon[2]	Feb. 15-Mar. 15	Feb. 15-Apr. 1	Feb. 15-Apr. 15	Feb. 1-Apr. 15	Mar. 15-Apr. 15	Apr. 1-May 1	Apr. 10-May 15

CROP	Apr. 10	Apr. 20	Apr. 30	May 10	May 20	May 30	June 10
Asparagus[1]	Mar. 10-Apr. 10	Mar. 15-Apr. 15	Mar. 20-Apr. 15	Mar. 10-Apr. 30	Apr. 20-May 15	May 1-June 1	May 15-June 1
Beans, lima	Apr. 1-June 30	May 1-June 20	May 15-June 15	May 25-June 15	May 15-June 30	May 25-June 15	
Beans, snap	Apr. 10-June 30	Apr. 25-June 30	May 10-June 30	May 10-June 30	Apr. 25-June 15	May 1-June 15	May 15-June 15
Beet	Mar. 10-June 1	Mar. 20-June 1	Apr. 1-June 15	Apr. 15-June 15	May 1-June 15	May 10-June 15	May 20-June 10
Broccoli, sprouting[1]	Mar. 15-Apr. 15	Mar. 25-Apr. 20	Apr. 1-June 15	Apr. 15-June 1	May 1-June 1	May 10-June 10	May 20-June 10
Brussels sprouts[1]	Mar. 15-Apr. 15	Mar. 25-Apr. 20	Apr. 1-June 15	Apr. 15-June 1	May 1-June 1	May 10-June 10	May 20-June 1
Cabbage[1]	Mar. 12-Apr. 1	Mar. 10-Apr. 1	Mar. 15-Apr. 10	Apr. 1-May 15	May 1-June 15	May 10-June 15	May 20-June 1
Cabbage, Chinese	(2)	(2)	(2)	Apr. 1-May 15	May 10-June 15	May 10-June 15	May 20-June 1
Carrot	Mar. 10-Apr. 20	Apr. 1-June 1	Apr. 10-June 1	Apr. 20-June 15	May 1-June 1	May 10-June 1	May 20-June 1
Cauliflower[1]	Mar. 1-Mar. 20	Mar. 15-Apr. 20	Apr. 10-May 10	Apr. 15-May 15	May 10-June 15	May 20-June 1	May 20-June 1
Celery and celeriac	Apr. 1-Apr. 20	Apr. 1-May 1	Apr. 15-May 1	Apr. 20-June 15	May 10-June 15	May 20-June 1	June 1-June 15
Chard	Mar. 15-June 1	Apr. 1-June 15	Apr. 15-June 15	Apr. 20-June 15	May 10-June 15	May 20-June 1	June 1-June 15
Chervil and chives	Mar. 1-Apr. 1	Mar. 15-Apr. 15	Mar. 20-Apr. 20	Apr. 1-May 1	Apr. 15-May 15	May 1-June 1	June 1-June 15
Chicory, witloof	June 10-July 1	June 15-July 1	June 15-July 1	June 1-20	June 1-15	June 1-15	May 15-June 1
Collards[1]	Mar. 1-June 1	Mar. 10-June 1	Mar. 10-June 1	Apr. 15-June 1	May 1-June 1	May 10-June 1	June 1-15
Corn, sweet	Apr. 10-June 1	Apr. 25-June 15	May 10-June 15	May 10-June 15	May 15-June 1	May 20-June 1	May 20-June 1
Cornsalad	Feb. 1-Apr. 1	Feb. 15-Apr. 15	Feb. 15-Apr. 15	Apr. 1-June 1	Apr. 15-June 1	May 15-June 1	
Cress, upland	Mar. 10-Apr. 15	Mar. 20-May 1	Apr. 10-May 10	Apr. 20-May 20	May 1-June 1	May 1-June 1	
Cucumber	Apr. 20-June 1	May 1-June 15	May 15-June 15	May 20-June 15	May 15-June 15	May 15-June 15	May 15-June 15
Eggplant[1]	May 1-June 1	May 10-June 1	May 15-June 10	May 20-June 15	June 1-15	June 1-15	May 15-June 15
Endive	Mar. 15-Apr. 15	Mar. 25-Apr. 15	Apr. 15-May 15	Apr. 15-May 15	May 1-30	May 1-30	May 15-June 1
Fennel, Florence	Mar. 15-Apr. 15	Mar. 25-Apr. 15	Apr. 1-May 1	Apr. 15-May 15	May 1-30	May 1-30	May 15-June 1

[1]Plants.
[2]Included in this section because no appropriate coverage is available elsewhere.

FIGURE 39-4 Earliest dates, and ranges of dates, for safe spring planting of vegetables in the open (USDA Home and Garden Bulletin 202) (continued)

SPRING PLANTING DATES FOR LOCALITIES IN WHICH AVERAGE DATE OF LAST FREEZE IS (continued)

CROP	Apr. 10	Apr. 20	Apr. 30	May 10	May 20	May 30	June 10
Garlic	Feb. 20–Mar. 20	Mar. 10–Apr. 1	Mar. 15–Apr. 15	Apr. 1–May 1	Apr. 15–May 15	May 1–30	May 15–June 1
Horseradish[1]	Mar. 10–Apr. 10	Mar. 20–Apr. 20	Apr. 1–30	Apr. 15–May 15	Apr. 20–May 20	May 1–30	May 15–June 1
Kale	Mar. 10–Apr. 1	Mar. 20–Apr. 10	Apr. 1–20	Apr. 10–May 1	Apr. 20–May 10	May 1–30	May 15–June 1
Kohlrabi	Mar. 10–Apr. 10	Mar. 20–May 1	Apr. 1–May 10	Apr. 10–May 15	Apr. 20–May 20	May 1–30	May 15–June 1
Leek	Mar. 1–Apr. 1	Mar. 15–Apr. 15	Apr. 1–May 1	Apr. 15–May 15	May 1–May 20	May 1–15	May 1–15
Lettuce, head[1]	Mar. 10–Apr. 1	Mar. 20–Apr. 15	Apr. 1–May 1	Apr. 15–May 15	May 1–June 30	May 10–June 30	May 20–June 30
Lettuce, leaf	Mar. 15–May 15	Mar. 20–May 15	Apr. 1–June 1	Apr. 15–June 15	May 1–June 30	May 10–June 30	May 20–June 30
Muskmelon[2]	Apr. 20–June 1	May 1–June 15	May 15–June 15	June 1–June 15			
Mustard	Mar. 10–Apr. 20	Mar. 20–May 1	Apr. 1–May 10	Apr. 15–June 1	May 1–June 30	May 10–June 30	May 20–June 30
Okra	Apr. 20–June 15	May 1–June 1	May 10–June 1	May 20–June 10	June 1–20		
Onion[1]	Mar. 1–Apr. 1	Mar. 15–Apr. 10	Apr. 1–May 1	Apr. 10–May 1	Apr. 20–May 15	May 1–30	May 10–June 10
Onion, seed	Mar. 1–Apr. 1	Mar. 15–Apr. 1	Mar. 15–Apr. 15	Apr. 1–May 1	Apr. 20–May 15	May 1–30	May 10–June 10
Onion, sets	Mar. 1–Apr. 1	Mar. 10–Apr. 1	Mar. 10–Apr. 10	Arp. 10–May 1	Apr. 20–May 15	May 1–30	May 10–June 10
Parsley	Mar. 10–Apr. 10	Mar. 20–Apr. 20	Apr. 1–May 1	Apr. 15–May 15	May 1–20	May 10–June 1	May 20–June 10
Parsnip	Mar. 10–Apr. 10	Mar. 20–Apr. 20	Apr. 1–May 1	Apr. 15–May 15	May 1–20	May 10–June 1	May 20–June 10
Peas, black-eye	May 1–July 1	May 10–June 15	May 15–June 1				
Peas, garden	Feb. 20–Mar. 20	Mar. 10–Apr. 10	Mar. 20–May 1	Apr. 1–May 15	Apr. 15–June 1	May 1–June 15	May 10–June 15
Pepper[1]	May 1–June 1	May 10–June 1	May 15–June 10	May 20–June 10	May 25–June 15	June 1–15	
Potato	Mar. 10–Apr. 1	Mar. 15–Apr. 10	Mar. 20–May 10	Apr. 1–June 1	Apr. 15–June 15	May 1–June 15	May 15–June 1
Radish	Mar. 1–May 1	Mar. 1–May 1	Mar. 20–May 10	Apr. 1–June 1	Apr. 15–June 15	May 1–June 15	May 15–June 1
Rhubarb[1]	Mar. 1–Apr. 1	Mar. 1–Apr. 1	Mar. 20–Apr. 15	Apr. 1–May 1	Apr. 15–May 10	May 1–20	May 20–June 1
Rutabaga			May 1–June 1	May 1–June 1	May 1–20	May 10–20	May 10–June 1
Salsify	Mar. 10–Apr. 15	Mar. 20–May 1	Apr. 1–May 15	Apr. 15–June 1	May 1–June 1	May 10–June 1	May 10–June 1
Shallot	Mar. 1–Apr. 1	Mar. 15–Apr. 15	Apr. 1–May 1	Apr. 10–May 1	Apr. 20–May 10	May 1–June 1	May 20–June 10
Sorrel	Mar. 1–Apr. 15	Mar. 15–May 1	Apr. 1–May 15	Apr. 15–June 1	May 1–June 1	May 1–June 1	
Soybean	May 1–June 30	May 10–June 20	May 15–June 15	May 25–June 10			
Spinach	Feb. 15–Apr. 1	Mar. 1–Apr. 1	Mar. 20–Apr. 20	Apr. 1–June 15	Apr. 10–June 15	Apr. 20–June 15	May 1–June 15
Spinach, New Zealand	Apr. 20–June 1	May 1–June 15	May 1–June 15	May 10–June 15	May 20–June 15	June 1–15	June 10–20
Squash, summer	Apr. 20–June 1	May 1–June 15	May 1–30	May 10–June 10	May 20–June 15	June 1–20	
Sweet potato	May 1–June 1	May 10–June 10	May 20–June 10		May 25–June 15	June 1–15	
Tomato	Apr. 20–June 1	May 5–June 10	May 10–June 15	May 15–June 10	May 25–June 15	June 5–20	June 15–30
Turnip	Mar. 1–Apr. 1	Mar. 10–Apr. 1	Mar. 20–May 1	Apr. 1–June 1	Apr. 15–June 1	May 1–June 15	May 15–June 15
Watermelon[2]	Apr. 20–June 1	May 1–June 15	May 15–June 15	June 1–15	June 15–July 1		

[1] Plants.

[2] Included in this section because no appropriate coverage is available elsewhere.

FIGURE 39-4 Earliest dates, and ranges of dates, for safe spring planting of vegetables in the open (USDA Home and Garden Bulletin 202) (continued)

SUMMER OR FALL PLANTING DATES FOR LOCALITIES IN WHICH AVERAGE DATE OF LAST FREEZE IS

CROP	Aug. 30	Sept. 10	Sept. 20	Sept. 30	Oct. 10	Oct. 20
Asparagus[1]				June 1–15	Oct. 20–Nov. 15	Nov. 1–Dec. 15
Beans, lima					June 1–15	June 15–30
Beans, snap	May 15–June 15	May 15–June 15	June 1–July 1	June 1–July 10	June 15–July 20	July 1–Aug. 1
Beet	May 1–June 1	May 1–June 15	June 1–July 1	June 1–July 10	June 15–July 25	July 1–Aug. 5
Broccoli, sprouting	May 1–June 1	May 1–June 1	May 1–June 15	June 1–30	June 15–July 15	July 1–Aug. 1
Brussels sprouts	May 1–June 1	May 1–June 1	May 1–June 15	June 1–30	June 15–July 15	July 1–Aug. 1
Cabbage[1]	May 1–June 1	May 1–June 1	May 1–June 15	June 1–July 10	June 1–July 15	July 1–20
Cabbage, Chinese	May 15–June 15	May 15–June 15	June 1–July 1	June 1–July 15	June 15–Aug. 1	July 15–Aug. 15
Carrot	May 15–June 15	May 15–June 15	June 1–July 1	June 1–July 10	June 1–July 20	June 15–Aug. 1
Cauliflower[1]	May 1–June 1	May 1–July 1	May 1–July 1	May 10–July 15	June 1–July 25	July 1–Aug. 5
Celery[1] and celeriac	May 1–July 1	May 15–July 1	May 15–July 1	June 1–July 5	June 1–July 15	June 1–Aug. 1
Chard	May 15–June 15	May 15–July 1	May 15–July 1	June 1–July 5	June 1–July 20	June 1–Aug. 1
Chervil and chives	May 10–June 10	May 15–June 15	May 15–June 15	(2)	(2)	(2)
Chicory, witloof	May 15–June 15	May 15–June 15	May 15–June 15	June 1–July 1	June 1–July 1	June 15–July 15
Collards[1]	May 15–June 15	May 15–June 15	May 15–June 15	June 15–July 15	July 1–Aug. 1	July 15–Aug. 15
Corn, sweet			June 1–July 1	June 1–July 1	June 1–July 10	June 1–July 20
Cornsalad	May 15–June 15	May 15–July 1	June 15–Aug. 1	July 15–Sept. 1	Aug. 15–Sept. 15	Sept. 1–Oct. 15
Cress, upland	May 15–June 15	May 15–July 1	June 15–Aug. 1	July 15–Sept. 1	Aug. 15–Sept. 15	Sept. 1–Oct. 15
Cucumber			June 1–15	June 1–July 1	June 1–July 1	June 1–July 15
Eggplant[1]				May 20–June 10	May 15–June 15	June 1–July 1
Endive	June 1–July 1	June 1–July 1	June 15–July 15	June 15–Aug. 1	July 1–Aug. 15	July 15–Sept. 1
Fennel, Florence	May 15–June 15	May 15–July 15	June 1–July 1	June 1–July 1	June 15–July 15	June 15–Aug. 1
Garlic	(2)	(2)	(2)	(2)	(2)	(2)
Horseradish[1]	(2)	(2)	(2)	(2)	(2)	(2)
Kale	May 15–June 15	May 15–June 15	June 1–July 1	June 15–July 15	July 1–Aug. 1	July 15–Aug. 15
Kohlrabi	May 15–June 15	June 1–July 1	June 1–July 15	June 15–July 15	July 1–Aug. 1	July 15–Aug. 15
Leek	May 1–June 1	May 1–June 1	(2)	(2)	(2)	(2)
Lettuce, head[1]	May 15–July 1	May 15–July 1	June 1–July 15	June 15–Aug. 1	July 15–Aug. 15	Aug. 1–30
Lettuce, leaf	May 15–July 15	May 15–July 15	June 1–Aug. 1	June 1–Aug. 1	July 15–Sept. 1	July 15–Sept. 1
Muskmelon			May 1–June 15	May 15–June 1	June 1–June 15	June 15–July 20
Mustard	May 15–July 15	May 15–July 15	June 1–Aug. 1	June 15–Aug. 1	July 15–Aug. 15	Aug. 1–Sept. 1
Okra			June 1–20	June 1–July 1	June 1–July 1	June 1–Aug. 1
Onion[1]	May 1–June 10	May 1–June 10	(2)	(2)	(2)	(2)
Onion, seed	May 1–June 1	May 1–June 10	(2)	(2)	(2)	(2)
Onion, sets	May 1–June 1	May 1–June 10	(2)	(2)	(2)	(2)
Parsley	May 15–June 15	May 15–June 15	June 1–July 15	June 1–July 15	June 15–Aug. 1	July 15–Aug. 15
Parsnip	May 15–June 1	May 15–June 15	May 15–June 15	June 1–July 1	June 1–July 10	(2)
Peas, black-eye				June 1–Aug. 1	June 1–July 1	June 1–July 1
Peas, garden	May 10–June 15	May 1–July 1	June 1–July 15	June 1–July 1	(2)	(2)
Pepper[1]			June 1–June 20		June 1–July 10	June 1–July 10
Potato	May 15–June 1	May 1–June 15	May 1–June 15	May 1–June 15	May 15–June 15	June 15–July 15

[1]Plants.

FIGURE 39-5   Latest dates, and ranges of dates, for safe fall planting of vegetables in the open (USDA Home and Garden Bulletin 202)

SUMMER OR FALL PLANTING DATES FOR LOCALITIES IN WHICH AVERAGE DATE OF LAST FREEZE IS (continued)

CROP	Aug. 30	Sept. 10	Sept. 20	Sept. 30	Oct. 10	Oct. 20
Radish	May 1–July 15	May 1–Aug. 1	June 1–Aug. 15	July 1–Sept. 1	July 15–Sept. 15	Aug. 1–Oct. 1
Rhubarb[1]	Sept. 1–Oct. 1	Sept. 15–Oct. 15	Sept. 15–Nov. 1	Oct. 1–Nov. 1	Oct. 15–Nov. 15	Oct. 15–Dec. 1
Rutabaga	May 15–June 15	May 1–June 15	June 1–July 1	June 1–July 1	June 15–July 15	July 10–20
Salsify	May 15–June 1	May 10–June 10	May 20–June 20	June 1–20	June 1–July 1	June 1–July 1
Shallot	(2)	(2)	(2)	(2)	(2)	(2)
Sorrel	May 15–June 15	May 1–June 15	June 1–July 1	June 1–July 15	July 1–Aug. 1	July 15–Aug. 15
Soybean				May 25–June 10	June 1–25	June 1–July 5
Spinach	May 15–July 1	June 1–July 15	June 1–Aug. 1	July 1–Aug. 15	Aug. 1–Sept. 1	Aug. 20–Sept. 10
Spinach, New Zealand						June 1–Aug. 1
Squash, summer	June 10–20	June 1–20	May 15–July 1	May 15–July 1	June 15–July 15	June 1–July 20
Squash, winter			May 20–June 10	June 1–15	June 1–July 1	June 1–July 1
Sweet potato					May 20–June 10	June 1–15
Tomato	June 20–30	June 10–20	June 1–20	June 1–20	June 1–20	June 1–July 1
Turnip	May 15–June 15	June 1–July 1	June 1–July 15	June 1–Aug. 1	July 1–Aug. 1	July 15–Aug. 15
Watermelon			May 1–June 15	May 15–June 1	June 1–15	June 15–July 20

FALL PLANTING DATES FOR LOCALITIES IN WHICH AVERAGE DATE OF FIRST FREEZE IS

CROP	Oct. 30	Nov. 10	Nov. 20	Nov. 30	Dec. 10	Dec. 20
Asparagus[1]	Nov. 15–Jan. 1	Dec. 1–Jan. 1				Sept. 1–Oct. 1
Beans, lima	July 1–Aug. 1	July 1–Aug. 15	July 15–Sept. 1	Aug. 1–Sept. 15	Sept. 1–30	Sept. 1–Nov. 1
Beans, snap	July 1–Aug. 15	July 1–Sept. 1	July 1–Sept. 10	Aug. 15–Sept. 20	Sept. 1–30	Sept. 1–Dec. 31
Beet	Aug. 1–Sept. 1	Aug. 1–Oct. 1	Sept. 1–Dec. 1	Sept. 1–Dec. 15	Sept. 1–Dec. 31	Sept. 1–Dec. 31
Broccoli, sprouting	July 1–Aug. 15	Aug. 1–Sept. 1	Aug. 1–Sept. 15	Aug. 1–Oct. 1	Aug. 1–Nov. 1	Sept. 1–Dec. 31
Brussels sprouts	July 1–Aug. 15	Aug. 1–Sept. 1	Aug. 1–Sept. 15	Aug. 1–Oct. 1	Aug. 1–Nov. 1	Sept. 1–Dec. 31
Cabbage[1]	Aug. 1–Sept. 1	Sept. 1–15	Sept. 1–Dec. 1	Sept. 1–Dec. 31	Sept. 1–Dec. 31	Sept. 1–Dec. 31
Cabbage, Chinese	Aug. 1–Sept. 15	Aug. 15–Oct. 1	Sept. 1–Oct. 15	Sept. 1–Nov. 1	Sept. 1–Nov. 15	Sept. 1–Dec. 1
Carrot	July 1–Aug. 15	Aug. 1–Sept. 1	Sept. 1–Nov. 1	Sept. 15–Dec. 1	Sept. 15–Dec. 1	Sept. 15–Dec. 1
Cauliflower[1]	July 15–Aug. 15	Aug. 1–Sept. 1	Aug. 1–Sept. 15	Aug. 15–Oct. 10	Sept. 1–Oct. 20	Sept. 15–Nov. 1
Celery[1] and celeriac	June 15–Aug. 15	July 1–Aug. 15	July 15–Sept. 1	Aug. 1–Dec. 1	Sept. 1–Dec. 31	Oct. 1–Dec. 31
Chard	June 1–Sept. 10	June 1–Sept. 15	June 1–Oct. 1	June 1–Nov. 1	June 1–Dec. 1	June 1–Dec. 31
Chervil and chives	(2)	(2)	Nov. 1–Dec. 31	Nov. 1–Dec. 31	Nov. 1–Dec. 31	Nov. 1–Dec. 31
Chicory, witloof	July 1–Aug. 10	July 10–Aug. 20	July 20–Sept. 1	Aug. 15–Sept. 30	Aug. 15–Oct. 15	Aug. 15–Oct. 15
Collards[1]	Aug. 1–Sept. 15	Aug. 15–Oct. 1	Aug. 25–Nov. 1	Sept. 1–Dec. 1	Sept. 1–Dec. 31	Sept. 1–Dec. 31
Corn, sweet	June 1–Aug. 1	June 1–Aug. 15	June 1–Sept. 1			
Cornsalad	Sept. 15–Nov.1	Oct. 1–Dec. 1	Oct. 1–Dec. 1	Oct. 1–Dec. 31	Oct. 1–Dec. 31	Oct. 1–Dec. 31
Cress, upland	Sept. 15–Nov. 1	Oct. 1–Dec. 1	Oct. 1–Dec. 1	Oct. 1–Dec. 31	Oct. 1–Dec. 31	Oct. 1–Dec. 31
Cucumber	June 1–Aug. 1	June 1–Aug. 15	June 1–Aug. 15	July 15–Sept. 15	Aug. 15–Oct. 1	Aug. 15–Oct. 1
Eggplant[1]	June 1–July 1	June 1–July 15	June 1–Aug. 1	July 1–Sept. 1	Aug. 1–Sept. 30	Aug. 1–Sept. 30
Endive	July 15–Aug. 15	July 15–Aug. 15	Sept. 1–Oct. 1	Sept. 1–Nov. 15	Sept. 1–Dec. 31	Sept. 1–Dec. 31

[1]Plants.

FIGURE 39-5   Latest dates, and ranges of dates, for safe fall planting of vegetables in the open (USDA Home and Garden Bulletin 202) (continued)

FALL PLANTING DATES FOR LOCALITIES IN WHICH AVERAGE DATE OF FIRST FREEZE IS (continued)

CROP	Oct. 30	Nov. 10	Nov. 20	Nov. 30	Dec. 10	Dec. 20
Fennel, Florence	July 1–Aug. 1	July 15–Aug. 15	Aug. 15–Sept. 15	Sept. 1–Nov. 15	Sept. 1–Dec. 1	Sept. 1–Dec. 1
Garlic	(2)	Aug. 1–Oct. 1	Aug. 15–Oct. 1	Sept. 1–Nov. 15	Sept. 15–Nov. 15	Sept. 15–Nov. 15
Horseradish[1]	(2)	(2)	(2)	(2)	(2)	(2)
Kale	July 15–Sept. 1	Aug. 1–Sept. 15	Aug. 15–Oct. 15	Sept. 1–Dec. 1	Sept. 1–Dec. 31	Sept. 1–Dec. 31
Kohlrabi	Aug. 1–Sept. 1	Aug. 15–Sept. 15	Sept. 1–Oct. 15	Sept. 1–Dec. 1	Sept. 15–Dec. 31	Sept. 15–Dec. 31
Leek	(2)	(2)	Sept. 1–Nov. 1	Sept. 1–Nov. 1	Sept. 1–Nov. 1	Sept. 15–Nov. 1
Lettuce, head[1]	Aug. 1–Sept. 15	Aug. 15–Oct. 15	Sept. 1–Nov. 1	Sept. 1–Dec. 1	Sept. 15–Dec. 31	Sept. 15–Dec. 31
Lettuce, leaf	Aug. 15–Oct. 1	Aug. 25–Oct. 1	Sept. 1–Nov. 1	Sept. 1–Dec. 1	Sept. 15–Dec. 31	Sept. 15–Dec.31
Muskmelon	July 1–July 15	July 15–July 30				
Mustard	Aug. 15–Oct. 15	Aug. 15–Nov. 1	Sept. 1–Dec. 1	Sept. 1–Dec. 1	Sept. 1–Dec. 1	Sept. 15–Dec. 1
Okra	June 1–Aug. 10	June 1–Aug. 20	June 1–Sept. 10	June 1–Sept. 20	Aug. 1–Oct. 1	Aug. 1–Oct. 1
Onion[1]		Sept. 1–Oct. 15	Oct. 1–Dec. 31	Oct. 1–Dec. 31	Oct. 1–Dec. 31	Oct. 1–Dec. 31
Onion, seed			Sept. 1–Nov. 1	Sept. 1–Nov. 1	Sept. 1–Nov. 1	Sept. 15–Nov. 1
Onion, sets		Oct. 1–Dec. 1	Nov. 1–Dec. 31	Nov. 1–Dec. 31	Nov. 1–Dec. 31	Nov. 1–Dec. 31
Parsley	Aug. 1–Sept. 15	Sept. 1–Nov. 15	Sept. 1–Dec. 31	Sept. 1–Dec. 31	Sept. 1–Dec. 31	Sept. 1–Dec. 31
Parsnip	(2)	(2)	Aug. 1–Sept. 1	Sept. 1–Dec. 1	Sept. 1–Dec. 1	Sept. 1–Dec. 1
Peas, black-eye	June 1–Aug. 1	June 15–Aug. 15	July 1–Sept. 1	July 1–Sept. 10	July 1–Sept. 20	July 1–Sept. 20
Peas, garden	Aug. 1–Sept. 15	Sept. 1–Nov. 1	Oct. 1–Dec. 1	Oct. 1–Dec. 31	Oct. 1–Dec. 31	Oct. 1–Dec. 31
Pepper[1]	June 1–July 20	June 1–Aug. 1	June 1–Aug. 15	June 15–Sept. 1	Aug. 15–Oct. 1	Aug. 15–Oct. 1
Potato	July 20–Aug. 10	July 25–Aug. 20	Aug. 10–Sept. 15	Aug. 1–Sept. 15	Aug. 1–Sept. 15	Aug. 1–Sept. 15
Radish	Aug. 15–Oct. 15	Sept. 1–Nov. 15	Sept. 1–Dec. 1	Sept. 1–Dec. 31	Aug. 1–Sept. 15	Aug. 1–Sept. 15
Rhubarb[1]	Nov. 1–Dec. 1					
Rutabaga	July 15–Aug. 1	July 15–Aug. 15	Aug. 1–Sept. 1	Sept. 1–Nov. 15	Oct. 1–Nov. 15	Oct. 15–Nov. 15
Salsify	June 1–July 10	June 15–July 20	July 15–Aug. 15	Aug. 15–Sept. 30	Aug. 15–Oct. 15	Sept. 1–Oct. 31
Shallot	(2)	Aug. 1–Oct. 1	Aug. 15–Oct. 1	Aug. 15–Oct. 15	Sept. 15–Nov. 1	Sept. 15–Nov. 1
Sorrel	Aug. 1–Sept. 15	Aug. 15–Oct. 15	Aug. 15–Oct. 15	Sept. 1–Dec. 15	Sept. 1–Dec. 15	Sept. 1–Dec. 31
Soybean	June 1–July 15	June 1–July 25	June 1–July 30	June 1–July 30	June 1–July 30	June 1–July 30
Spinach	Sept. 1–Oct. 1	Sept. 15–Nov. 1	Oct. 1–Dec. 1	Oct. 1–Dec. 31	Oct. 1–Dec. 31	Oct. 1–Dec. 31
Spinach, New Zealand	June 1–Aug. 1	June 1–Aug. 15	June 1–Aug. 15			
Squash, summer	June 1–Aug. 1	June 1–Aug. 10	June 1–Aug. 20	June 1–Sept. 1	June 1–Sept. 15	June 1–Oct. 1
Squash, winter	June 10–July 10	June 20–July 20	July 1–Aug. 1	July 15–Aug. 15	Aug. 1–Sept. 1	Aug. 1–Sept. 1
Sweet potato	June 1–15	June 1–July 1	June 1–July 1	June 1–July 1	June 1–July 1	June 1–July 1
Tomato	June 1–July 1	June 1–July 15	June 1–Aug. 1	Aug. 1–Sept. 1	Aug. 15–Oct. 1	Sept. 1–Nov. 1
Turnip	Aug. 1–Sept. 15	Sept. 1–Oct. 15	Sept. 1–Nov. 15	Sept. 1–Nov. 15	Oct. 1–Dec. 1	Oct. 1–Dec. 31
Watermelon	July 1–July 15	July 15–July 30				

[1]Plants.

FIGURE 39-5 Latest dates, and ranges of dates, for safe fall planting of vegetables in the open (USDA Home and Garden Bulletin 202) (continued)

planting have been taken care of. The next step is to harrow or rake the garden in preparation for planting. The soil should be worked to break all large clods of soil and to form a firm seedbed. For a seed to germinate, there must be good contact with the soil so that moisture can be pulled from the soil into the seed. Large, open air spaces result in poor germination. *The smaller the seed the more important it is to have a fine, firm seedbed.*

To be sure that the rows of seeds are planted straight, use a string stretched across the garden. To get the proper number of seed per foot or row, a wooden marker measured off in 1 foot lengths may be used, figure 39-6. If there is no problem estimating distance, use of the marker pole is not necessary. To establish the proper distance between rows, and the depth at which to plant seed, see figure 39-7.

The planting row should be made at least deep enough to cover the seed at the proper depth. Small seeds, such as carrot and lettuce, should be covered with only ¼ to ½ inch of soil. The hoe handle can be used to form these shallow rows. For deeper planting, use the corner of the hoe blade or a wheel hoe. Read the directions on the seed package for the proper depth and distance apart to plant the seed, or use figure 39-7 as a guide. Fertilizer may be banded on each side of the row.

A hoe or garden rake, figure 39-8, page 422, is used to cover the seed. Try to cover the seed with fine soil that is free of lumps, stones, and clods. Cover all seeds in the row with the same depth of soil. Firm the soil around the seeds with the rake or hoe so that close contact is made between soil and seed. If the soil is a little wet when planting, do not pack or press soil down on seeds. This causes a hard crust to form and seeds will have difficulty breaking through the soil surface. Do not plant when the soil is wet.

When using the wide row system, small, easily covered seeds are planted by spreading them out over the top of the bed to the desired width and seed spacing and pulling dirt in from the edges to cover seeds. Several rows side by side can also be opened for larger seeds and covered the same as with the single row system, the only difference being that rows are closer together. Figure 39-9, page 422, shows a wide row being prepared for planting.

PLANTING VEGETABLE PLANTS

SEEDING

Some vegetables are not direct seeded in the garden. These seeds are planted in flats or other containers before the frost-free date and later transplanted to the garden. Figure 39-10, page 422, shows a flat with rows marked off being seeded with vegetables for later transplanting. Soil in the flat should be a fine, carefully screened mixture of sand, soil, and peat moss or one of the soilless mixes. Seeds may also be planted in individual peat pots or Jiffy 7 peat pellets. Figure 39-11, page 422, shows a Jiffy 7 peat pellet planted with a single seed.

The length of time needed to reach transplanting size varies from vegetable to vegetable. A good general rule is to plant the seed in the flat or other container six to eight weeks before plants are to be transplanted to the garden.

After the seeds are planted, the flat or pot should be watered to settle the medium around the seeds. Close contact between the soil and seed is important for germination. Cover the flat or pot with newspaper, glass, or plastic to hold in moisture. The seedbed container should never be allowed to dry out, since newly germinating seeds are easily killed by lack of moisture. Uncover the seedlings as soon as shoots appear through the soil. Water as often as necessary to keep the growing medium moist but not wet. Fertilize the seedlings as recommended on the fertilizer container label. Do not use too much fertilizer—new plants are easily burned. (Some soilless mixes have slow-release fertilizer in the mix, making additional fertilizer unnecessary.)

FIGURE 39-6 Using a line and marker pole to establish a straight row and the proper number of seeds per foot of row. Notice the even texture of the soil and the bean seeds. (Ed Reiley, Photographer)

Crop	REQUIREMENT FOR 100 FEET OF ROW		Depth for Planting Seed (Inches)	DISTANCE APART		Plants in the Row
	Seed	Plants		Rows		
				Horse- or Tractor-Cultivated (Feet)	Hand-cultivated	
Asparagus	1 ounce	75	1–1½	4–5	1½ to 2 feet	18 inches
Beans:						
Lima, bush	½ pound		1–1½	2½–3	2 feet	3 to 4 inches
Lima, pole	½ pound		1–1½	3–4	3 feet	3 to 4 feet
Snap, bush	½ pound		1–1½	2½–3	2 feet	3 to 4 inches
Snap, pole	4 ounces		1–1½	3–4	2 feet	3 feet
Beet	2 ounces		1	2–2½	14 to 16 inches	2 to 3 inches
Broccoli:						
Heading	1 packet	50–75	½	2½–3	2 to 2½ feet	14 to 24 inches
Sprouting	1 packet	50–75	½	2½–3	2 to 2½ feet	14 to 24 inches
Brussels sprouts	1 packet	50–75	½	2½–3	2 to 2½ feet	14 to 24 inches
Cabbage	1 packet	50–75	½	2½–3	2 to 2½ feet	14 to 24 inches
Cabbage, Chinese	1 packet		½	2–2½	18 to 24 inches	8 to 12 inches
Carrot	1 packet		½	2–2½	14 to 16 inches	2 to 3 inches
Cauliflower	1 packet	50–75	½	2½–3	1 to 2½ feet	14 to 24 inches
Celeriac	1 packet	200–250	⅛	2½–3	18 to 24 inches	4 to 6 inches
Celery	1 packet	200–250	⅛	2½–3	18 to 24 inches	4 to 6 inches
Chard	2 ounces		1	2–2½	18 to 24 inches	6 inches
Chervil	1 packet		½	2–2½	14 to 16 inches	2 to 3 inches
Chicory, witloof	1 packet		½	2–2½	18 to 24 inches	6 to 8 inches
Chives	1 packet		½	2½–3	14 to 16 inches	in clusters
Collards	1 packet		½	3–3½	18 to 24 inches	18 to 24 inches
Corn, sweet	2 ounces		2	3–3½	2 to 3 feet	drills, 14 to 16 inches; hills, 2½ to 3 feet
Cornsalad	1 packet		½	2½–3	14 to 16 inches	1 foot
Cress, upland	1 packet		⅛–¼	2–2½	14 to 16 inches	2 to 3 inches
Cucumber	1 packet		½	6–7	6 to 7 feet	drills, 3 feet; hills, 6 feet
Dasheen	5 to 6 pounds	50	2–3	3½–4	3½ to 4 feet	2 feet
Eggplant	1 packet	50	½	3	2 to 2½ feet	3 feet
Endive	1 packet		½	2½–3	18 to 24 inches	12 inches
Fennel, Florence	1 packet		½	2½–3	18 to 24 inches	4 to 6 inches
Garlic	1 pound		1–2	2½–3	14 to 16 inches	2 to 3 inches
Horseradish	Cuttings	50–75	2	3–4	2 to 2½ feet	18 to 24 inches
Kale	1 packet		½	2½–3	18 to 24 inches	12 to 15 inches
Kohlrabi	1 packet		½	2½–3	14 to 16 inches	5 to 6 inches
Leek	1 packet		½–1	2½–3	14 to 16 inches	2 to 3 inches
Lettuce, head	1 packet	100	½	2½–3	14 to 16 inches	12 to 15 inches
Lettuce, leaf	1 packet		½	2½–3	14 to 16 inches	6 inches
Muskmelon	1 packet		1	6–7	6 to 7 feet	hills, 6 feet

FIGURE 39-7 Quantity of seed and number of plants required for 100 feet of row, depths of planting, and distances apart for rows and plants (USDA Home and Garden Bulletin 202)

REQUIREMENT FOR 100 FEET OF ROW				DISTANCE APART		
				Rows		
Crop	Seed	Plants	Depth for Planting Seed (Inches)	Horse- or Tractor-Cultivated (Feet)	Hand-cultivated	Plants in the Row
Mustard	1 packet		1/2	2 1/2–3	14 to 16 inches	12 inches
Okra	2 ounces		1–1 1/2	3–3 1/2	3 to 3 1/2 feet	2 feet
Onion:						
Plants		400	1–2	2–2 1/2	14 to 16 inches	2 to 3 inches
Seed	1 packet		1/2–1	2–2 1/2	14 to 16 inches	2 to 3 inches
Sets	1 pound		1–2	2–2 1/2	14 to 16 inches	2 to 3 inches
Parsley	1 packet		1/8	2–2 1/2	14 to 16 inches	4 to 6 inches
Parsley, turnip-rooted	1 packet		1/8–1/4	2–2 1/2	14 to 16 inches	2 to 3 inches
Parsnip	1 packet		1/2	2–2 1/2	18 to 24 inches	2 to 3 inches
Peas	1/2 pound		2–3	2–4	1 1/2 to 3 feet	1 inch
Pepper	1 packet	50–70	1/2	3–4	2 to 3 feet	18 to 24 inches
Physalis	1 packet		1/2	2–2 1/2	1 1/2 to 2 feet	12 to 18 inches
Potato	5 to 6 pounds, tubers		4	2 1/2–3	2 to 2 1/2 feet	10 to 18 inches
Pumpkin	1 ounce		1–2	5–8	5 to 8 feet	3 to 4 feet
Radish	1 ounce		1/2	2–2 1/2	14 to 16 inc hes	1 inch
Rhubarb		25–35		3–4	3 to 4 feet	3 to 4 feet
Salsify	1 ounce		1/2	2–2 1/2	18 to 26 inches	2 to 3 inches
Shallots	1 pound (cloves)		1–2	2–2 1/2	12 to 18 inches	2 to 3 inches
Sorrel	1 packet		1/2	2–2 1/2	18 to 24 inches	5 to 8 inches
Soybean	1/2 to 1 pound		1–1 1/2	2 1/2–3	24 to 30 inches	3 inches
Spinach	1 ounce		1/2	2–2 1/2	14 to 16 inches	3 to 4 inches
Spinach, New Zealand	1 ounce		1–1 1/2	3–3 1/2	3 feet	18 inches
Squash:						
Bush	1/2 ounce		1–2	4–5	4 to 5 feet	drills, 15 to `8 inches; hills, 4 feet
Vine	1 ounce		1–2	8–12	8 to 12 feet	drills, 2 to 3 feet; hills, 4 feet
Sweet potato	5 pounds, bedroots	75	2–3	3–3 1/2	3 to 3 1/2 feet	12 to 14 inches
Tomato	1 packet	35–50	1/2	3–4	2 to 3 feet	1 1/2 to 3 feet
Turnip greens	1 packet		1/4–1/2	2–2 1/2	14 to 16 inches	2 to 3 inches
Turnips and rutabagas	1/2 ounce		1/4–1/2	2–2 1/2	14 to 16 inches	2 to 3 inches
Watermelon	1 ounce		1–2	8–10	8 to 10 feet	drills, 2 to 3 feet; hills, 8 feet

FIGURE 39-7 Quantity of seed and number of plants required for 100 feet of row, depths of planting, and distances apart for rows and plants (USDA Home and Garden Bulletin 202) (continued)

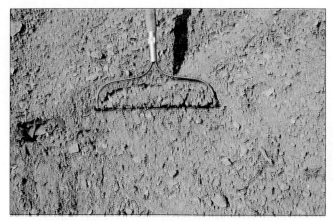

FIGURE 39-8 A garden rake is used to cover seeds evenly. (Ed Reiley, Photographer)

FIGURE 39-9 A wide row being prepared for planting vegetables (Courtesy Gardenway, Inc.)

FIGURE 39-10 Seeds being planted in a flat to be grown indoors for later transplanting to the garden. Notice the labels placed for seed identification. (Ed Reiley, Photographer)

FIGURE 39-11 Jiffy 7 pellet used to start individual plants from seed. (Left) pellet before soaking in water; (right) plant ready for transplanting to the garden (Ed Reiley, Photographer)

To determine which vegetable crops are usually planted in the garden as seeds and which ones may be transplanted to the garden as plants, see figure 39-7. The column titled *Requirement for 100 Feet of Row* lists both the amount of seed and the number of plants for those crops that are transplanted to the garden as plants.

FIRST TRANSPLANTING

If seeds are sown as directed on the package, two transplantings will be necessary. The first transplanting must be made soon after the seedlings develop the first pair of true leaves. At this first transplanting, each plant should be spaced about 2 inches apart in a flat or planted in individual peat pots or Market Packs. The soil mix used is the same mix the seeds were planted in—one-third soil, one-third sand or perlite, one-third peat moss, or an artificial soil mix. A good soluble fertilizer should be used within a few days after transplanting and repeated according to directions. Some of the soilless planting mixes contain fertilizer, making the addition of fertilizer unnecessary.

HARDENING OFF

As the date nears for the plants to be transplanted to the garden, the atmosphere must be gradually changed to resemble as closely as possible the conditions of the outside garden. If plants are in a greenhouse in which temperature and humidity are high, the following should be done.

PROCEDURE

HARDENING OFF IN THE GREENHOUSE

1. Place the plants where the temperature can gradually be lowered over a period of ten days to two weeks. This slows the growth rate and toughens the plant.

2. Lower the humidity and allow the soil to become dryer. This also slows growth and toughens the plant. Do not allow the plants to wilt, however.

3. If planted in flats, cut a square around the plants with a sharp knife about ten days before transplanting. Use all the area of the flat, and make the blocks as big as possible. In the period of ten days to two weeks, the plant will grow new roots in a tighter root ball and be able to withstand transplanting shock better.

These three steps slow the growth rate to prepare the plant for chilling weather, drying winds, a shortage of water, or high temperatures. Tender, succulent growth must be hardened for best growth of plants.

The three steps of hardening off can be done in a greenhouse if all the plants it contains are being hardened off, or in a special cold frame. Cold frames are usually used for hardening off. Perma-

FIGURE 39-12 Temporary cold frame used for hardening off flower and vegetable plants. (Courtesy Gardener's Supply Company)

nent cold frames are usually constructed of concrete block or wood with a glass sash on top. A temporary cold frame is shown in figure 39-12. Cold frames are never heated—the frame is opened enough during the day to keep the sun from raising the temperature too high and closed at night to hold heat in and prevent freezing of the plants. The cold frame is gradually opened more and more each day until the plants are hardened off and living in the outdoor environment. Most plants are then ready for transplanting. (Some plants, such as tomatoes, must not be transplanted to the garden until all danger of frost is past. On the other hand, other crops, such as cabbage, are more tolerant of cold weather and can be transplanted to the garden earlier.)

SECOND TRANSPLANTING

The vegetable plants are now properly hardened off and ready to transplant to the garden.

PROCEDURE

TRANSPLANTING PLANTS TO THE GARDEN

1. Water the plants to be transplanted the evening before, or at least several hours before transplanting.

2. If plants are grown in peat pots, peel off the top half of the pot and break off the bottom. If roots are already growing through the pot, the bottom need not be broken. This practice helps to speed the spread of roots into the garden soil, especially if dry weather follows transplanting.

3. If plants are grown in flats, cut the soil between the plants and lift them out with a putty knife or other small tool. Leave as much soil around the roots as possible. Handle carefully.

4. Check the distance apart at which the plants are to be placed and set them immediately in the garden soil. Place the plants a little deeper in the garden soil than they grew in the original container. Some tall, spindly plants, especially tomatoes, may be planted deeper.

Tips: Do not allow plant roots to be exposed to the sun or drying wind.

5. Pull soil toward the plant and firm slightly. Fill the hole only two-thirds full. Stop and water enough to settle the soil around the roots. Pull in enough dry soil to level the hole and cover the wet area. This prevents moisture loss and baking of the soil.
   Figure 39-13 shows tomato plants being transplanted from pots to the garden.

FIGURE 39-13 Transplanting a tomato plant to the garden. Notice the plant is in a peat pot so that the roots are not disturbed at planting time. Plant as deep as possible, placing the plant down in the hole with just the top leaves exposed. The peat pot should be completely covered to prevent drying out. (Ed Reiley, Photographer)

If possible, transplant on a cloudy day or in late afternoon or evening. Plants will not wilt as much if they are not exposed to bright sunlight immediately after transplanting, resulting in a higher percentage of healthy, growing plants.

A strip of cardboard, aluminum foil, or newspaper wrapped around the stem and extending about 1 inch underground and about 2 inches aboveground will prevent cutworms from cutting off and killing the new plant, figure 39-14.

EARLY PLANTING

Some new products on the market now allow earlier outdoor planting of vegetables. These products provide both protection from the cold and protection against some insects, promoting earlier maturity and higher yields.

WATER-FILLED TEPEE The first of these products is a small item called the water-filled tepee, figure 39-15. It is double-walled, made of plastic, measures 18 inches tall and 18 inches across, and holds ³⁄₁₀-gallon (38 ounces) of water. It is open at the top. Placed around a newly planted plant, a tepee will control the temperature around that

FIGURE 39-14 Placement of a cardboard collar around a tomato plant to prevent cutworms from cutting off the plant just above the soil level. (Ed Reiley, Photographer)

FIGURE 39-16   Lightweight floating row covers are very effective in protecting plants from heat, cold, insects, and rabbits. (Courtesy Gardener's Supply Company)

FIGURE 39-15   A water-filled tepee around a tomato plant. Tepees are very effective in protecting plants from both heat and cold. (Courtesy Gardener's Supply Company)

plant to a great degree since water is an excellent storage element for heat. In the daytime, the water warms up, while keeping the plant inside from getting too hot. At night, the heat is released to keep the plant from freezing when temperatures are below freezing. Planting can be done a month earlier than normal, resulting in earlier maturity of crops and higher yields, especially on crops like tomatoes.

FLOATING COVER   Another new gardening product is the floating row cover, figure 39-16. Made of very lightweight polyester material, the cover or blanket is placed over a row of plants and sealed with soil on both edges. As the plants grow, they lift up the lightweight blanket. The white material allows sunlight and rain to penetrate, holds in the heat at night, allows for only minor temperature changes, and may be left on until the plants get too large for the cover.

Frost protection down to 27°F is provided by the covers. Temperatures under these covers are from one to four degrees warmer than outside temperatures, so planting can be done one to two weeks sooner. This is especially important if early market prices are higher, resulting in more profits.

Some plants remain covered until harvest if sprays for insects are not needed. Some insects are kept away from plants by these blankets, and sprays may not be needed. However, strong winds may be a problem and a rye crop or other windbreak in narrow rows may be necessary.

The lightweight row covers have been tested at many state universities. Some of the results are:

1. Increased yields on beans by as much as 40 percent.

2. Increased yields, and delayed bolting (see glossary) of lettuce, spinach, and Chinese cabbage.

3. Protection against insects for cabbage, cucumbers, and melons.

4. Protection from virus disease spread by aphids.

Floating covers are being used both by home gardeners and by large commercial growers.

Many gardeners are now using floating covers in wide row planting. A row 1 to 3 feet wide is planted with closely spaced plants of the same variety. There is no wide spacing between rows; in fact, there are no rows in many cases. Seeds or plants are planted as close together as practical in a single wide row.

There are also other row covers that are placed over wire hoops stuck in the ground. They are applied in the same manner as the floating covers, but they also have a wire support to hold them up. For more information, contact your local agricultural experiment station.

DOUBLE CROPPING

Since some of the vegetables reach maturity earlier than others, consideration should be given to planting something in their place as they are harvested. Replanting the same crop or planting other crops are both possibilities. Vegetables may be grouped according to how long they grow in the

CROPS OCCUPYING THE GROUND ALL SEASON

asparagus	onions (from seeds)
beans, pole lima	parsnips
beans, pole snap	peppers
beets, late	pumpkins
carrots, late	rhubarb
corn, late	rutabagas
cucumbers	salsify
eggplant	squash
melons	tomatoes
okra	

SUCCESSIVE CROPS WHICH MAY BE REPLANTED OR FOLLOWED BY OTHER CROPS

beans, dwarf	peas
lettuce	radishes
mustard	spinach
parsley	turnips

EARLY CROPS WHICH MAY BE FOLLOWED BY OTHERS

beets, early	mustard
cabbage, early	onion sets
carrots, early	spinach
corn, early	turnips, early

LATE CROPS WHICH MAY FOLLOW OTHERS

beets, late	kale
Brussels sprouts	mustard
cabbage, late	peas, late
cauliflower	spinach
endive	turnips

FIGURE 39-17 By grouping vegetables in the garden according to their use as replanted or successive crops, harvesting is simplified.

garden and their use as replanted or successive crops, figure 39-17.

GARDEN TOOLS

Very few tools are necessary for cultivation of a small garden, figure 39-18. In most cases, these include:

- ◆ a spade or spading fork for digging up soil
- ◆ a steel bow garden rake for leveling and preparing the seed bed
- ◆ a garden hoe for laying off rows and controlling weeds
- ◆ a strong cord for laying off rows
- ◆ a wheelbarrow for hauling fertilizer and mulch
- ◆ a garden hose long enough to water the garden

For large gardens measuring about 2,000 to 4,000 square feet, a wheel hoe is helpful in saving labor.

For gardens measuring 4,000 square feet, a rotary tiller rather than a rake is useful for preparing the soil for planting and cultivating to control weeds.

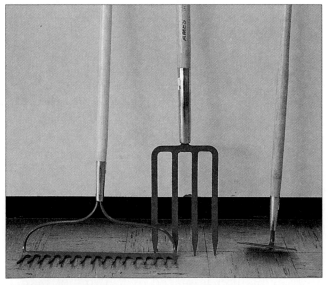

FIGURE 39-18 Essential tools for the home garden. From left to right: garden rake, fork spade, and garden hoe. (Ed Reiley, Photographer)

Student Activities

1) Plant seeds of vegetable crops in flats for sale or for use in a home or school garden.

2) Give a demonstration on the proper planting techniques of a home garden. Be sure to mention that the class has plants that are for sale.

3) Construct a temporary cold frame structure.

Self-Evaluation

1) Specify whether the following vegetables are direct seeded in the garden or transplanted to the garden as small plants.

beans	corn	peas	spinach
beets	cucumbers	peppers	squash
cabbage	eggplant	potatoes	sweet potatoes
carrots	onions	radishes	tomatoes

2) Specify whether the following vegetables are cold-tender or cold-hardy.

broccoli	cucumber	peas	sweet corn
beans	eggplant	radishes	tomatoes
cabbage	onions		

3) List the five steps in transplanting to the garden.

4) List five crops which

   (a) occupy the soil all season.

   (b) are successive crops.

   (c) are late crops which may follow others.

5) List the biggest advantage of the new row covers for vegetable plants.

Caring for the Vegetable Garden

OBJECTIVE

To care for a vegetable garden, producing a marketable-quality crop.

Competencies to Be Developed

After studying this unit, you should be able to

- ◆ determine watering needs of vegetable plants and list three ways in which water is applied.
- ◆ determine the type and amount of fertilizer for a specific crop by using a soil test recommendation.
- ◆ establish weed control programs using mulches, cultivation, and herbicides.
- ◆ list five vegetables grown in the area, a pest which commonly attacks each vegetable, and one method of control for each.

MATERIALS LIST

- ✔ a rain gauge
- ✔ garden hose and sprinkler
- ✔ garden fertilizer
- ✔ garden tools (listed in Unit 39)
- ✔ weed killers if used
- ✔ mulching material
- ✔ insecticides (malathion, Sevin)
- ✔ fungicides (captan, zineb)
- ✔ sprayer and/or duster
- ✔ rubber gloves and respirator

Planting the garden is only a small part of the work involved in growing vegetables. In planning the size of the garden, the time available for its care should be considered. It is better to have a small, well-kept garden than a large, weedy one. The best preparation, planning, and planting is of little value if the garden is not cared for properly.

WATERING

All plants must be watered when they are transplanted to the garden. There may also be other times during the growing season when lack of rainfall makes watering desirable for best yield, or even necessary for best plant growth.

When plants wilt for any length of time, production is reduced; watering helps to prevent serious damage to the plants at these times. If the plants do not recover from wilting soon after the sun is down, the need for water is even more serious.

APPLICATION OF WATER

If the garden is watered, it must be done properly—most plants are actually harmed by shallow watering. Soak the soil to a depth of at least 6 inches. This requires a great deal of water. To be sure the proper depth has been reached, dig down into the soil to see how far the water has penetrated. Another way to tell if enough water has been added is to place a can with straight sides or a rain gauge in the garden and measure how much water has been applied as a result of the sprinkling. The plants should receive no less than 1 inch of water per week. This amount includes any rainfall which may occur. Frequent, light waterings encourage shallow root systems and may actually reduce crop yield. Wide row plantings of vegetables need watering less often, since shading by plants keeps the soil cooler.

Water may be applied with sprinklers, a hose, by trench, or by ditch irrigation. Ditch or trench irrigation can be used if the garden is fairly level.

The use of natural mulches 3 to 4 inches thick can reduce the need for water in the garden. A good mulch shades the soil and keeps it cool, reducing the rate of evaporation of water from the soil. Porous mulches also slow down runoff of rainwater. Plastic mulch does a good job in moisture retention and also warms the soil during the early season, encouraging growth.

FERTILIZERS (ADDITIONAL)

By this time, some fertilizer should have been applied to the garden by broadcasting and plowing or digging in. Additional fertilizer application after the plants are growing is made by spreading the fertilizer around the plants (band application) or along the row (sidedress application), figure 40-1. If earlier application of fertilizer was not made to the garden, the amount of fertilizer applied after the plants have established growth will be increased slightly. In these cases, refer again to the soil test recommendations for the amount to apply. Never apply more than 3 pounds of fertilizer per 100 square feet in a single application.

Plants requiring high amounts of nitrogen, such as corn and the leafy vegetables, respond best to applications of fertilizer to the soil surface, such as by sidedressing or band application. This is because nitrogen leaches down to the roots much faster than phosphorus or potash.

Use of organic fertilizers such as cow manure can have a significant effect on insect control in the garden. Where cow manure rather than chemical fertilizers is used, populations of flea beetles, aphids, and caterpillars are much lower. Better plant nutrition through minor elements or other

FIGURE 40-1 Fertilizer being placed around a newly transplanted tomato plant after planting (Ed Reiley, Photographer)

materials added by the organic fertilizers apparently results in healthier plants. These results definitely lend support to the organic gardeners' claims that organic fertilizers can promote resistance to insect pests.

WEED CONTROL

Weed control is the single most important cultural practice in most gardens. All other effort is useless if weeds are allowed to overtake the garden. Weeds rob garden vegetables of water, plant food, and light. Some weeds help spread diseases, insects, and nematodes. Weeds are most easily controlled before they break through the ground or while they are quite small.

CULTIVATION

When weeds are small (less than 1 inch tall) shallow cultivation (turning the soil over 1 inch deep) will control the weeds and prevent damage to crop roots. Deep cultivation damages many crop roots and brings moist soil and more weed seeds to the surface which dries out and causes moisture loss from the soil. If weeds are allowed to grow taller than 1 inch, deep cultivation is needed to kill the weeds, and in the process, many vegetable plant roots are damaged.

The secret to good weed control is to kill the weeds when they are very small by using a hoe or other shallow cultivator to kill them. The oscillating hoe is an excellent tool for shallow cultivation. By sliding it back and forth just under the soil surface, small weeds are cut off and killed. This shallow cultivation is quicker, requires less muscle, and does not damage vegetable crop roots. As mentioned in Unit 39 a special type of hoe is used for wide row planting.

As soon as the soil is dry enough to work after a rain, it should be hoed or cultivated to kill any weed seeds that have sprouted. This also leaves the soil surface loose to absorb later rainfall. Another advantage to the loose surface soil is that the weed seeds do not germinate and grow as quickly. Hoeing and cultivation should be done often enough to eliminate weeds while they are less than 1 inch tall. Hoeing or cultivation is unnecessary if weeds are not present, unless the soil has a hard crust on it that would prevent rainwater from soaking in.

MULCHING

Mulching is another excellent method of weed control in the garden, figure 40-2. Organic material such as straw, leaves, sawdust, grass clippings, ground corn cobs, and wood chips make excellent mulch. Apply several inches of mulch around plants to cover the entire soil surface. The application of materials that pack tightly, such as grass clippings, should be more shallow (2 inches deep) than coarse materials such as straw.

If weeds grow up through the mulch, the mulch is either not thick enough or not applied uniformly. It is important that the mulch keep all light from the soil surface to prevent weed seeds from germinating. For warm-season crops, such as tomatoes, melons, and beans, organic mulches are best applied only after the soil has warmed in late spring. Cool-season crops, such as cabbage, may be mulched earlier. Too early an application shades the soil, thereby preventing it from warming up in the spring and early summer.

> Tips: Be sure the material used for mulch does not contain weed seeds.

When plant material is used for mulch, it is often necessary to apply extra nitrogen fertilizer. As the soil organisms rot the mulch, the soil is robbed of nitrogen, which is used as food by microorganisms. Sawdust uses a great deal of nitrogen in this rotting process, and fresh sawdust is best allowed to rot down for a year before use. To

FIGURE 40-2 Straw mulch being applied around a tomato plant. Notice that the mulch is thick enough to completely shade the soil. (Ed Reiley, Photographer)

FIGURE 40-3 Black polyethylene plastic used as a mulch around tomato plants. (Ed Reiley, Photographer)

compensate, add extra nitrogen around the plants as a sidedress when organic mulches are used.

Black polyethylene plastic makes an excellent mulch for vegetable gardens. The black color warms the soil and may be used early in the spring when plants are first planted in the garden. Polyethylene is available in rolls which may be rolled down the garden row on each side of the plants. Soil is placed over the edge of the plastic to hold it in place, figure 40-3. Plants may be planted by punching a hole through the plastic and placing the plant in the soil below.

Mulches also conserve moisture in the soil, reducing the need for watering. If properly mulched, the garden need not be cultivated, saving a great deal of work in weed control.

Mice may become a problem under mulch.

Recent research shows that the color of a mulch affects yield of certain crops. For example, USDA research has shown that the use of red mulch can increase tomato production from 32,921 pounds per acre to 37,057 pounds. Different colors of light, either reflected or directly applied, affect phytochrome, a light-sensitive pigment in plants. Red light increases yield in tomatoes and white mulch increases potato yields as much as 25 percent. No difference in fertilizer or water use is required.

Straw mulch can help to control Colorado potato beetle on either potato or tomato crops. For effective control, mulch before the plants come up in the case of potatoes or at transplanting time with tomatoes.

Hairy vetch is a nitrogen-fixing winter legume which makes an excellent mulch for vegetable crops. It offers many advantages over plastic mulches, such as higher yields, reduced fertilizer needs (especially nitrogen), better soil aeration and thus deeper root development, cooler soil in hot weather which conserves moisture, deterring the Colorado potato beetle, and not having to be removed at the end of the season as plastic does.

Vetch is planted in the fall and is allowed to grow until spring when it is mowed to a height of 1 to 2 inches and left on the ground as a mulch. Tomatoes or other vegetables are planted through the mulch. The soil is not tilled or plowed before planting.

According to Dr. Aref Abdul Baki, a USDA plant physiologist who researched use of hairy vetch as a mulch for tomatoes, yields are increased 24 percent compared with black plastic mulch. The vetch rots down during the growing season, adding plant nutrients to the soil.

CHEMICAL WEED CONTROL

Modern herbicides provide dependable weed control where hand labor was once required. Herbicides require careful application, since too much may injure the crop and too little may not control the weeds.

Contact your local extension service or horticulture teacher for materials to use.

APPLICATION

A sprinkling can may be used to apply these herbicides. Determine the rate of application by applying only water first; then add the chemical and apply to the measured area. Generally, it is more desirable to make two trips over the area and walk in a different direction each time while applying one-half of the total amount of chemical. This technique provides a more uniform application. A similar application technique should be used when applying herbicides with hand sprayers. It is probably best not to use weed killing chemicals in the small home garden.

> *Caution: Weed control chemicals are considered poisonous even though many of them are relatively nontoxic to humans. Read the label on the container and follow all directions and safety precautions. Avoid contact with the skin and be sure to wash thoroughly after use. Store herbicides out of reach of children and pets and away from foodstuffs.*

THINNING

Plants that are allowed to grow too close together compete with one another for such things as food and water. For this reason, it is sometimes necessary to remove a number of plants. Check directions on the seed package for thinning instructions. Never plant transplanted plants more closely together than recommended.

INSECT AND DISEASE CONTROL

Insects and diseases can cause a serious decrease in production of the garden if not controlled.

PREVENTION

Some of the best control measures are preventives rather than cures.

- **Dispose of crop residue.** Dispose of all crop residue that may carry diseases or insects over to the next year's planting. Compost this material so that it rots before the next season.

- **Rotate crops.** Rotation of crops is another good preventive. Avoid planting the same crop in the same area each year. A three- or four-year rotation is most effective.

- **Plant treated seeds.** Plant seeds which are treated to control disease as a good step in prevention of certain diseases.

- **Purchase resistant varieties.** Buy resistant varieties whenever possible. Many new varieties are disease resistant; some are insect or nematode resistant. A good example is the tomato Better Buy. It is labeled VFN, which means that it resists verticillium wilt, fusarium wilt, and nematodes.

- **Purchase healthy plants.** Plants that are vigorously growing because of good fertilizer and other cultural practices are more able to survive disease and insect attacks and produce good crops.

If disease-resistant hybrids are planted and good cultural practices are followed, many garden vegetables will need no further insect or disease control. Corn, tomatoes, peas, beets, carrots, lettuce, onions, and turnips rarely need much additional help. Other vegetables, especially melons, cucumbers, beans, and cabbage, generally need some additional insect control. This varies with location and local problems. *Use biological control agents wherever possible* (see Unit 16).

CHEMICAL CONTROL

Most garden insects can be effectively controlled with two relatively safe insecticides, malathion and Sevin. Most fungus diseases can be controlled with zineb or maneb and captan fungicides. These insecticides and fungicides can be mixed together and applied at the same time as a spray. Captan should not be mixed with an emulsifiable concentrate of malathion, but is compatible with the wettable powder of malathion. Read the label for mixing, use, and safety precautions. Check to see how close to harvest date the sprays or dusts can be applied. The chemical must have time to break down and become harmless before the vegetables are eaten.

> *Caution: Always apply pesticides only to those crops for which they are recommended. Apply only the amount that is specified on the label. Wash all treated vegetables before eating. USE ALL SAFETY PRECAUTIONS.*

APPLICATION

Many garden pesticides are applied as dusts using a hand duster. Dusts are available in diluted form and ready to apply. Sprays are often applied with a knapsack sprayer. Material that is sprayed is always diluted with water. Application should be made on a quiet day when the wind is not blowing so that the chemical does not drift to other areas.

A spray schedule for the vegetable garden can be obtained at your local extension agent's office or from a State Agriculture College Extension Service office. Another guide, *Insects and Diseases of Vegetables in the Home Garden* (Bulletin 46), is available from the USDA.

VEGETABLE SPRAY SCHEDULE

Figure 40-4 lists vegetables that most often require spraying or dusting, the pests that commonly attack the plant, and recommended chemicals for control. Remember that many vegetables never require chemical sprays or dusts. Many others that do require the use of chemicals respond to a

PLANT	PEST AND DESCRIPTION (Only the most damaging insects are listed for each crop in the chart.)	WHEN TO SPRAY	CHEMICAL (PER GALLON OF WATER)
Asparagus	**Asparagus Beetle.** Metallic blue to black beetle with orange markings. Larvae olive green, 1/3 inch long. Feeds on leaves and shoots.	When larva or adult first begins to feed. Check label for days to wait before harvesting crop.	carbaryl (Sevin) (50% WP) 2 tablespoons or malathion (50% EC) 2 level tablespoons or rotenone (5% WP) 4 level tablespoons
Bean	**Mexican Bean Beetle.** Feeds on underside of leaves. Adult is orange to yellow with 16 black spots on back. Larvae are fuzzy, 1/3 inch long.	When larva or adult begins to feed; repeat in 7 to 10 days	malathion (50% EC) 2 level tablespoons or carbaryl (Sevin) 2 level tablespoons or rotenone (5% WP) 4 level tablespoons
Beet Swiss Chard	**Leaf Miner.** Adult is slender, gray, black-haired fly 1/4 inch long. Larvae pale green or whitish. Leaves have blotches.	When blisters first appear on leaves; repeat in 7 days	malathion (50% WP) 2 level teaspoons or pyrethrum spray mixed with water according to directions
	Beet Webworm. Larvae are yellow to green, with black stripe and numerous black spots on back. Up to 1 1/4 inches long.	When leaves start to roll or fold	
Broccoli Brussels sprouts Cabbage Cauliflower Kale Kohlrabi	**Cabbage Looper.** Pale green worm, white stripes on back, which loops as it crawls. Up to 1 1/2 inches long. Chews holes in leaf.	When worms first begin to feed; repeat in 7 days. Continue up to 1 week before harvest if necessary.	carbaryl (Sevin) (50% WP) 2 level tablespoons or Dipel or Thuricide (biological) See label for amount to use.
	Imported Cabbage Worm. Velvet green, size up to 1 1/4 inches long. Chews holes in leaves.		
	Aphid. Small, green to powdery blue, soft. Causes curling of leaves.	When aphids first appear. Do not apply within 7 days of harvest.	malathion (25% WP) 4 level tablespoons
	Club Root Disease. (swelling on roots causing wilting of plants)	At time of planting	Terrachlor, 6 level tablespoons per gallon of water. Apply 1 cupful of diluted chemical in each planting hole.

Asparagus Beetle

Mexican Bean Beetle

Beet Webworm

Cabbageworm

FIGURE 40-4 Spray schedule for vegetables most often requiring spray

PLANT	PEST AND DESCRIPTION (Only the most damaging insects are listed for each crop in the chart.)	WHEN TO SPRAY	CHEMICAL (PER GALLON OF WATER)
Celery	**Leaf Tier.** Larvae are greenish caterpillars ¾ inch long. Eats holes in leaves and stalks and ties leaves together with webs.	When webs are first seen. Repeat in ½ hr. First dust drives them out of webs; second application kills them.	pyrethrum dust (0.2%) comes ready to use (no water) or Dipel or Thuricide (biological) See label for amount to use.
Corn (Sweet)	**Corn Earworm.** Larvae are green, brown, or pink worms; light stripes along sides and on back. Up to 1¾ inches long. Feeds on early corn shoot in season. Later feeds on kernels near top of ear.	For shoot damage, spray early in season. For ear damage, spray silks the day after silks appear. Spray until silk is wet. Repeat 4 times at 2-day intervals.	carbaryl (Sevin) (50% WP) 3 level tablespoons or Dipel or Thuricide (biological) See label for amount to use.
Cucumber Pumpkin Squash	**Cucumber Beetle, Flea Beetle, Leaf Hopper, Pickle Worm.** Cucumber Beetle adult is yellow with black dots; Flea Beetle adult is black; Leaf Hopper adult is light green or brown; Pitch Worm adult is green.	When insect damage is first seen. Repeat in 7 days. Do not spray open blossoms.	malathion (50% WP) 2 level teaspoons or methoxychlor (50% WP) 2 level tablespoons
	Anthracnose. Reddish-brown spots on leaves, tan cankers on stems; round sunken spots on fruit.	When damage is first seen. Repeat every 7 to 10 days.	zineb or captan according to directions on label. Plant rotation of not less than 3 years is helpful.
Eggplant	**Colorado Potato Beetle.** Adult is yellow, black-striped, ⅜ inch long. Larvae are brick red, hunchbacked, ⅗ inch long. Eats leaves. Resistant to many chemicals.	When insect is seen and leaf damage is first noticed.	methoxychlor and biological (see Unit 16) See label for amount to use.
	Eggplant Lacebug. Adult is grayish to light brown; flat lacelike wings, 1/16 inch long. Nymph: yellowish, louselike; up to 1/10 inch long. Feeds on underside of leaves; leaves turn yellow and brown.	Apply as soon as insect or damage is noticed. Repeat in 7 to 10 days. Do not apply within 7 days of harvest.	malathion (25%WP) 4 level tablespoons

Celery Leaf Tier

Corn Earworm

Cucumber Beetle

Colorado Potato Beetle

Eggplant Lacebug

FIGURE 40-4 Spray schedule for vegetables most often requiring spray (continued)

PLANT	PEST AND DESCRIPTION (Only the most damaging insects are listed for each crop in the chart.)	WHEN TO SPRAY	CHEMICAL (PER GALLON OF WATER)
Lettuce	Aphid. Same as broccoli.	When insect is first seen. Repeat in 7 days. Use up to 14 days before harvest if needed.	malathion (57% EC) 2 level teaspoons or Dipel or Thuricide (biological)
	Cabbage Looper. Same as broccoli. Feeds on underside of leaves, leaving ragged holes.	When insect or damage is first seen. Repeat in 7 days. Do not apply within 7 days of harvest.	
Muskmelon	Aphid. Same as broccoli.	Same as for lettuce	Same as for lettuce.
	Downy Mildew and Leaf Spot. Brownish spots on leaves; leaves die. Fruits are not affected.	When injury is first noticed. Repeat in 7 to 10 days.	zineb or captan as recommended on the label. Use resistant varieties.
	Mosaic. (viruses) Mottled green and yellow curled leaves.	Control aphids and striped cucumber beetles to prevent spread. See control already given.	See control for aphids and striped cucumber beetles. Destroy weeds and diseased plant. Do not plant seed from infected plants. Use resistant varieties.
Okra	Corn Earworm. Same as corn. Eats holes in pods.	As pods form; repeat at least twice at 7-day intervals	carbaryl (Sevin) (50% WP) 2 or 3 level tablespoons
	Stink Bug. Black, green, or brown shield-shaped bug; 5/8 inch long.	As pods form; repeat at least twice at 7-day intervals	malathion (52% EC) 2 level teaspoons
	Wilt. Causes yellow and wilted leaves.		Rotate crops at least 3 years apart. Use resistant varieties.
Onion	Maggot. White legless worm 3/8 inch long. Burrows into young bulbs.	At planting time	diazinon (25% EC) 2 level teaspoons. Apply in row before planting onion sets.
	Thrip. Adult is tiny, active yellow or brownish, winged insect. Larvae are white, same appearance as adult, only smaller. Sucks juices and causes blotches on leaves. Tips may turn brown and die.	When injury is first noticed. Repeat 2 or 3 times at 7- to 10-day intervals.	malathion (25% WP) 4 level tablespoons

FIGURE 40-4   Spray schedule for vegetables most often requiring spray (continued)

PLANT	PEST AND DESCRIPTION (Only the most damaging insects are listed for each crop in the chart.)	WHEN TO SPRAY	CHEMICAL (PER GALLON OF WATER)
Pea	Aphid. Same as broccoli.	When insects first appear	malathion (57% EC) 2 teaspoons
	Aphid		
Potato	Colorado Potato Beetle. Same as eggplant.	Same as eggplant	Same as eggplant
	Flea Beetle. Black, brown, or striped jumping beetles; about 1/16 inch long. Leaves appear to have been shot full of small holes.	As soon as beetle is seen	carbaryl (Sevin) (50% WP) 2 level tablespoons See Unit 16 for biological control.
Colorado Potato Beetle	Early Blight. Leaves show small, irregular, dark brown spots. Fungus is carried in soil.	As soon as damage is seen	zineb or captan according to directions Plant only disease-free tubers.
	Late Blight. Dark irregular dead areas on leaves and stems. Infected tubers rot in storage; plants may die in season.	Same as for early blight	Same as for early blight. Use blight-resistant varieties.
Potato Flea Beetle			
Tomato	Flea Beetle. Same as potato.	Same as for potato	Same as for potato
	Tomato Fruit Worm. Larvae are green, brown, or pink; light green stripes along sides and on back; up to 1 3/4 inches long. Same as corn earworm. Eats holes in fruit and buds.	When fruits first begin to set. Repeat 3 times at 2-week intervals	carbaryl (Sevin) (50% WP) 2 level tablespoons
Flea Beetle	Fusarium and Verticillium Wilts.		Plant resistant varieties.

FIGURE 40-4 Spray schedule for vegetables most often requiring spray (continued)

mixture of malathion and Sevin for insect control and captan or zineb for control of fungus diseases.

> *Tips: If a chemical must be used, read the recommendation on the pesticide label before purchase.*

HARVESTING

Harvest vegetable crops at the peak of quality, figure 40-5. Some crops, such as sweet corn, peas, and asparagus, lose quality rapidly. With these vegetables, just a few days make a great difference in quality. Other crops, such as beets, cabbage, and pepper, retain their quality over a long period of time. See Unit 41 for more detail on harvesting.

GARDENING NOTES— WHAT'S NEW

- ◆ **Tomatoes.** New varieties set fruit in hot or cold weather. "Oregon Pride" and "Oregon Star" are two examples for cool areas. "Solar Set," "Sunfire," "Arkansas Traveler," and "Big Rainbow" for hot areas.

    Tickle your tomato plants for 25 to 35 percent shorter, stockier plants. Brushing the top of tomato seedlings for 1½ minutes twice a day results in shorter plants, according to a study at the University of Georgia.

    Mulch with red plastic or hairy vetch.

- ◆ **Potatoes.** 25 percent more production with white mulch.

    Straw mulch can help to control the Colorado potato beetle if applied as soon as plants emerge.

- ◆ **Leeks.** Grow from bulblets faster than from seed.

- ◆ **Mixed plantings** of vegetables reduces damage by insects. Mixing confuses insects because of the mixed smell of chemicals given off by the different plants. They have difficulty finding their favorite.

- ◆ **Broccoli.** "Emperor" variety resistant to black rot.

- ◆ **Watermelon.** Seedless varieties are sweet and easier to eat.

Some of the most destructive vegetable insects are pictured in figures 40-6 through 40-9.

FIGURE 40-5 A well-cared for garden planted using the wide row system (Courtesy of Gardenway, Inc.)

FIGURE 40-6A The Colorado potato beetle larvae (Courtesy USDA)

FIGURE 40-6B The Colorado potato beetle adult (Courtesy USDA)

FIGURE 40-7 The Mexican Bean beetle, eggs, larvae, pupae, and adult (Courtesy USDA)

FIGURE 40-8 Corn ear worm (Courtesy USDA)

FIGURE 40-9 Tomato hornworm. This one is parasitized by the Branconid wasp. The white eggs attached to the worm will hatch and the small wasps will enter the worm and kill it. (Courtesy USDA)

SUMMARY

1. Choose fertile, well-drained soil and apply water and proper fertilizer to produce vigorous plants.

2. Plant crops adapted to the area.

3. Keep weeds under control.

4. Use only disease-free seed, certified if possible.

5. Treat seed with chemicals to protect against decay and damping-off.

6. Purchase disease-free plants from a reputable dealer.

7. Plant disease-resistant varieties whenever possible.

8. Destroy all plant refuse as soon as harvest is over to prevent carryover of diseases and insects. Compost if possible.

9. Use insecticides as a last resort, and then only according to directions.

10. Select insecticides that do the least harm to helpful insects and the environment.

Student Activities

1) Actually plant a garden. Demonstrate the techniques of planting and caring for the garden.

2) Work along with an experienced gardener.

3) Mulch tomatoes with a red mulch and check for increased yield.

4) Demonstrate how to compost garden refuse.

Self-Evaluation

A) Select the best answer from the choices offered to complete the statement or answer the question.

1) How much rain does the average plant require per week?

a) 1 foot
b) 1 inch

c) ½ inch
d) ¾ inch

2) A signal which indicates that plant production is being reduced and water should be added is

a) the plants turn brown.
b) the plants fall over.

c) the plants wilt.
d) all of these

3) Organic mulches prevent water loss from the soil by reducing

a) evaporation of water.
b) water runoff.

c) soil temperature.
d) all of these

4) Plants requiring more nitrogen than most plants respond best to extra application of fertilizer to the soil surface because

a) they are heavy feeders.
b) nitrogen leaches down to the roots quickly.
c) they are grasses.
c) all of the above

5) Fertilizer should not be applied heavier than

a) 10 pounds per 100 square feet.
b) 7 pounds per 100 square feet.
c) 5 pounds per 100 square feet.
d) 3 pounds per 100 square feet.

6) The single most important cultural practice in most gardens is

a) weed control.
b) watering.

c) insect control.
d) disease control.

7) A garden should be cultivated or hoed for weed control

a) every 10 days.
b) after every rain.
c) while weeds are less than 1 inch tall.
d) just before harvest time.

8) Cultivation of the garden should be shallow (1 or 2 inches deep) because

a) shallow cultivation controls small weeds.
b) deep cultivation damages vegetable crop roots.
c) soil moisture is brought to the surface and lost in deep cultivation.
d) all of the above

9) Organic mulches used for weed control, such as straw, should be applied to the soil in a heavy enough layer so that

a) mud does not come through.
b) no light can reach the soil.

c) it is soft to the touch.
d) none of these

10) When sawdust mulch is used

    a) it tends to pack too tightly.
    b) good weed control is assured.
    c) extra nitrogen fertilizer must be applied to the soil around the plants.
    d) the soil becomes acidic.

11) Chemical pesticides should be applied

    a) to all garden vegetable crops.
    b) with sprayers.
    c) early in the morning.
    d) only to those crops recommended on the label.

12) Straw mulch is best applied late in the spring because

    a) it shades the soil and keeps it from warming up.
    b) that is when the weeds are worst.
    c) there is plenty of moisture in the early spring.
    d) it is easier to obtain straw then.

13) Two disease control methods which do not involve the use of chemicals are

    a) crop rotation and the use of resistant varieties.
    b) seed treatment and the use of resistant varieties.
    c) burning crop residue and seed treatment.
    d) none of the above

B) Provide a brief answer for each of the following.

1) List the ten key points for insect and disease control.

2) Select two vegetables and list the pest which most commonly attacks them, a description of the pest, and the chemicals used to control the pest.

3) If fertilizer recommendations call for 50 pounds of actual nitrogen per acre, how many pounds of a 10-10-10 fertilizer would you need to apply to 1 acre?

Favorite Garden Vegetables

OBJECTIVE

To cultivate and harvest five vegetables commonly grown in the local area.

Competencies to Be Developed

After studying this unit, you should be able to

- list the requirements for planting, fertilizing, harvesting, and storing five garden vegetables that grow successfully in the local area.
- select five locally grown vegetables and plant and grow them to marketable quality.

MATERIALS LIST

- ✔ seed catalogs
- ✔ seed
- ✔ planting site
- ✔ fertilizer
- ✔ minimal garden tools (spade, rake, hoe)
- ✔ duster or sprayer
- ✔ chemical sprays or dusts if necessary

NOTES ON THE UNIT

In making fertilizer recommendations, the two preplant methods of application mentioned for each vegetable are broadcasting and digging in, and band application. The standard fertilizer application consists of about 3 pounds of a 10-10-10 fertilizer per 100 square feet or 75 feet of row area. These applications were mentioned in previous units. Additional needs are given for each vegetable in this unit.

ASPARAGUS

Asparagus is among the earliest of spring vegetables to appear. It grows best in areas where the soil freezes to a depth of at least a few inches in winter. Asparagus, a perennial, may live for many years.

The asparagus crop may be grown on almost any well-drained fertile soil. First dig a trench 16 inches wide, laying aside the topsoil. Plow or spade lots of manure, leaf mold, rotted leaves, or peat into the subsoil to a depth of 14 to 16 inches (below the normal soil surface.) Mix in 5 to 10 pounds of a complete fertilizer per 75 feet of row (length). Fill the trench with topsoil to within 6 inches of the natural soil level. Set the asparagus crowns and cover with 1 to 2 inches of soil. As the shoots grow during the first season, gradually replace the topsoil in the trench until it reaches the natural soil level. Set the plants at least 1½ feet apart in the row. Rows should be 4 to 5 feet apart.

FERTILIZER

The first year after planting, sidedress with 1 pound of 5-10-10 fertilizer per 100 square feet of bed or 75 feet of row at the first cultivation. In established beds, apply 3 or 4 pounds of a 5-10-5 fertilizer per 100 square feet of bed or 75 feet of row just before growth begins in the spring.

HARVEST

Remove no shoots the first two years after planting. Each following year, remove all the shoots until July 1, when harvesting must be stopped to allow the plant to replenish root reserves of plant food. Harvest when the spears are 6 to 10 inches long. To harvest, break the spears off above soil level. Cutting below ground level reduces yield. Harvest frequently to prevent tough spears. *In the fall, remove the dead plant tops and work them into the soil.*

USE

Most asparagus is cooked fresh. It may be frozen or canned.

STORAGE

Store by freezing or canning. It may be kept fresh for short periods of time at 32°F (0°C) and 85 to 90 percent relative humidity.

LIMA BEANS

Lima beans, figure 41-1, are a warm-season crop and should not be planted until the soil and the

FIGURE 41-1 Burpee's Fordhook lima beans (Courtesy W. Altee Burpee Company)

air have warmed sufficiently. Limas are able to grow in most soils. In heavy soils that tend to turn crusty on top, the seed should be covered with sand rather than soil so that the seeds can push through the soil more easily.

Lima beans require a growing season of about 4 months at a relatively high temperature. They need a richer soil than snap or green beans.

Limas grow both on poles and in bush form. The pole limas need some means of support such as poles or trellises. When poles are used, plant three or four plants per hill and space the hills 2 to 3 feet apart. When planted to grow on trellises, space the plants 6 to 12 inches apart. Bush varieties are planted 3 to 4 inches apart in rows spaced 3 feet apart.

FERTILIZER

The ordinary application of fertilizer at planting time is usually enough. Lima beans should not be treated with a high nitrogen fertilizer and should not be sidedressed in late season.

HARVEST

Harvest lima beans when the seeds are close to full size but before the color changes from green to white.

USE

Fresh cooked, frozen, canned, dried.

STORAGE

Limas remain fresh at 32°F (0°C) and 85 to 90 percent relative humidity. They can also be frozen, canned, or dried.

SNAP BEANS

Snap beans, figure 41-2, should not be planted until the ground is warm. Successive plantings can be made every 2 to 3 weeks until 8 weeks before the fall frost date. Snap beans can be grown in a wide variety of soils, providing that the soil is well drained and not so heavy that it crusts, preventing seeds from emerging from the soil.

FERTILIZER

The general garden fertilizer program before planting is usually sufficient. On light sandy soils, a sidedressing of a 5-10-5 fertilizer may be beneficial when the plants bloom.

HARVEST

Pick when seeds in the pods are one-third their mature size and when the pods are still young and tender.

USE

Fresh cooked, frozen, canned.

FIGURE 41-2 Burpee's Tender Pod snap beans (Courtesy W. Altee Burpee Company)

STORAGE

Ideal fresh storage should be at 45° to 50°F (7° to 10°C) and 85 to 90 percent relative humidity. They can also be frozen or canned.

BEETS

Beets, figure 41-3, are very hardy vegetables and grow under a wide range of climatic conditions. They will not, however, tolerate severe freezing. Beets require a loose, loamy soil for best results.

Beets are planted in rows as close as 16 inches apart for hand hoeing. Plants are spaced 3 inches apart in the row. Successive plantings may be made at three-week intervals during the season. Seeds planted in hot, dry weather may not germinate well. Mulching or watering to keep the soil moist will help to resolve this problem. Beets do not like acid soils. Check the soil pH and add lime if needed to keep soil pH above 6.5.

FERTILIZER

Beets should be encouraged to maintain a steady growth pattern. A 5-10-10 fertilizer should be applied at the time of planting and another application made at one of the early cultivations.

FIGURE 41-3 Burpee's Red Ball beet (Courtesy W. Altee Burpee Company)

HARVEST

For tender beets, harvest when they are 1½ to 2 inches in diameter. A fall crop which is to be stored may be allowed to grow a little larger.

USE

Fresh cooked, canned, pickled.

STORAGE

Store fresh at 32°F (0°C) and 90 to 95 percent relative humidity for best results.

BROCCOLI

If a spring crop of broccoli is desired, plant seeds indoors and set plants in the garden in March or April. Set the plants 12 to 18 inches apart in rows 3 feet apart. Plants begin to yield sprouts about 10 weeks after planting, figure 41-4. A fall crop may be direct seeded by sowing seed in June.

FIGURE 41-4 Broccoli (Courtesy Southern States Corporative)

Tips: Do not plant in the same area in which cabbage, cauliflower, or Brussels sprouts have been planted, since the same pests attack all these plants.

FERTILIZER

Being a leafy vegetable, broccoli responds well to additional fertilizer. In addition to the fertilizer applied at the time of planting, broccoli responds to a sidedressing soon after planting. The sidedressing should be done with a high nitrogen fertilizer such as 10-10-10 at a rate of 1 or 2 pounds per 100 feet of row.

HARVEST

The broccoli head, which is a cluster of flower buds, should be harvested before the buds begin to open. Small side clusters develop after the main head is cut, and may be harvested over a long period of time.

USE

Fresh (as a greens crop), frozen.

STORAGE

For best results, store fresh broccoli at 32°F (0°C) and at 90 to 95 percent relative humidity.

BRUSSELS SPROUTS

Being hardier than cabbage, Brussels sprouts can survive outdoors over winter in all the mild sections of the country. They may be grown as a winter crop in the southern regions of the United States and the same way as early and late cabbage is grown in southern sections. As the heads begin to form, break the lower leaves from the stem to give them more room. Do not remove top leaves.

FERTILIZER

See Broccoli.

HARVEST

Remove the sprouts when they are 1 to 1¼ inches in diameter by cutting them close to the stem.

USE

Fresh, frozen.

STORAGE

See Broccoli.

CABBAGE

Cabbage, figure 41-5, is one of the most important of the home garden crops. It can be grown throughout almost the entire United States. For a spring crop, set the plants in the garden as early as the soil can be prepared. For fall crops, plant again in late summer.

Cabbage may be planted after early potatoes, peas, beets, spinach, or other early crops, or it may be set between rows of crops before these crops are harvested. Set plants about 1 foot apart in rows 3 feet apart. Plantings must be very shallow.

Tips: Be sure to employ methods for control of the cabbageworm.

FERTILIZER

Fast growth of plants is necessary for good quality cabbage. Being a leaf crop, cabbage requires more nitrogen than many other plants. An additional sidedressing of a high nitrogen fertilizer, such as a 10-10-10, when the heads are half grown is very helpful.

HARVEST

Harvest when the heads are firm, but before they split.

USE

Fresh (in salads and coleslaw), cooked as sauerkraut.

STORAGE

Fall cabbage may be stored in an outside pit or trench or in a cool vegetable cellar. Storage at 32°F (0°C) and at 90 to 95 percent relative humidity is best.

CARROT

Carrots, figure 41-6, are cold-hardy. Seed may be planted as early in spring as the ground can be worked. Carrot seed does not germinate well in hot, dry weather. The soil requirements for carrots are very similar to those for beets. They prefer a deep, loose, sandy loam which is free of stones.

Space plants no closer than 1 inch apart in the row. Cultivate by pulling loose soil in toward the plant to keep the crown covered. This prevents the top of the carrot from turning green.

FERTILIZER

Carrots respond to nitrogen fertilizer. Apply 2 pounds of 10-10-10 fertilizer per 100 feet of row by band application beside the seed row before planting. Sidedress with 1 pound of a 10-10-10 or 10-6-4 fertilizer after the second or third cultivation.

HARVEST

Harvest carrots when the plants are 1 to 1½ inches in diameter, depending on the variety.

USE

Raw (alone or in salads), cooked fresh, canned.

STORAGE

Carrots must be dug and stored before frost kills the roots. Fresh carrots are stored best at 32°F (0°C)

FIGURE 41-5 Cabbage (variety Earliana) (Courtesy W. Altee Burpee Company)

FIGURE 41-6 Carrots, six different varieties (Courtesy W. Altee Burpee Company)

and at 90 to 95 percent relative humidity. They may also be stored outdoors covered with straw or other material to prevent freezing.

CAULIFLOWER

Cauliflower, figure 41-7, requires cultural conditions similar to those of cabbage. Cauliflower will not tolerate as much cold weather as cabbage, however. High temperatures cause poor quality heads. Cauliflower grows on any type of well-drained, fertile soil. Set the plants 1½ to 2 feet apart in 3-foot rows. When the heads are 1 to 2 inches in diameter, tie the outer leaves loosely around the heads to *blanch* them (remove color).

FERTILIZER

Sidedress with a high nitrogen fertilizer. Add Borax to the fertilizer to supply the minor element boron, which cauliflower uses in large amounts. (Boron is needed in very small amounts and may be purchased already mixed in the fertilizer.)

HARVEST

Harvest when the heads are firm and compact. Do not allow the heads to become "ricey" appearing.

USE

Cooked, fresh.

STORAGE

Fresh storage at 32°F (0°C) and 85 to 90 percent relative humidity is best. May also be frozen and canned.

FIGURE 41-7 Cauliflower (variety Early White) (Courtesy W. Altee Burpee Company)

CELERY

Celery is a cool-season crop grown as a spring or fall crop in the upper southern and in northern sections of the country. It is a good winter crop in lower southern regions.

Soil which is rich, moist, well-drained, deeply prepared, and loose is a must for celery. Plenty of organic material is also necessary. About ten weeks are required to grow healthy plants from seed to transplanting size. Set plants 18 to 24 inches apart for hand cultivation. For tractor cultivation, space plants 6 inches apart in rows 3 feet apart.

FERTILIZER

At planting time, dig in 5 pounds of a 10-10-10 fertilizer per 100 feet of row. Mix thoroughly with the soil.

HARVEST

Harvest when large enough to eat.

USE

Fresh, alone, in salads and other dishes.

STORAGE

For early winter use, bank soil around the plants and cover the tips with straw or leaves to prevent freezing. May also be stored in a cold frame or cellar with high humidity. Storage at 34°F (2°C) with 90 to 95 percent relative humidity is best.

CELERIAC

The cultivation of celeriac, known as turnip-rooted celery, is the same as that of celery. The root, rather than the top, is eaten.

FERTILIZER
See Celery.

HARVEST

Use the roots any time after they are big enough.

STORAGE

Store in the ground with straw or in cold storage.

CHARD

See Beets. The only difference is that the plants grow larger and require spacing of at least 6 inches apart in the row. Chard needs a rich, loose soil. It is sensitive to soil acidity.

CHERVIL

Salad chervil is grown as parsley is. The seed must be bedded in damp sand for a few weeks before planting, or germination is very slow.

Turnip-rooted chervil is planted in the fall in southern regions and may not appear until spring. In northern regions, it is either planted in the fall or seeds are started in flats to grow transplants for spring planting. For cultivation, see Beets.

CHICORY, WITLOOF

Witloof chicory is grown for both its roots and tops. It is a hardy plant which tolerates both hot and cold climates. A rich, loamy soil without much organic matter is best.

Sow seeds in rows in late spring. Thin so that they are kept 6 to 8 inches apart in rows. If sown too early, the plants go to seed, preventing later forcing.

FERTILIZER

Work a good garden fertilizer into the soil at planting time. No additional fertilizer is needed.

HARVEST

The tops are sometimes harvested when young.

USE

Top is used as cooking greens.

STORAGE

The roots are lifted in the fall and placed in moist soil in a warm cellar for forcing. Roots are covered with a few inches of sand. Under this covering, the leaves form a solid head known as a *witloof* (Dutch for "white leaf").

THE ONION GROUP (CHIVES, GARLIC, LEEK, SHALLOT, ONION)

Onions thrive under a variety of climatic and soil conditions. A moist, warm climate is best.

The soil should be fertile, moist, and loose. It should also contain plenty of compost, have a fine texture, and be free of clods and foreign matter. Soil pH should be high for best results.

Onions are generally started as sets or as small dry onions grown the previous year from seed.

Seedling onions have an advantage over sets in that they rarely go to seed and thus produce better bulbs.

Plant onion sets 3 inches apart in rows 1 foot apart. Plant 2 inches deep. Onion plants should have soil pulled away from them during cultivation to expose the onion; leeks should have soil pulled up around them.

FERTILIZER

A fertilizer high in phosphorus and potash, such as 5-10-10, should be applied at the rate of 5 pounds per 100 square feet. Sidedress if the tops do not appear to be healthy, green, and actively growing.

HARVEST

Onion stalks may be harvested young for salad greens or left in the ground to mature. The mature onions are pulled for storage after the tops have died down.

USE

Fresh (in salads or alone), cooking, seasoning.

STORAGE

Store mature onions at 32°F (0°C) and in a relatively dry, well-aerated location at 70 percent relative humidity for best results.

The following is an explanation of members of the onion group only as they differ from onions.

TYPES OF ONIONS

CHIVES Chives are perennials and should be planted where they can be left for more than one year. Divide and reset if plants become too thick. They are used primarily for seasoning.

GARLIC Large bulbs must be separated before planting, as each contains about ten small bulbs. Garlic is used primarily for seasoning.

LEEK Rather than forming a bulb, leek produces a thick, fleshy, stemlike onion, figure 41-8. Leek is started from seed or small bulblets and harvested any time it is large enough to eat. Store like celery.

SHALLOT The shallot is a small onion. Its bulbs have a milder flavor than that of most onions. They seldom form seeds and are propagated by small divisions into which the plant splits during growth. The shallot should be lifted and separated, and the smaller ones replanted, each year.

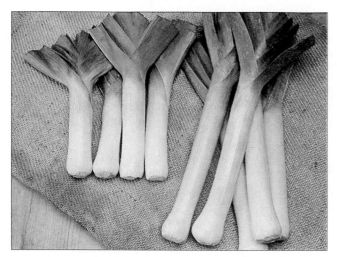

FIGURE 41-8 Leeks harvested young for fresh consumption (Courtesy W. Altee Burpee Company)

COLLARDS

Collards are grown and used similarly to cabbage. They withstand heat better than other members of the cabbage group.

Collards do not form a true head, but rather a rosette of leaves which may be tied together for blanching.

CORN, SWEET

Sweet corn, figure 41-9, requires a great deal of garden space and is best used only in larger gardens. It is planted in 3-foot rows, 12 to 14 inches apart in the row, or in hills of three plants per hill spaced 3 feet apart.

FIGURE 41-9 Sweet corn (varieties Breeders Choice Bicolor, Silver Queen, and Breeders Choice) (Courtesy W. Altee Burpee Company)

Corn is a tender crop and should not be planted until the danger of frost has passed. The new supersweet varieties require warmer soil temperatures to germinate. Soaking seed in warm water increases percent of germination. It should not be planted in single rows. Plantings four rows wide are desirable, even if rows must be short to aid in pollination. Sweet corn grows in almost any well-drained garden soil.

FERTILIZER

Corn is a grass and has a high nitrogen requirement. About half the fertilizer used (3 pounds per 100 square feet of 10-10-10) should be dug into the soil before planting or banded beside the row. Apply the rest of the 10-10-10 fertilizer as a sidedress every 3 weeks until silking of ears. This greatly increases yields.

HARVEST

Sweet corn is ready to eat when milk squirts from the kernel as the thumbnail is pushed into the kernel. The kernels should be full size. Pull and cook immediately. Sugar content of corn decreases rapidly after it is pulled from the plant. New, high-sugar varieties stay sweet much longer; the sugar in them is not converted to starch.

USE

Fresh as roasting ears, frozen, canned.

STORAGE

Storage of sweet corn is by freezing and canning. Fresh ears are best held for a short time at 32°F (0°C) and at 85 to 90 percent relative humidity.

CUCUMBER AND MUSKMELON (CANTALOUPE)

Cucumbers, figure 41-10, and muskmelons are warm-season crops. Hot weather may be too severe and greatly lower yields. In southern sections, they may be grown in the fall. A very fertile, loose soil high in rotted organic matter is necessary for best yields.

Seeds are planted in hills 6 feet apart, followed by thinning to four plants per hill, or in rows 6 feet apart with plants spaced 2 to 3 feet apart in the row. Cover seed with 1½ inches of soil. If soil

FIGURE 41-10 Cucumbers used in pickling (Courtesy W. Altee Burpee Company)

is heavy and becomes crusty, cover seed with sand. Plants may be started indoors in peat pots and transplanted to the garden for earlier crops.

> *Tips: Control of the cucumber beetle is necessary for good crop yields. The beetle spreads bacterial wilt, which kills the plants.*

FERTILIZER

Heavy application of fertilizer is necessary. Apply 5 pounds of a 10-10-10 fertilizer for every 50 feet of row or for every ten hills. Mix the fertilizer into the top 8 to 10 inches of soil before planting.

HARVEST

Cucumbers are harvested for fresh slicing while they are crisp and fresh and before any yellowing begins. Cucumbers used for pickling may be harvested at any size desired but before any yellowing begins. Pull all mature fruit off the plant, even if it is not wanted. Yield is reduced if fruit is allowed to remain on the vine.

Cantaloupe or muskmelon is harvested when the fruit separates from the stem with a slight tug or push with the thumb.

USE

Cucumber is used in salads and pickling. Cantaloupe is eaten as fresh fruit.

STORAGE

Fresh storage should be at 45°F (7°C) and 85 to 90 percent relative humidity for best results.

EGGPLANT

The eggplant, figure 41-11, demands warm weather and a warm soil. A growing season of from 100 to 140 days is needed for crop maturity.

Culture and fertilization are the same as those for tomatoes, except that eggplant may be planted closer together—18 inches apart in the row. Eggplant is a highly productive plant; six plants will furnish enough for the average family.

HARVEST

Harvest any time after proper size is reached and while the fruit is still glossy.

USE

Cooked.

STORAGE

Should be stored like tomatoes.

FIGURE 41-11 Eggplant (Courtesy W. Altee Burpee Company)

ENDIVE

For cultural requirements, see head lettuce under Lettuce. Endive is less sensitive to heat than head lettuce. In the south, it is a winter crop and in the north, a spring, summer, and fall crop. Tie leaves over the heart or shade and blanch the plant.

FERTILIZATION, HARVEST, AND USE

See Lettuce.

STORAGE

Dig plants with a root ball attached and place in a cold frame or cellar where they will not freeze. Use as needed.

KALE

Kale is a member of the cabbage family. It is a hardy plant and lives over winter as far north as southern Pennsylvania. It is also resistant to heat and may be grown in summer. It is best grown as a cool-weather crop.

Plant in rows 12 inches apart and keep plants thinned to 6 inches apart in the row. Sow seed in early spring or in early August for a fall crop. Seed may also be broadcast and raked in. Cover seed with only ½ inch of soil.

FERTILIZER

Dig about 3 pounds per 100 feet of row of a 10-10-10 fertilizer into the soil when seeding. A sidedress of fertilizer high in nitrogen halfway through the growing season helps to produce a tender and larger crop.

HARVEST

Kale may be harvested by pulling the entire plant from the soil or by cutting off the outside leaves as the proper size is reached. Old leaves are tough and stringy.

USE

Cooked as greens, fresh, frozen.

STORAGE

Keep cool (40°F) and moist.

KOHLRABI

Kohlrabi is grown for its swollen stem. For cultural requirements and uses, see Cabbage.

HARVEST

Harvest kohlrabi when it is young and tender.

STORAGE

Keep cool (40°F) and moist.

LETTUCE

Lettuce, a cool-season crop, is very sensitive to heat. In the southern United States, the growing of lettuce is continued into late fall, winter, and spring. In most other areas, it is a spring and fall crop.

Lettuce grows in any good garden soil. The soil should not be highly acidic, however; check soil pH before planting. Head lettuce plants are grown indoors and transplanted to the garden. Leaf lettuce may be direct seeded. It may be planted very early since temperatures as low as 28°F (−2°C) do not harm plants that have been properly hardened off. Set head lettuce plants 12 inches apart in rows 3 feet apart. Leaf lettuce plants may be planted much closer to each other.

FERTILIZER

A fertilizer high in nitrogen and phosphorus such as 10-10-10 should be dug into the soil at planting time or sidedressed along the rows at a rate of 3 pounds per 100 feet of row.

HARVEST

Harvest head lettuce when the heads are firm or slightly earlier. Leaf lettuce is picked before the leaves become tough. The older leaves should be used first.

USE

Fresh, in salads and other dishes.

STORAGE

Lettuce is stored at 32°F (0°C) and 90 to 95 percent relative humidity for best results.

MUSTARD

Mustard grows in almost any type of soil. It is a cool-season crop and matures rapidly. Prepare same as a fresh-cooked greens dish, such as kale.

FERTILIZATION AND STORAGE

See Lettuce.

OKRA

See Tomato.

PARSLEY

Parsley is hardy in cold weather, but sensitive to heat. It thrives under the same conditions as kale and lettuce. It is important that parsley be planted early in the season. It grows in any good, loose garden soil. Since young parsley plants are slow to start, it is best to start them indoors in flats and transplant to the garden. Soaking seed overnight in water helps to speed up germination.

Plant parsley in rows 14 inches apart with plants 6 inches apart in the row. No special fertilizer is needed.

HARVEST

Harvest in late fall after frost.

STORAGE

Store in the soil with mulch or in cool, moist conditions same as other root crops such as carrots.

TURNIP-ROOTED PARSLEY This type of parsley is grown for the root and has much the same flavor as celeriac. The cultural conditions are the same as for parsley. It is stored as other root crops are.

PEAS

Peas, figure 41-12, are a cool-season crop and can be seeded in the spring as soon as the ground can be prepared. A soil high in organic matter is best.

Plant in double rows 6 to 8 inches apart with just enough space between double rows to allow cultivation. Sow seeds 1 inch apart in the rows.

FERTILIZER

Peas respond to high applications of fertilizer. Apply 3 pounds of a 10-10-10 fertilizer per 100 feet on each side of the row before planting and 2 pounds of the same fertilizer at the early blossom stage.

HARVEST

Harvest when the pods are well filled but just before the seeds reach full size. Cook or process immediately after picking for best results.

USE

Fresh cooked.

STORAGE

Peas are especially good fresh from the garden. If they must be stored, they remain freshest at 32°F (0°C) and 85 to 90 percent relative humidity. Freeze or can for permanent storage.

PEPPERS

The cultural practices for peppers, figure 41-13, are the same as those for tomatoes, except that

FIGURE 41-12 Pea (variety Bush Snapper) (Courtesy W. Altee Burpee Company)

FIGURE 41-13 Sweet peppers (Courtesy Southern States Cooperative)

warmer soil is required before planting. Set plants 1½ to 2 feet apart in rows 3 feet apart.

FERTILIZER

Apply 2 pounds of a 10-10-10 fertilizer per 100 feet of row and dig it into the soil, or band it beside rows at planting. Apply a sidedress of fertilizer at first blossom with a 10-10-10 fertilizer. Use enough to keep plants actively growing.

HARVEST

Harvest peppers when they reach the desired size.

USE

Fresh (in salads and other dishes) and pickled.

STORAGE

Store at 45° to 50°F (7° to 10°C) and at 85 to 90 percent relative humidity for best results.

POTATO

The potato is one of the highest producing vegetables in terms of food per area of land. A cool-season crop, potatoes are planted as early in spring as possible or in mid to late summer for a fall crop. Any loose, well-drained, fertile soil is suitable. The soil should be moderately acidic.

Plant small (1½ to 2 ounces), chunky rather than thin, cut pieces of potato with one or more eyes. Place the seed pieces 3 inches deep in the soil and space them 10 to 12 inches apart in rows spaced 3 feet apart.

FERTILIZER

Apply 10 to 15 pounds of 10-10-10 fertilizer per 100 feet of row. The fertilizer can be broadcast and worked into the soil or applied as a sidedressing beside the row at planting time, but should never touch the seed pieces. Sidedress applications of fertilizer may be made later if the tops do not appear dark green.

HARVEST

Potatoes may be dug for immediate use as soon as they are large enough to eat. Do not dig potatoes that are to be stored until the tops are dead. Handle them carefully and avoid skinning. Do not leave potatoes exposed to light or they will turn green and be unfit for use.

USE

Cooked fresh.

STORAGE

Store potatoes at 40° to 45°F (4° to 7°C) in a well-ventilated area. Keep the storage area dark to prevent greening. If possible, curing at 50° to 60°F (10° to 16°C) in moist air before storage is an advantage.

RADISH

Radishes, figure 41-14, are hardy in cold weather but cannot tolerate heat. They may be grown in spring or fall. The crop matures from 3 to 6 weeks after planting, so a number of plantings can be made. A rich, loose, well-drained soil is best.

There are two types of radishes—the mild, small, quickly maturing type which reaches edible size in twenty to forty days, and the large winter radishes which require seventy-five days for growth.

FERTILIZER

Fertilize by digging 2 pounds of a 10-10-10 fertilizer per 100 feet of row into the soil at planting time. Sidedress if needed to maintain steady growth.

FIGURE 41-14 Cherry Belle and White Icicle radishes (Courtesy W. Altee Burpee Company)

HARVEST

The early, fast-growing radishes are harvested as soon as they have reached a desirable size for storage.

USE

Fresh (in salads).

STORAGE

Store as root crops are stored.

RUTABAGA AND TURNIP

Rutabagas and turnips, cool-season crops, are similar in terms of soil and fertilizer requirements and ways in which they are used as food items. They are both grown in southern regions chiefly in the fall, winter, and spring; in northern parts of the country, they are grown mostly in the spring and autumn. Rutabagas grow best in northern regions; turnips are better for gardens south of Indianapolis, Indiana, or northern Virginia. A half ounce of seed of these vegetables allows for the broadcasting of 300 square feet.

Turnips require sixty to eighty days to mature; rutabagas require 90 to 110 days.

FERTILIZER

If planted after early potatoes or other early crops which have been fertilized, no additional fertilizer is needed.

HARVEST

Harvest as soon as the roots are large enough to eat.

USE

Fresh cooked.

STORAGE

Storage is the same as for the other root crops— 32°F (0°C) at 90 to 95 percent relative humidity.

SPINACH

Spinach, figure 41-15, is a hardy, cool-weather crop. It is able to withstand winters in the southern regions and is an early spring and late fall crop in the northern regions. Seed should be planted as early in the spring as possible. Sow the seed in rows 12 to 30 inches apart (depending upon the method of cultivation). Plants should be thinned so that they stand 2 to 4 inches apart in the row.

FIGURE 41-15 Spinach (variety Melody Hybrid) (Courtesy W. Altee Burpee Company)

Spinach will grow in almost any well-drained soil. Check the requirement for spinach and test the soil.

FERTILIZER

Apply 3 or 4 pounds of a 10-10-10 fertilizer per 100 square feet at planting time and work into the soil. Sidedress with nitrogen to keep the plants actively growing.

HARVEST

Harvest as leaves become mature enough to eat.

USE

Cook fresh (as greens) or fresh (in salads).

STORAGE

Spinach keeps freshest at 32°F (0°C) and 90 to 95 percent relative humidity.

NEW ZEALAND SPINACH

New Zealand spinach is not related to common spinach. It is generally grown as a substitute for common spinach in hot weather. The plants are larger than common spinach and must be spaced 18 inches apart in rows that are 3 feet apart. Plant seed as soon as frost danger is passed.

SQUASH

The squash, figure 41-16, is a hardier crop than melons and cucumbers. It is, however, a warm-season crop.

Summer squash, which includes all bush types, should be eaten while young and tender. This group consists of scallops, yellow straightneck, yellow crookneck, and the vegetable marrows such as cocozelle and zucchini. Summer squash plants are heavy producers.

Winter squash consists principally of vine types. They have hard rinds, making them easy to store. Winter squash includes such varieties as Boston marrow, table queen, Hubbard, delicious, and butternut.

Plant bush squash 2 to 3 feet apart in rows 3 to 4 feet apart. Plant vine squash 4 to 6 feet apart in rows 6 to 8 feet apart.

FIGURE 41-16 Summer squash (variety Butterstick Hybrid) (Courtesy W. Altee Burpee Company)

FERTILIZER

Squash responds well to applications of manure and fertilizer. Dig manure and liberal amounts of fertilizer high in potash, such as 5-10-15, into the soil before planting. Sidedress at the first blossom with a 10-10-10 fertilizer.

HARVEST

Harvest summer squash while the skin is still tender. Winter squash must be allowed a longer time to mature and is harvested just before the first killing frost.

USE

Fresh cooked.

STORAGE

Store summer squash at 32° to 40°F (0° to 4°C) and at 85 to 90 percent relative humidity. Winter squash is stored at 50° to 55°F (10° to 13°C) and at 70 to 75 percent relative humidity.

SWEET POTATO

Sweet potatoes, figure 41-17, grow best in warm weather. They are grown in gardens as far as southern New York and southern Michigan. A frost-free period of 150 days with relatively high temperatures is needed for their cultivation. Sprouts or slips should not be set out in the garden until all danger of freezing is passed. The home gardener should purchase the sprouts for planting, and not attempt to grow them.

FIGURE 41-17 Sweet potatoes (Ed Reiley, Photographer)

Plant in rows 3 or 3½ feet apart and space plants 12 to 15 inches apart in the row. Sweet potato plants are generally set on top of ridges. A well-drained, sandy loam soil which is moderately deep is best.

FERTILIZER

Moderate use of a fertilizer high in potash, such as 5-10-15, is best. Dig and mix half the fertilizer into the soil at planting time (1 pound per 50 feet of row) and apply the rest as a sidedress during July.

HARVEST

Harvest sweet potatoes just before the first frost. Allow the roots to dry before storage. Be careful not to bruise or injure the potatoes in any way.

USE

Fresh cooked.

STORAGE

Sweet potatoes should be allowed to cure for one to two weeks at a relatively high temperature (85°F). Storage is best at 50° to 55°F (10° to 13°C) and 85 to 90 percent relative humidity.

TOMATO

The tomato is probably the most popular garden vegetable, figure 41-18. It is a warm-weather crop and a highly productive plant. Tomatoes are planted in the garden only after the danger of frost is passed. Space the plants 3 feet apart in rows which are 3 feet apart. Set the plants deeper in the garden than they were growing before transplanting. Plants are sometimes staked or caged for protection. A mulch is of great value in conserving moisture. Installation of a cutworm collar around each plant is necessary. Try the new wire caging method along with a heavy mulch for almost care-free tomato production. Tomatoes grow best in a well-drained loam soil.

FERTILIZER

Before planting, dig compost or other organic matter and 5 pounds of a 5-10-10 fertilizer per 100 square feet of area into the soil. Sidedress when early fruit sets with a 10-10-10 fertilizer. Limestone placed in the hole at planting time may help to prevent blossom-end rot (caused by a calcium deficiency).

FIGURE 41-18 Tomatoes (Courtesy Dr. Frank Gouin)

HARVEST

Harvest tomatoes as they turn red, but before the fruit becomes soft. For canning, the fruit should be fully red and fully ripe.

USE

Fresh (in salads), cooked fresh, canned, and for making catsup and tomato paste.

STORAGE

Ripe fruit is stored ideally at 50°F (10°C) and at 85 to 90 percent relative humidity. Green mature fruit stores best at 55° to 70°F (13° to 21°C).

WATERMELON

Watermelons, figure 41-19, require a great deal of garden space. They are more sensitive to cold than cucumbers or cantaloupes. Soil conditions are the same as for cucumbers.

Plant watermelons in hills 8 feet apart. Plant four to six seeds per hill and thin to two to three

FIGURE 41-19 Watermelon (variety Bush Sugar) (Courtesy W. Altee Burpee Company)

plants per hill. Mix about a bushel of organic material in the soil of each hill. (New bush types take much less garden space and seedless varieties are easier to eat.)

FERTILIZER

Mix a handful of a 10-10-10 fertilizer with the organic matter applied to each hill.

HARVEST

Watermelons are ready to harvest when the tendrils nearest the fruit die and the side of the fruit nearest the ground changes from white to creamy yellow.

USE

Fresh (as dessert or in salads); rinds may be pickled.

Student Activity

1) Plant a vegetable garden as a class project. Examine a seed catalog and select the varieties best suited for the area. (Local extension and experiment station bulletins give recommendations of varieties to plant.) Meet as a class the following fall to discuss reasons for the success or failure of the garden.

2) Visit a local farmers market.

Self-Evaluation

Select five vegetables that are grown in the local area from the vegetables discussed in this unit. For each vegetable, list cultural requirements, including soil preparation and special needs of the plant, and how the vegetable is harvested.

Select the best answer from the choices offered to complete the statement or answer the question.

1) Asparagus is

    a) an annual vegetable.
    b) a bi-annual fruit.

    c) a perennial.
    d) harvested all year.

2) Beans are

    a) a warm-weather crop.
    b) a cool-season crop.

    c) heavy users of nitrogen.
    d) none of the above

3) Broccoli is

    a) a warm-weather crop.
    b) a cool-season crop.

    c) a perennial.
    d) a root crop.

4) Carrots

    a) do not need nitrogen fertilizer.
    b) are a warm-season crop.

    c) prefer heavy clay soils.
    d) are a root crop.

5) Celery

    a) is a warm-season crop.
    b) prefers a rich, loose, well-drained soil.

    c) prefers a heavy clay soil.
    d) is a perennial.

6) Which of the following is not a type of onion?

    a) chives
    b) leek

    c) shallot
    d) chicory

7) The new supersweet sweet corn requires

    a) a warm soil to germinate.
    b) a high nitrogen fertilizer.

    c) a lot of garden space.
    d) all of the above

8) The potato is

    a) one of the highest producing vegetables.
    b) a cool-season crop.
    c) a good storage vegetable.
    d) all of the above

9) The tomato is

    a) a warm-weather crop.
    b) a root crop.

    c) planted in narrow rows.
    d) a cool-season crop.

10) Watermelons

    a) require a lot of garden space.
    b) are a warm-weather crop.

    c) are now available in seedless varieties.
    d) all of the above

The Small Fruit Garden

Strawberries

OBJECTIVE

To grow strawberries of marketable quality in the home garden and to become familiar with commercial strawberry-producing areas in the United States.

Competencies to Be Developed

After studying this unit, you should be able to

- ◆ select a site for a strawberry planting and properly prepare the soil for planting.
- ◆ diagram four planting systems for strawberries.
- ◆ list the seven points to consider when selecting a strawberry variety to plant.
- ◆ list the seven steps in planting strawberries.
- ◆ explain the cultural practices of weed control, mulching, and fertilizing as they relate to care of strawberry plantings.
- ◆ follow the sixteen steps to better strawberry yields.

MATERIALS LIST

- ✔ extension bulletins on strawberry culture in the local area
- ✔ strawberry plants for identification of plant parts

If a planting site is available:

- ✔ plants
- ✔ fertilizer
- ✔ string
- ✔ shovel
- ✔ plant labels
- ✔ mulch or cultivator or hoe
- ✔ tape measure
- ✔ measuring sticks cut to proper plant spacing length

The strawberry, figure 42-1, is a perennial plant. The leaves die back to the crown each fall in areas where winter temperatures drop below 23°F (–5°C). The roots live from year to year and new leaves

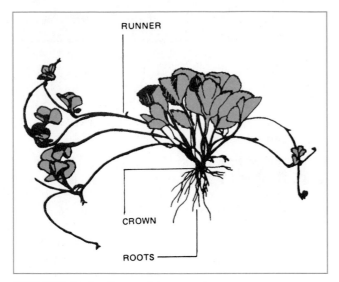

FIGURE 42-1 Parts of the strawberry

and blossoms grow from the crown each spring. The new fruit is produced from these blossoms.

Runners are the means by which strawberries naturally propagate. Plants normally send out runners from early summer until fall. Each runner roots and forms new plants. Everbearing plants that produce berries all summer have fewer runners.

STRAWBERRY PRODUCTION

The strawberry is one of the first fresh fruits to be marketed in spring and is usually in high demand at this time. Strawberry plants are high producers on a per acre basis and are considered to be a high cash crop, figure 42-2.

Strawberries are grown nationwide, but there are two large areas: (1) the Pacific Coast states and (2) the eastern United States. Approximately 65 percent of all strawberries grown in the United States are grown on the Pacific Coast. There are some major differences in cultural requirements between the two growing areas and within the Pacific Coast area.

GROWING AREAS ON THE PACIFIC COAST

CALIFORNIA

Strawberries are grown in California by means of a hill system. The plants are kept in cold storage until summer, when they are set in the field. They are generally set in single or double rows. Production the first year is heavy.

Summer plantings are irrigated to keep salt accumulation in the soil minimal. All plantings are mulched with clear polyethylene.

CENTRAL COAST

This area includes Santa Cruz, Monterey, and Santa Clara counties. Plants are planted in early August and do not produce runners after the first year because of warm winters. Harvest season is from April until November.

CENTRAL VALLEYS (NEAR FRESNO)

This area experiences much cold weather in winter and hotter than average weather in summer. The harvest season is April, May, and part of June. Plants are set in the fields in mid-July.

SANTA MARIA

This area includes Santa Barbara and San Luis Obispo counties. Harvest begins in April and continues through the summer and fall. Plants are set in the field in both winter and summer.

SOUTHERN CALIFORNIA

This area includes the Oxnard coastal plain of Ventura County and the coastal regions of the counties of Los Angeles, Orange, and San Diego. Plants are harvested for only one year. Plants are set in the field both in summer and winter.

OREGON

The Northern Willamette Valley has the largest acreage of strawberries of any single area in the United States. Harvest is from May until early July. Many berries are also grown in the Hood River Valley.

WASHINGTON

The main strawberry-producing area in Washington is the Puget Sound region. Harvest is from early June until mid-July. Southwestern Washington also produces strawberries commercially.

Strawberry varieties grown in the Pacific Coast states are the spring- or June-bearing varieties. These varieties produce flower buds in late summer and bear flowers and fruit the following spring.

Since cultural requirements differ from area to areas, specific recommendations should be obtained from the local extension service or a horticulturist. For a specific description of strawberry varieties, consult Bulletin 1043 *Strawberry Varieties in the United States*, available from local extension offices.

FIGURE 42-2 Strawberries ready for market (Courtesy Dr. Harry J. Swartz)

GROWING AREAS IN THE EASTERN UNITED STATES

This growing area includes all regions east of the Great Plains and north of the Southern Coastal Plains. Cultural practices are generally the same for this entire area.

In the eastern United States, dormant plants are usually set in early spring. Runners grow from the original plants starting in June and a succession of new runners grow and take root around the original plant. Flowers are removed from the plants the first year, causing more vigorous growth. Fruit buds are formed in the fall and, by the end of October, can be seen in the crown. Fruit buds also develop in leaf axils of vigorous plants.

The number of leaves on a plant in the fall is a good indication of what the following year's production will be like. The more leaves there are, the more berries will result. The crop is generally harvested in late May and June. Most of these berries are sold fresh.

The following information applies to both the Pacific Coast areas and the eastern United States.

SELECTING THE PLANTING SITE

SOIL

Strawberries grow best on a loam or sandy loam soil. Clayey loam or silty loam soils are very good to excellent for yields if well drained. If planted on a very sandy soil, plantings should be fruited (allowed to bear fruit) for only one season.

It is very important to supply the soil with an abundance of organic matter. This can be added by plowing under a crop such as rye or by applying barnyard manure. Care should be taken not to plow under weed seeds. (Chemical weed control has helped to decrease this problem.)

DRAINAGE

Drainage is very important for good growth of strawberry plants. Raised beds help to improve drainage. The fungus disease red stele is almost always present in wet soils. It attacks the plant roots and kills the plant. There are now many excellent varieties that are resistant to red stele disease.

FROST

Strawberries cannot be grown in low areas where frost is a problem during the blossoming time. Cold air collects in low areas in the same way that water does; frost damage in these pockets can be severe. Irrigation and row covers offer some protection.

DISEASES

Avoid soils infested with red stele root rot or areas below infected fields where water may wash the disease into the planting site. Surecrop, Earliglow, and Allstar are strawberry varieties that are resistant to the common types of red stele. These varieties may be grown where the disease is a problem. Check for other, more recently developed varieties for local areas.

If tomatoes, potatoes, or peppers have been grown in a soil recently, verticillium wilt is probably present. If this is the case, plant only a disease-resistant variety, such as the ones mentioned.

SOIL PREPARATION

Soil preparation for planting strawberries should begin one or two years before the actual planting date. Soil tests should be made to determine soil pH or acidity. Strawberries grow best at a pH of 6.0 to 6.5. If it is necessary, lime or sulfur should be added to correct the soil pH at least six months or, better yet, one year before planting. This allows time for the lime or sulfur to dissolve and correct the soil pH before plants are set in the field. If magnesium is low, apply dolomitic limestone.

Fertile soils high in organic matter are necessary for top crop yield. Large quantities of animal or green manure should be plowed under to raise the organic matter level one to two years before planting. However, heavy green manure crops should not be plowed under just before planting strawberries—they should have time to partially decay so that large loose air pockets are not left in the soil.

In preparation for planting, the soil is plowed to a depth of 7 to 9 inches. If the soil test shows phosphorus and potassium to be low, both should be added to correct the deficiency before planting. Any phosphorus or potash should be either plowed under or mixed into the soil after plowing. The soil should be worked until it is firm and free of large clods and air pockets. The soil must pack firmly around the new plants.

Land that was in sod for years should not be directly planted with strawberries. Such soils may contain root-eating grubs which destroy new plants. It is best to plant a sod area with a cultivated crop, such as corn, for a year before planting strawberries.

Areas which contain nutgrass, quackgrass, Bermuda grass, Johnson grass, or other perennial weeds should be avoided. Chemical weed control can help solve this problem.

Strawberries are being grown on raised beds more now than in the past. Because of this, the soil warms earlier in spring, drainage problems are lessened, and picking is easier. Use of the new floating row covers is also easier with raised beds.

CHOOSING VARIETIES TO PLANT

The selection of varieties to plant depends on climate, soil, and the purpose for which the crop is grown. The following are other factors to be considered.

◆ A variety that produces firm berries is needed for shipping to market.

◆ Large, softer textured varieties for freezing have a bright red color, firm flesh, and a tart flavor.

◆ The fruit must ripen when the best market prices are available.

◆ If frost protection is not available, use plants that bloom late in the season to miss spring frosts that could destroy the crop.

◆ If planting for the wholesale market, varieties that are well known are planted, yielding enough volume to provide truckload lots. This insures a better price for the fruit.

◆ Disease-resistant varieties should be used wherever possible.

◆ For the home garden and local roadside market, three or more varieties that ripen at different times are planted. This supplies fresh fruit over a longer picking season.

Day neutral everbearing strawberries are now a reality. These new varieties bear throughout the growing season, not just in spring and fall as many of the old everbearers did. They produce fewer runners and should be planted 6 inches apart each way, in double rows spaced 1 foot apart. For roadside markets, pick-your-own operations, and the home garden, this continuous fruiting is a big advantage.

Varieties now available are Tristar, Tribute, Tillikum, Hecker, Aptos, and Brighton. Tribute and Tristar are also disease resistant to red stele, verticillium wilt, and powdery mildew.

Day neutral varieties require different management practices. They should be planted as early in spring as possible or in November or December in mild coastal and southwestern areas. Flowers should be pinched off for six weeks after planting and all runners should be removed for best production. First fruit will be ripe about thirty days after blossom pinching stops.

Because they are more shallow rooted and the roots are more sensitive to air temperature than most other strawberries, day neutral varieties should be mulched to keep the soil cool and moist. This will maintain yields better in hot weather. The variety Tristar is the best selection for hot weather.

Some growers treat day neutral varieties as annuals, harvesting for one year only. They can, however, be wintered over in many areas with adequate mulch.

PURCHASING PLANTS

Purchase plants from a reputable nursery and be certain they are disease and nematode free. Nursery plants should be certified true to variety and free of insects and signs of disease. Be sure that the plants are fresh upon arrival. The roots should appear bright and plump. If roots appear dry, soak the roots in water for an hour before planting or heeling in.

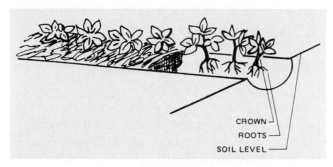

FIGURE 42-3 Heeling in. The plants are placed in the trench with the crown at soil level. The trench is then filled with soil and packed down.

If plants cannot be planted immediately, either heel them in the soil outside or store them in cold storage with the roots packed in moist sphagnum moss or other packing.

Heeling in is done by making a V-shaped trench deep enough for the roots to spread out with the crowns at ground level. The plants are then placed along one edge of the trench so that the roots are separated, and the trench is filled in and packed around the roots. The plants can be kept in the trench until planted to the field, figure 42-3.

If cold storage is to be used, store plants that are not dormant at 40°F (4°C) and dormant plants at 32° to 36°F (0° to 2°C). When stored at the proper temperature, dormant plants can be kept for a longer period of time. When using cold storage, first wrap the plants in polyethylene plastic to keep them from drying out.

PLANTING TIME

In the eastern United States, the best time to plant is early spring. Moisture is usually available from natural rainfall and the temperatures are cool at this time. Pacific Coast regions are planted at differing times, as discussed earlier.

PLANTING STRAWBERRIES

PLANTING SYSTEMS

Five planting systems will be discussed in this section. The first three are generally used in the eastern United States. The other two systems are used by both eastern and western U.S. growers.

HILL SYSTEM (No runners are allowed to grow.) In the hill system, plantings are made in

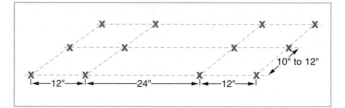

FIGURE 42-4 Double-row hill system. In this system, a 24-inch alley separates two rows that are 12 inches apart. Plants in the rows are 10 to 12 inches apart.

double rows. Rows are spaced 12 inches apart and plants in the row are 10 to 12 inches apart. A 24-inch space is left between the double rows, figure 42-4. The double-row system allows for 29,000 plants per acre.

The unwanted runners are generally cut off by machine then thrown into the alley between rows where they are cut by sharp discs.

SPACED MATTED-ROW SYSTEM (Some runners are allowed to grow to fill the rows in.) In this system, plants are set 18 to 24 inches apart in the row, and a 42-inch space is left between rows. An acre allows for 6,225 plants if they are spaced 24 inches apart in the row and 8,300 plants if the 18-inch spacing is used, figure 42-5.

MATTED-ROW SYSTEM (Most runners are allowed to grow.) Plant spacing for the matted-row system is the same as for the spaced matted-row system. The difference is that all runners are allowed to grow freely and fill in the rows. Plants become crowded in this system, and individual fruit size may be smaller than under the spaced matted-row system. This system is losing favor as narrow rows which can be mechanically tilled show advantages of lower cost and higher quality fruit.

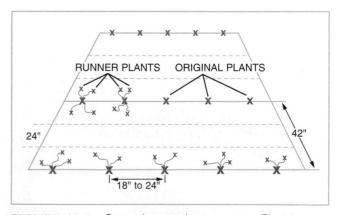

FIGURE 42-5 Spaced matted-row system. Plants are spaced 18 to 24 inches apart in rows 42 inches apart. Runners are allowed to grow so that plants are eventually spaced 6 inches apart. A 24-inch alley is left between runner plants.

FOUR ROWS, DOUBLE ROWS, AND SINGLE ROW ON RAISED BEDS This is the most common method of planting in California. The system consists of raised beds spaced 2 feet apart. The two or four rows in each bed are spaced 8 to 12 inches apart and plants are set in the rows 8 to 12 inches apart. See figure 42-6.

In the Pacific Northwest, plants are set in single rows on raised beds 39 to 42 inches apart with plants spaced 12 to 14 inches apart in the rows. Newer plantings are being made with 8-inch spacing in the row. Plants are often replanted each year with raised-bed plantings.

SINGLE-ROW PLANTING Recent research indicates that most strawberries will be grown on raised beds in single rows, with plants close together in the rows and rows 3 feet apart center to center. About 60,000 plants per acre will be planted.

In the East, a grower may plant in early spring and harvest a first crop in sixty days. Since runners are not needed with this type of close planting, plants will be allowed to flower the first year and bear fruit. The following spring will produce the largest crop and several years of production may be expected form the planting. The introduction of new varieties, which produce few runners, will make this system more practical. These plantings often become narrow matted rows in time.

STEPS IN PLANTING

Regardless of the system of planting the general steps in the procedure are the same.

FIGURE 42-6 Strawberries planted in raised bed, four rows wide (Courtesy Dr. Harry J. Swartz)

PROCEDURE

PLANTING STRAWBERRIES

1. Cultivate the soil immediately before planting to kill germinating weed seeds. Be sure all clods are broken up and that the soil is firm. Fumigation of the soil or other chemicals may be used.

2. Lay out rows according to spacing requirements. Rows must be opened deep enough to plant the roots straight down.

3. Keep the plant roots moist before planting. Never expose roots to the drying sun or wind.

4. Set the plant in the row at the proper depth with the crown just at ground level, figure 42-7. This is very important. Plants set too deep smother and die; plants set too shallow dry out.

5. Fan the roots out and straighten them. If roots are too long, they may be trimmed to a length as short as 4 inches.

6. Pack the soil around the roots firmly. Do not cover the crown and do not leave any roots exposed.

7. If possible, water immediately after planting.

Many plants are set by machine in large operations. For small operations, a two-person team with a spade and buckets to carry plants is sufficient. One person inserts the spade in the ground and pushes it forward. The second person places a plant in the hole, after which the first person withdraws the spade and firms the soil around the plant with the foot.

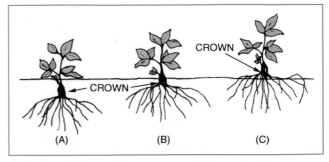

FIGURE 42-7 Plants are set at different depths: (A) plant set too deep; (B) plant set at correct depth; (C) plant set too shallow.

Further watering will be needed in the East, and frequent irrigation during the summer is generally needed in all areas of California.

CARE OF THE PLANTING

FLOATING ROW COVERS

One of the newest growing techniques for strawberries is the use of row covers. Made of lightweight materials (from .5 ounce to 2.0 ounces per square yard) row covers allow from 40 percent to 80 percent light transmission. Rain water passes through and temperature is modified under the covers. The covers are rolled out over rows of plants and edges are covered with soil to hold them in place. No support is needed to hold them up off the plants.

These covers can help to get crops to market as much as two weeks earlier than normal in northern regions, when prices are higher. California growers also find that the covers provide some frost protection.

Dr. James Pollard, an associate professor of Plant Science at the University of New Hampshire, has tested row covers on perennial strawberries. He has found that floating row covers provide the following benefits:

1. They induce plants to set more flower buds in the fall (more fruit produced).

2. Winter protection.

3. Spring frost protection (Dr. Pollard recommends irrigation in addition, since the degree of frost protection by row covers *alone* is not predictable (uncover during flowering).

4. Protection from virus diseases spread by aphids.

RESULTS OF TESTS BY DR. POLLARD

◆ 1982–1983: Not much snow, low temperatures −15°F to −20°F.

Production of strawberries:

With row covers—11,500 pounds per acre.

Mulched with hay—6,500 pounds per acre.

There were three severe frosts while the covers were still on. This yield with the covers is 75 percent greater than yield for the hay-mulched crop.

◆ 1983–1984: Mild winter, no spring frosts.

Production of strawberries:

With row covers—23,000 pounds per acre.

Mulched with hay—17,000 pounds per acre.

The row covers produced 6,000 pounds of 35 percent more strawberries.

Profit was calculated as follows:

Row cover cost:	$750.00 per acre
Cost to apply:	150.00 per acre
Total cost:	$900.00 per acre

At $.70 per pound for pick-your-own strawberries, even in the mild winter a profit was made.

6,000 pounds at $.70 per pound:	$4,200.00 per acre
Minus cost of row covers:	− 900.00 per acre
Profit:	$3,300.00 per acre

Earliness to market with better prices per pound and reduced mulch costs from using hay or straw should result in even more profits.

Row covers must be removed at full bloom for fruit pollination, so frost protection is not a major benefit.

It appears that these new row covers show great promise. Contact your local agricultural college for more details.

CULTIVATION

Plantings should be cultivated within six weeks after they are placed in the soil. This loosens the soil and kills germinating weed seeds. Cultivation should be shallow (no more than 1 or 2 inches) and far enough away from the plant that the crowns are not exposed or covered too deeply by soil thrown by the cultivator. After this first cultivation, all commercial plantings should be treated with weed killers. Additional cultivation may be needed when the chemical weed control begins to fail, usually in about two months.

Any plants that die may be replaced by placing runners in the location and allowing them to take root.

WEED CONTROL

To obtain good yield of quality fruit, some method of weed control must usually be employed. Hand weeding is too expensive a method to use in strawberry production. The combination of mechanical cultivation and use of weed killers is the most practical method. Black plastic mulch is effective in controlling weeds, and it also conserves moisture and helps to control fruit rot.

In California, soil fumigation before planting is used to control weeds and nematodes. Because of the many different climatic and soil conditions that affect weed control with chemicals, no specific control measures are given here. Contact a local extension service or horticulturist for recommendations.

MULCHING

Mulching is extensively used in eastern growing regions. Mulching helps to control freezing and thawing of the soil, which pushes plants out of the soil and exposes roots to drying.

Mulching with natural material also helps in weed control if it is applied so that it is thick enough (3 or 4 inches of straw), figure 42-8. It helps to conserve moisture and keeps the fruit clean. Small-grain straw makes the best mulch, but almost anything that does not contain weed seeds and stays in place is suitable. Black plastic also makes an excellent mulch except where excess heat builds up.

Clear polyethylene bed mulch is essential to successful strawberry growing in California. Plastic mulching takes much of the danger out of winter planting by increasing soil temperature as much as 10 degrees. This results in earlier ripening of fruit, larger harvests, and longer harvest seasons. As mentioned earlier, new floating row covers are also being used.

FIGURE 42-8 A strawberry patch planted with single rows of plants, mulched with straw for weed control. (Ed Reiley, Photographer)

FERTILIZING AFTER PLANTING

After planting, fertilizers should be used only in the late summer or early fall to increase the number of fruit buds in spring-bearing plants, $1\frac{1}{4}$ pounds of 10-10-10 per 100 square feet is usually adequate. Fertilizer containing nitrogen, applied in the spring, generally results in soft berries lacking in flavor, and should not be used.

FROST CONTROL

Planting on high ground prevents most frost damage to spring blossoms. If frost should occur during blooming time, an irrigation sprinkler system turned on during the time freezing temperatures exist will prevent the blossoms from freezing. Plants mulched with straw can be protected from frost by pulling the mulch back over the plants on nights when frost is expected. Row covers are also effective.

WINTER KILL

In the eastern regions of the United States, freezing and thawing of the soil can heave newly transplanted plants out of the ground, resulting in drying of roots and death of the plants. A heavy mulch applied after the plants have been hardened off to winter conditions (usually after a temperature of 20°F has been reached) will greatly reduce this loss. The plants should be completely covered to a depth of 3 or 4 inches. Floating row covers are also helpful.

HARVESTING

All strawberries are picked by hand since no machine is currently available that can do the job properly. Poor harvesting techniques can ruin even the best berries. If handled roughly, berries are bruised. If stems are too short, they puncture other fruit in the container. In eastern growing areas, strawberries are almost always picked with a ½-inch stem attached. The berry should not be pulled from the vine; the stem should be pinched off instead. Pulling breaks the skin on the berry and quality drops rapidly. In California, many berries are picked without stems for the freezing market. The fruit should be ripe for best flavor, but not too soft. Pick-your-own operations, where the customer picks the fruit, have been the answer to many labor problems.

CARE OF THE PLANTING AFTER HARVEST

WEED CONTROL

Weeds must be controlled after harvest either by mechanical cultivation and hand pulling, by mulching, or by chemical weed control. Failure to control weeds results in reduced crop production the following year.

RENEWAL (REVITALIZING THE PLANTING)

The root system of strawberries is often weakened while the fruit crop is maturing. To correct this situation, many growers mow the tops off the plants immediately after harvest. As soon as the tops dry, the area between the rows is dug up so that the row of plants is back to the narrow row width desired. Plants are thinned out within the rows with a hoe or spike-toothed harrow. Proper weed control must follow this operation.

Some strawberry plantings are allowed to produce fruit for only one year, and are then plowed up and replanted.

BIOCONTROL OF FRUIT ROT

Gray mold is the most common cause of fruit rot of strawberries. A novel new control method is to use honeybees to spread another beneficial mold (Gliocladium roseum) to the blossoms of strawberries as the bees pollinate the flowers. A dispenser, fit in the beehive entrance, fixes fungus spores to hair on the bee's body as she leaves the hive. This control has been as effective as the chemical captan in research tests. This beneficial fungus is not yet commercially available.

STEPS FOR BETTER STRAWBERRY YIELDS

1. Build up the level of organic matter in soils (a year ahead if possible).
2. Fumigate the soil for diseases, insects, and weeds.
3. Select locally adapted varieties. *This is very important.*
4. Plant as early as possible.
5. Plant on raised beds, especially in heavy soils.
6. Plant at the proper depth.
7. Space plants for maximum light (narrow rows).
8. Remove excess runners.
9. Fertilize in the early fall to encourage flower-bud formation.
10. Protect plants over the winter with mulch or row covers.
11. Protect plants from spring frost.
12. Provide bees for good pollination of blossoms.
13. Supply adequate moisture.
14. Spray for good disease and insect control.
15. Pick only ripe fruit and, in pick-your-own operations, insist that the customers pick only ripe fruit.
16. Renovate beds as soon as possible after the harvest.

 # Student Activities

1) Plan and care for a patch of strawberries at home or on the school grounds.
2) Visit a local strawberry grower.
3) Write in your notebook the sixteen steps to better strawberry yields.

Self-Evaluation

A) Select the best answer from the choices offered to complete the statement.

1) Most of the strawberries grown in the eastern regions of the United States are sold

    a) on the fresh market.
    b) for processing.
    c) in summer and fall.
    d) in carload lots.

2) Strawberries grow best on a loamy soil that is high in

    a) calcium.
    b) phosphorus.
    c) organic matter.
    d) sand content.

3) If strawberries must be planted in a soil infested with red stele disease, the grower should

    a) drain the soil to remove excess water.
    b) add as much organic matter as possible.
    c) sterilize the soil.
    d) plant only disease-resistant varieties.

4) Strawberries should not be planted in soil recently planted with tomatoes, potatoes, peppers, or eggplant because

    a) these plants spread verticillium wilt which would attack the strawberry plants.
    b) these plants spread red stele disease which would attack the strawberry plants.
    c) these plants leave a toxic chemical in the soil.
    d) none of the above

5) Strawberries grow best at a soil pH of

    a) 5.5 to 6.0.
    b) 6.5 to 7.0.
    c) 6.0 to 6.5.
    d) 5.0 to 5.5.

6) If lime is needed to raise the pH of soil in which strawberries are grown, it should be added

    a) two years before planting strawberries.
    b) from six months to a year before planting strawberries.
    c) at the time of planting.
    d) as a sidedress after plants are set.

7) If land that has been sodded for years is plowed and planted directly with strawberries,

    a) too much organic matter would be present.
    b) there is a danger of root damage from grubs.
    c) weed control would be almost impossible.
    d) the plant food supply will require a great increase.

8) In the hill system of planting strawberries, runners are

    a) hand placed to widen the hills.
    b) allowed to grow in limited numbers.
    c) allowed to grow freely.
    d) allowed to root and grow.

9) In the matted-row system of planting strawberries in the East, the planting

    a) tends to become overcrowded with plants.
    b) requires a great deal of hand placement and removal of runners.
    c) usually produces the largest berries of any system.
    d) remains productive for more years than any other system.

10) The most important reason for cultivating or harrowing the soil just before planting strawberries is

a) to level the soil.

b) to mix in fertilizer and lime.

c) to firm the soil.

d) to kill germinating weed seeds.

11) After the first cultivation of newly set strawberry plants, all commercial plantings should be treated with

a) a soil fumigant.

b) a weed killer.

c) an insecticide.

d) a fungicide.

12) Mulching of strawberry plantings with straw or other material is used in eastern regions of the United States to

a) prevent winter freezing and thawing and to control weeds.

b) keep fruit clean and control fungus diseases.

c) control insects and field mice.

d) force early blooming and protect against frost.

13) Nitrogen or a complete fertilizer is applied to spring-bearing strawberry plantings in early fall to

a) produce larger berries.

b) encourage runner formation.

c) encourage setting of a greater number of fruit buds.

d) none of the above

14) Strawberries for the fresh fruit market are generally picked with

a) the cap and a ½-inch piece of stem connected.

b) machines.

c) the cap and stem removed.

d) all of the above

15) In the renewal process, the tops are mowed off all plants in the strawberry planting and the plants are

a) fertilized.

b) sprayed for disease control.

c) thinned out.

d) none of these

B) Provide a brief answer for each of the following.

1) What type of use can best be made of the new day neutral everbearing strawberries?

2) What planting system seems to be the system of the near future?

Blueberries

OBJECTIVE

To plant and grow marketable blueberries.

Competencies to Be Developed

After studying this unit, you should be able to

- list the six areas of the country in which blueberries grow naturally and the type of berry grown in each area.

- identify the two types of commercially grown blueberries and select the one best suited for growing in the local area.

- determine the soil conditions needed for best highbush blueberry production by listing the three main soil requirements.

- draw to scale a diagram of a blueberry planting.

- plant, mulch, fertilize, spray, and prune one variety of blueberry.

MATERIALS LIST

- ✔ rooting frame to propagate blueberry plants
- ✔ blueberry cutting wood
- ✔ propagating knife, pruner, sprayer
- ✔ rooting hormone
- ✔ rooting medium (sphagnum moss and sand or perlite)
- ✔ soil test kit or soil sample boxes
- ✔ graph paper

Blueberries are one of the easiest and most rewarding small fruits to grow. If the soil pH can be adjusted so that it is low enough to support the acid soil requirement of blueberries, cultural practices should

(1) The Evergreen, or Box, Blueberry *(Vaccinium ovatum)*

(2) The Mountain Blueberry *(Vaccinium membranaceum)*

(3) The Dryland, or Low, Blueberry *(Vaccinium pallidum)*

(4) The Rabbiteye Blueberry *(Vaccinium ashei)*

(5) The Highbush Blueberry *(Vaccinium austcale)*

(6) The Lowbush Blueberry *(Vaccinium lamarckii)*

FIGURE 43-1 Map of the United States, showing areas in which blueberries are extensively harvested (Courtesy USDA)

not be difficult. In addition to being a fruit plant, blueberries make an attractive landscape plant either as single plants or planted in a hedge. The spring blossoms are very fragrant and the fall foliage is a beautiful red. The berries attract birds who come to feed on them.

It is much easier to cultivate blueberries in areas of the country where they grow naturally. Any commercial production is usually restricted to these areas. Figure 43-1 is a map showing the areas in the United States where blueberries grow in the wild. Many blueberries are harvested for sale from wild plants growing in the regions shown in the map. There are two major cultivated species of blueberries, the highbush and the rabbiteye.

HIGHBUSH BLUEBERRIES

LOCATION

The highbush, figure 43-2, grows naturally in an area from southeastern North Carolina to southern Maine and westward to southern Michigan.

Highbush blueberries are planted where the soil is moist and acid. The berries have better flavor where the days are long and the nights are cool during the ripening season. Since blueberries bloom early, plantings should not be made in low areas where late spring frosts may kill the blossoms. Most plants are not winter-hardy to tem-

FIGURE 43-2 A fruit cluster of the highbush blueberry (Courtesy Stark Brothers Nurseries and Orchards, Co.)

peratures below −20°F (−29°C). New hybrids, recently developed, are more cold-hardy and extend the growing area northward.

Highbush blueberries do require some winter cold to break the winter rest period and to blossom normally, about 1,200 hours of temperature below 40°F (5°C). They are not planted farther south than northern Georgia and northern Louisiana for this reason.

SOIL

The highbush blueberry grows best in moist, acidic soil. A pH range of 4.6 to 5.0 is best. If soil pH is lower (more acidic) than this, add ground limestone to raise the pH. If the pH is higher than 5.0, add finely ground sulfur or fertilize with ammonium sulfate to lower the pH (make the soil more acid). Clay soils require four times as much as sandy soils. Digging in sphagnum peat moss also lowers soil pH.

The soil must have good drainage. If water stands on the soil surface for several hours after a hard rain or for extended periods of time in winter or early spring, the area is not right for blueberries.

The best indication that blueberries may succeed in a soil is the presence of other native plants that grow best in acidic soil. If wild blueberries, huckleberries, azaleas, or laurel grow in an area, chances are good that cultivated blueberries will grow well there. A soil water table between 14 and 30 inches is ideal.

The soil should be high in organic matter. The plowing under of large amounts of humus or green manure crops a year before planting is very helpful. This material helps in moisture retention and in the development of good soil structure. Natural sand-peat mixtures make excellent blueberry soils. They drain well and the peat content holds moisture. The soil should be plowed and harrowed and rows established with a layoff plow for easier planting. (A layoff plow opens a trenchlike row in the soil.)

PLANTING

Plants are spaced 4 or 5 feet apart in rows that are spaced 9 or 10 feet apart. Spacing plants as such means that there are about 1,100 plants per acre. Two-year-old plants are generally used and are planted at the same depth at which they grew in the nursery row.

The planting is usually done as early in the spring as the soil can be worked. In New Jersey, some plants are set in the fall, and in eastern North Carolina, planting is done in late fall and winter.

Plant roots must not be exposed to drying wind or sun during the planting process. To prevent this, keep plants covered with moist burlap or some other covering. When planted, the roots of the plant are placed in holes big enough for them to fit in without bending or twisting around in the hole. Firm the soil around the plant roots and water if possible. Plants are cut back one-third to one-half at planting time by thinning stems, not cutting off terminal ends.

FERTILIZER

Soil pH or acidity should be adjusted, if necessary, six months before planting by mixing either lime or sulfur into the soil, depending on which way the pH is to be adjusted. If the soil tests low in phosphorus or potash, these elements should also be plowed down or mixed into the soil prior to planting. Use a soil test to determine how much fertilizer is needed.

The following fertilizer schedule is suggested after planting.

1. After buds start to swell: 400–500 pounds of a 10-10-10 fertilizer per acre (*not neutralized*—it should be acidic). Acidic fertilizer helps to keep the soil acidic.

2. Six weeks later: If soil pH is 4.8 or higher, 100 pounds of ammonium sulfate per acre adds nitrogen and helps to keep the

soil acidic. If plants are growing vigorously, this second application and additional fertilizer use may not be necessary. On very sandy soils where nitrogen leaches rapidly, a second application of ammonium sulfate six weeks after the application described earlier is recommended.

The fertilizer is broadcast around the plant to within 6 inches of the plant and out as far as the roots extend. This is usually a foot or more beyond the spread of the outer branches. Older plant roots extend farther out than younger plants. New shoot growth of 14 to 18 inches indicates proper fertilizer use.

IRRIGATION

The highbush blueberry is not drought resistant. During periods of dry weather, 1 to 2 inches of water per acre should be added every ten days to two weeks during the picking season. Do not water any less than this amount at a single application. Shallow watering encourages shallow-rooted plants which fail to survive during periods of stress and winter months. Trickle irrigation under the mulch works well.

MULCHES

Mulching with straw, sawdust, wood chips, or other material helps to conserve soil moisture and control weeds and erosion. These materials should be added deeply enough to shade the ground and to prevent the surfacing of weeds. A fine material such as sawdust should be applied only 3 or 4 inches thick, but coarse, open material such as straw should be up to 6 inches thick.

When organic mulches are used, especially sawdust, additional nitrogen fertilizer must be applied to maintain good growth. This is because the rotting process of organic mulches robs the soil of nitrogen. Two pounds of ammonium sulfate or 6 pounds of 10-10-10 should be added for each 100 pounds of sawdust mulch added.

Mulches are the best weed control for blueberries since cultivation damages the shallow root system.

POLLINATION

Blueberries are partially self-fruitful or self-pollinating. But two or more varieties should be planted within two rows of each other to ensure good pollination. Honeybees should be placed within 200 yards of the plants to carry out the pollination. Wild insect pollinators may not be present in sufficient numbers to assure proper set of the fruit.

PRUNING

Fruit is produced on wood of the previous season's growth. The largest fruit is produced on the most vigorous wood. This means that good vigorous growth (14 to 18 inches) must be maintained each year to produce a crop of large berries year after year.

Most varieties tend to bear too much fruit. Unless many of the fruit buds are removed, the berries are small and there is not enough vigorous growth for the next year's crop. To prune varieties that are growing erect, thin out the plants in the center. To prune spreading varieties, remove lower branches.

Heavy pruning should not be done in any one year. The heavier the pruning is, the smaller the crop will be. Too heavy pruning causes fruit to ripen earlier and over a shorter period of time. It is sometimes done to ripen berries earlier to obtain the highest price market.

Light pruning consists of thinning out the smaller, weaker branches and cutting some old stems to ground level each year. It is usually sufficient to insure vigorous growth for the next year's crop and good-sized fruit during the current year, figure 43-3. *A vigorous plant requires less pruning than a weak, slow-growing plant. Vigorous plants tend to produce large berries without much pruning.*

Very little pruning is needed until the third year after planting. After that, pruning should be done each year. Pruning done in the fall delays flowering in the spring by about one week. This provides protection against frost, and does not change the harvesting date. Fall pruning should be done from Rhode Island south.

(A)

(B)

FIGURE 43-3 (A) A highbush blueberry bush before pruning (B) The same blueberry bush after pruning. Notice that a number of large canes were removed to soil level. This stimulates new, vigorous shoots that are more productive. (Ed Reiley, Photographer)

SUMMARY

1. The center of the bush is kept open by removing old and weak wood.

2. Low-hanging branches are removed.

3. Most of the small slender branches are cut off at ground level, leaving only strong vigorous shoots and branches.

4. Some old branches are removed to ground level each year.

PROPAGATION

Propagation of highbush blueberries is generally by hardwood cuttings, figure 43-4. Cuttings 4 to 5 inches long are made from dormant shoots of the previous season's growth in winter. Only wood with leaf buds is used. If there are blossom buds on the tips of shoots, cut the ends off and discard them. Cuttings are usually grown in the nursery

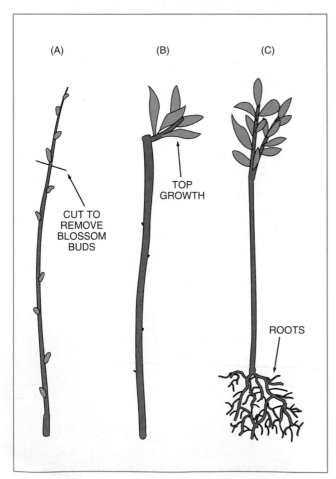

(A) (B) (C)

TOP GROWTH

CUT TO REMOVE BLOSSOM BUDS

ROOTS

FIGURE 43-4 Propagation by hardwood cuttings. (A) shows the cutting wood and where it is cut to remove blossom buds. (B) shows the cutting as it begins to callus and establish top growth. (C) shows well-rooted cutting.

for a year after rooting and sold as two-year-old plants for planting to the field.

VARIETIES

For use in western North Carolina, Maryland, and New Jersey:

Bluetta	Bluecrop	Lateblue	Jersey
Collins	Berkeley	Blueray	Coville

For use in eastern North Carolina:

Berkeley	Croatan	Jersey
Morrow	Murphy	Wolcott

For use in Michigan and New England:

Bluecrop	Blueray	Collins	Coville
Earliblue	Jersey	Lateblue	

For use in western Oregon and Washington:

Berkeley	Bluecrop	Blueray	Collins
Coville	Darrow	Dixi	
Herbert			

Most of these varieties are hybrids which tend to produce large fruit on vigorous plants.

Florida and Georgia are now getting into the early blueberry market with such varieties as Sharpblue and Floridablue, ripening about May 1. Fruit is being shipped to Michigan and other northern markets.

A fall-bearing variety is being developed in Minnesota. If the program produces a good commercial variety, a whole new blueberry market could be opened.

RABBITEYE BLUEBERRIES

The rabbiteye blueberry has a natural range in southern Georgia, southern Alabama, and northern Florida. Rabbiteye blueberries are important mainly because they are able to grow in dry upland areas where the highbush will not, and they can be grown farther south because they require a very short rest period in winter. They are not as sensitive to soil conditions as the highbush, and resist heat and drought better. The rabbiteye grows in areas from eastern North Carolina to central Florida and west to Arkansas and east Texas.

CULTURE

Rabbiteye varieties are planted in midwinter in rows 12 feet apart with plants spaced 6 to 12 feet apart in the row. Small plants are sensitive to fertilizer; none should be applied the first year.

Pruning is not generally done but some pruning on older bushes is desirable. The older stems and smaller young shoots may require removal to prevent the bushes from becoming too thick. The rabbiteye blueberry is vigorous enough that pruning is not needed to produce strong new growth. The fruit ripens later and over a longer period of time than the highbush blueberry.

Plants are propagated by offshoots, or suckers, which grow up from the roots as far as 8 feet from the plant. they are dug up and cut free of the main root and grow in a nursery for one year before planting to the field. Propagation by softwood cuttings is also very successful.

VARIETIES

- ◆ Woodward is the earliest and largest, and has good flavor.

- ◆ Tifblue has the lightest blue color and excellent flavor.

- ◆ Homebell exhibits large fruit, and very vigorous growth.

Many wild rabbiteye blueberries are harvested in the United States.

FIGURE 43-5   A highbush blueberry plant loaded with fruit. (Ed Reiley, Photographer)

HARVESTING (ALL VARIETIES)

For the fresh market, blueberries are generally picked by hand. In eight hours, a picker can harvest 60 to 80 pints. Machine harvesting is used in some areas, employing machines that shake the berries into a net or hopper. Fruit harvested by machine must be sorted and cleaned. Most machine-harvested fruit is used for fruit that is processed before being sold.

Yields of blueberries range from 50 pints per acre on a two-year-old planting to 6,000 pints on a six-year-old planting. Blueberries are becoming a popular pick-your-own fruit. See figure 43-5.

DISEASES

North Carolina suffers the most severe losses of blueberries due to disease. Stem canker causes the most damage. Foliage diseases, stunt virus, stem blight, and root rot are also serious enough to consider a control program.

In New Jersey, the most serious diseases are stunt virus and mummy berry. In Michigan, fusicoccum canker, phomopsis canker, and mummy berry are serious problems. In the Pacific Northwest, mummy berry is difficult to control and is very serious.

> *Tips: A well-drained soil and good sanitation help to control many diseases.*

PEST CONTROL

WEEDS

Weeds may be controlled by mulching, shallow cultivation, or use of weed killers. For weed killer recommendations for your area, see a local horticulturist or extension agent.

INSECTS (SEE SPRAY SCHEDULE)

A spray schedule for the home blueberry grower is provided in figure 43-6. More detailed information for commercial growers is available from the local extension service in USDA Bulletin No. 2254.

DISEASE OR INSECT AND WHEN TO CONTROL	DESCRIPTION	CONTROL	REMARKS
Scale insects (delayed dormant)	Small round spots on stem.	Superior oil applied in delayed dormant stage.	Plants must be completely covered with spray.
Mummy berry (when buds first break)	Stems and blossoms blighted. As berries approach maturity, they become mummified and drop to the ground.	Clean cultivation in the spring after each rain to disturb the mummies. Spray with ferbam 1 lb. and 2 oz. per 100 gal. of water.	Fungus lives over winter in the mummified berries and spreads spores in early spring.
Other fungus diseases (continue use of ferbam every two weeks)	Leaf, fruit, and stem fungus diseases.	Ferbam at 2-week intervals. Prune and burn diseased plant parts.	Sanitation is very important.
Plum curculio (2 weeks after bloom)	Dark brown beetles—1/4 inch long, larvae grayish white, legless, curved body, small brown head.	Malathion 25% WP 6 lbs. per 100 gal. of water 2 weeks after bloom.	Weed in areas where the adult lives during winter months.
Fruit worms (2 weeks after bloom)	Small gray moths lay eggs after bloom has fallen. Larvae are brownish red on top. Some are green on sides and underneath.	Same as plum curculio, or use Sevin as directed.	More than one species of fruit worms may be present.
Leaf hoppers (late May to June 5; repeat in 10 days)	Small, tent-shaped insects.	Malathion 57% EC 1 qt. per 100 gal. of water.	Spread virus. Spray dates given are for Mid-Atlantic states.
Blueberry maggot (June 22; repeat every 10 days until harvest)	Adults look like small houseflies and are 3/16 inch long with a brown face and shiny black body, white on the side and rear of thorax. Maggots hollow berry by eating the flesh.	Malathion 57% EC 1 qt. per 100 gal. of water.	Check label for days before harvest and cease application as necessary.

FIGURE 43-6   Blueberry spray schedule

Student Activities

1) Propagate blueberry plants from ten hardwood cuttings. See Unit 8 for information on propagation of hardwood cuttings.

2) Take a soil test of a possible site for a blueberry planting at home or on the school grounds. Determine if the pH is suitable for growing blueberries. Report to the class on the findings. If suitable soil can be located, prepare the site for a blueberry planting, using the plants propagated in Activity 1.

3) If blueberries are cultivated locally, visit a local grower for tips on growing the crop.

Self-Evaluation

Select the best answer from the choices offered to complete the statement or answer the question.

1) Blueberries prefer a moderately acid soil. The best pH range for cultivation of blueberries is

  a) 6.5 to 7.0.
  b) 6.0 to 6.5.
  c) 3.5 to 4.0.
  d) 4.6 to 5.0.

2) If the soil is too acidic for blueberries, what material should be added to raise the pH?

  a) lime
  b) sulfur
  c) phosphorus
  d) potash

3) Raising the soil pH from 4.6 to 5.0 makes the soil

  a) less acidic.
  b) more acidic.
  c) neutral.
  d) alkaline.

4) The highbush blueberry prefers a soil that is

  a) moist and acidic.
  b) dry and acidic.
  c) moist, acidic, and well drained.
  d) moist, alkaline, and well drained.

5) Fertilizer for blueberries should not be neutralized so that it will

  a) help keep the soil acidic.
  b) be more readily available.
  c) be more economical to use.
  d) be easier to spread.

6) If highbush blueberries need irrigation during the harvest season, what is the least amount of water which should be applied at one time?

  a) 1 acre foot
  b) ½ inch
  c) 3 inches
  d) 1 inch

7) If a soil test indicates a need for phosphorus and potash, the chemicals are applied

  a) by plowing and mixing them into the soil before planting.
  b) in bands beside the plants at planting time.
  c) as a sidedress after planting.
  d) two weeks before picking the first harvest.

8) Ammonium sulfate is a good source of nitrogen for blueberries. Knowing that nitrogen leaches through the soil rather rapidly, the best application method is to

    a) plow the soil and make the application.
    b) place it in bands beside the plants at planting.
    c) spread it on the soil surface around the plants.
    d) place it in the bottom of the hole at planting time.

9) When organic or natural mulches are used around blueberry plants, the rotting of these materials robs the soil of the plant food element

    a) nitrogen.   c) phosphorus.
    b) calcium.   d) potash.

10) Blueberry varieties must be planted near each other to

    a) lengthen the picking season.
    b) insure pollination and fruit set.
    c) help to control the spread of disease.
    d) all of these

11) Blueberries are pruned to

    a) remove some of the fruit buds and promote vigorous new growth.
    b) increase production per acre.
    c) let the sun shine inside the plant.
    d) help to control fungus diseases.

12) Blueberry plants that grow erect are pruned by

    a) pruning off lower branches.
    b) cutting back the tips of branches.
    c) thinning out the center of the plant.
    d) removing only tall growing tops.

13) Heavy pruning of blueberry plants results in

    a) large crops of large berries.
    b) fruit ripening earlier and over a shorter period of time.
    c) too much new growth.
    d) a reduction in plant vigor.

14) Highbush blueberries are propagated by

    a) root suckers.   c) budding.
    b) layering.   d) hardwood cuttings.

15) Rabbiteye blueberries are propagated by

    a) root suckers and softwood cuttings.
    b) hardwood cuttings and layering.
    c) grafting.
    d) budding.

16) Root rot of blueberries, a fungus disease, can largely be prevented by

    a) planting on a well-drained soil.
    b) planting only rot-resistant varieties.
    c) spraying with a fungicide.
    d) shallow cultivation.

17) Since blueberries bloom early, a frost-free site is important. Which site would give the best protection from spring frost?

    a) a high, level area
    b) a high, slightly sloping area
    c) a low, level area
    d) a low area surrounded by hills

The Bramble Fruits

OBJECTIVE

To outline a program for the propagation, cultivation, and maintenance of bramble fruit planting.

Competencies to Be Developed

After studying this unit, you should be able to

- list the five key points to consider when selecting a site for planting of bramble fruits.
- outline a soil preparation plan for a bramble fruit planting.
- demonstrate the planting procedure for bramble fruits, including proper spacing of plants.
- outline a fertilizer and pruning program for bramble fruits.
- recognize insect and disease pests which infest brambles in the local area.
- list the seven measures for disease prevention in bramble fruit plantings.

MATERIALS LIST

One of each of the following bramble fruits for identification purposes:

✔ red raspberry
✔ black raspberry
✔ erect blackberry
✔ trailing blackberry

The bramble fruits are some of the most delicious and useful of the small fruits, figure 44-1. Although this unit concerns the blackberry, red raspberry, and black raspberry, most of the information also

FIGURE 44-1 A blackberry branch with a typical fruit cluster ready for picking (Courtesy Stark Brothers Nurseries and Orchards, Co.)

applies to the boysenberry, dewberry, loganberry, and tayberry, which are other brambles. Any difference in pruning or other practices concerning brambles will be noted.

LOCATION OF THE PLANTING

Good air circulation is important for bramble plantings. Because of this, a location that is higher than the surrounding area or with a slight slope should be chosen. A site that is above the level of the surrounding ground is not as subject to spring frosts that could kill the blossoms or cause winter injury of plants. However, brambles are not as subject to spring frost damage as strawberries and blueberries.

Do not plant in an area close to wild raspberry and blackberry plants. These wild plants may be diseases and will spread disease to the cultivated crop. Wild plants may be destroyed or removed from the area prior to planting.

Red raspberries should be planted at least 1,000 feet from black raspberries. Reds carry a virus disease which can be passed on to the black raspberries, causing severe damage to the blacks.

Brambles should not be planted in a soil that has been planted with potatoes, tomatoes, peppers, eggplant, or tobacco in the last four or five years. These plants may leave a fungus wilt disease in the soil that affects the planting of bramble fruits. The wilt organism can live only four or five years in the soil away from its host plant (a plant on or in which the disease lives).

SOIL FOR PLANTING SITE

The soil should have good drainage capacity and be high in organic matter. Subsoil drainage is more important than soil type. Organic matter increases the moisture-holding capacity of the soil. This promotes good growth, but does not interfere with good drainage. The subsoil must be well drained with no hard layer that will stop roots from penetrating the soil or slow water movement through the soil. Plants that cannot establish a deep root system cannot produce a crop during dry weather. The brambles grow in a wide variety of soils, from sandy loams to clay loams. Red raspberry roots are finer and more sensitive to high soil temperatures.

SOIL PREPARATION

If possible, begin preparing the soil for blackberries or raspberries one or two years before planting. Plow under green manure crops or animal manure to build up the organic matter level of the soil. The year the plants are to be set, plow the soil to a depth of 7 to 9 inches. Condition the soil by harrowing to break up all large clods and to level the soil surface.

A soil test should be taken prior to plowing. If lime or phosphorus and potash fertilizer are needed, they are applied and either plowed under or harrowed into the surface. Do not apply lime and fertilizer in the same application, however. It is probably best to plow under the fertilizer and harrow the lime into the surface after plowing. Lime may not be needed since bramble fruit grows well in a pH as low as 6.0 to 6.5.

THE PLANTING PROCESS

Always start with plants that are certified to be disease free and true to variety. Do not allow the plant roots to dry out during the planting process. Start with wet roots and keep the roots covered with wet burlap or plastic until planted. Do not place plants in the row so far ahead of the planters that they dry.

Establishing rows with a layoff plow at the proper distance apart makes the planting process easier. Place the plant in the row and cover with soil, firming around the roots. The plants are set at the same depth in the ground at which they

FIGURE 44-2 Black raspberry tip plant being set in the ground. A spade is being used to make the hole in which the plant is placed. Note that the plant is held by the cane from the parent plant. (Ed Reiley, Photographer)

PLANT	DISTANCE BETWEEN ROWS*	DISTANCES BETWEEN PLANTS IN THE ROW
Black raspberry	7 to 10 feet	3 to 4 feet
Blackberry and Tayberry	7 to 10 feet	5 feet
Boysenberry (trailing types)	7 to 10 feet	8 feet
Purple raspberry	7 to 10 feet	3 to 4 feet
Red raspberry	7 to 10 feet	2 to 3 feet
Yellow raspberry	7 to 10 feet	2 to 3 feet

*Distance between rows varies with the method of cultivation that will be used. A small garden tractor works well at 7 feet apart; a large tractor requires 10 feet.

FIGURE 44-3 Planting distances for bramble fruits

grew before. Red raspberries are planted 1 to 2 inches deeper than they grew in the nursery. Do not cover new sprouts that may have started to grow. Instead, pull soil up around them later, after the sprouts have had a chance to grow more. A spade may be used to plant the brambles, figure 44-2. To plant, push the spade in the ground, pull it forward, place the plant in the opened slit spreading and keeping the roots straight, pull the spade out, step on the soil to force the slit closed, and firm the soil around the roots.

The 6-inch piece of cane on the new plant that is used as a handle is cut level with the ground and removed from the field. This is a disease control measure. Tissue culture plants are disease free.

PLANTING SYSTEMS

Most of the bramble fruits are best grown in rows, either freestanding, on wire trellises, or staked to hold them up. Figure 44-3 gives recommended planting distances between rows of brambles.

The most effective trellis system for erect-growing brambles is one having two parallel wires, one on each side of the post. After harvesting fruit the wires can be lowered to the ground until pruning of old canes is complete. The wires are then lifted to 24 to 30 inches high and fastened to each side of posts which are spaced 15 to 20 feet apart in the row. The new plant canes are placed inside

the two wires. Wires may be wired together at intervals to prevent outward movement.

Trailing plants such as the boysenberry should have a two-wire trellis with the first wire 30 inches from the ground and the second wire 5 feet from the ground. The canes are tied to the wire with string. Some growers prune erect black-berries and raspberries back to an initial heading of 24 inches. Most canes of this height will stand without support. This practice is used commercially because of the cost of trellis construction. Some canes fall and fruit is lost. If a trellis is not used, plants must be pruned shorter so that the canes are able to stand.

If trellis rows are 400 to 500 feet long, it is best to break the trellis system to provide a cross-lane through the center. Figure 44-4 illustrates a trellis system for erect blackberries or black

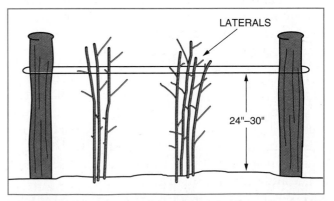

FIGURE 44-4 Two-wire trellis for black raspberry and erect-growing blackberry. The posts are set 15 to 20 feet apart.

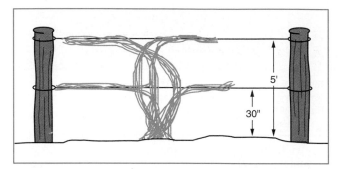

FIGURE 44-5 Two-wire trellis system for trailing blackberries and boysenberries

raspberries. Figure 44-5 shows a two-wire trellis system for trailing blackberries or boysenberries.

PRUNING AND TRAINING

The top growth of brambles lives for two years. New canes are produced each year from the crown or roots. Fruit is formed on wood of the previous season; these canes die soon after the fruit is harvested. The roots of plants are perennial and continue to live for many years.

PRUNING AND TRAINING FREESTANDING PLANTS

PRUNING BLACKBERRIES DURING GROWING SEASON
The first-season plants are pruned to a height of about 15 inches. After the first year, the height is increased to 24 inches since the plants are more vigorous and are able to produce larger, stronger canes that stand better. Not all canes reach the correct size at the same time, so the planting should be checked once a week and all canes 24 inches or longer headed back. Heading back causes the plants to form many side branches or laterals, which increases fruit production. Prune when no rain is expected for forty-eight hours as a disease prevention practice.

PRUNING BLACKBERRIES IN EARLY SPRING
Before buds swell in the spring, remove all canes that are ½ inch or less in diameter by cutting off at ground level. At least two canes must be left, but most plants should have four or five canes. Shorten all of the laterals to 18 to 24 inches in length. Fruit from pruned laterals is larger and of better quality.

PRUNING BLACKBERRIES AFTER HARVEST
Immediately after the crop is harvested, prune all old canes that fruited by cutting off at ground level. This promotes new cane growth from buds at the base of the plant and helps to control diseases. At this time, continue the growing cycle by cutting back all new canes to the 24-inch height as soon as this height is reached. Purple raspberries are pruned the same way except that the new canes are headed back to 30 to 36 inches instead of 24 inches. Commercial growers in Texas leave old canes for four or five years, then mow off all tops and remove them from the field. This is a labor-saving practice.

If plants are trained to grow on a trellis, the heading back of new canes should be kept about 12 inches higher than recommended for non-supported plants. Other pruning practices are the same.

New genetically thornless blackberries are now available. These varieties are easier to handle and pick fruit from and are nearly as productive as the thorny types, for example, Black Satin and Smoothstem are hardy to 5°F, Thornfree 0°F., Hull Thornless −10°F, and Chester −12°F. Check with your local agricultural research center for varieties best suited to your area.

PRUNING RED RASPBERRIES DURING GROWING SEASON
Red raspberries may be tied to a trellis or left freestanding. The canes are thinned to stand no closer than 4 to 6 inches apart, leaving about seven or eight strong canes per hill.

PRUNING RED RASPBERRIES AFTER HARVEST
All old canes are removed as with black raspberries. New canes are thinned and all root suckers growing into the rows or between rows are removed.

The new Heritage and Cherokee fall-bearing red raspberries now make it profitable to grow red raspberries as annuals to produce a fall crop. Rows 20 to 24 inches wide are mowed to a height of 6 inches each spring. All canes are removed. New canes come up the same spring and produce a fall crop.

PRUNING UPRIGHT-GROWING BLACKBERRIES DURING GROWING SEASON
All suckers that sprout between the rows should be pulled as soon as possible to prevent the planting from becoming too thick.

PRUNING UPRIGHT-GROWING BLACKBERRIES AFTER HARVEST
All old canes that fruited are cut off at ground level and removed immediately after harvest. Commercial growers may want to leave old canes for four or

five years and mow off everything for removal at that time. This saves labor costs. These canes should be burned or otherwise destroyed to help to control diseases. Thin any new canes that surface, leaving four of the most vigorous canes per plant.

Blackberry canes with a great deal of fruit often fall due to lack of support. A wire trellis support may be needed. Tie canes to the wire with soft string or place canes between wires in a two-wire trellis.

PRUNING SEMITRAILING BLACKBERRIES AND BOYSENBERRIES DURING GROWING SEASON   Trailing blackberries are neither headed back nor summer pruned in any manner except to remove the old canes after fruiting. The new canes are allowed to grow along the ground.

PRUNING SEMITRAILING BLACKBERRIES AND BOYSENBERRIES IN EARLY SPRING   Laterals are shortened on all canes to 10 inches and the long trailing vines are wound into a cable and tied to a two-wire trellis with soft string.

PRUNING SEMITRAILING BLACKBERRIES AND BOYSENBERRIES AFTER HARVEST   Remove the old canes that have fruited immediately after harvest and destroy them.

CULTIVATION

Cultivate the bramble plants often enough to control weeds. Cultivation should be shallow—no more than 2 inches deep. Stop cultivation during harvest time to prevent fruit loss. Brambles should not be cultivated in the fall since this stimulates late growth and plants do not harden off properly for winter.

Weeds may also be controlled with mulches and weed killers. Contact the local extension service for chemical weed control recommendations.

FERTILIZING

It is necessary to apply lime, phosphorus, and potash before planting, as mentioned earlier. After planting, sidedress a nitrogen fertilizer 6 inches from the plant at a rate of 40 pounds of actual nitrogen per acre.

Each year thereafter, apply a 5-10-5 fertilizer as a sidedress just before blossoming time at a rate of about 600 pounds per acre, or about one-half cupful around each plant. Keep fertilizer at least 6 inches away from the crown of the plant. After harvesting, apply ammonium nitrate (a nitrogen fertilizer) at a rate of 80 to 100 pounds per acre.

HARVESTING

Berries are picked when they are fully ripe but still firm. To keep ahead of the fruit so that none gets too ripe, raspberries should be picked every two or three days. Hot weather ripens fruit more quickly and could cause picking dates to change. Blackberries should be picked every three or four days.

Only the firm berries are picked; all injured or soft berries are thrown away. Do not hold handfuls of berries during picking. This practice causes bruising of fruit which results in early spoilage and a product that cannot be marketed.

Place the fruit in the shade or some other cool spot as soon as it is picked. Heating in the container causes rapid spoilage. Berries picked in the morning when they are cool remain fresh longer than berries picked after the sun has heated the fruit.

Many growers are operating pick-your-own operations. This is one of the most economical systems for harvesting fruit if organized properly. Hand picking for the fresh market is very expensive, making up 50 percent of the total cost to produce the berries. Processed berries are often picked by machines. This process calls for special pruning and good disease control.

Average yields of fruit are about 6,000 pounds of blackberries per acre. Raspberries yield about 2,000 quarts per acre, figure 44-6.

FIGURE 44-6   A quart box of red raspberries (Courtesy Dr. Harry J. Swartz)

PROPAGATION

Black and purple raspberries are propagated by tip layering. Red raspberries are propagated from root suckers and root cuttings. The rooted suckers are dug up and separated from the parent plant and planted in early spring.

Root cuttings from red raspberries are taken in early spring. Pieces of root 2 or 3 inches in length are cut from the parent plant and covered with 2 inches of soil and placed in the nursery. The roots send up shoots which are set in the field the following spring. When old plants are dug up, roots left in the ground send up many shoots which form new plants. This is an easy way to propagate red raspberries.

Upright blackberries are propagated from root suckers or root cuttings, as red raspberries are. Trailing blackberries are propagated by tip layering, just as black raspberries are. Tissue culture allows rapid propagation of virus-free plants of all the bramble fruits.

VARIETIES

There are many varieties of the bramble fruits and new varieties are constantly being developed. Some varieties grow best in colder climates, others survive better in southern regions of the United States. Recommended varieties for the local area may be obtained from the local agricultural extension service.

INSECTS AND DISEASES

Insects are not usually as destructive to the bramble fruits as diseases are. If control measures are used, attack by disease organisms can be kept to a minimum without the use of an extensive spray schedule. Some spraying is necessary, however, depending on local conditions and the needs of that particular year.

The following are recommended disease prevention measures.

1. Start with the proper site.
   a. Do not plant black or purple raspberries in fields that were planted with tomatoes, potatoes, peppers, or eggplant.
   b. Keep plantings of black and red raspberries separated by at least 1,000 feet.
   c. Destroy all wild raspberry plants in or near the planting site one year before planting.
2. Choose disease-resistant varieties.
3. Buy and plant only disease-free plants. Tissue culture plants are safer for freedom from diseases.
4. Remove all diseased plants and burn them.
5. Remove old canes after harvest.
6. Keep the planting free of weeds and fallen leaves.
7. Use pesticides when necessary. Obtain a spray schedule from the local extension service or the USDA Farmers' Bulletin 2208, *Controlling Diseases of Raspberries and Blackberries.*

 # Student Activities

1) Visit a small fruit grower and discuss all aspects of the operation with him or her.
2) Propagate and plant at least one bramble fruit. Include proper pruning techniques in the program.

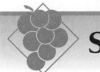

Self-Evaluation

Select the best answer from the choices offered to complete each statement.

1) Wild raspberry and blackberry plants are removed from areas planted with commercial plantings of brambles because

    a) the plants cross-pollinate, resulting in small fruits.
    b) disease is spread from the wild plants to the new planting.
    c) new plants selected from the planting may become mixed with the wild planting.
    d) all of the above

2) A fungus wilt disease is generally present in soils that have been planted with potatoes, tomatoes, peppers, eggplant, or tobacco. This disease also attacks bramble fruits. It is necessary to wait _____ years after any of these disease-carrying plants are grown on a soil before planting brambles.

    a) three or four   c) five to ten
    b) four or five   d) two or three

3) Subsoil type is more important than soil type for bramble fruits because

    a) brambles grow in a wide variety of soils.
    b) an open subsoil allows deep rooting.
    c) soil drainage through the subsoil is important.
    d) all of the above

4) If it is found that potash or phosphorus fertilizer is needed as a result of a soil test, it is

    a) applied as a band application on each side of the plants.
    b) plowed under or harrowed into the soil before planting.
    c) spread in a circle around the plants.
    d) applied after picking the crop.

5) When planting black raspberry plants, they are set

    a) at least 2 inches deeper than the depth at which they grew in the nursery.
    b) at least 1 inch above the top of the soil level.
    c) at the same depth at which they grew in the nursery.
    d) at about 3 inches deeper than the depth at which they grew in the nursery.

6) The planting depth of red raspberry plants is

    a) 1 or 2 inches deeper than the depth at which they grew in the nursery.
    b) at the same depth at which they grew in the nursery.
    c) 3 inches deeper than the depth at which they grew in the nursery.
    d) at least 2 inches deeper than the depth at which they grew in the nursery.

7) The 6-inch piece of plant used as a handle to hold the new plant during planting is cut off black raspberry and blackberry plants after planting and removed from the field to

    a) help control disease.
    b) make the planting more attractive.
    c) prevent the cultivator from hooking the new plant.
    d) all of the above

8) Organic matter is important in the soil in which brambles are planted because it

    a) stimulates earthworm activity.
    b) is cheaper than fertilizer in promoting plant growth.
    c) helps in weed control.
    d) helps the soil to hold moisture.

9) If a trellis is not used to support black raspberry plants, the plants must be cut back or topped

    a) more often.
    b) to a shorter total length.
    c) at a higher total height.
    d) twice a year instead of once a year.

10) When purchasing bramble fruit plants, be sure that the plants are certified

    a) to be proper planting size.
    b) free of insects.
    c) to grow.
    d) free of disease and true to variety.

11) Pruning during the growing season to cut back the canes on raspberry plants results in

    a) the formation of more side branches.
    b) the production of more fruit.
    c) canes that can support themselves better.
    d) all of the above

12) Pruning of black raspberries and blackberries in early spring consists of

    a) shortening all laterals and removing small, weak canes.
    b) topping or cutting back canes.
    c) removing the old canes that fruited the previous summer.
    d) thinning the number of canes.

13) Immediately after harvest, pruning of all bramble fruits consists primarily of

    a) thinning out the number of canes.
    b) heading back or shortening canes.
    c) removing the old canes that just finished fruiting.
    d) shortening laterals to promote the following year's fruiting.

14) During the growing season, red raspberry canes are thinned to leave about

    a) four strong canes per hill.
    b) seven or eight strong canes per hill.
    c) five or six strong canes per hill.
    d) as many strong canes over ½ inch in diameter as possible.

15) Pruning of trailing blackberries during the growing season consists of

    a) shortening the canes to promote lateral growth.
    b) cutting back lateral branches.
    c) no pruning at all.
    d) thinning out weak canes.

16) Fall cultivation of brambles should not be done because

    a) it is not necessary for weed control.
    b) many small roots are destroyed and do not have the time to grow back.
    c) weed seeds are spread over a wide area.
    d) it stimulates late growth and plants do not harden off for winter.

17) Raspberries should be picked every

    a) two or three days.
    b) one or two days.
    c) two days.
    d) any of the above, depending on how hot the weather is.

18) Black raspberries, purple raspberries, and trailing blackberries are propagated by

    a) root cuttings.   c) shoots.
    b) tip layering.   d) suckers.

19) Red raspberries and upright blackberries are propagated by

    a) tip layering.   c) root suckers.
    b) stem cuttings.   d) stem layering.

20) Black raspberries are planted at least 1,000 feet from red raspberries because

    a) red raspberries carry a disease that is harmful to black raspberries.
    b) cross-pollination results in a poor red raspberry crop.
    c) black raspberries carry a disease that is harmful to red raspberries.
    d) none of the above

Grapes

OBJECTIVE

To outline a program for the establishment and cultivation of a vineyard.

Competencies to Be Developed

After studying this unit, you should be able to

- ◆ select a planting site for a vineyard and give reasons for its selection.
- ◆ prepare the soil and plant grapes in the selected planting site.
- ◆ describe a fertilizer program for a vineyard.
- ◆ list the characteristics of ripe grapes.
- ◆ list the procedures for weed control in vineyards.

MATERIALS LIST

- ✔ pruners
- ✔ canes from a grapevine for demonstration of proper pruning techniques
- ✔ If a hands-on activity involving the planting and care of a grape planting is possible: all planting tools and materials for the construction of a trellis.

Grapes are rapidly becoming a popular homegrown fruit. Grapes have many uses and are very nutritious. They are consumed fresh, as juices and wines, as raisins, jam and jelly, and as frozen products. Grapes are native to the United States. Many of the domesticated varieties have native wild grape ancestry in at least one parent. American fine wine grapes are native to central Asia. Grapes are grown commercially in most areas of the United States, except where the frost-free season is less than 150 days, in arid sections, and in areas having extremely high temperatures and very humid conditions. Some newer cultivars can be grown in central Florida and other warm areas.

CHOOSING A PLANTING SITE

A level site or a steep slope that is higher than the surrounding area is important to reduce danger of frost damage, especially to fruit in the fall. The soil should have good drainage capacity and be high in organic matter.

A location to the south or east of a large body of water is favorable since water has a modifying effect on temperature. It protects against fall frost, delays early spring blooming, and thus holds plants until the danger of spring frost is passed and keeps wood and buds from losing hardiness in spring.

SOIL

Grapes grow well on most soil types, providing the subsoil allows deep rooting. *Hardpan* (compacted clayey soil), rock, or wet subsoil close to the surface are not suitable. A deep, fertile, well-drained loamy soil with a moderate amount of organic matter is best. A pH range from 5.5 to 7.0 is satisfactory.

Soils that are overfertilized, rich, or too high in organic matter produce a late-maturing crop with a low sugar content. Sandy or light soils tend to produce small crops of early-maturing fruit that is high in sugar content. A soil that falls somewhere between these two examples is best.

SOIL PREPARATION

Grapes should not be planted directly into soil that has been in sod for years. Such land should be planted with a cultivated crop such as corn for at least one year before planting grapes. The soil is plowed deeply and disc harrowed until it is in good planting condition.

After a soil test is done, any phosphorus or potash fertilizer needed should be plowed or disced into the soil before planting.

PLANTING

In southern regions of the United States, grapevines are planted in the fall as soon as they be-

come dormant. North of Arkansas, Tennessee, and Virginia, vines are planted in early spring.

The ideal grape plant at planting time is one year old. The vines are planted at the same depth at which they grew in the nursery. Do not allow roots to dry out during the planting producature. Immediately after planting, prune to a single stem with two or three buds remaining. The plants are spaced 8 to 10 feet apart in the row, with the rows about 10 feet apart. The plants are placed directly under the trellis wire.

Grapes respond best to shallow cultivation (not over 2 inches deep). The vines should be cultivated soon after planting to kill germinating weeds. Cultivation is repeated as needed for weed control. The grape hoe, consisting of a manually operated blade mounted on a tractor, is an excellent tool for keeping the rows clear of weeds.

TRELLIS CONSTRUCTION

Grapes are trained to grow on a trellis so that the vines are given proper support. Generally, a two-wire trellis is used. Wooden, concrete, or steel posts are placed about 20 feet apart. The end posts must be securely planted at least 3 feet deep to remain firm. Line posts can be set 2 feet deep. Earth anchors set outside end posts are strongly recommended.

Number 9 steel wire is used for the top wire and smaller number 12 steel wire for the bottom wire. The bottom wire is 30 inches from ground level and the top wire 30 inches from the bottom wire, figure 45-1. The wires are fastened to the posts so that they slide through the staples or clips for easy tightening. The wires are tightened each spring before the vines are tied.

New planting and trellis systems are being evaluated as researchers look for higher production

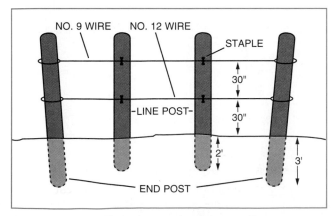

FIGURE 45-1 Typical two-wire grape trellis

per acre and trellis systems which make it possible to prune mechanically.

TRAINING AND PRUNING

Grapevines are trained to fit the trellis construction described above. Although there are several systems of training, the *four-arm kniffin system* will be explained here. This system is used for training bunch grapes and works very well.

After the first growing season, the most vigorous cane is selected and tied to the top wire. It is then cut off above the top wire and all other canes are removed, figure 45-2.

When pruning during the dormant season of the second or third year, select four vigorous canes for arms. The most fruitful canes are darkest in color, mid-size, and on the outside where most light was received. Prune each of these canes to eight or ten buds in length. Select four other canes located as close to the arms as possible and cut each back to two buds each. These short canes are called *renewal spurs*, and the buds on them will develop new canes to serve as the new arms for the vine the following year. All other canes are removed. Each year, this process of pruning is repeated. New arms are formed from spurs each year for the following year's crop. The old arms are removed during the pruning process. Fruit is always produced on new growth from the canes that are one year old.

The vines are pruned each year late in winter or early spring before growth begins. Most of the canes are removed each year, with the canes left as arms and renewal spurs greatly shortened, figure 45-3. This may seem to be very severe pruning, but it is essential for continued high yields and quality fruit.

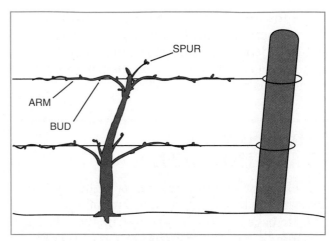

FIGURE 45-3   Pruning and training the second and third year

The more vigorous a vine is, the more fruit it is able to produce. The number of buds left to produce fruit on the four arms as described earlier is from thirty-two to forty buds total (eight to ten on each arm). This is considered average. Weak vines have slightly fewer buds and strong vines more. A good rule of thumb is to leave twenty to thirty buds and then add ten buds for each pound of one-year-old wood pruned from the vine.

After pruning, the canes are tied to the trellis wire with string. The end of the arm is tied lightly just behind the last bud. That bud is rubbed off so that it does not grow and break the string as the cane grows. Other ties are made as needed to hold the arm to the wire. These are tied loosely so that the cane has room to grow and expand in diameter, figure 45-4.

The four-arm kniffin system is only one training system. There are many others used. Mechani-

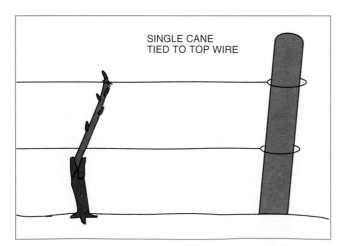

FIGURE 45-2   Pruning and training at the end of the first growing season

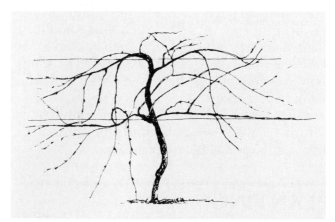

FIGURE 45-4   Grapevine trained to the four-arm kniffin system and tied to a two-wire trellis.

FIGURE 45-5A   Vines trained to the Geneva Double Curtain system shown before pruning. Wooden spacers extend out on both sides of posts to hold the two wires about 3 feet (1 meter) apart. Every other plant in the row is trained to the wire on the right side of the row and alternate plants to the wire on the left side. This gives a double curtain plant foliage. (Ed Reiley, Photographer)

FIGURE 45-5B   Vines pruned to shorten canes using the Geneva double curtain system (Ed Reiley, Photographer)

cal harvesting requires different pruning and trellis construction. Plants must be cordon trained for mechanical harvesting. Canes are cut in horizontal and vertical planes, leaving many short canes and spurs instead of the long arms explained using the four-arm kniffin system.

A new system called the Geneva Double Curtain system allows more light to the plants and gives higher production since it has two fruiting canopies per row.

Figure 45-5A and B shows grape vines trained to the Geneva Double Curtain system, where two wires tied to arms spaced about 3 feet (1 meter) apart are used to spread the grape canopy. Many short fruiting canes are left on main branches of the vine tied to these wires. Plants are spaced 8 feet apart with this system and alternate plants in the row are tied to the same wire (one plant to the wire on the right, the next plant to the wire on the left of the row, etc.). Contact your local agricultural college for the latest on grape pruning for your area.

FERTILIZER

Fertilize grapes in early spring with 500 pounds of 10-10-10 fertilizer. Apply according to soil test recommendations. The fertilizer should be broadcast because grape roots tend to spread over a wide area.

HARVESTING

Grapes are harvested when they are ripe. This stage is often difficult to detect since some grapes change color and appear to be ripe a month before they are actually ready to pick. The best test for table grapes is to taste them for sweetness or check the color of the seeds. Seed color changes from green to brown as the fruit ripens.

In grapes used in juices and wines, the percentage of soluble solids or sugar content is important. A refractometer or Balling hydrometer is used to determine ripeness. Proper acidity also insures best wine flavor and storagability.

It is not always possible to harvest fruit when it is fully ripe. Some grapes crack or split open as ripening progresses and must be picked before they are fully ripe to prevent loss of fruit. Rotting in rainy weather or danger of early frost may also make early picking necessary.

Table grapes for fresh consumption are generally picked by hand while juice and wine grapes are picked by machine. Figure 45-6 shows a grapevine with an abundant grape crop.

WEED CONTROL

Weed control is generally accomplished through shallow cultivation and the use of the grape hoe.

FIGURE 45-6 Grapevine with abundant crop of fruit

cultural experiment station. Since the grape canopy is above the weed canopy, weed killers can be used more easily than with other small fruits.

INSECTS AND DISEASES

Grapes are sprayed for both insects and fungus diseases. At least three spray applications are needed for good control. Commercial growers spray up to eight times during the season. Consult your local county agricultural agent for the USDA bulletin on spraying grapes.

Major diseases which require control are black rot, powdery mildew, and downy mildew. Major insect pests are the flea beetle, leaf hopper, berry moth, and Japanese beetle.

PROPAGATION

Grapes are propagated commercially by hardwood cuttings. See the section of the text on propagation by hardwood cuttings for a description of this technique. If a disease-resistant rootstock is needed for certain varieties, plants are grafted or budded to a disease-resistant seedling rootstock.

In one hoeing, the soil is pulled away from the wires toward the center of the rows. About two weeks later, the hoe is turned the other way and the soil is pulled back toward the vines and under the trellis. The grape hoe removes most of the weeds and leaves only a few for hand hoeing. Chemical weed killers are also sometimes used. Check local recommendations of the state agri-

Student Activities

1) Complete the following activity as a class project, using a section of school grounds or a local vineyard.

   ◆ Plant and prune one-year-old grapevines.

   ◆ Fertilize a producing grapevine.

   ◆ Prune the grapevine and tie it to the trellis.

   ◆ Discuss the site selected.

   ◆ Discuss weed control for the planting and carry out, if necessary.

   ◆ Observe the application of pesticides.

   ◆ Determine when the fruit is ripe and demonstrate proper picking techniques.

2) Ask a local grower to come and speak to the class about the culture of grapes.

Self-Evaluation

Select the best answer from the choices offered to complete each statement.

1) Grapes grow throughout most of the United States except in those areas where the frost-free season is less than

   a) 200 days.   c) 100 days.
   b) 150 days.   d) 120 days.

2) A planting site that is _____ the surrounding area helps to reduce frost danger to grapes.

   a) lower than   c) higher than
   b) level with   d) protected from

3) Grapes grow well on most types of soil providing that the subsoil allows

   a) deep rooting and good drainage.
   b) nutrients to leach through.
   c) hardpan to hold nutrients and water.
   d) deep cultivation.

4) In areas north of Arkansas, Tennessee, and Virginia, grapes are planted

   a) in the fall.   c) in the early spring.
   b) in August.   d) none of these

5) The proper depth at which to plant one-year-old grape plants is

   a) 2 inches deeper than they grew in the nursery.
   b) 1 inch deeper than they grew in the nursery.
   c) 3 or 4 inches deeper than they grew in the nursery.
   d) the same depth at which they grew in the nursery.

6) The two-wire grape trellis has support wires spaced

   a) 30 inches apart.   c) 48 inches apart.
   b) 36 inches apart.   d) 24 inches apart.

7) In the four-cane system of training grapes, each of the four arms has _____ buds left on it after pruning.

   a) five or six   c) eight to ten
   b) ten to fifteen   d) twenty to thirty

8) Short two-bud spurs are left on the grapevine after pruning. The purpose of these spurs is to

   a) produce extra fruit close to the main stem.
   b) produce canes for new arms the following year.
   c) provide arms in case the selected arms die.
   d) provide foliage to protect the main stem from sunscald.

9) A very vigorous grapevine should be pruned to leave _____ canes and _____ buds than are left on a less vigorous vine.

   a) longer, more
   b) shorter, fewer
   c) the same number of, the same number of
   d) none of the above, it depends on the variety

10) A complete fertilizer should be broadcast around grapevines

    a) after the fruit is picked.
    b) in early fall.
    c) after the fruit has set and has started to develop.
    d) in early spring.

11) Grapevines are pruned

    a) just after fruiting to remove old canes.
    b) just after new fruit has set.
    c) in early fall.
    d) in late winter.

12) The best test for ripeness in table grapes is

    a) color of the fruit.
    b) sweetness and change in seed color.
    c) softness of the fruit.
    d) ease of separation of the grapes from the vines.

13) Grapes are generally propagated commercially by

    a) hardwood cuttings.   c) budding.
    b) grafting.   d) layering.

14) Grapevines are pruned to remove most of the canes each year because

    a) fruit is produced on current year's wood.
    b) heavy pruning is needed to produce quality fruit.
    c) heavy pruning is needed to produce high fruit yields.
    d) all of the above

15) Weed control in vineyards is primarily through use of

    a) shallow cultivation and the grape hoe.
    b) weed killers.
    c) hand pulling.
    d) a disc harrow.

Holiday Crafts and Floral Designs

Wreaths and Door Swags

OBJECTIVE

To construct various seasonal decorative pieces, including the evergreen wreath, Della Robbia wreath, door swag, and kissing ball.

Competencies to Be Developed

After studying this unit, you should be able to

- identify ten different plant materials used in holiday decorations.
- construct an evergreen wreath, Della Robbia wreath, door swag, and kissing ball.
- list five different fruit and nut pods which can be used in the construction of a Della Robbia wreath.

MATERIALS LIST

- ✔ evergreen trees and shrubs used in wreath construction (white pine, Scotch pine, English boxwood, Douglas fir, noble fir, balsam fir, Norway spruce, blue Colorado spruce, eastern red cedar, holly, rhododendron, and yew)
- ✔ seed pods of plants, pine cones, spruce cones, oak acorns, walnuts, shellbarks, pecans, filberts
- ✔ box wreath wire frame measuring 12, 14, 16, or 18 inches
- ✔ styrofoam ball, 4" in diameter
- ✔ 16-gauge and 22-gauge florist wire
- ✔ straw and sphagnum moss
- ✔ Hillman wreath frame, 10"-36" in diameter

Holiday decorations, materials for their construction, and holiday plants play an important part in the wintertime business of nurseries, greenhouses, and garden centers. Some of the more popular holiday items are wreaths, door swags, and table centerpieces. These decorations usually consist of a base, some type of moisture-holding material, greenery, and decorative accessories which contrast with the plant material.

EVERGREEN BOX WREATHS

To construct a foundation for an evergreen foliage wreath, a box wreath wire frame is used, figure 46-1. After an appropriate frame has been chosen, it is filled with straw or sphagnum moss. This material is dampened to help to hold it in place and keep the evergreen foliage fresh. After the straw or moss is packed in the wire frame, the frame is wrapped with a special plastic wreath wrap material. See figure 46-2. The wrap holds the straw or moss in place, keeps moisture in, and supports the greenery when it is punched into the frame.

The next step in the construction of a wreath is to select the foliage. The selection of foliage depends upon personal preference, but the final design should be considered. Boxwood has been used in the construction of the wreath shown in this unit, but any type of greenery can be used in basically the same way.

Sprigs of the greenery in lengths of 4 to 6 inches are then punched individually into the wreath. This operation moves in one direction

FIGURE 46-2A The box wreath wire frame is filled with sphagnum moss and then wrapped tightly with plastic wreath wrap.

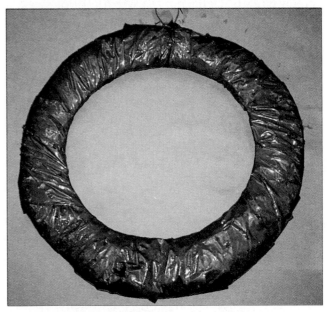

FIGURE 46-2B A completed wreath form packed with sphagnum moss, with wire hanger.

and is continued around the frame until the entire base is covered, figure 46-3. The material should flow smoothly and evenly for the best finished product. In other words, there should be no open spaces or sprigs which extend farther than the others. The completed wreath is shown in figure 46-4.

HILLMAN WREATH FRAME SYSTEM

The Hillman wreath frame system consists of a 9-gauge wire wreath frame and a Hillman wreath-

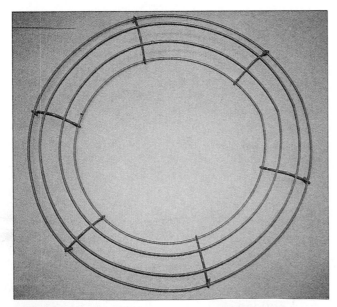

FIGURE 46-1 A box wreath frame. Frames are available from most florist supply companies in sizes of 10 to 36 inches in diameter.

FIGURE 46-3   Sprigs of greenery (in this case, boxwood) are pushed into the wrapped sphagnum moss base until it is completely covered.

FIGURE 46-4   A completed boxwood wreath

making machine, figure 46-5. The wreath frame is available from most florist supply companies in sizes from 10 to 36 inches in diameter. Unlike the box wreath frame, the Hillman wreath frame requires no filler material (straw or sphagnum moss). The greenery is held in place by a system of crossbars that are clamped together using the Hillman wreath-making machine.

Evergreens such as white pine, blue Colorado spruce, Douglas fir, noble fir, balsam fir, or Scotch pine are generally chosen to construct the Hillman wreath. The greenery is pruned to 6- to 8-inch lengths and is gathered into bunches of four or five sprigs. For fullness, only those sprigs with terminal as well as lateral bud growth are used. Each bunch of sprigs is then clamped firmly into the wreath frame with the Hillman machine. This process is continued until the entire wreath is completed. After all of the greenery has been attached, accessories such as bows, cones, or ornaments may be added.

(A) The Hillman wreath frame system.

(B) A student using a Hillman wreath-making machine clamping into place the correct size bundles of greenery to give the wreath an even flow.

FIGURE 46-5   Hillman wreath making

(C) Placing the finishing touches to the Hillman wreath.

(D) A completed wreath with a bow for a special accent.

FIGURE 46-5 Hillman wreath making (continued)

DOOR SWAGS

Door swags are a popular decoration for entranceways to homes during the holiday season. Their construction is less complicated than that of the wreath. To make a door swag, several evergreen branches are simply grouped in an attractive arrangement that can be fastened to a door. The evergreen branches should be 24 to 36 inches long. If more than one branch is used, they may be tied together at the top with 16-gauge florist wire. As this is wired to the greenery, a small loop is left at the end to serve as a hanger.

Pines cones, lotus pods, or fruit may be fastened to the branches as accent items, figure 46-6. These, also, are attached with florist wire. The most dramatic highlight for a swag is a large bow attached to the base where the hanger is fastened, figure 46-7.

KISSING BALL

Mistletoe, very popular in holiday decorating, is combined with other greenery to create the kissing ball. A kissing ball is made by covering a ball of sphagnum moss with a piece of chicken wire measuring 8 inches square. The wire is rolled into a ball, with the moss held in the center. The final size of the ball should be about the size of a softball. This serves as the foundation or base for the

greenery. Other materials may be used, such as a styrofoam ball or a large potato (tuber).

A piece of 16-gauge florist wire is pushed through the center of the ball to serve as a hanger. The ball can then be hung at about eye level or slightly lower. This makes insertion of the greenery much easier. The greenery is cut into pieces about 4 to 5 inches long and inserted into the ball. For best results, start at the side of the ball and work around the bottom in a circular pattern. The

FIGURE 46-6 Door swags are constructed of several boughs of greenery.

FIGURE 46-7 Bows, cones, or other ornaments are added for interest. The bow covers the cut ends of the boughs.

FIGURE 46-8 To line the inside and the outside of the wreath frame, the pine cones are pushed between the wires of the frame and secured with florist wire.

sprigs of greenery must be even in length for the best appearance. A small piece of mistletoe is then fastened to the bottom of the ball with an attractive piece of ribbon or a bow.

DELLA ROBBIA WREATH

A Della Robbia wreath is a wreath consisting of cones, pods, nuts, and fruits. It is constructed on a box wire wreath frame or a styrofoam ring. A 10-inch ring makes an attractive candle-ring display, while the 16- to 18-inch rings form interesting door decorations.

The frame is placed face down on a table. The cones to be used for the wreath are then chosen and soaked in water for thirty minutes. This will cause the cones to close, making it easier to insert them into the frame. The cones will open again after they dry out. The cones which are used

for the foundation of the wreath must be equal in size. The cones are inserted into the frame by pushing them in between the wires. Each cone is then wired in place with 22-gauge florist wire, figure 46-8. The wire is attached to the cone and then woven through the frame and secured on the inside or back of the frame, concealing the wire from view. This process is continued until the outside and inside edges of the frame are lined with cones.

The center of the frame, which has been left open to this point, can be filled with cones placed upside down. This provides a contrast to the other two rows of cones. If a center row is formed with cones, clusters of seeds and nuts can be added to accent the wreath. The entire middle row can be filled with seed pods and nuts if desired, figure 46-9.

LOCATING CONES, SEED PODS, NUTS, AND FRUITS

The best source of cones, seed pods, nuts, and fruit is a local florist supply house. However, field trips to the woods can often supply different types of seeds and nuts which are native to a particular area. For example, sweet gum balls, beechnuts, horse chestnuts, acorns, and shellbarks can be collected from trees in some areas. Rose hips, hollyhock seeds, and bamboo fruit might be found in your own backyard. A walk in the woods or a vacant lot might produce teasel, lotus, or milkweed pods.

FIGURE 46-9 A completed pine cone wreath as this one can be used for several years.

FASTENING SEED PODS AND NUTS

To fasten seed pods or nuts to a wreath, they must first be drilled. The drilling must be done carefully since nuts often have a tendency to slip out of the driller's hand. Once the nuts or seeds have been drilled, they are fastened to the wreath with 22-gauge florist wire. Or, they may be glued into place with a hot glue gun. Observe caution when handling the glue gun.

The completed wreath can be sprayed with a clear lacquer for a shiny appearance. Clear lacquer is available in aerosol cans at most hardwood stores. If a more natural or dull finish is desired, no lacquer is needed.

Student Activities

1) Take a field trip to local woods or fields to collect supplies for wreaths. Collect seed pods from trees and shrubs in the area. Cut branches of Scotch pine, white pine, and Douglas fir. (Note the distinct fragrance of white pine and Douglas fir.)

2) Visit a local florist supply house and examine materials available for holiday decorations.

3) Ask a local florist to speak with the class about special preparations for the holiday season and construction of holiday decorations.

4) Visit a library and examine magazines and pamphlets containing ideas for holiday decorations.

5) Construct and sell evergreen and Della Robbia wreaths as a special fund-raising project for the class or school.

6) Attend a holiday craft fair to gain ideas to be used in the class.

Self-Evaluation

Provide a brief explanation for each of the following.

1) What are five types of greenery that can be used in wreath construction?

2) How is the box wreath wire frame which holds the greenery prepared?

3) What lengths of greenery are used in the construction of a wreath, and how are they attached to the frame?

4) What are the two main differences between constructing an evergreen wreath and a door swag?

5) What materials are used to construct the foundation of a kissing ball? What is the finished size of the foundation?

6) What is the function of the moss or straw used in the wreath and kissing ball?

7) How long are the pieces of greenery used in the construction of the kissing ball?

8) What materials are used in a Della Robbia wreath?

9) How are cones and nuts attached to the frame of a Della Robbia wreath?

10) What are five different cones, fruit pods, and nuts that can be used in Della Robbia wreaths?

Creating Holiday Centerpieces

OBJECTIVE

To construct and care for holiday centerpieces.

Competencies to Be Developed

After studying this unit, you should be able to

- ◆ construct a holiday centerpiece.
- ◆ list ten plant materials which may be used for centerpieces.
- ◆ describe the procedures for keeping a centerpiece fresh.

MATERIALS LIST

- ✔ pieces of evergreen trees and shrubs
- ✔ floral foam (Oasis)
- ✔ one candle
- ✔ 18-gauge florist wire
- ✔ florist bowl

CONSTRUCTION OF CENTERPIECES

In many homes centerpieces are an important part of the holiday season as they are used to decorate and create interest throughout the home. They are often found on dining room tables, dry sinks, and hutches. Candles, which have special meaning for many people, are usually part of the centerpiece design.

A variety of evergreens are suitable for use in centerpiece construction. Spruce, white pine, Scotch pine, Douglas fir, noble fir, balsam fir, juniper, yew, holly, and boxwood are some of the greens which may be chosen.

PROCEDURE

CREATING A PINE AND HOLLY CENTERPIECE

1. *Preparing the base.* Select a half block of oasis, a spongelike material, to be used as a base and to hold moisture for the greenery, figure 47-1. Soak it in a pan of water until there are no longer any air bubbles coming out of the oasis. This insures that it is thoroughly saturated.

2. *Inserting the pine.* Place the block on the design table. Take small clippings of white pine 4 to 5 inches long and push them in around the base of the oasis. Be certain that a complete line of greenery is formed and that the pieces of pine are even in length.

> *Caution: Never pull the greenery out of the oasis after it is in place; this causes air pockets which will prevent the greenery from absorbing the available moisture.*

Start the next line of greenery about ½ inch above the first line. This allows the design to flow smoothly. Continue until the line is completed.

Add additional rows until the greenery reaches the top of the oasis block, figure 47-2. Care must be taken not to insert the last row so that it is too close to the top of

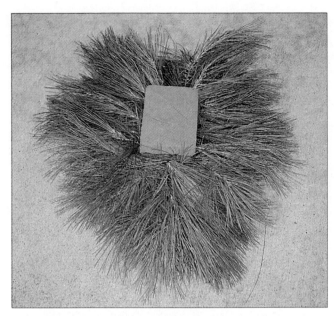

FIGURE 47-2   Pieces of pine 4 to 5 inches long are inserted into the sides of the oasis block.

the block, since oasis easily crumbles and breaks apart.

3. *Positioning the candle and holly.* Place a candle in the center of the block. To be sure the candle stays in place, push a toothpick about ¼ inch into the center of its bottom. This acts as a balance and holds the candle firmly in the oasis. Insert sprigs of variegated English holly 4 to 5 inches long into the top portion of the oasis. This particular type of holly has spines on the leaves and is very prickly, so this process must be done carefully. Add enough holly to cover the top of the block and make the centerpiece appear complete, figure 47-3.

PROCEDURE

CREATING A MINIATURE CHRISTMAS TREE CENTERPIECE

1. *Preparing the base.* Select a full block of oasis and soak it in water until there are no longer any air bubbles coming out of the oasis.

2. *Shaping the base.* Firmly set the full block of oasis vertically into a standard florist bowl, figure 47-4. Using a piece of 18-gauge florist wire, cut the oasis to form a tapered cone by removing the corners as shown in figure

FIGURE 47-1   To begin construction of the centerpiece, select half a block of oasis. Soak the oasis in water until no air bubbles form. Set the soaked oasis into a florist bowl container.

FIGURE 47-3 Placement of a candle and variegated foliage completes the centerpiece to create a special accent for the arrangement.

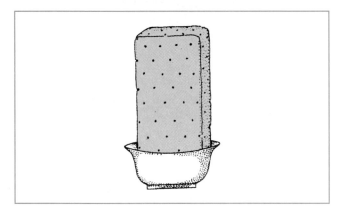

FIGURE 47-4 Set the oasis firmly in the florist bowl.

47-5. Hold the florist wire firm while making these cuts. When finished, the block should appear as shown in figure 47-6.

3. *Preparing the greenery.* Select your greenery from any of a number of narrowleaf or broadleaf evergreens. Prune sprigs to about 4 to 6 inches long.

4. *Inserting the greenery.* Beginning at the base of the oasis cone, insert 4- to 6-inch sprigs evenly around the base, covering the container. As you move up the cone, you will have to keep pruning the sprigs of greenery so that the shape of your "mini-tree" tapers with its intended design.

5. Your finished tree should look like the one in figure 47-7. Fill in with sprigs anywhere necessary, and decorate as you desire.

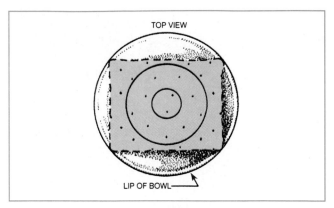

FIGURE 47-5 Top view of oasis

FIGURE 47-6 The cone-shaped oasis

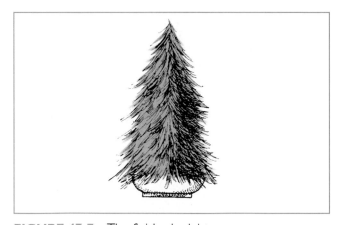

FIGURE 47-7 The finished mini-tree

KEEPING THE ARRANGEMENT FRESH

To keep the centerpiece fresh during the entire holiday season, it must be watered. To water it, submerge the centerpiece in a tub of water deep enough to cover the entire arrangement. Allow it to stand in the water until there are no longer any air bubbles coming out of the oasis. It is necessary to perform this procedure several times during the holidays to keep the greenery looking fresh.

Student Activities

1) Take a field trip in a local area to collect greenery for the construction of centerpieces.
2) Design and make a centerpiece.
3) Set up a demonstration on how to water centerpieces after their construction.

Self-Evaluation

Provide a complete answer for each of the following.

1) List five types of evergreens that can be used for centerpiece construction.
2) How is oasis used in the construction of centerpieces, and how is it prepared before it is used?
3) How long are the evergreen pieces that are used in the construction of centerpieces?
4) Why should evergreen pieces never be pulled out of the block after they are in place?
5) Describe the way in which pieces of greenery are inserted in the base.
6) What material is used to fill in the top of the centerpiece, and how long are the pieces that are used?
7) Explain how to water a centerpiece.

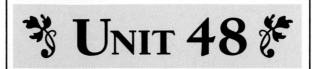

Creating Bows for Floral Designs

OBJECTIVE

To create attractive bows for decorative arrangements.

Competencies to Be Developed

After studying this unit, you should be able to

- ◆ select ribbon of appropriate size and color for a specific arrangement.
- ◆ construct a bow to be used on a wreath or door swag.

MATERIALS LIST

- ✔ 22-gauge florist wire
- ✔ wire cutters
- ✔ florist scissors
- ✔ one bolt #40 red velvet ribbon
- ✔ one bolt #9 red velvet ribbon
- ✔ one bolt #5 red velvet ribbon
- ✔ one bolt #3 red velvet ribbon
- ✔ one bolt #1 red velvet ribbon

The bow is an important accessory to many floral projects; it is used to lend a graceful touch to the design. Color, material, size, and quantity must be considered when selecting the ribbon used to make a bow.

Bow color will depend upon the arrangement, fashion trends, and individual tastes. Color should be a pleasing accent to the overall design. An almost infinite variety of ribbon colors are available. Patterns and styles, like color, will vary with fashion trends and individual tastes.

Many different materials are used to make ribbon today. Ribbon varies in texture, pattern, style, weight, and ease of workability. Because of these factors, certain designers have distinct preferences in the types of ribbons they use. Texture, for example, ranges from fine to coarse, such as a satin ribbon compared to a heavy burlap ribbon. Ease of workability varies with the texture and weight of the ribbon. Satin is regarded as the easiest ribbon with which to work. In addition, today's designer can choose from velvet, plastic, cotton, flocked, and burlap ribbon. Each has a different application in the designer's world.

Ribbon size follows an industry standard. The basic sizes used in this unit are Number 9 (1⅜ inches wide), and Number 40 (2¾ inches wide). These sizes are shown in figure 48-1. Ribbon is either single-faced or double-faced. The single-faced has a shiny side and a dull side. The double-faced is the same texture on both sides. Ribbon is available at local florist or craft shops. It is sold by the yard or by the roll, also called a bolt.

Many arrangements, in addition to those used during holiday celebrations, require decorative bows. There are two basic types of decorative bows to use for holiday arrangements. The first type is a pinch bow that is made by loops of ribbon in a circular fashion. The other, a decorator's bow, is in the shape of a bow tie. A well-made bow adds an attractive finishing touch and can be made easily with practice.

DECORATOR'S BOW

The bow demonstrated here is known as a *decorator's bow.* It is made of velvet ribbon which is 2¾ inches wide (Number 40 ribbon).

The materials needed for the sample bow are Number 40 velvet ribbon and 22-gauge florist wire. From the velvet ribbon, cut one piece 6 inches long (for center of bow), one piece 18 inches, one piece 20 inches, and one piece 30 inches long (for loops). Use three pieces of florist wire, each 18 inches long.

PROCEDURE

MAKING A DECORATOR'S BOW

1. Make a circle with the 6-inch piece of ribbon, allowing the ends to overlap about an inch, figure 48-2. Pinch the ribbon together at the center where the two ends overlap, figure 48-3.
2. Feed a piece of wire through the circle, pulling it tight around the crushed, overlapped side of the loop, figure 48-4.
3. Secure the wire by turning the ribbon two or three times while holding the ends of the wire, figure 48-5. The loop formed acts as the center of the bow and covers all exposed wires when the bow is completed.

FIGURE 48-2

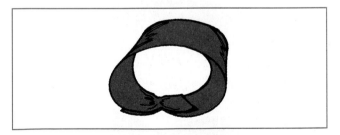

FIGURE 48-3

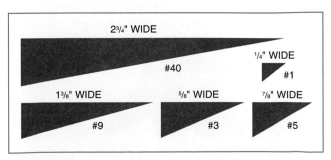

FIGURE 48-1   Ribbon sizes

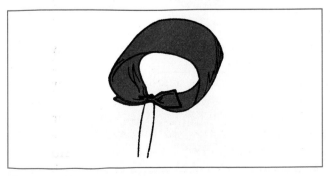

FIGURE 48-4

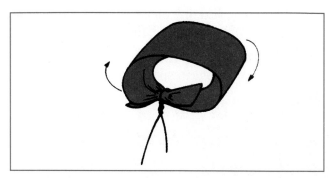

FIGURE 48-5

4. Make the bottom loop of the bow using the 20-inch piece of velvet. Overlap the ends as shown for the center loop. By pinching the middle of the loop together, form a figure eight with the ribbon, figure 48-6. Take a piece of florist wire and wire the loop in the center, holding the bow tight. Twist the wire by turning the ribbon two or three times, figure 48-7.

5. Form the top loop of the bow with the 18-inch piece of ribbon. Overlap the ends as before, and then form a figure eight as in Step 4. Crush the loop in the center and wire it with the last piece of wire. Twist the ribbon two or three times to secure the wire.

6. Take the two larger loops and place the smaller of the two on top of the larger one, figure 48-8.

FIGURE 48-6

FIGURE 48-7

FIGURE 48-8

7. Use the 30-inch piece of ribbon to form the tails of the bow. Fold the ribbon in half to determine the middle. Crush the ribbon together at this point so that it can be easily wired, figure 48-9. Place the tails on top of the two loops from Step 6.

8. With the wire attached to the center loop formed in Steps 2 and 3, wire the entire bow together. Pull the center tight to form the completed bow, figure 48-10.

9. The tails of the bow should be two different lengths to create interest. The ends of the ribbon may be cut on a diagonal or in a fishtail design for an interesting variation, figure 48-11.

FIGURE 48-9

FIGURE 48-10

FIGURE 48-11

PINCH BOW

To make the pinch bow use about 3 yards of Number 9 florist ribbon. Refer to figure 48-12 while following the procedure.

PROCEDURE

MAKING A PINCH BOW

Note: Take a piece of 22-gauge florist wire and tape the center 3 inches of the wire with florist tape. Folding the wire in half, work the folded wire between the thumb, index finger, and third finger, making an inverted "U" shape. Put the wire aside.

1. Hold the ribbon between the thumb and index finger with the shiny side of the ribbon facing you.

2. Take one end of the ribbon and make a loop about 3 inches long, shiny side up.

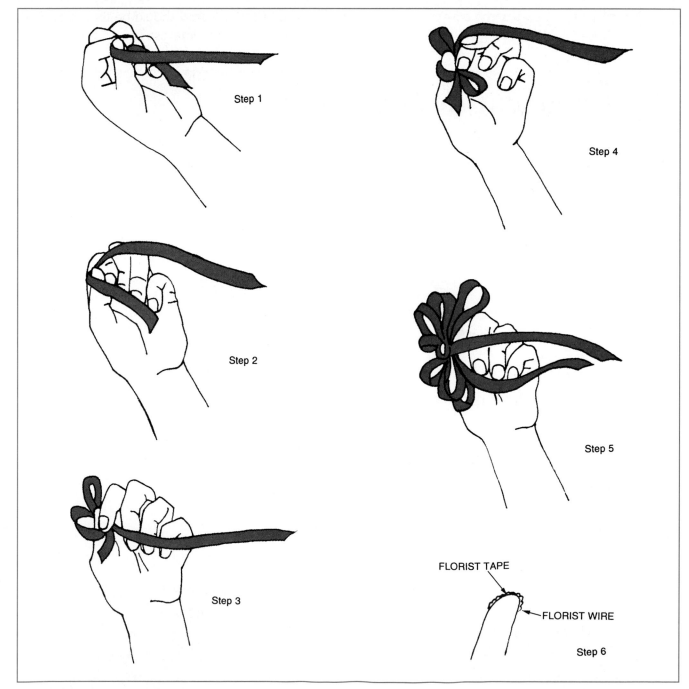

FIGURE 48-12   Pinch bow

3. Pull ribbon through and make a loop on the opposite side the same size as the first loop.

4. As the second loop is made, pinch the bow in the center. Continue the same procedure until there are three loops on one side and two loops on the opposite side. As the ribbon is turned to make the third loop, pull the ribbon up into the center of the bow to make the bow appear fuller. Repeat previous steps until the entire amount of the ribbon is used.

5. Now, hold the bow firmly between the index finger and the thumb. The florist wire that you made ready earlier is used to fasten the bow together. This wire is 18 inches long and coated green to blend better with arrangements.

6. With your free hand, pick up the piece of previously folded wire. Pull the wire up around the pinched center of the bow. Take the opposite end of the florist wire and run it through the loop in the wire, pulling it tightly to form a completed bow.

Student Activities

1) Visit a local florist and ask to observe their procedure for making bows.

2) Construct decorative bows for use on holiday decorations. Consider selling the bows as a fund-raising project.

3) Participate in the National FFA Floriculture Contest.

Self-Evaluation

Select the best answer from the choices offered to complete each statement.

1) A Number 40 florist ribbon is

    a) 1 inch wide.
    b) 1½ inches wide.
    c) 2¾ inches wide.
    d) 3 inches wide.

2) Ribbon that is 1⅜ inches has a florist number of

    a) 40.
    b) 3.
    c) 9.
    d) 60.

3) The best gauge florist wire to use to make a bow is

    a) 18 gauge.
    b) 22 gauge.
    c) 20 gauge.
    d) 14 gauge.

4) What part of the decorator's bow does the 6-inch piece of ribbon form?

    a) tail of the bow
    b) covers the center of the bow
    c) top loop of the bow
    d) none of these

5) To make a pinch bow _____ of Number 9 ribbon is needed.

a) 1 yard

b) 2 yards

c) 3 yards

d) 4 yards

6) Single-faced ribbon

a) has a shiny side and a dull side.

b) is dull on both sides.

c) is shiny on both sides.

d) none of these

7) Double-faced ribbon has a _____ side both top and bottom.

a) dull

b) shiny

c) red

d) none of these

Floral Designs

OBJECTIVE

To create a floral design.

Competencies to Be Developed

After studying this unit, you should be able to

- ◆ identify the types of flowers used in floral designing.
- ◆ list and describe the nine basic principles of floral design.
- ◆ identify the six basic floral design shapes.
- ◆ design a circular floral arrangement.

MATERIALS LIST

- ✔ fresh floral material (carnations, baby's breath, baker's fern)
- ✔ floral container
- ✔ floral foam (Oasis)
- ✔ florist wire and wire cutters
- ✔ sharp florist knife
- ✔ florist shears (see figure 49-1)

Basic floral design is an art that has played an important role throughout history and is a part of many functions in our lives. Floral design is used at weddings, funerals, and other formal and informal functions, figure 49-2. Flowers help to adorn the function where they are being used.

Today, the florist industry has broadened opportunities for those people who are interested in a career in the florist business. The industry has expanded to malls, supermarkets, airports, and hotels in order to meet the ever-growing needs of individuals who recognize the value of flowers in their everyday lives. The florist trade

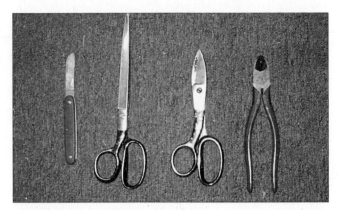

FIGURE 49-1 Florist tools

association and wholesalers of floral products offer employment opportunities. It is important that people seeking employment in these fields have a basic skill and knowledge of the florist industry.

BASIC FLORAL DESIGN

In basic floral design, one must be familiar with the factors that affect the design. When designing an arrangement it is necessary to use the basic principles of design which serve as the guidelines for floral design. These principles are balance, focal point, proportion, scale, accent, repetition, rhythm, harmony, and unity.

DESIGN PRINCIPLES

Balance is achieved in a floral design when flower size and container flow together and complement each other. The arrangement has visual weight on each side of the axis to make it stand alone. If the arrangement is symmetrical the floral design could be divided in half and have equal visual weight on each side of the arrangement. This type of design is used to create a more formal arrangement. In asymmetrical design the arrangement is offset, creating a visual weight of being heavier to one side of the axis. This design allows the individual to use a more natural effect or informal approach to designing.

The **focal point** is important in some designs because it creates the accent and interest. This will catch the eye of the viewer. Focal points are created by the use of unusual flowers, by grouping flowers of one color, or by using a larger flower. The focal point must complement the entire arrangement.

Proportion in an arrangement is important to keep relative size, color, and texture of the flowers to create a pleasing arrangement. The general rule of thumb for floral designs is that they be at

(A)

(B)

(C)

FIGURE 49-2 Some floral designs. (A) Cascading wedding bouquet. (B) Asymmetrical right angle informal design. (C) Symmetrical design.

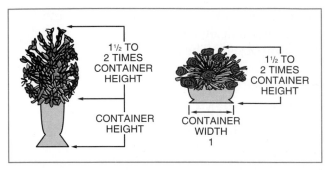

FIGURE 49-3 Height of the floral arrangement. The height of the arrangement is 1½ to 2 times the height or width of the container.

least one-and-one-half to two times the height or width of the container, figure 49-3. To create the proper proportion in the arrangement, the points farthest from the focal point should have the flower buds. In designing toward the focal point, flowers in full bloom are used.

Scale is a principle that must be taken into consideration. The size of the flowers, where the arrangement is to be used, placement of the arrangement, and overall size of the completed design are all factors of scale. The design must be in proportion to the surrounding area.

Accent is used in a design to draw attention to the design. Many designers use the focal point to create this emphasis. Accent is that point of interest that catches one's attention and enhances the design.

Repetition is used to accentuate the flowers and colors used in a design. Repeating the use of material or color throughout the design can be very effective.

Rhythm creates a sense of continuity among the shapes, colors, and textures. It creates a feeling in the viewer that everything flows together in the arrangement, thus giving a feeling of motion.

Harmony is created in a floral design when all the parts flow together to give a total, completed look to the arrangement.

Unity is created when all of the arrangement flows together and is not segmented. For a look of unity the flowers in an arrangement must complement each other through color, size, and shape.

DESIGN MATERIALS

In floral design there are four basic types of design materials: line, mass, form, and filler flowers.

Line flowers are ones such as snapdragons, delphinium, and gladiolus. In most cases these flowers are used to establish the outer framework of an arrangement.

Massing flowers are those which have the flower head on the terminal end of the stem. These flowers are used in the arrangement to create all of the design principles and to accent the focal point. Some examples are roses, carnations, chrysanthemums, and asters.

Form flowers are those which have a unique shape or form. They may be used to create the accent or the focal point. Bird of paradise, potea, and orchid are a few.

Filler flowers are used to tie the arrangement together. They help to fill any empty spaces in the design. They are also used to cover the mechanics of the arrangement. Some examples are baker's fern, asparagus fern, baby's breath, statice, heather, huckleberry, and palm.

BASIC FLORAL DESIGN SHAPES

When designing floral arrangements it is necessary to determine what shape arrangement is desired. The person designing the arrangement must be able to visualize what this arrangement will look like when it is completed.

Basic floral design shapes include round (circular), triangle, horizontal, crescent, hogarth (S) curve, and right angle, figure 49-4.

DESIGNING A FLORAL ARRANGEMENT

A good, basic design to start with is the circular design that has a rounded form. It has many applications and can be used in various settings such as a centerpiece, coffee-table arrangement, or in a foyer in front of a mirror where it is viewed from all sides.

In the construction of the circular design, select massing flowers to establish the outline and form the circular shape of the design. A circular design does not have a focal point because it is viewed from all sides. To have the correct proportion of the design use the general rule of thumb of one-and-one-half to two times the height or width of the container. The stem length

OVAL

INVERTED-T

VERTICAL

ASYMMETRICAL TRIANGLE

HORIZONTAL

HOGARTH CURVE

RIGHT ANGLE

ROUND

CRESCENT

DIAGONAL

EQUILATERAL TRIANGLE

FAN

FIGURE 49-4   Basic floral design shapes

of the massing flowers will be determined by this rule. Arrange the massing flowers to establish the basic outline of the design. With the basic outline established, arrange the other massing flowers so they are within the outline to give a rounded appearance.

Filler flowers have several purposes: to tie the arrangement together, to fill in any voids in the arrangement, and to provide the greenery to cover the floral foam and container. Filler flowers and greenery provide more fullness and give the arrangement a completed appearance.

PROCEDURE

DESIGNING A CIRCULAR ARRANGEMENT

1. Select a round container that will hold a half-block of floral foam. Other materials needed are an 18-inch piece of florist wire, shears, a block of floral foam, and a sharp knife.

> *Caution: Sharp florist knife—use safely.*

The knife is needed to cut the stems of the floral material and the shears are needed to cut wire. The florist wire is used to cut the dry floral foam. Floral foam will hold the flowers in place and supply moisture to the floral material. See figure 49-5.

2. In this design, a half-block of floral foam will be used. Set a block of dry floral foam firmly on the container. Take the 18-inch florist wire, hold it firmly, and slice down through the block of floral foam, as shown in figure 49-6.

3. Place the floral foam into the container and push down firmly to hold it in place. The cleats in the bottom of the container will hold it securely, figure 49-7. Set the container with the floral foam into a bucket of water for twenty minutes to allow the foam to absorb the water. It must be completely soaked through the entire block to help keep the floral arrangement fresh.

4. Select baker's fern (leatherleaf) for the greenery material. Arrange the material to

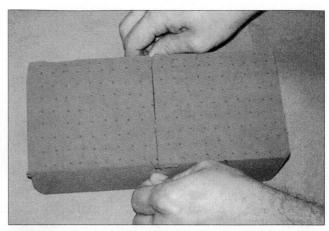

FIGURE 49-6   Cutting dry floral foam with florist wire to fit the container.

FIGURE 49-7   This container has cleats in the bottom to hold the floral foam firmly in place. The floral foam is placed in the container dry, and then soaked with water.

cover the entire block of floral foam, having it extend one-and-one-half times the width of the container, figure 49-8.

5. Using the rule of thumb of one-and-one-half to two times the width of the container, first arrange the carnation to establish the outline of the arrangement. This will be the height of the arrangement, figure 49-9.

6. Establish the circular outline at the base of the arrangement and maintain the same distance for this carnation as for the first one in Step 5. See figure 49-10.

7. Turn the design so the third carnation may be placed in the arrangement, figure 49-11. Remember to maintain one-and-one-half to two times the width of the container.

FIGURE 49-5   The mechanics of a floral arrangement include the following materials: container, floral foam, shears.

FIGURE 49-8   Baker's fern used as a greenery material extends 1½ times the width of the container to form a foundation.

FIGURE 49-9   Establishing the height of the arrangement with the first carnation added to the arrangement.

FIGURE 49-10   Establishing the width of the floral design at the base of the arrangement.

FIGURE 49-11   Placing the carnations at the base establishes the circular design. Continue the circular pattern, maintaining space between each flower, to finish the arrangement.

8. Arrange the next carnation to establish the floral outline. With these points set, the basic foundation of the circular arrangement is organized.

9. Arrange two carnations in between the top and base of the design. Be sure to maintain the outline of the arrangement.

10. Add the remaining carnations to the arrangement until the circular shape of the arrangement is filled. It is important to keep space in between each carnation. This allows their natural beauty to radiate. The greenery should extend out around each carnation.

11. To complete the design, filler flowers (baby's breath) are arranged in between the carnations. This will give a completed arrangement, figure 49-12.

FIGURE 49-12 Baby's breath is added to tie the arrangement together.

Student Activities

1) Visit a local florist and observe the designing of various arrangements.

2) Attend a flower and garden trade show to observe competitive events of floral designers.

3) Have a local florist come to class as a guest speaker.

4) Participate in the FFA Floriculture Contest.

5) Put on floral demonstrations at local fairs, malls, shopping centers, and garden club meetings.

6) Make a bulletin board showing the basic floral design shapes.

7) Read florist trade association magazines to keep up-to-date on current trends in the florist industry.

8) Design floral arrangements for fund-raising projects.

9) Attend local floral design schools for a career demonstration at the college level.

10) Visit museums and mansions and observe the floral designs used there.

Self-Evaluation

Select the best answer from the choices offered to complete each statement.

1) A floral design must use the basic principles of design which include

   a) balance and harmony.   c) scale and focal point.

   b) unity and rhythm.   d) all of these

2) What type of balance can be used to create and design with an informal approach?

    a) symmetrical   c) asymmetrical
    b) circular   d) isometrical

3) The basic principle of design that creates interest and accent in a floral design is

    a) balance.   c) scale.
    b) harmony.   d) focal point.

4) An accent is used in a design to

    a) draw attention to the design.   c) emphasize a point of interest.
    b) create a focal point.   d) all of the above

5) To create the proper scale in a floral design, flowers at the focal point should be at

    a) bud stage.   c) full bloom.
    b) tight-bud stage.   d) none of these

6) The design of a floral arrangement should be _____ times the height or width of the container.

    a) 2 to 3   c) 1 to 2
    b) 3 to 4   d) 1½ to 2

7) The basic types of floral material include

    a) line.   c) form.
    b) mass.   d) all of these

8) Which one of the following flowers is an example of line flowers?

    a) roses   c) snapdragon
    b) baby's breath   d) aster

9) Filler flowers include

    a) delphinium.   c) statice.
    b) roses.   d) tulips.

10) The basic floral design shapes include

    a) circular and triangle.   c) hogarth.
    b) horizontal and crescent.   d) all of these

11) A circular design does not have

    a) balance.   c) harmony.
    b) focal point.   d) none of these

Corsages and Boutonnieres

OBJECTIVE

To create a corsage and a boutonniere.

Competencies to Be Developed

After studying this unit, you should be able to

- ◆ list the various floral materials used to make corsages and boutonnieres.
- ◆ select the correct wiring procedure for the different flower types.
- ◆ tape the flowers correctly.
- ◆ demonstrate the four wiring procedures used with the appropriate flowers.
- ◆ select the correct size ribbon for the construction of a corsage.
- ◆ feather a carnation flower.

MATERIALS LIST

✔ floral shears	✔ ribbon
✔ floral tape	✔ corsage pins
✔ floral wire	✔ boutonniere pins
✔ floral knife	✔ cut flowers

Corsages and boutonnieres have been a special part of the florist business for years. They are popular for many occasions, such as school proms, dances, holiday parties, and weddings.

Many special techniques that require more dexterity than other forms of floral design are used in the construction of corsages and boutonnieres. With practice, a designer can create unique arrangements that will be charming and beautiful.

CORSAGES

Corsages should be created with the wearer in mind; they must be complementary to the individual. The size of a corsage is a major consideration. For example, a bird of paradise worn by a flower girl in a wedding would be totally out of scale and proportion to her size. Sweetheart roses or miniature carnations would be better choices. Therefore, the designer must use good judgment when selecting the flowers for a corsage.

The corsage should be coordinated with the individual's apparel. In many situations the wearer will have a consultation with a designer regarding the type and color of flowers and ribbon to use, and the position at which the corsage will be worn. For example, a wristlet corsage could be used when a pin-on shoulder corsage would not be appropriate because the wearer's dress is strapless. When constructing a corsage, the designer should keep in mind that standard practice is that flowers be displayed in their natural growing position.

CONSTRUCTION OF A CORSAGE

In order to make an attractive corsage, the designer must first select a high-quality flower. Next, the designer must wire the flowers. Wiring is necessary because the wire allows the designer to place the florets in the correct position in the design, while natural stems make the corsage too bulky.

When wiring corsages, always use florist wire. This wire is available in a standard length of 18 inches and a wide range of gauges (thicknesses). The lower the gauge number, the thicker the wire will be. A range of 22- to 32-gauge wire is good for corsage construction.

The florist wire is used for several reasons:

◆ to hold the flowers in the correct position in the design

◆ to make arranging the flowers in the corsage easier

◆ to prevent the flower heads from breaking off the natural stems

◆ because it is not as bulky to use as natural stems

◆ to make adding accessories such as a bow, puff, or flocked leaves easier.

METHODS OF WIRING FLOWERS

There are four methods of wiring flowers to make a corsage or boutonniere: the hook, piercing, clutchwire, and hairpin methods.

PROCEDURE

HOOK METHOD

This method is used for composite (daisy-like) florets. Typical flowers wired this way are asters, black-eyed Susans, daisy chrysanthemums, and any flower with a short tough calyx, figure 50-1.

1. Cut the stem to within 1/2 inch of the calyx. Strip any leaves off the stem.

2. Insert the end of a 6-inch piece of 22-gauge wire in the stem and up through the flower.

3. Bend the end of the wire above the flower into a hood 1/4 inch long.

4. Slowly and gently pull the wire back through the flower until it catches in the top of the calyx and is not visible in the flower center.

5. Tape the flower, starting high on the calyx and continuing down the stem and wire, using a smooth, twisting motion. (The procedure for taping flowers is given in detail later in this unit.)

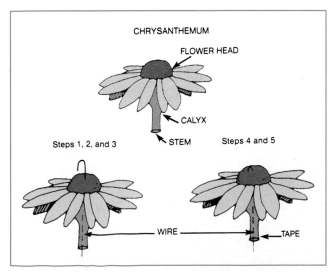

FIGURE 50-1   The hook method of wiring flowers

PROCEDURE

PIERCING METHOD

This method is used for flowers with a large calyx, such as roses and carnations, figure 50-2.

1. Remove the stem 1/2 inch below the calyx and pierce the calyx halfway down, pulling the wire through to one-half the length of the wire.

2. Bend the wire down on both sides. Do not wrap the wire around the stem. Wrapping will make the stem too bulky.

3. Tape the flower, starting high on the calyx, pulling and taping just under the petals. Continue taping both the flower and the wire until all of the wire is covered. If the flower is not evenly taped, retape to get a smooth, tight wrap around the flower and the wire.

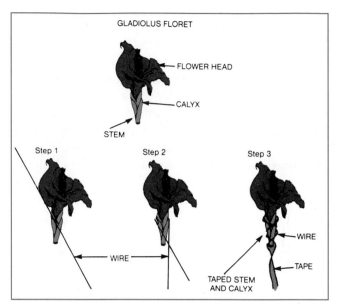

FIGURE 50-3 The clutchwire method of wiring flowers

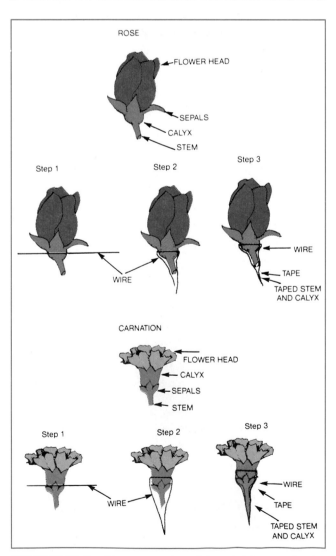

FIGURE 50-2 Two examples of the piercing method of wiring flowers

PROCEDURE

CLUTCHWIRE METHOD

This method is used on gladiolus florets, camellias, and garden, football (standard), and spider mums, figure 50-3.

1. Take 8 inches of wire and fold it in half. Place the fold around the calyx and/or stem of the flower.

2. Wrap one-half of the wire around the flower's stem or calyx, and around the other half of the wire several times, until all the wire is used.

3. Tape the flower, starting at the calyx and working down until all the wire is covered.

PROCEDURE

HAIRPIN METHOD

This method is used with flowers that have a tubular floret, such as the stephanotis, figure 50-4.

1. Take 8 inches of florist wire and bend it in half. Place a small wad of moist cotton in the bend to support the flower.

2. Insert the wire down through the floret, keeping the cotton wad in place.

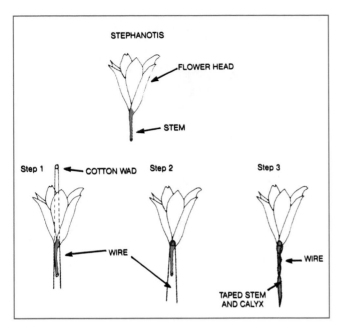

FIGURE 50-4   The hairpin method of wiring flowers

> **3.** Gently pull the florist wire down until the cotton wad is firmly fixed in the flower trumpet. Twist the florist wire around the calyx. Tape the floret into place.

TAPING THE FLOWER

Florist tape is also used in the construction of corsages and boutonnieres. It is used to cover all exposed florist wire, stems, and calyxes in a corsage. Florist tape is available in ¼-inch, ½-inch, and 1-inch widths. The most commonly used width is ½ inch. Florist tape also comes in a variety of colors. Although green is the most common color, tape also comes in orchid, brown, black, blue, white, red, yellow, purple, and orange.

The tape is applied to the stem by holding the wired flower between the thumb and forefinger of one hand and applying the tape to the stem by twisting the wired flower with the other hand, pulling the tape tightly around the stem. This will create a stickiness, causing the tape to adhere to the stem as the wired floret is wrapped. The tape should be applied evenly, covering all wire and exposed stem. The excess wire on the stem should be left for the time being, as it will be helpful when the construction process begins.

Florist tape is sold under such brand names as Floraltape and Parafilm.

TECHNIQUES OF CORSAGE MAKING

Corsage construction is an area of the florist business that tests one's creative ability and dexterity if an attractive corsage is to be produced. The following steps should be used to make a single carnation corsage.

PROCEDURE

MAKING A SINGLE CARNATION CORSAGE

1. Select a top-quality standard carnation. The color choice is based on the individual's preference and desire. Remove the stem ½ inch below the calyx.

2. Using a piercing method, wire the flower with an 8-inch piece of 20- to 22-gauge florist wire. Tape the flower by holding the flower in one hand, twisting it as the tape is applied to the calyx stem and the wire. Using ½-inch florist tape, cover all exposed stem and wire. Remember to keep the tape tight while taping with the twisting motion. Leave the stem about 4 inches long.

3. Select a piece of greenery for the backing of the corsage. Baker's fern, asparagus fern, sprengeri fern, salal leaves, podocarpus tips, carnation foliage tips, or ribbon leaves may be used. After selecting the greenery, tape it to the back of the carnation with the ½-inch tape. Be sure it is in the proper proportion to the flower.

4. Make a pinch bow as explained in Unit 48, figure 48-12. In making the bow, be sure the ribbon is the same width as the flower when making the loops. Personal taste will also dictate how large the bow is made.

5. Take the taped carnation with the baker's fern and add the bow to the underside of the carnation. Tape the bow into place with a short piece of florist tape. Using florist shears, remove about 1 inch of the taped stem.

6. Place a corsage pin on the back of the corsage by inserting it vertically into the stem. This way no one will be injured with the corsage pin.

PROCEDURE

MAKING A FEATHERED CARNATION CORSAGE

1. To create this corsage, select two large carnations. To feather a carnation bloom, use a sharp florist knife and cut **one** of the blooms in half. Using 9-inch pieces of 24-gauge florist wire, wrap the remaining calyx of one of the flower halves. (**Note:** hold the flower firmly as the wire is twisted around the calyx.) Tape the carnation calyx and wire. Repeat the same process for the other half of the flower.

2. Take the other standard carnation flower and with a sharp florist knife, split the calyx about ½ inch on each side. This will allow the flower to open more widely. This flower will be used as the focal point in the corsage. Using the piecing method at right angles to the cuts in the calyx, wire and tape this flower.

3. For greenery in this arrangement, choose three short tips of carnation foliage. Using three pieces of 4-inch, 28-gauge florist wire, tape a foliage tip to each of the individual wires.

4. Lay out the three flowers as taped along with the three individual carnation foliage tips.

5. Design the corsage, figure 50-5.

6. Take one of the feathered carnations and one of the carnation foliage tips and press them together. The stickiness of the tape will hold them sufficiently until you can put them together with a short piece of florist tape.

FIGURE 50-6   Completed feather carnation corsage

7. Add the next feathered carnation, moving down the main stem of the corsage. Press this flower into place and tape it in the main stem.

8. Add the two remaining foliage tips below the last feathered carnation and on each side of the stem. Again, press the taped foliage tips to the stem and tape them into place with a short piece of florist tape.

9. Now add the focal point, the standard carnation, taping it at the base of the two feathered carnations.

10. Select a bow that will harmonize with the corsage and tape it below the last flower. Trim the long stems, but be sure to leave them straight to balance the corsage. See figure 50-6.

BOUTONNIERES

The boutonniere is a single flower or small bouquet worn for numerous occasions by a man. It is used to add a touch of color to the man's clothing, and is worn pinned to his lapel or in a lapel buttonhole.

Flowers for boutonnieres are selected based on the individual's preference. Flowers most commonly used for boutonnieres are the carnation, rose, and stephanotis. Other flowers available on a seasonal basis are the bachelor's button and lily of the valley.

The technique of making a boutonniere is simple. The first step is to select a flower of the highest quality and freshness. Wire the flower with the proper florist wire and tape it smoothly and evenly. Use a 20- to 22-gauge wire for roses, carnations, and chrysanthemums, and a 24-gauge wire for stephanotis. Following are the steps to use for making a rose boutonniere, figure 50-7.

FIGURE 50-5   Design for a feathered carnation corsage

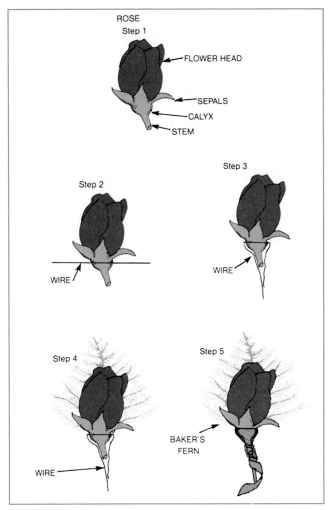

FIGURE 50-7   Steps in making a rose boutonniere

PROCEDURE

MAKING A ROSE BOUTONNIERE

1. Select a high-quality flower.

2. Using the piercing method, attach a 6-inch piece of 20- to 22-gauge wire to the stem.

3. Bend the florist wire down next to the stem.

4. Select a piece of greenery to use on the back of the boutonniere. In this case, use baker's fern. Be sure to keep the greenery in proportion to the size of the rose, as it is used to complement the rose.

5. Tape the baker's fern, rose stem, and wire all in one step. Be sure the tape is applied tightly and evenly down the stem. After taping, cut off the excess stem. The stem must be kept in proportion to the flower. Leave the stem straight because it gives a more natural appearance on the lapel.

6. Mist the boutonniere and attach a boutonniere pin with the sharp point parallel to the stem. Be sure it will not injure someone. Place the boutonniere in a small bag or box.

Student Activities

1) Visit a local florist and observe the making of corsages and boutonnieres.

2) Consider as a class project the making of corsages and boutonnieres to sell as a fund-raising project, perhaps around a holiday or for a formal school dance (e.g., the prom).

Self-Evaluation

A) Provide a brief answer for each of the following.

1) What are the five floral materials used to make corsages and/or boutonnieres?

2) There are four methods of wiring flowers for boutonnieres or corsages. Give an example of flowers for each method.

B) Select the best answer from the choices offered to complete the statement.

1) Florist tape is used to

    a) cover exposed florist wire.
    b) hold the flowers more tightly.
    c) cover exposed flower stems.
    d) all of the above

2) The most commonly used size of florist tape is the

    a) 1-inch width.   c) ½-inch width.
    b) ¼-inch width.   d) none of the above

C) Match the correct wiring procedure with each flower listed below.

    a) Rose   1) Hook method
    b) Carnation   2) Hairpin method
    c) Gladiolus   3) Piercing method
    d) Camellia   4) Clutchwire method
    e) Chrysanthemum (football)
    f) Daisy mum
    g) Stephanotis

GLOSSARY

Absorb – to take in

Absorption – the process of taking in a liquid or gaseous substance

Accent – an item used to create interest in a landscape

Adventitious buds – buds which occur at sporadic and unexpected places on a vegetative structure

Aeration – movement of air into and through the soil

Aggregation – the clinging together of soil particles to form larger crumblike particles

Allelopathy – the toxic effect of one plant on another

Annual – a plant which grows, flowers, produces seed, and dies in one year

Anther – saclike structure at the top of the stamen which contains pollen (the male sex cell)

Apical dominance – a form of regulating growth in plants where the terminal bud secretes chemicals which inhibit or prevent the growth of lateral buds on the same shoot

Arboriculture – cultivation of trees and shrubs

Arm – cane of a vine left after pruning which is tied to a trellis and produces fruit; replaced each year with new wood

Asexually – without the union of male and female sex cells

Asymmetrical – having unequal visual weight on each side of the axis

Auxins – plant hormones that accelerate growth by stimulating cell enlargement

Available capillary water or field capacity – the water left after capillary movement stops

Balance – a basic principle of design

Banding fertilizer – applying fertilizer in trenches on each side of a row of plants

Bare root – plants with no soil, or soilless mix on the roots; roots are washed or shaken clean of all soil

Basal plate – base of a bulb

Berm – a ridge of soil placed around a newly planted tree to retain water

Biennial – a plant which produces vegetation in one year, flowers the next, and then dies

Biological – descriptive of natural living organisms, or from living organisms

Biological insect control – control of insects by the use of the natural enemies of certain insects rather than chemical substances

Biostimulant – natural, organic products, from living organisms, which stimulate soil microbial activity, stimulate plant growth, and promote disease resistance in plants

Bolting – occurs when the plant sends out a seed stalk and stops growing edible leaves

Broadcast – to spread over an entire area

Bud scale scar – a scar located where a terminal, or end, bud has been the previous year

Bud scales – leaves which cover the outside of a bud

Budstick – a small shoot of current season's growth used to cut buds for budding

Bulb – a vegetative structure which consists of layers or fleshy scales overlapping each other, such as the onion

Bulb scales – leaflike parts of a bulb which surround the flower bud and are attached to the basal plate

Bulblet – immature bulb which develops at the base of the bulb

Burlap – a coarse cloth made of jute or hemp often used to protect plants from the weather; holds root ball together on B&B plants

Callus – mass of cells which forms around the wounded area of a plant to start the healing process

Cambium – thin, green, actively growing tissue located between the bark and wood of a plant; in grafting, the cambium of the scion must touch the cambium of the stock

Cane – mature shoot of a vine's previous season's growth

Capillary action – movement of water upward through narrow spaces in the soil

Capillary water – water held against the force of gravity

Chemical retardant – a chemical used to slow down, shorten, or dwarf plant growth

Chlorophyll – the green substance which gives many plants their green color and is necessary for photosynthesis

Chloroplast – small green particle containing chlorophyll found in leaves

ClandoSan – a biological control method which controls nematodes with a by-product of crab and shellfish industries

Cold frame – an outside structure covered on top with glass or plastic used to harden off plants or protect tender plants during the winter

Common name – the English name of a plant which may differ in various localities

Compatibility – ability to unite with, grow, and live together as the scion and rootstock in the grafting and budding processes

Complete flower – flower with both male and female parts; sexually complete

Container growing – growing nursery crops to marketable size in a container

Corm – swollen underground stem which grows upright; is a food storage organ and a means of reproduction

Cormel – a new corm produced by a larger corm

Cotyledons – the first leaves to appear on a plant; seed leaves

Crescent – basic shape of a floral design with a half-moon curve

Cross-pollination – a process in which pollen (male sex cell) of one plant unites with the egg (female sex cell) of a different plant

Cultivars – another name for a specific plant; same as *variety*

Cutting (noun) – a section of stem or root used for propagation of plants

Cytokinins – plant hormones that stimulate cell division. They work along with auxins (will not work without auxins present)

Deciduous plant – a plant which loses it leaves during certain seasons

Dermal – through the skin

Dicot – a plant having two cotyledons, or seed leaves

Dilute – to make weaker or thinner

Direct seeding – planting seeds in a permanent growing site

Division – a method of propagation requiring the cutting and dividing of plants

Dolomitic limestone – limestone with a high magnesium content

Dormant – in a resting, or nongrowing, state

Drip line – the imaginary circle which indicates the outer edge or farthest extension of a tree's branches

Durability – length of use or how long something like a container will last

Embryonic Plant – the entire plant before germination; embryo

Endophytes – small organisms, fungi or bacteria usually, which live in or on other plants and protect them from insects and disease

Endosperm – the food supply for the young developing seedling which is contained in the seed

Environment – the surrounding area

Epidermis – the skin of the leaf

Erosion – washing or blowing away the soil caused by water or strong winds

Evergreen plant – a plant which has leaves or needles throughout the year

Filament – the stalklike part of the stamen, or male part of the flower

Filler flowers – a type of flower used to tie together an arrangement, create interest, and give the arrangement a completed appearance

Flat – a wooden box with slotted bottom used to start seedlings

Floral foam – a mechanism to hold flowers and water in an arrangement

Floriculture – cultivation of ornamental flowering plants

Flower stalk – the stem of the plant which supports the flower

Focal point – an accent point in a flower arrangement where attention is focused

Forcing – growing plants to flower at other than their normal season

Form flower – type of flower that has a unique shape, such as Bird of Paradise, protea, or orchid

Free moving capillary water – water that moves in all directions in the soil

Fungicide – any substance which destroys or prevents the growth of fungi; usually applied to plants as a spray or dust to control fungus diseases

Genetic engineering – moving or changing the genes of plants

Germinate – to sprout or begin to grow

Gibberellins – plant hormones that stimulate growth in stem and leaf by cell elongation; also stimulate premature flowering, growth of young fruits, and breaking of dormancy

Grafting – uniting two different plants so that they grow as one

Grafting wax – a pliable, sticky, waterproof material made of bee's wax, resin, and tallow

Gravitational water – water that the soil is unable to hold against the force of gravity

Green manure – plants plowed or mixed into the soil to rot and add organic matter to the soil

Guard cells – cells on the underside of the leaf which open and close the stomata, or leaf pores

Hardening off – gradually subjecting plants to more difficult growing conditions by withholding water and decreasing temperature; prepares plants for transplanting

Hardiness – ability of a plant to withstand the minimum temperature of an area

Hardwood cutting – a cutting made from current season's stem tissue which is mature or hard

Harrow – a farm implement pulled behind a tractor which breaks up clods of earth, levels the soil surface, and compacts the seed bed

Herbicide – a substance which destroys weeds

Hogarth curve – basic shape of a floral design that has the appearance of an S curve

Hormones – growth-regulating substances in plants

Horticulture – a field which includes growing of fruits, nuts, vegetables, ornamental plants, and flowers, and the sale and processing of these items; from the Latin word meaning "garden cultivation"

Host plant – the plant on which a disease or insect lives

Humidity – the amount of moisture in the atmosphere

Hybrid – an offspring of two different varieties of one plant which possesses certain characteristics of each parent

Inconspicuous – small and unnoticeable (as *inconspicuous* flowers)

Inhalation – the process of taking in air through the lungs; breathing in

Inhibitors – (abscisic acid), plant hormones that inhibit seed germination, stem elongation, and hasten ripening of fruit (ethylene gas)

Initial – the first of something (for example, initial growth)

Initiate – to start; begin

Insert – to place into an opening

Irrigation – the addition of water to plants to supplement natural rainfall

Laminate (tunicate) bulb – a bulb having dry membranous outer scales which protect it from rough handling and drying

Larvicide – a material that kills the larval stage of an insect

Lateral – side branch

Layering – a type of asexual reproduction in which roots are developed on the stem of a plant while it is still attached to the parent plant

Leach – to wash through or out of soil

Leaf axil – the angle made where the leaf leaves the stem; usually contains a bud

Leaf blade – the flat part of the leaf which extends from either side of the leaf stalk

Leaf scale scar – a scar on a stem where a leaf has been attached in a previous year

Lenticels – breathing pores in the bark of woody stems

Lethal – deadly

Limestone – a natural rock used to reduce soil acidity

Line flowers – a spike-appearing flower, such as gladiolus, delphinium, or snapdragon

Lining out – planting cuttings in rows in the field

Loam – soil with approximately equal amounts of sand, silt, and clay; generally considered the best type of soil for plant growth

Major elements – plant food elements required in large amounts: nitrogen, phosphorus, and potash; they must be added to plant media by means of fertilizers

Manure – animal wastes used to improve soil; low in plant food but high in bacteria and other elements used for plant growth

Market pack – plant container which usually holds 6 to 12 plants and is sold as a unit

Massing flower – a flower that has a rounded appearance, such as a carnation, rose, or chrysanthemum

Maturity – age; plant tissue hardens as it matures

Medium – a material which is used to start and grow seeds and plants (plural: media)

Microbes – microscopic organisms

Minor elements – elements required by plants in small amounts; they may or may not need to be added to media for the best growth

Misting system – a piped water system used in greenhouses which uses nozzles to spray fine droplets of water on plants

Molluscicide – a chemical used to kill snails and slugs

Monocot – a plant having only one cotyledon, or seed leaf

Morphology – the classification of plants by form and structure

Mulch – any material used to cover the soil for weed control and moisture retention

Nematocide – a substance used to kill nematodes (small worms)

Neutral – neither acidic nor alkaline; having a pH of 7

Node – the joint of a stem; the swollen place where leaves and buds are generally attached

Nomenclature – a system of naming used to classify a group, such as the botanical names of plants

Nontunicate (scaly) bulb – a bulb which has no outer covering of scales and is much more sensitive to drying or bruising

Nutrients – plant food elements

Oasis (floral foam) – spongelike material used as a base for flower arrangements

Optimum growth – the best possible growth

Orally – through the mouth

Organic matter – dead and decaying plant parts, such as green manure crops, peat moss, or animal manures which improve the water-holding and fertilizer-holding capacity of the soil

Organic mulch – material composed of decayed plant life, such as straw, peat moss, or wood chips used as a mulch

Oscillator – a type of sprinkler head used in overhead irrigation

Ovary – the lower part of the pistil in which the eggs are fertilized and develop; becomes the seed coat or fruit

Ovules – eggs; the female sex cells

Parasitize – to live on or in, often resulting in death of the host

Peat pellets – compressed peat moss disks used to start seeds or root cuttings which are transplanted with the plant

Peat pots – pot-shaped plant containers made of compressed peat moss instead of clay or plastic

Perennial – a plant that grows year after year without replanting; plant whose roots live from year to year

Perlite – a white granular material used to help loosen or open up spaces in rooting media

Pest – an unwanted animal, plant, bacterium, or fungus

Petal – one type of leaf on flowers; usually considered the most striking part of the flower

Petiole – stalk or stem which supports the blade of the leaf

pH level – the measure of a soil's acidity or alkalinity on a scale of 1–14; 1–6 indicates an acid, 7 is neutral, and 8–14 indicates a base

Phloem – the tubes in plant stems which conduct food from leaves through the stem to the roots

Photoperiodism – response of plants to different periods of light and darkness in terms of flowering and reproduction cycle

Photosynthesis – manufacture of food by green plants in which carbon dioxide and water are combined in the presence of light and chlorophyll to form sugar and oxygen

Pistil – the female reproductive part of a flower; contains the female sex cells in the ovary

Polarity – the tendency of tuberous roots to grow shoots on one end, or the top, and roots on the other end, or bottom

Pollen – male sex cell in plants

Pollination – the process in which the pollen cell is transferred to the female plant part and the eggs are fertilized

Polyethylene – a clear plastic that holds in moisture but allows the passage of light and a small amount of air

Pot-bound – roots growing in a tight pattern around the walls of the container

Precooled – cooled for forcing

Propagate – to increase in number, to reproduce

Pruning collar – the rings at the base of a branch which is attached to the trunk of the tree

Regenerate – to reproduce or grow anew

Relative humidity – the amount of moisture in the air compared with what it can hold at that temperature; expressed as a percent of the amount of moisture the air could hold if saturated

Renewal spur – a cane of a vine pruned to two buds and left to grow new canes the following year

Reproduce – to increase in number

Resist – to withstand or prevent

Rhizome – underground stem which produces roots on the lower surface, and extends leaves and flowering shoots above the ground

Rodenticide – a substance used to kill rodents

Root hair – a very tiny rootlike extension on small feeding roots; absorbs minerals and water from the soil

Rooting – the development or growth of new roots

Rooting hormone – a chemical, in powder or liquid form, that helps cuttings to root faster and have a greater number of roots

Rootstock – the lower portion of a graft which becomes the stem and roots of the new plant

Scientific name – the Latin name of a plant giving its genus and species

Scion – a short piece of shoot containing several buds which becomes the new top of a grafted plant

Seed coat – the outer covering of the seed

Seedlings – young plants which have been germinated several days

Self-fruitful – able to pollinate itself

Self-pollination – fertilization of a plant by its own pollen

Sepal – green, leaflike part of the flower that covers and protects the flower bud before it opens

Separation – method of propagation that occurs naturally in which reproductive organs of a plant detach from the parent plant to become new plants

Set – to develop fruit

Sexual – reproduction involving male and female sex cells (pollen and egg)

Shallow – not deep, generally less than 2 inches

Shoot – current season's growth of a plant; rapid, new stem growth

Sidedress – to fertilize along the side of rows near plants

Site – location or growing area

Skew – a tool used to level media in flats before planting

Slow-release fertilizer – fertilizer that releases plant nutrients over a long period of time

Soilless – no soil in the growing mix

Soluble – able to be dissolved in water

Sphagnum peat moss – a material formed from decayed bog plants; used in growing media

Stamen – the male reproductive part of the flower containing the male sex cells or pollen

Sterile – free from any living organisms

Stigma – part of the pistil which catches pollen on its sticky surface

Stomata – small pores or holes in the leaf which allow the plant to breathe and give off moisture

Style – the tube in the pistil which leads from the stigma to the ovary and through which pollen reaches the egg

Susceptible – weak, unable to resist (disease)

Symmetrical – having equal visual weight on each side of the axis

Symptoms – warning signals of plant growth problems

Systemic – inside and moving throughout a plant

Taxonomy – the study of plant names and the identification of plants

Tendrils – structures on a plant vine or stem which allow the plant to attach itself to a trellis or other structures

Terrarium – a covered or closed container for growing plants

Texture – the coarseness or fineness of plant leaves

Thatch – dead grass clippings and roots which build up in lawns, causing growth problems

Thermo-blanket – an insulating blanket with white plastic on the outside and microfoam insulation on the inside

Tissue culture and micropropagation – new method for rapidly reproducing plants using terminal bud tissue

Translocated – moved from one part of the plant to another, such as root to leaf

Transpiration – loss of water through the leaves or stems of plants

Transplant – to move plants from one growing location to another, thus giving them more space in which to develop

Trellis – a supporting structure made of posts and wire which is used to tie up grapes and some of the bramble fruits

Trickle tube irrigation – a system for watering container-grown plants using a small tube running to each container

True to seed – reproducing offspring from seed which are nearly exact duplicates of the parent plant

Trunk – the main stem of a plant which lives on for years, supporting new growth each year from arms, spurs, or limbs

Tuber – fleshy root which reproduces by growing roots from an eye, or bud

Tuberous root – a thick root containing large amounts of stored food; has roots located at one end and stem at the other

Turf grasses – grasses that act as vegetative ground cover

Turgid – swollen, filled with moisture

Unavailable capillary water – water held as a film around soil particles; can only be moved as vapor

Vectobac – a biological insecticide for fungus gnats applied as a soil drench

Ventilation – the movement and exchange of air in a greenhouse

Vermiculite – a light mineral with a neutral pH used to increase the moisture-holding capacity of media

Weed – any plant growing where it is not wanted

Wetting agent – a chemical used to cause material like sphagnum peat moss to soak up water and get wet much more quickly

Windbreak – any natural or manmade screen to slow wind speed

Xeriscaping – a landscaping technique used to practice water conservation

Xylem – conducting tissue (tubes) in plants which transports water and nutrients from the roots to the stem and leaves

Abelia grandiflora – Abelia
Abies balsamea – Balsam Fir
Abies concolor – White Fir
Abies nordmanniana – Nordman Fir
Acer ginnala – Amur Maple
Acer palmatum 'Atropurpureum' – Japanese Maple
Acer platanoides – Norway Maple
Acer platanoides 'Crimson King' – Crimson King Maple
Acer platanoides 'Emerald Queen' – Emerald Queen Maple
Acer rubrum – Red Maple
Acer rubrum 'October Glory' – October Glory Maple
Acer rubrum 'Red Sunset' – Red Sunset Maple
Acer saccharinum – Silver Maple
Acer saccharum – Sugar Maple
Achillea millefolium – Yarrow
Adiantum cuneatum 'Croweanum' – Maidenhair Fern
Aechmea fascitata – Bromeliads
Aeschynanthus lobbianus – Lipstick Vine
Aesculus carnea – Red Horsechestnut
Aesculus parviflora – Bottlebrush Buckeye
Ageratum houstonianum – Ageratum
Aglaonema simplex – Aglaonema
Agropyron cristatum – Crested Wheatgrass
Agrostis palustris – Creeping Bentgrass
Agrostis tenius – Colonial Bentgrass
Ajuga reptans – Bugleweed
Aloe vera – Medicine Plant
Alyssum saxatile – Golddust
Ananas comosus – Pineapple
Anchusa italica – Alkanet
Anemone pulsatilla – Windflower
Anthemis tinctoria – Golden Daisy
Antirrhinum sp – Snapdragon
Aphelandra squarrosa – Zebra Plant
Aquilegia alpina – Columbine
Aquilegia sp – Columbine
Arabis alpina – Rockcress
Araucaria excelsa – Norfolk Island Pine

Araucaria heterophylla – Norfolk Island Pine
Arctostaphylos uva-ursi – Bearberry
Armeria maritima – Sea Pink
Asarum caudatum – Wild Ginger
Asparagus sprengeri – Asparagus Fern
Aspidistra elatior – Aspidistra
Asplenium nidus – Birds Nest Fern
Aster alpinus – Hardy Aster
Astilbe arendsii – False Spirea
Aucuba japonica – Japanese Aucuba
Axonopus affinis – Carpet Grass
Beaucarnea recurvata – Ponytail Palm
Begonia evansiana – Hardy Begonia
Begonia socotrana – Begonia
Berberis thunbergi – Japanese Barberry
Betula pendula – Weeping Birch
Bougainvillea spectabilis – Bougainvillea
Boutelous gracilis – Blue Grama
Browallia sp – Browallia
Buchloe dactyloides – Buffalo Grass
Buddleia davidii – Butterfly Bush
Buxus sempervirens – Common Boxwood
Calendula sp – Pot Marigold
Callistephus chinensis – Aster
Calluna vulgaris – Heather
Camellia japonica – Camellia
Campanula calycanthema – Bellflower
Campanula medium – Canterbury Bells
Campsis radicans – Trumpet Vine
Canna indica – Canna Lily
Carex morrowii variegata – Sedge
Cedrus deodara – Deodar Cedar
Celastrus scandens – American Bittersweet
Celosia sp – Cockscomb
Cerastium tomentosum – Snow-in-Summer
Cercidyphyllum japonicum – Katsura Tree
Cercis canadensis – Redbud
Chaenomeles japonica – Flowering Quince
Chamaecyparis obtusa 'Nana' – Dwarf Hinoki Cypress
Chamaecyparis pisifera 'Plumosa' – Plume False-cypress
Chamaecyparis pisifera 'Squarrosa' – Moss False-cypress

Chasmanthium latifolium – Wild Oats
Chionanthus virginicus – Old Man's Beard
Chlorophytum comosum – Spider Plant
Chrysanthemum maximum – Shasta Daisy
Chrysanthemum morifolium – Chrysanthemum
Cissus rhombifolia – Grape Ivy
Clematis hybrida – Clematis
Cleome speciosa – Spiderflower
Codiaeium variegatum – Croton
Coleus blumei – Coleus
Convallaria majalis – Lily of the Valley
Coreopsis lanceolata – Tickseed
Coreopsis sp – Coreopsis
Cornus canadensis – Bunchberry
Cornus florida – Flowering Dogwood
Cornus kousa – Chinese Dogwood
Coronilla varia – Crown Vetch
Cosmos sp – Cosmos
Cotoneaster adpressus 'Praecox' – Cotoneaster
Cotoneaster congesta – Cotoneaster
Cotoneaster dammeri – Lowfast Cotoneaster
Cotoneaster dammeri 'Skogholmen' –
　Skogholmen Cotoneaster
Cotoneaster horizontalis – Rock Spray
　Cotoneaster
Crassula argentea – Jade Plant
Crataegus crus-galli – Cockspur Hawthorne
Crataegus phaenopyrum – Washington
　Hawthorne
Crataegus oxycantha 'Pauli' – Paul's Scarlet
　Hawthorne
Cryptomeria japonica – Japanese Cryptomeria
Cyclamen persicum – Cyclamen
Cynodon dactylon – Bermuda Grass
Cypressocyparis Leylandi – Leyland Cypress
Cytisus x 'Praecox' – Warminster Broom
Dahlia variabilis – Dahlia
Daphne odora – Daphne
Delphinium elatum – Delphinium
Delphinium sp – Larkspur
Deutzia gracilis – Slender Deutzia
Dianthus sp – Pinks
Dianthus barbartus – Sweet William
Dianthus caryophyllus – Carnation
Dieffenbachia exotica – Dieffenbachia
Digitalis purpurea – Foxglove
Dimorphotheca sp – Cape Marigold
Dizygotheca elegantissima – Aralia
Dracaena fragrans 'Massangeana' – Corn Plant
　Dracaena
Echeveria secunda – Echeveria (Cat and Kittens)
Echinacea purpurea – Cone Flower

Elaeagnus umbellata – Autumn Olive
Epimedium grandiflorum – Barrenwort
Eremochloa ophiuroides – Centipede Grass
Erica carnea – Heath
Euonymus alatus – Winged Euonymus
Euonymus fortunei 'Argenteo marginatus' –
　Wintercreeper Euonymus
Euonymus fortunei 'Carrieri – Wintercreeper
　Euonymus
Euonymus fortunei 'Colorata' – Wintercreeper
　Euonymus
Euonymus fortunei 'Vegetus' – Wintercreeper
　Euonymus
Euonymus japonicus – Euonymus
Euonymus kiautschovicus – Spreading Euonymus
Euphorbia pulcherrima – Poinsettia
Fagus grandifolia – American Beech
Fagus sylvatica – Beech
Fatshedera lizei – Fatshedera
Festuca arundinacea – Tall Fescue
Festuca glauca – Blue Fescue
Festuca rubra – Creeping Red Fescue
Ficus elastica 'Decora' – Rubber Plant
Fittonia ssp – Fittonia
Fothergilla monticola – Fothergilla
Franklinia alatumala – Franklinia
Fraxinus americana – White Ash
Fraxinus pennsylvanica lanceolata 'Marshall' –
　Green Ash
Fritillaria imperialis – Fritillaria
Fuchsia excorticata – Fuchsia
Gaillardia grandflora – Gaillardia
Gaillardia sp – Gaillardia
Gardenia jasminoides – Gardenia
Gaultheria procumbens – Wintergreen
Ginkgo biloba – Ginkgo
Gladiolus grandiflorus – Gladiolus
Gleditsia triacanthos 'Shademaster' –
　Shademaster Honeylocust
Gomphrena globosa – Globe amaranth
Gynura sarmentosa – Velvet Plant
Gypsophila paniculata – Baby's Breath
Hedera helix – English Ivy
Hedera helix 'Baltic' – Baltic English Ivy
Helianthemum nummularium – Sunrose
Helianthus sp – Sunflower
Helichrysum sp – Strawflower
Hermerocallis sp – Daylily
Hibiscus moscheutos – Mallow Rose
Hibicus syriacus – Rose of Sharon
Hippeastrum sp – Amaryllis
Hosta undulata – Hosta

Hoya carnosa – Hoya
Hyacinthus orientalis – Hyacinth
Hydrangea quercifolia – Oakleaf Hydrangea
Hypericum ssp – St. Johns wort
Iberis sp – Candytuft
Ilex aquifolium – English Holly
Ilex crenata – Japanese Holly
Ilex cornuta 'Nellie Stevens' – Nellie Stevens Holly
Ilex crenata 'Convexa' – Convexleaf Japanese Holly
Ilex crenata 'Helleri' – Helleri Holly
Ilex crenata 'Hetzi' – Hetz Holly
Ilex crenata 'San Jose' – San Jose Japanese Holly
Ilex crenata 'Stokes' – Stokes Holly
Ilex glabra – Inkberry Holly
Ilex x meserveae – Meservea Holly
Ilex opaca – American Holly
Ilex verticillata – Winterberry
Impatiens wallerana – Impatiens
Ipomoea sp – Morning Glory
Iris sp – Iris (Flag)
Juglans nigra – Black Walnut
Juniperus chinensis 'Hetzii' – Hetz Juniper
Juniperus chinensis 'Pfitzeriana' – Pfitzer Juniper
Juniperus chinensis 'Sargentii' – Sargent Juniper
Juniperus conferta – Blue Pacific Juniper
Juniperus horizontalis 'Douglasii' – Waukegan Juniper
Juniperus horizontalis 'Plumosa' – Andorra Juniper
Juniperus horizontalis 'Wiltoni' – Blue Rug Juniper
Juniperus squamata – Meyeri Juniper
Juniperus virginiana – Eastern Redcedar
Kalanchoe blossfeldiana – Kalanchoe
Kalanchoe pinnata – Kalanchoe
Kalmia latifolia – Mountain Laurel
Kerria japonica – Mt. Vernon Shrub
Koelreuteria paniculata – Golden Rain Tree
Kolkwitzia amabilis – Beauty bush
Lantana camara – Lantana
Larix decidua – European Larch
Lathyrus sp – Sweet Pea
Leucothoe catesbaei – Leucothoe
Leucothoe fontanesiana – Leucothoe
Liatris pycnostachya – Gayfeather
Ligustrum japonicum – Privet
Ligustrum lucidum – Glossy Privet
Ligustrum ovalifolium – California Privet
Lilium longiflorum – Lily
Limonium sp – Statice

Lippia citriodora – Lemon Verbena
Liquidambar styrciflua – Sweet Gum
Liriodendron tulipifera – Tulip Tree
Liriope muscari – Liriope
Lobelia sp – Lobelia
Lobularia sp – Sweet Alyssum
Lolium multiflorum – Annual Rye Grass
Lolium perenne – Perennial Rye Grass
Lonicera japonica – Japanese Honeysuckle
Lonicera tatarica – Tatarian Honeysuckle
Lupinus polyphyllus – Lupine
Lysimachia nummularia – Moneywort
Magnolia acuminata – Cucumber Magnolia
Magnolia grandiflora – Southern Magnolia
Magnolia stellata – Star Magnolia
Magnolia virginiana – Sweet Bay Magnolia
Magnolia x 'Soulangeana' – Saucer Magnolia
Mahonia aquifolium – Oregon Holly-Grape
Mahonia bealei – Leatherleaf Mahonia
Mahonia repens – Creeping Mahonia
Malus floribunda – Japanese Flowering Crabapple
Maranta leuconeura – Prayer Plant
Matteuccia modulosa – Ostrich Fern
Matthiolia sp – Stock
Metasequoia glyptostroboides – Dawn Redwood
Mirabilis sp – Four-o-Clock
Miscanthus floridulus – Giant Miscanthus
Mitchella repens – Partridgeberry
Musa maurelii – Banana
Myosotis sp – Forget-me-not
Nanidina domestica – Heavenly Bamboo
Narcissus jonquilla – Narcissus (Daffodils)
Nephrolepis exaltata – Boston Fern
Nicotiana sp – Flowering Tobacco
Nyssa sylvatica – Black Gum
Ophiopogon japonicus – Dwarf Lilyturf
Oxydendrum arboreum – Sourwood
Pachysandra terminalis – Japanese Pachysandra
Paeonia sp – Peony
Panicum virgatum – Switch Grass
Papaver nudicaule – Iceland Poppy
Papaver oriental – Oriental Poppy
Papaver sp – Poppy
Parthenocissus tricuspidata – Boston Ivy
Paspalum notatum – Bahia Grass
Paxistima canbyi – Paxistima
Pelargonium hortorum – Geranium
Peperomia obtusifolia – Peperomia
Peris japonica – Japanese Andromeda
Petunia hybrida – Petunia
Phalaris arundinaccea 'Picta' – Ribbon Grass
Philadelphus tomentosus – Mock Orange

Philodendron cordatum – Heartleaf Philodendron
Phlox paniculata – Summer Phlox
Phlox sp – Phlox
Phlox sublata – Moss Phlox
Physalis alkekengi – Chinese Lantern
Picea abies – Norway Spruce
Picea abies 'Nudiformis' – Bird's Nest Spruce
Picea omorika – Serbian Spruce
Picea pungens – Colorado Spruce
Picea pungens 'Glauca' – Blue Colorado Spruce
Pieris floribunda – Mountain Andromeda
Pieris japonica – Andromeda
Pilea cadierei – Pilea
Pinus griffithi – Himalayan Pine
Pinus mugo 'Mugo' – Mugho Pine
Pinus nigra – Austrian Pine
Pinus resinosa – Red Pine
Pinus rigda – Pitch Pine
Pinus strobus – White Pine
Pinus sylvestris – Scotch Pine
Pinus taeda – Loblolly Pine
Pinus virginiana – Virginia Pine
Pittosporum tobira – Pittosporum
Platanus occidentalis – Sycamore
Platcerium bifurcatum – Staghorn Fern
Plectrantus australis – Swedish Ivy
Poa annua – Annual Blue Grass
Poa pratensis – Kentucky Blue Grass
Poa trivialis – Rough Blue Grass
Podocarpus macrophyllus – Yew Podocarpus
Polygonum aubertii – Silverlace Vine
Populus nigra – Lombardy Poplar
Portulaca sp – Rose Moss
Potentilla fruticosa – Cinquefoil
Primula polyantha – Primrose
Prunus cerasifera 'Thundercloud' –
   Thundercloud Flowering Plum
Prunus grandulosa – Flowering Almond
Prunus serrulata – Oriental Cherry
Prunus subhirtella 'Pendula' – Weeping Cherry
Pseudotsuga menziesii – Douglas Fir
Pyracantha coccinea 'Lalandi' – Firethorn
Pyrethrum roseum – Painted Daisy
Quercus alba – White Oak
Quercus borealis – Red Oak
Quercus coccinea – Scarlet Oak
Quercus palustris – Pin Oak
Quercus virginiana – Live Oak
Rhododendron minus 'Carolinianum' –
   Rhododendron Carolinianum
Rhododendron nudiflorum – Azalea Nudiflora
Rhododendron 'Exbury Hybrid' – Deciduous
   Azalea

Rhododendron 'Gibraltar' – Deciduous Azalea
   Gibraltar
Rhododendron 'Hino Crimson' – Hino Crimson
   Azalea
Rhododendron 'Hinodegiri' – Hinodegiri Azalea
Rhododendron 'P.J.M.' – P.J.M. Rhododendron
Rhododendron 'Purple Gem' – Purple Gem
   Rhododendron
Rhododendron prunifolium – Deciduous Azalea
Rhododendron schlippenbachi – Schlippenbachi
   Rhododendron
Rhododendron ssp – Azalea
Rhododendron x 'Catawblinse' – Catawba
   Rhododendron
Rosa hybrida – Hybrid Tea Rose
Rudbeckia fulgida 'Goldstrum' – Black-Eyed
   Susan
Saintpaulia ionantha – African Violet
Salpiglossis sp – Giant Velvet Flower
Salvia sp – Scarlet Sage
Sansevieria hahni – Hahn Snake Plant
Sansevieria trifasciata – Sansevieria (Snake
   Plant)
Sarcococca hookeriana 'Humilis' – Sarcococca
Scabiosa sp – Pincushion Flower
Schefflera actinophylla – Schefflera
Schefflera digitata – Umbrella Tree
Scindapsus aureus – Pothos
Sedum sp – Stonecrop
Senecio cruentus – Cineraria
Sinningia speciosus – Gloxinia
Sorbus aucuparia – European Mountain Ash
Spiraea bumalda 'Anthony Watereri' – Anthony
   Waterer Spiraea
Spiraea prunifolia – Bridalwreath Spirea
Spiraea vanhouttei – Vanhoutte Spirea
Stenotaphrum secundatum – St. Augustine Grass
Stephanandra incisa – Cutleaf Stephanandra
Strelitzia reginae – Bird of Paradise
Syngonium ssp – Nephthytis
Syringa chinensis – Chinese Lilac
Syringa vulgaris – Common Lilac
Tagetes ssp – Marigold
Taxodium distichum – Bald Cypress
Taxus baccata 'Repandens' – Spreading English
   Yew
Taxus cuspidata aurencens – Dwarf Japanese Yew
Taxus cuspidata 'Capitata' – Japanese pyramidal
   yew
Taxus x *media* 'Hicksi' – Hicks Yew
Taxus x *media* 'Nigra' – Anglojap Yew
Thuja occidentalis – American Arborvitae

Thuja orientalis – Oriental Arborvitae
Thunbergia sp – Black-eyed Susan Vine
Tilia americana American Linden
Tilia cordata – Little Leaf Linden
Tropaeolum sp – Nasturtium
Tsuga canadensis – Canada Hemlock
Tulipa hybrida – Tulip
Ulmus americana – American Elm
Verbena x *hortensis* – Verbena
Veronica spicata – Speedwell
Viburnum carlesii – Korean Spice Viburnum
Viburnum dentatum – Arrowwood
Viburnum macrocephalum – Chinese Snowball
Viburnum opulus – Cranberry Bush

Viburnum plicatum 'Tomentosum' – Doublefile Viburnum
Vinca major – Large Leaf Vinca
Vinca minor – Periwinkle
Viola tricolor – Pansy
Weigela sp – Weigela
Wisteria floribunda – Wisteria
Wisteria sinensis – Wisteria
Yucca filamentosa – Desert Candle (Yucca)
Zebrina pendula – Wandering Jew
Zinnia sp – Zinnia
Zoysia japonica – Zoysia Grass
Zygocactus truncatus – Christmas Cactus

❧ INDEX ❧

Numbers followed by f indicate figures.

Numbers followed by f indicate figures.

Numbers followed by f indicate figures.

Numbers followed by f indicate figures.

Numbers followed by f indicate figures.